NEAR-FIELD COSMOLOGY WITH DWARF ELLIPTICAL GALAIXES

IAU COLLOQUIUM No. 198

COVER ILLUSTRATION:

The Yellow Emperor Huáng Dí (2696-2598 BC) said "The principle of Yin and Yang is the foundation of the entire universe. It underlies everything in creation. Heaven was created by the concentration of Yang, the force of light, earth was created by the concentration of Yin, the forces of darkness. Yang stands for peace and serenity; Yin stands for confusion and turmoil. Yang stands for destruction; Yin stands for conservation. Yang brings about disintegration; Yin gives shape to things ...".

Image credits: the cover illustration shows a Yin-Yang arrangement of two images, the cosmic large-scale structure as derived from a N-body simulation and the dwarf elliptical galaxy NGC185, companion of M31.

The cosmic web image was produced by Joerg Colberg as part of the Virgo consortium for cosmological supercomputer simulations. The reference to the Virgo simulations is Jenkins *et al.* 1998, ApJ, 499, 20.

The CCD image of NGC 185 was obtained with the 24-inch telescope of the Grasslands Observatory. It is reproduced with the permission of Tim Hunter and James McGaha (Tucson, Arizona, USA).

INTERNATIONAL ASTRONOMICAL UNION

UNION ASTRONOMIQUE INTERNATIONALE

International Astronomical Union

NEAR-FIELD COSMOLOGY WITH DWARF ELLIPTICAL GALAXIES

PROCEEDINGS OF THE 198th COLLOQUIUM OF THE INTERNATIONAL ASTRONOMICAL UNION HELD IN LES DIABLERETS, SWITZERLAND MARCH 14–18, 2005

Edited by

HELMUT JERJEN
Research School of Astronomy & Astrophysics, Australian National University, Australia

and

BRUNO BINGGELI
Astronomical Institute, University of Basel, Switzerland

CAMBRIDGE
UNIVERSITY PRESS

CAMBRIDGE UNIVERSITY PRESS
The Edinburgh Building, Cambridge CB2 2RU, UK
40 West 20th Street, New York, NY 10011 4211, USA
477 Williamstown Road, Port Melbourne, VIC 3207, Australia
Ruiz de Alarcón 13, 28014 Madrid, Spain
Dock House, The Waterfront, Cape Town 8001, South Africa

First published 2005

Printed in the United Kingdom at the University Press, Cambridge

Typeset in System $\LaTeX\ 2_\varepsilon$

A catalogue record for this book is available from the British Library

Library of Congress Cataloguing in Publication data

ISBN 0 521 85204 8 hardback
ISSN 1743-9213

Table of Contents

CHEMICAL EVOLUTION OF DWARF ELLIPTICAL GALAXIES

PHYSICAL PROCESSES IN DWARF ELLIPTICAL GALAXIES

DWARF ELLIPTICALS IN GROUPS AND CLUSTERS

GLOBULAR CLUSTERS, NUCLEI AND ULTRA COMPACT DWARFS

Preface

Though of humble appearance and easily overlooked because of their modest luminosities, dwarf galaxies are the most numerous type of galaxy known. It is recognised today that the phenomenon "dwarf galaxies" represents a cornerstone in our understanding of the physical processes that govern galaxy formation and evolution at high redshift. These systems are fundamental in both major formation scenarios. In hierarchical clustering theories they constitute the building blocks from which larger systems have been created by merging, while in monolithic collapse scenarios they can represent the debris of massive galaxies unable to start forming stars until $z \sim 1$. Furthermore, dwarf galaxies yield important information on the chemical enrichment of the cosmos, the star formation processes, and the characteristics of dark matter.

In spite of a rapidly growing dwarf galaxy community (as indicated, e.g., by the electronic newsletter "Dwarf Tales"), international conferences dedicated to dwarf galaxies are held on rare occasions. Since it seems to be easier for dwarf irregular topics to be accomodated in other conferences, e.g., on star formation, ISM, chemical evolution, radio astronomy, we felt the need for a meeting on *dwarf elliptical* galaxies (including dwarf spheroidals) with an emphasis on their role in cosmology. Great effort has been made over the past decade in delineating the properties of both Local Group satellite dEs (dwarf spheroidals) and the large dE populations of nearby groups and clusters. Only there the starformation history, gas and dark matter contents of stellar systems can be studied with highest resolution that is impossible to achieve at high redshifts. The mismatch of the faint-end galaxy luminosity function with the current cosmological model (ΛCDM) predictions, the puzzling diversity of star-formation histories among Local Group dwarf ellipticals, and the growing evidence that dwarfs are repositories of a lot of dark matter demonstrate the importance of these optically elusive stellar systems.

The first ideas on the range of topics, speakers to invite, and conference venue were discussed in July 2003 when we met in Sydney on the occasion of the IAU General Assembly. A Scientific Organizing Committee was quickly established, as many of our colleagues shared our wish for such a conference. Our choice for the conference location was *Les Diablerets*, a picturesque mountain resort in the French speaking part of Switzerland near Geneva, which had already hosted a number of successful "Saas-Fee Advanced Courses" of the Swiss Society of Astronomy and Astrophysics.

Following endorsement by IAU Commissions 28 (Galaxies) and 47 (Cosmology) as well as IAU Division VII (Galaxies and the Universe), our proposal for an IAU Colloquium entitled "Near-Field Cosmology with Dwarf Elliptical Galaxies" was finally approved and sponsored by the IAU Executive Committee in March 2004. Soon thereafter we started to invite speakers and to inform and invite the community at large. The overwhelming, positive response materialized in 92 astronomers from 23 countries who participated at the conference. IAU Colloquium No. 198 took place in Les Diablerets from 14–18 March, 2005. It was a most fruitful and enjoyable meeting, scientifically and socially.

This book contains all but 7 of the 69 oral contributions and all but 2 of the 23 poster contributions presented at the conference. The order of the written contributions slightly deviates from the schedule of the conference program, as the written material is easier to bring into a logical flow than living people! We have structured the contributions, including the poster contributions, in a number of main topics. Questions, answers, and comments following the talks were hand-written by the participants involved and later transcribed and added to the written contributions. The social photographs scattered throughout the book are our attempt to capture some of the conference spirit.

It is a great pleasure to acknowledge the financial support by the IAU, and IAU Assistant General Secretary, Dr. Karel van der Hucht, for his efficiency in dealing with all IAU matters and his sympathy with the organizers. We thank our colleagues from the SOC for their important advice in putting together the program, and our colleagues from the LOC for their invaluable, unwavering help in confining the chaos shortly before and during the conference week. Last not least it is a pleasure to thank the management and crew of the hotel EUROTEL Les Diablerets, who contributed very much to make the conference a memorable experience.

Helmut Jerjen and Bruno Binggeli, co-chairs SOC
Canberra and Binningen, June 2005

THE ORGANIZING COMMITTEES

Scientific

B. Binggeli (co-chair, Switzerland)
J. Davies (UK)
H. Jerjen (chair, Australia)
M. Mateo (USA)
J. Silk (UK)
B.R. Tully (USA)
N. Caldwell (USA)
E. Held (Italy)
I. Karachentsev (Russia)
B. Moore (Switzerland)
E. Tolstoy (The Netherlands)

Local

K. Ammon
F. de Lorenzi
T. Hascher
A. Kayser
S. Kautsch
B. Binggeli (chair)
E. Grebel
H. Jerjen (co-chair)
A. Koch
T. Lisker

Acknowledgements

The colloquium is sponsored and supported by the IAU Divisions VIII (Galaxies and the Universe) and by the IAU Commissions No. 28 (Galaxies) and No. 47 (Cosmology).

The Local Organizing Committee operated under the auspices of the Astronomical Institute, University of Basel.

Funding by the International Astronomical Union, is gratefully acknowledged.

Participation List

1. Bruno Binggeli
2. Wyn Evans
3. Matthew Walker
4. Matthew Coleman
5. Ben Moore
6. Eleanor K. Ferguson
7. Harry Ferguson
8. Stefano Zibetti
9. Philippe Prugniel
10. Jürg Diemand
11. Igor Karachentsev
12. Justin Read
13. Maren Hempel
14. Liese van Zee
15. Fabio Barazza
16. Paola Popesso
17. Joanna Woo
18. Alan McConnachie
19. James E. Taylor
20. Andrés Jordán
21. Sergey Mashchenko
22. Brent Tully
23. Rami Rekola
24. Jeremy Bailin
25. Pavel Kroupa
26. Dominik Argast
27. Carsten Weidner
28. Igor Chilingarian
29. Jay S. Gallagher, III
30. Andrea Marcolini
31. Andrea Kayser
32. Ovidiu Vaduvescu
33. Daniel Harbeck
34. Chiara Mastropietro
35. Gary Da Costa
36. Bettina Ogbuagu-Poledna
37. Karl Rakos
38. Monica Hasegan
39. Carme Gallart
40. HongSheng Zhao
41. Marina Rejkuba
42. Michael Hilker
43. David Silva
44. Steffen Mieske
45. Sangmo Sohn
46. Bryan Miller
47. Andreas Koch
48. Eva Grebel
49. Nobuo Arimoto
50. Chris Conselice
51. Lucio Mayer
52. Thorsten Lisker
53. Ivo Saviane
54. Stefan Kautsch
55. Helmut Jerjen
56. Dolf Michielsen
57. Gerhard Hensler
58. Laura Dunn
59. Myung Gyoon Lee
60. Victor Debattista
61. Alister Graham
62. Enrico Held
63. Ana Matković
64. Narae Hwang
65. Sven DeRijcke
66. Patricia Knezek
67. Michael Drinkwater
68. Doug Lin
69. Hajime Susa
70. Manuel Metz
71. Andreea Font
72. Rachel Smith
73. Antoine Bouchard
74. Sergio Colafrancesco
75. Max Pettini
76. Analia Smith Castelli
77. Patrick Côté
78. Marco Gullieuszik
79. Evan Skillman
80. Luca Rizzi
81. Ekaterina Evstigneeva
82. Margarita Sharina

Participants missing on the photo (in alphabetic order): Nelson Caldwell, Avishai Dekel, Rodrigo Ibata, Katarina Kovač, David Martinez-Delgado, Mario Mateo, Massimo Ricotti, Eline Tolstoy, Mark Wilkinson, Irina Yegorova, and Simone Zaggia.

Conference photo was taken by St. Scholz. Other photos in the proceedings were kindly provided by Andreas Koch, Manuel Metz, Rami Rekola, Ovidiu Vaduvescu, and Stefano Zibetti.

Participants

Dominik **Argast**, Institut für Physik, Universität Basel, Switzerland	argast@quasar.physik.unibas
Nobuo **Arimoto**, National Astronomical Observatory, Japan	arimoto@optik.mtk.nao.ac
Jeremy **Bailin**, Swinburne University of Technology, Melbourne, Australia	jbailin@astro.swin.edu.
Fabio **Barazza**, Space Telescope Science Institute, Baltimore, MD, USA	barazza@stsci.e
Bruno **Binggeli**, Astronomical Institute Basel, Switzerland	binggeli@astro.unibas
Antoine **Bouchard**, Research School of Astronomy and Astrophysics, ANU, Australia	bouchard@mso.anu.edu.
Nelson **Caldwell**, Smithsonian Astrophysical Observatory, USA	caldwell@cfa.harvard.e
Igor **Chilingarian**, Sternberg Astronomical Institute, Moscow, Russia	chil@sai.msu
Sergio **Colafrancesco**, INAF-Osservatorio Astronomico di Roma, Italy	cola@mporzio.astrc
Matthew **Coleman**, Research School of Astronomy and Astrophysics, ANU, Australia	coleman@mso.anu.edu.
Chris **Conselice**, Caltech, USA	cc@astro.caltech.e
Patrick **Côté**, Rutgers University, USA	patrick.cote@nrc-cnrc.gc
Gary **Da Costa**, Research School of Astronomy and Astrophysics, ANU, Australia	gdc@mso.anu.edu.
Victor **Debattista**, University of Washington, USA	debattis@astro.washington.e
Avishai **Dekel**, The Hebrew University, Israel	dekel@phys.huji.a
Sven **De Rijcke**, Ghent University, Belgium	sven.derijcke@UGent.
Jürg **Diemand**, UCO/Lick Observatory, USA	diemand@ucolick.
Michael **Drinkwater**, University of Queensland, Australia	mjd@physics.uq.edu.
Laura **Dunn**, Research School of Astronomy and Astrophysics, ANU, Australia	laurad@mso.anu.edu.
Wyn **Evans**, Institute of Astronomy, Cambridge, UK	nwe@ast.cam.ac.
Ekaterina **Evstigneeva**, University of Queensland, Australia	katya@physics.uq.edu.
Harry **Ferguson**, Space Telescope Science Institute, Baltimore, MD, USA	ferguson@stsci.e
Andreea **Font**, Wesleyan University, USA	afont@astro.wesleyan.e
Jay **Gallagher**, University of Wisconsin-Madison, USA	jsg@astro.wisc.e
Carme **Gallart**, Instituto de Astrofisica de Canarias, Spain	carme@iac
Alister **Graham**, Research School of Astronomy and Astrophysics, ANU, Australia	graham@mso.anu.edu.
Eva **Grebel**, Astronomical Institute Basel, Switzerland	grebel@astro.unibas
Marco **Gullieuszik**, Dipartimento di Astronomia, Padova, Italy	gullieus@pd.astrc
Daniel **Harbeck**, UC Berkeley, USA	harbeck@ssl.berkeley.e
Monica **Hasegan**, Rutgers University, USA	mhasegan@physics.rutgers.e
Enrico **Held**, Osservatorio Astronomico di Padova, Italy	held@pd.astrc
Maren **Hempel**, Michigan State University, USA	hempel@pa.msu.e
Gerhard **Hensler**, Institute of Astronomy, Vienna, Austria	hensler@astro.univie.ac
Michael **Hilker**, University of Bonn, Germany	mhilker@astro.uni-bonn
Narae **Hwang**, Seoul National University, Korea	nhwang@astro.snu.ac
Rodrigo **Ibata**, Observatoire de Strasbourg, France	ibata@newb6.u-strasbg
Helmut **Jerjen**, Research School of Astronomy and Astrophysics, ANU, Australia	jerjen@mso.anu.edu
André **Jordán**, European Southern Observatory, Garching, Germany	ajordan@eso.
Igor **Karachentsev**, Special Astrophysical Observatory, Russia	ikar@luna.sao
Stefan **Kautsch**, Astronomical Institute Basel, Switzerland	kautsch@astro.unibas
Andrea **Kayser**, Astronomical Institute Basel, Switzerland	akayser@astro.unibas
Patricia **Knezek**, National Optical Astronomy Observatory, USA	pknezek@noao.e
Andreas **Koch**, Astronomical Institute Basel, Switzerland	koch@astro.unibas
Katarina **Kovač**, Kapteyn Instituut, Groningen, The Netherlands	kovac@astro.rug
Pavel **Kroupa**, University of Bonn, Germany, Germany	pavel@astro.uni-bonn
Myung Gyoon **Lee**, Seoul Natonal University, Korea	mglee@astrog.snu.ac
Doug **Lin**, UCO/Lick Observatory, USA	lin@ucolick.
Thorsten **Lisker**, Astronomical Institute Basel, Switzerland	tlisker@astro.unibas
Andrea **Marcolini**, Universita die Bologna, Italy	andrea.marcolini@bo.astr
David **Martinez-Delgado**, IAA, Spain	ddelgado@iaa
Sergey **Mashchenko**, Mc Master University, Canada	syam@physics.mcmaster
Chiara **Mastropietro**, Universität Zürich, Switzerland	chiara@physik.unizh
Mario **Mateo**, University of Michigan, USA	mmateo@umich.e
Ana **Matković**, University of Florida, USA	matkovic@astro.ufl.e
Lucio **Mayer**, Universität Zürich, Switzerland	lucio@physik.unizh
Alan **McConnachie**, Institute of Astronomy, Cambridge, UK	alan@ast.cam.ac.
Manuel **Metz**, University of Bonn, Germany	mmetz@astro.uni-bonn
Dolf **Michielsen**, Ghent University, Belgium	dolf.michielsen@UGent
Steffen **Mieske**, University of Bonn, Germany	smieske@astro.uni-bonn
Bryan **Miller**, Gemini Observatory, Chile	bmiller@gemini.e
Ben **Moore**, Universität Zürich, Switzerland	moore@physik.unizh
Bettina **Ogbuagu-Poledna**, Institute of Astronomy, Vienna, Austria	poledna@astro.univie.ac
Max **Pettini**, Institute of Astronomy, Cambridge, UK	pettini@ast.cam.ac.
Paola **Popesso**, European Southern Observatory, Garching, Germany	ppopesso@eso.
Philippe **Prugniel**, Lyon Observatory, France	prugniel@obs.univ-lyon
Karl **Rakos**, Institute of Astronomy, Vienna, Austria	karl.rakos@chello
Justin **Read**, Institute of Astronomy, Cambridge, UK	jir22@ast.cam.ac
Marina **Rejkuba**, European Southern Observatory, Garching, Germany	mrejkuba@eso.
Rami **Rekola**, Tuorla Observatory, Finland	rareko@ut
Massimo **Ricotti**, University of Colorado at Boulder, USA	ricotti@ast.cam.ac
Luca **Rizzi**, University of Hawaii, USA	rizzi@ifa.hawaii.e
Ivo **Saviane**, European Southern Observatory, Chile	isaviane@eso.
Margarita **Sharina**, Special Astrophysical Observatory, Russia	sme@sao
David **Silva**, European Southern Observatory, Garching, Germany	dsilva@eso.
Evan **Skillman**, University of Minnesota, USA	skillman@astro.umn.e
Rachel **Smith**, Bristol University, UK	rachel.smith@bristol.ac
Analia **Smith Castelli**, National University of La Plata, Argentina	asmith@fcaglp.fcaglp.unlp.edu
Sangmo **Sohn**, Korea Astronomy Observatory, Korea	tonysohn@kao.re
Hajime **Susa**, Department of Physics, Rikkyo University, Japan	susa@rikkyo.ac
James E. **Taylor**, California Institute of Technology, USA	jet@astro.caltech.e
Eline **Tolstoy**, Kapteyn Institute Groningen, The Netherlands	etolstoy@astro.rug
Brent **Tully** University of Hawaii, USA	tully@willick.ifa.hawaii.e
Ovidiu **Vaduvescu**, York University, Canada	ovidiuv@yorku
Liese **van Zee**, Indiana University, USA	vanzee@astro.indiana.e
Matthew **Walker**, University of Michigan, USA	mgwalker@umich.e
Carsten **Weidner**, University of Bonn, Germany	cweidner@astro.uni-bonn
Mark **Wilkinson**, Institute of Astronomy, Cambridge, UK	markw@ast.cam.ac
Joanna **Woo**, The Hebrew University, Israel,	joaw@phys.huji.a
Irina **Yegorova**, SISSA, Italy	yegorova@siss
Simone **Zaggia**, Oss. Astronomico di Trieste, Italy	zaggia@ts.astr
Hong Sheng **Zhao**, University of St Andrews, UK	hz4@st-and.ac
Stefano **Zibetti**, Max Planck Institute for Extraterrestrial Physics, Germany	szibetti@mpe.mpg

Address by the Organizing Committees

Dear colleagues and friends!

On behalf of both the SOC and the LOC I have the great pleasure to welcome you to this IAU conference on dwarf elliptical galaxies in Les Diablerets. We are happy that the response to our call was so positive and that we are now such a large crowd of participants! This certainly shows the timelyness of, and the need for, a conference with this topic. Indeed, the last major meeting on dwarf galaxies dates back 12 years. That occasion was the memorable ESO/OHP workshop in Southern France entitled simply "dwarf galaxies".

We felt that this time, after an explosion of knowledge in the field, the general title "dwarf galaxies" would be too broad for a conference of a suitable, managable and pleasant size, and so we narrowed it down by focussing on early-type dwarf galaxies (which are obviously the pet objects of the organizers!). By this we discouraged large parts – but fortunately not all! – of the communities working on dwarf irregulars, star formation, the Magellanic Clouds, interstellar matter, and radio astronomy.

By early-type dwarfs we mean dwarf ellipticals and dwarf spheroidals. Most of us believe that these are actually one and the same species of galaxies, differing only in luminosity, dwarf spheroidals being the very faintest dEs that until recently were known only in the local neighbourhood, while the brighter dwarf ellipticals are known as the dominant population of nearby clusters of galaxies, such as Virgo. Merely for practical reasons, two communities developped: the dwarf spheroidal community essentially studying very faint stars, and the dwarf elliptical community studying very low-surface brightness "clouds". Fortunately, both of these communities are well represented here. So, I believe this conference offers a unique opportunity to realize, i.e. to make real, the unity of dSphs and dEs.

Today, nobody would dare to claim that dwarf galaxies are irrelevant for astronomy, but we feel there is a kind of overestimation of high-redshift extragalactics that again tends to shade the importance of (low-redshift) dwarf galaxies. The breakthrough for cosmology by the Hubble Deep Field and other deep fields has been, and is, indeed spectacular. But it has to be borne in mind that a correct interpretation of those marvelous deep images can only rest on a complete knowledge of the nearby universe. *The local volume is the only fair sample of the universe we have!* And this is not going to change with the advent of JWST and other great facilities planned for the future.

To emphasize the cosmological significance of the study of the nearby universe and, by implication, dwarf galaxies, we like to use the term "Near-Field Cosmology" – hence the title of the conference. Near-field cosmology is, however, not meant to be in competition with, but rather to complement, Far-Field or Deep-Field Cosmology. This complementary character is nicely expressed by the Yin-Yang-Symbol of Helmut's conference logo. Looking more closely, you will of course notice that the nearby dwarf (NGC 185) is not complemented by a deep field image but by a snapshot from a CDM numerical simulation. Hence we might also regard this as a symbol for the complementary character of observation versus theory, and of visible matter versus dark matter.

Dark matter is of course another, maybe the primary reason why dwarf galaxies are important for cosmolgy. Faint dwarf galaxies, by their extremely low surface brightness, have indeed always given us a taste of dark matter. As it turned out, dwarfs are among the darkest objects in the universe and therefore they are ideal test beds for dark matter theories. Remarkably, both the discovery of the first dwarf spheroidals (Sculptor and Fornax) by Harlow Shapley, and the first hint to the existence of what later was called dark matter by Fritz Zwicky, date back to the 1930ies (70 years ago). For decades there was little follow-up on these observations; e.g. before 1970, Paul Hodge was working almost alone in the field of dwarf galaxies.

The most exciting development began in the 1980ies, when, again more or less coinciding in time, Marc Aaronson, who later died in a tragic accident on Kitt Peak, boldly announced an unusually high velocity dispersion of Draco based on the spectra of only three stars (!), while the CDM model set out to become the standard scenario of structure formation. The latest twist in the close relationship between dwarfs and dark matter is what people like to call the "Missing satellite problem", which undoubtedly will be one of the hot topics of this conference. Another fundamental feature of the favoured CDM scenario is its bottom-up direction of structure formation, rendering the present-day dwarf galaxies the closest relatives to those pregalactic lumps that merged together to form Milky-Way type spiral galaxies.

But I should stop here – there is no need to convince ourselves that dwarf galaxies, and perhaps dwarf ellipticals in particular, are key objects for the study of galaxy formation and evolution. Nor do I want to put up a list of the most pressing open questions around dEs, only to add that we probably won't answer them at this conference, anyway. The questions and possible answers will be presented by you in the following days. What has been achieved will be summarized on Friday morning by Ben Moore and Jay Gallagher.

Let me close by wishing you simply a very fruitful and enjoyable time here – and thank you very much for coming!

Bruno Binggeli, co-chair SOC, chair LOC
Les Diablerets, 14 March 2005

Near-Field Cosmology With Dwarf Elliptical Galaxies
Proceedings IAU Colloquium No. 198, 2005
H. Jerjen and B. Binggeli, eds.

doi:10.1017/S174392130500339X

Near-field cosmology with Local Group dwarf spheroidals

Eva K. Grebel

Astronomical Institute, University of Basel, CH-4102 Binningen, Switzerland
email: grebel@astro.unibas.ch

Abstract. The Local Group offers an excellent laboratory for near-field cosmology by permitting us to use the resolved stellar content of its constituent galaxies as probes of galaxy formation and evolution, which in turn is an important means for testing cosmological models of hierarchical structure formation. In this review, we discuss the the least massive, yet most numerous type of galaxy in the Local Group, the dwarf spheroidal galaxies, and compare their properties to cosmological predictions. In particular, we point out problems found with a simple building block scenario and with effects expected from reionization. We show that the star formation histories of dSphs are inconsistent with the predicted cessation of star formation after reionization; instead, extended star formation episodes are observed. The Galactic dSphs contain in part prominent intermediate-age populations, whereas the Galactic halo does not. Conversely, the M31 dSphs are almost entirely old, while the M31 halo contains a substantial intermediate-age population. These differences in the population structure as well as the differences in the modes of star formation inferred from [α/Fe] ratios make dSphs unlikely major contributors to the build-up of the Galactic and M31 halo unless most of the accretion occurred at early epochs. On the other hand, there is clear evidence for ongoing harassment and accretion of a number of dSphs.

Keywords. galaxies: Local Group, galaxies: dwarf, galaxies: evolution, galaxies: structure, galaxies: interactions, cosmology: observations, Galaxy: formation, Galaxy: evolution, stars: late-type, stars: abundances.

1. Introduction

The Local Group is an excellent laboratory for studies of galaxy evolution at the highest possible resolution in that it provides us with a wide range of different galaxy types and a variety of environments. Yet the Local Group is a poor group, contains relatively few galaxies, and lacks very massive galaxies, such as large ellipticals. Similar to many other nearby groups, the mass and the luminosity of the Local Group are dominated by two large spirals, the Milky Way and M31. Most of the other Local Group members are dwarf galaxies, and the majority of them are found in close proximity to the two large spirals.

Dwarf galaxies are often considered building blocks of more massive galaxies in models of hierarchical structure formation. Dwarf galaxies come in many different flavors and cover a range of masses, luminosities, morphologies, gas content, star formation histories, etc. The distinction between dwarf galaxies and larger galaxies is somewhat fuzzy. The difference is primarily a luminosity difference – it is customary to call galaxies with absolute magnitudes of $M_V > -18$ dwarf galaxies. Gas-rich dwarfs include dwarf spirals, dwarf irregulars (dIrrs) and blue compact dwarf galaxies, which usually show differing levels of ongoing star formation. Gas-poor dwarfs are primarily dwarf ellipticals (dEs). These can be further subdivided into subtypes such as the more massive, strongly centrally concentrated dwarf ellipticals with higher surface brightness, and the less massive, faint, fairly diffuse dwarf spheroidals (dSphs) (see also Gallagher & Wyse 1994; Grebel, Gallagher, & Harbeck 2003).

What makes the Local Group special (apart from its being our home) is that here we have the possibility to actually resolve its constituent galaxies into individual stars and to study the properties of these stars. We can use these stars as probes of the past – they permit us to uncover the evolutionary histories of their host galaxies. Moreover, they permit us to study these evolutionary histories at a level of detail and accuracy that is unmatched by any more distant galaxy, where only the integrated light can be studied. The Local Group is the only place where we can analyze even *ancient* stars and uncover the *early* formation history of individual galaxies beyond our own, particularly of the dwarf companions of the Milky Way. To summarize, the Local Group is ideally suited for studies of "near-field cosmology", i.e., for studies of galaxy evolution over cosmological epochs based on their resolved stellar fossil record, and for tests of the corresponding cosmological models.

2. Local Group Dwarf Spheroidals

The galaxy census of the Local Group remains uncertain. Within the Local Group's volume as defined by its zero velocity surface of $\sim$ 1 Mpc (Karachentsev *et al.* 2002), we currently know of some 38 probable member galaxies. Some were only recently discovered, and additional faint candidates continue to be found (e.g., Zucker *et al.* 2004a,b). All of the newly discovered dwarfs are dSphs, the least massive, least luminous galaxies known, and thus contribute to the faint end of the galaxy luminosity function. For reviews on Local Group dwarfs, see, e.g., Grebel (1997, 1999, 2000, 2001, 2005), Mateo (1998), and van den Bergh (1999, 2000).

DSphs are usually the most numerous type of galaxy in galaxy groups and are characterized by $M_V < -14$ mag, $\mu_V < 22$ mag arcsec^{-2}, $M_{\rm HI} < 10^5$ M$_\odot$, and $M_{tot} \sim 10^7$ M$_\odot$. Often their stellar populations are purely old, but mixtures of old and intermediate-age populations are found as well. In dSphs where several populations can be distinguished the younger and/or more metal-rich populations are more centrally concentrated, indicating extended star formation episodes in the centers of the shallow potential wells of their parent galaxies (Harbeck *et al.* 2001). The gas deficiency of dSphs remains an unsolved puzzle – dSphs typically contain even less gas than expected from red giant mass loss over time scales of several Gyr. The metallicity–luminosity relations of dSphs and dIrrs show the usual trend of increasing metallicity with increasing galaxy luminosity, but the relations are offset from each other: DSphs have higher mean stellar metallicities at a given optical luminosity, which may indicate more rapid star formation and enrichment at early times as compared to dIrrs (Grebel *et al.* 2003).

DSphs do not seem to be supported by rotation and appear to contain large amounts of dark matter. The latter is inferred from the high velocity dispersion and the resulting high mass-to-light ratios derived under the assumption of virial equilibrium. Indirectly, a high dark matter content is also supported by the morphology of some nearby dSphs (Odenkirchen et al. 2001) and by the observed lack of a significant depth extent (Klessen, Grebel, & Harbeck 2003). The radial velocity dispersion profiles of dSphs fall off large radii (Wilkinson *et al.* 2004), possibly indicating the presence of a kinematically cold stellar population at the outermost radii.

If dwarf galaxies in general and dSphs in particular are indeed building blocks of larger galaxies, then today's dwarf population may be considered to be the surviving population of satellites that has not yet been accreted. The most numerous type of dwarfs in galaxy groups, the dSphs, may then be the most pristine members of the original building block population. Studying dSphs may teach us about the properties of objects that presumably were once accreted in large numbers to form galaxies like the Milky Way. Alternatively, it is conceivable that dSphs are in fact not fossil building blocks, but stripped remnants of disrupted and originally much more massive galaxies that have since merged. To find out more about the nature of dSphs and their cosmological significance we need to understand their past and present-day properties.

3. Dwarf Spheroidals: The Earliest Measurable Epoch of Star Formation and Its Cosmological Implications

Cold dark matter models predict that low-mass systems were the first sites of star formation, possibly as early as at a redshift of 30 (e.g., Barkana & Loeb 2001). Since larger systems form through hierarchical merging of smaller systems, they should contain surviving populations of these early epochs of star formation. Furthermore, several models predict that small galaxies should have formed most of their stars prior to reionization, while reionization would have suppressed further star formation activity. In fact, galaxies less massive than 10^9 $M_\odot$ should have lost their star-forming material during reionization (e.g., Susa & Umemura 2004). Hence one would expect that low-mass galaxies contain ancient populations, while star formation should have ceased after reionization.

These are predictions that can be tested in the dwarfs in our immediate surroundings. The least massive dwarfs, the dSphs, should have been most severely affected. Deep color-magnitude diagram data of these dwarfs that reach below the oldest main-sequence turn-offs permit us to carry out relative age dating of their old populations with internal accuracies of fractions of $\sim$1 Gyr, the highest accuracy currently attainable for any method for old stars. Note that this method can only be applied to populations sufficiently numerous to form detectable main-sequence turn-offs. This holds only for old Population II stars, while potential Population III stars are far too few even in our Milky Way. The differential ages of old populations in dwarf galaxies – either field populations or globular clusters – can then be compared to the ages of the oldest Galactic globular clusters of similar composition.

Importantly, this method reveals that (1) old populations are ubiquitous (but their fractions vary) and (2) the oldest ages in all of the galaxies studied so far are indistinguishable within the measurement accuracy (see Grebel 2000 and Grebel & Gallagher 2004 for details). All nearby dwarf galaxies studied in sufficient detail were shown to contain ancient populations. Moreover, these nearby Milky Way companions and the Milky Way itself share a common epoch of star formation for their ancient Population II (within $\sim$ 1 Gyr). These observations are consistent with the expectations from the building block scenario.

However, the predicted cessation of star formation after reionization, expected to have affected particularly dSphs owing to their low mass, is not observed (Grebel & Gallagher 2004). Instead, even dSphs entirely dominated by old populations show evidence for star formation extending over many Gyr (Harbeck *et al.* 2001; Ikuta & Arimoto 2002). In dSphs with a mixture of populations we usually find star formation episodes that lasted many Gyr without being interrupted by a pronounced, long hiatus to the extent that we can measure the duration of star formation (exception: Carina with its episodic star formation). This may mean that the above quoted cosmological models are incorrect and do not properly consider other effects that might prevent complete photoevaporation (e.g., Susa & Umemura 2004). On the other hand, photoionization is one plausible way to circumvent the substructure problem (e.g., Somerville 2002). Alternatively, it is conceivable that dSphs were once considerably more massive (by roughly a factor of 100), which could also have prevented photoionization squelching. In this case the galaxies observed today as dSphs must have undergone substantial mass loss.

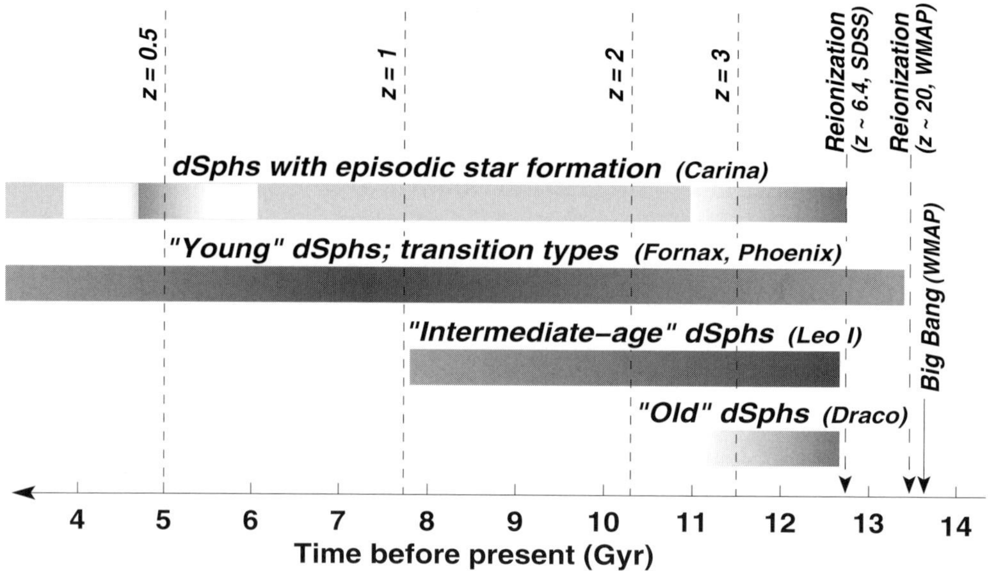

Figure 1. Sketch indicating the approximate duration of star formation episodes in dSph galaxies ($\sim 10^7$ $M_\odot$). The adopted beginning and end of the reionization epoch are based on results from WMAP and from the Sloan Digital Sky Survey. The predicted cessation of star formation after reionization is not observed. For details, see Grebel & Gallagher (2004).

4. Dwarf Spheroidal Star Formation Histories and Abundances

Thanks to deep, high-resolution photometry and synthetic color-magnitude diagram techniques, fairly detailed knowledge of the star formation histories of Local Group dSphs is now available. The resulting picture is one of high complexity: No two dSphs exhibit the same star formation history (Grebel 1997). As mentioned already, all dSphs studied in detail so far were found to contain old Population II stars. Some dSphs are dominated by ancient stars, others only have a small old population and a dominant intermediate-age population, and there is one example of a dSph that experienced star formation as recently as a few hundred Myr ago (Fornax, see Grebel & Stetson 1999). Generally, star formation has proceeded continuously in these galaxies, although the amplitude varied and eventually declined at intermediate or younger ages (e.g., Grebel *et al.* 2003). Only one dSph with clearly episodic star formation is known (Carina, Smecker-Hane *et al.* 1994 and Monelli *et al.* 2003). DSphs with several populations typically show population gradients in the sense that more metal-rich and/or younger populations are more centrally concentrated (Harbeck *et al.* 2001). Substructure of this kind is not necessarily symmetrically distributed (e.g., Stetson *et al.* 1998).

While the past decade was mainly one of photometrically derived star formation histories, we are now entering an era where the age-metallicity degeneracy, which is inherent to purely photometric determinations, can be broken by adding spectroscopic abundance information (e.g., Tolstoy *et al.* 2001; Pont *et al.* 2004, Cole *et al.* 2005, Koch et al. in these proceedings). This will ultimately permit us to derive detailed age-metallicity relations for these galaxies.

4.1. *Comparing stellar populations and star formation histories*

Comparing star formation histories of dwarf galaxies in general and dSphs in particular (e.g., Grebel 1997, 1999), one finds variations in the duration of star formation, in the star formation rates as a function of time, and in the enrichment. In spite of being overall metal-poor, all dSph galaxies that were studied spectroscopically so far show a spread of metallicities of typically 1 dex in [Fe/H] or more (e.g., Shetrone *et al.* 2001, 2003; Bonifacio *et al.* 2004). There appears to be a trend of increasing intermediate-age population fractions with increasing distance from the Milky Way among the Galactic dSphs (van den Bergh 1994; Grebel 1997), which may be due to the environmental impact of the Milky Way.

If environment was indeed the governing factor determining the evolution of these low-mass galaxies, then one should expect to find a similar trend among the dSph companions of M31. However, this is not observed. Although M31's dSphs cover a comparable range of distances as their Galactic counterparts, they all appear to be dominated by old populations and lack the indicators of prominent intermediate-age populations present in the more distant Milky Way dSphs (Harbeck *et al.* 2001, 2004, 2005).

Considering what we can infer from present-day dSphs about their star formation histories, how do they fit in as potential building blocks? With respect to stellar populations, dSphs dominated by old populations are compatible with the stellar content of the Galactic halo. DSphs with substantial intermediate-age populations seem less likely to have made a major contribution to the build-up of the halo of our Milky Way (Unavane et al. 1996). On the other hand, this problem would be diminished if most of the minor merger events took place at very early epochs. Comparing the old, metal-poor stellar populations in M31's dSphs to M31's halo indicates that the dSphs cannot have been primary building blocks of M31's halo since it was found to contain a substantial contribution from intermediate-age, comparatively metal-rich populations (Brown *et al.* 2003). An old, metal-poor halo population, however, has been detected as well (Brown et al. 2004), and again the population differences would be less severe if most of the dSph accretion had taken place at very early epochs, whereas the remainder of the younger halo of M31 would have been formed through the later accretion of more massive and more evolved galaxies. – These statements assume that dSphs have not changed appreciably over time (e.g., did not lose substantial amounts of mass) and that their observed stellar content permits one to arrive at a fair representation of their evolutionary history.

4.2. *Are DSph Abundance Patterns Consistent with the Building Block Scenario?*

During the past few years more and more detailed, high-resolution abundance analyses of individual red giants in dSphs have become available, leading to a growing, yet still limited body of knowledge about their detailed element abundance ratios. In particular, [α/Fe] ratios, r- and s-process abundances are being measured.

If dSphs were dominant contributors to the build-up of the Galactic halo, their abundance patterns should match those of the halo. However, the existing measurements show pronounced differences to the abundance ratios in our Galactic halo: Dwarfs are characterized by slower star formation rates, leading to reduced [α/Fe] ratios at lower metallicity ([Fe/H]) than found in the Galactic halo. This can be interpreted as a signature of a larger contribution of supernovae of Type Ia early on, such that a solar [α/Fe] is reached sooner (e.g., Matteucci 2003). In contrast, the Galactic halo experienced comparatively rapid star formation accompanied by gas removal, leading to low metallicities with higher α element ratios. These different properties lead to the conclusion that dSphs cannot have contributed in a major way to the build-up of the Galactic halo (Shetrone *et al.* 2001), unless the majority of the minor merger events occurred at very early epochs when the abundance ratios in the Milky Way and in the dSphs were still very similar.

6. Concluding remarks

What is the cosmological role of dSphs? With regard to the oldest measurable ages and the earliest epoch of star formation, we find consistency with expectations from cosmological modes. There appears to be a common epoch of early star formation in the Milky Way and in its dwarf companions. In contrast to model predictions, the expected cessation of star formation after reionization is, however, not observed in dSphs. The observed population structure in the (very different) halos of M31 and of the Milky Way, makes it seem unlikely that (present-day) dSphs played a major role in the build-up of the halo of the two spirals. The large variations in the star formation histories of dSphs and the presence of younger populations than in the Galactic halo can be reconciled with the building block scenario if most of the accretion events occurred very early on.

The global metallicities of the Milky Way dSphs are well-matched to those observed in the Galactic halo, but this is not the case for the M31 dSph companions. With regard to detailed chemical element abundance ratios, it is emerging that dSphs cannot have been dominant contributors to halo build-up unless – again – the merger events would have taken place at very early times. The differences in the metallicity-luminosity relation of different types of dwarfs seem to exclude that dSphs are simply stripped dIrrs.

Both disruption and accretion of dwarf companions are still occurring today, demonstrating that dSphs must have played some role in the growth of larger galaxies. Unfortunately, the number and importance of accretion events remains unclear for either of the two large spirals in the Local Group.

In spite of admirable progress, dwarfs remain an evolutionary puzzle. They are excellent and important test cases of cosmological predictions. Regardless of their cosmological importance, however, dwarf galaxies are also interesting in their own right!

Acknowledgements

Many thanks to Helmut Jerjen and Bruno Binggeli for a wonderful conference and for their patience while this contribution was finished. I am also indebted to Jay Gallagher for a critical reading of this text.

References

Barkana, R., & Loeb, A. 1999, *Phys. Rep.* 349, 125

Bonifacio, P., Sbordone, L., Marconi, G., Pasquini, L., & Hill, V. 2004, *A&A* 414, 503

Brown, T.M., Ferguson, H.C., Smith, E., Kimble, R.A., Sweigart, A.V., Renzini, A., Rich, R.M., & VandenBerg, D.A. 2003, *ApJ* 592, L17

Brown, T.M., Ferguson, H.C., Smith, E., Kimble, R.A., Sweigart, A.V., Renzini, A., & Rich, R.M. 2004, *AJ* 127, 2738

Choi, P.I., Guhathakurta, P., & Johnston, K.V. 2002, *AJ* 124, 310

Cole, A.A., Tolstoy, E., Gallagher, J.S., & Smecker-Hane, T.A. 2005, *AJ* 129, 1465

Gallagher, J.S., & Wyse, R.F.G. 1994, *PASP* 106, 1225

Gallagher, J.S., Madsen, G.J., Reynolds, R.J., Grebel, E.K., & Smecker-Hane, T.A. 2003, *ApJ* 588, 326

Grebel, E.K. 1997, *Reviews in Modern Astronomy* 10, 27

Grebel, E.K. 1999, in: P. Whitelock & R. Cannon (eds.), *The Stellar Content of the Local Group*, IAU Symp. 192, (San Francisco: ASP), p. 17

Grebel, E.K. 2000, in: F. Favata, A.A. Kaas, & A. Wilson (eds.), *Star formation from the small to the large scale*, 33rd ESLAB Symposium, ESA-SP 445 (Noordwijk: ESA), p. 87

Grebel, E.K. 2001, *A&SSS* 277, 231

Grebel, E.K. 2005, in: J. Mikolaewska & A. Olech (eds.), *Stellar Astrophysics with the World's Largest Telescopes*, AIP Conf. Proc. Vol. 752 (New York: AIP), p. 161

Grebel, E.K., Gallagher, J.S., & Harbeck, D. 2003, *AJ* 125, 1926

Grebel, E.K., & Gallagher, J.S. 2004, *ApJ*, 610 L89
Grebel, E.K., & Stetson, P.B. 1999, in: P. Whitelock & R. Cannon (eds.), *The Stellar Content of the Local Group*, IAU Symp. 192, (San Francisco: ASP), p. 165
Harbeck, D. *et al.* 2001, *AJ* 122, 3092
Harbeck, D., Gallagher, J.S., & Grebel, E.K. 2004, *AJ* 127, 2711
Harbeck, D., Gallagher, J.S., Grebel, E.K., Koch, A., & Zucker, D.B. 2005, *ApJ* 623, 159
Ikuta, C., & Arimoto, N. 2002, *A&A* 391, 55
Ibata, R.A., Gilmore, G., & Irwin, M.J. 1994, *Nature* 370, 194
Ibata, R., Irwin, M., Lewis, G., Ferguson, A.M.N., & Tanvir, N. 2001, *Nature* 412, 49
Karachentsev, I.D., et al. 2002, *A&A* 389, 812
Klessen, R.S., Grebel, E.K., & Harbeck, D. 2003, *ApJ* 589, 798
Martin, N.F., Ibata, R.A., Bellazzini, M., Irwin, M.J., Lewis, G.F., & Dehnen, W. 2004, *MNRAS* 348, 12
Mateo, M. 1998, *ARA&A* 36, 435
Matteucci, F. 2003, *Ap&SS* 284, 539
Mayer, L., Governato, F., Colpi, M., Moore, B., Quinn, T., Wadsley, J., Stadel, J., & Lake, G. 2001, *ApJ* 547, L123
Monelli, M., et al. 2003, *AJ* 126, 218
Newberg, H.J., et al. 2002, *ApJ* 569, 245
Odenkirchen, M., et al. 2001, *AJ* 122, 2538
Palma, C., Majewski, S.R., Siegel, M.H., Patterson, R.J., Ostheimer, J.C., & Link, R. 2003, *AJ* 125, 1352
Pont, F., Zinn, R., Gallart, C., Hardy, E., & Winnick, R. 2004, *AJ* 127, 840
Rocha-Pinto, H.J., Majewski, S.R., Skrutskie, M.F., Crane, J.D., & Patterson, R.J. 2004, *ApJ* 615, 732
Shetrone, M.D., Côté, P., & Sargent, W.L.W. *ApJ* 2001, 548, 592
Shetrone, M., Venn, K.A., Tolstoy, E., Primas, F., Hill, V., & Kaufer, A. 2003, *AJ* 125, 684
Skillman, E.D., & Bender, R. 1995, *RMxAC* 3, 25
Smecker-Hane, T.A., Stetson, P.B., Hesser, J.E., & Lehnert, M.D. 1994, *AJ* 108, 507
Somerville, R.S. 2002, *ApJ* 572, L23
Stetson, P.B., Hesser, J.E., & Smecker-Hane, T.A. 1998, *PASP* 110, 533
Susa, H., & Umemura, M. 2004, *ApJ* 600, 1
Tolstoy, E., Irwin, M.J., Cole, A.A., Pasquini, L., Gilmozzi, R., & Gallagher, J.S. 2001, *MNRAS* 327, 918
Unavane, M., Wyse, R.F.G., & Gilmore, G. 1996, *MNRAS* 278, 727
van den Bergh, S. 1999, *A&ARv* 9, 273
van den Bergh, S. 2000, *The Galaxies of the Local Group*, Cambridge Astrophysics Series Vol. 35 (Cambridge: Cambridge University Press)
Wilkinson, M.I., Kleyna, J.T., Evans, N.W., Gilmore, G.F., Irwin, M.J., & Grebel, E.K. 2004, *ApJ* 611, L21
Yanny, B., et al. 2003, *ApJ* 588, 824
Zucker, D.B., *et al.* 2004a, *ApJ* 612, L117
Zucker, D.B., *et al.* 2004b, *ApJ* 612, L121

Discussion

CONSELICE: What fraction of the stellar mass in Local Group dSphs formed before reionization?

GREBEL: In order to answer that question we need to determine *absolute* ages as opposed to *relative* ages, which forces us to rely on isochrone models whose oldest ages are inconsistent with the age of the Universe as inferred from WMAP results. Furthermore, we would need the ability to carry out highly accurate age determinations for individual stars in the old Population II to be able to establish when exactly they were formed – before, during, or after reionization. We lack such accurate age indicators.

FERGUSON: A pretty clear prediction is that the pre-reionization stellar population ought to have high [α/Fe]. Do the lowest metallicity stars in dSphs have high [α/Fe]? How well do we know high [α/Fe] for the lowest-metallicity stars in dSph galaxies?

GREBEL: Metal-poor giants in dSphs have high [α/Fe], but their values lie on average below those of Galactic stars at the same [Fe/H] (see, e.g., the compilation in Venn *et al.* 2004). Very few candidates for very metal-poor giants in dSphs have been identified so far, and even fewer have had detailed high-resolution spectroscopic abundance determinations. Those for which metallicities exist tend to like below the Galactic [α/Fe]. Clearly, more data are needed.

TAYLOR: Are there well-established age-metallicity relations within individual dwarfs?

GREBEL: This depends on your definition of "well-established". Synthetic color-magnitude diagram techniques provide age-metallicity relations, but purely photometric data may be affected by age-metallicity degeneracies. If instead one has the means to determine one of the ingredients through an independent method, this knowledge can be used as a fixed parameter when fitting the photometric data points. More specifically, when chemical abundances can be determined independently via spectroscopy, this knowledge can be taken into account when fitting synthetic color-magnitude diagrams to photometric data. Such studies are currently underway by several groups. Depending on the number of stars measured photometrically and spectroscopically and on the quality of the models used, well-established age-metallicity relations should result.

TAYLOR: Are there any dwarfs whose old populations look like/are consistent with the halo?

GREBEL: In terms of age, yes. As I pointed out, the old Population II stars in dwarfs and in the Galactic halo are coeval within the measurement accuracy. The mean metallicities are consistent as well. With regard to [α/Fe], there are the mentioned discrepancies. Note that this does not *exclude* dSph accretion; this merely means that they were not the *dominant* contributors to the halo build-up.

GALLART: I don't think we have enough information to exclude substantial amounts of intermediate-age populations in the Andromeda dSphs. We do not have color-magnitude diagrams for them that reach the oldest main-sequence turnoffs. My impression is that you would find at least similar amounts of intermediate-age populations in them as in M31's halo.

GREBEL: I agree that very deep color-magnitude diagrams that reach below the oldest main-sequence turnoffs are highly desirable for the M31 dSph companions. The presently available HST data only reach the horizontal branches. Hence attempts to carry out modeling with synthetic color-magnitude diagrams will be affected by substantial uncertainties. My statement that the M31 dSphs are predominantly old is, however, not based on such modeling. Instead, I am using the following indicators: (1) the absence of a red clump. (2) The absence of a clearly recognizable population of luminous AGB stars. (3) The lack of carbon stars. (4) The lack of other indicators such as a blue main sequence and a blue plume. From this I conclude that intermediate-age populations, if present at all, cannot be very numerous.

Near-Field Cosmology with Dwarf Elliptical Galaxies
Proceedings IAU Colloquium No. 198, 2005
H. Jerjen and B. Binggeli, eds.

doi:10.1017/S1743921305003406

The stellar populations of local dwarfs

Enrico V. Held

Osservatorio Astronomico di Padova, INAF, vicolo dell'Osservatorio 5, I-35122 Padova, Italy

Abstract. Recent progress in our knowledge of stellar populations in local dwarf spheroidal galaxies is briefly discussed. A few results are summarized including wide field observations of stellar populations and their spatial variations, studies of AGB and variable stars, extension to near-infrared wavelengths, and the interpretation effort based on synthetic color-magnitude diagrams and chemical evolution models.

Keywords. galaxies: dwarf, Local Group, galaxies: stellar content, galaxies: evolution

1. Introduction

The study of stellar populations of nearby dwarf galaxies provides a powerful means of learning, albeit indirectly, about the evolution of low mass galaxies at large look-back times. Galaxies in the Local Group are close enough that they can be resolved in the constituent stars and studied using the classical tools of stellar population work, the HR diagram in the first place. Diagnostics derived from the color-magnitude diagram of local dwarf galaxies are used to pinpoint and, to some extent, quantify the presence of young, intermediate-age (1–9 Gyr), and old ($\gtrsim$ 10 Gyr) stellar populations and to reconstruct their histories of star formation. Recent advances with spectroscopy of individual stars and nebulae provide complementary information to understand their chemical enrichment histories. I will focus here on the resolved stellar populations of dwarf spheroidal (dSph) galaxies, with the exclusion of nearest Milky Way companions and the dwarf galaxies around M 31 and in nearby groups, which are the subject of other contributions to this conference. I refer the reader to existing reviews (Mateo 1998, Da Costa 1998, van den Bergh 2000, Grebel 2000, Tolstoy 2003, Gallagher & Grebel 2004) for earlier work and a comprehensive account of stellar populations in Local Group dwarfs as emerged from the last decade of HST and ground-based studies.

2. The stellar populations of dwarf spheroidal galaxies

2.1. *Stellar populations: recent results*

Direct age measurements of the main-sequence turnoff in nearby dSph, as well as indirect evidence from horizontal branch stars in more distant dwarfs, have shown that all dwarf spheroidals in the Local Group are characterized by a common early epoch of star formation approximately coeval to the Galactic globular clusters (10–13 Gyr ago) (e.g., Held *et al.* 2000; Grebel 2000). This initial star formation episode is predominant in some dSph, but the general evidence is for multiple episodes or a continuous star formation until very recent times.

Two examples of the color-magnitude diagrams (CMDs) of "young" and "old" dwarf spheroidal galaxies are shown in Fig. 1: in both cases, the stellar populations show a complexity typical of a range of age and metallicity. Fornax (see Fig. 1, left panel; also

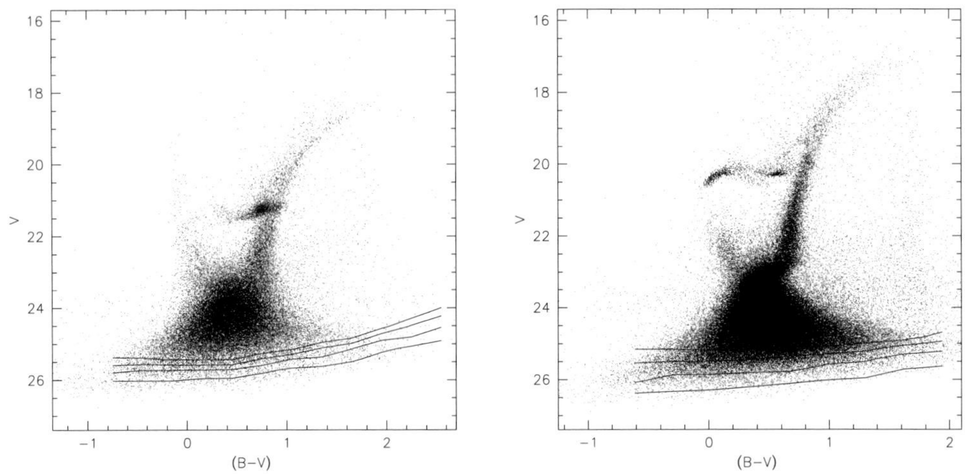

Figure 1. Wide-field color-magnitude diagrams of Fornax (*left*) and Sculptor (*right*) dwarf spheroidals based on ESO/WFI observations (Rizzi *et al.* 2005, in prep.).

Saviane et al. 2000) is an example of a "young" dwarf spheroidal: its CMD is characterized by a "blue plume" of main-sequence stars indicating an extended period of star formation, until a few hundred Myr ago. The very wide red giant branch clearly shows the presence of a mix of old and intermediate-age stars with different metal abundances. Pont *et al.* (2004) have recently studied the star formation and chemical enrichment in a central region of Fornax by combining VLT photometry and stellar spectroscopy. Their observations suggest an increase in star formation rate and metallicity in the last few Gyr: the metallicity distribution of Fornax is centered at [Fe/H]$= -0.9$ with a long tail of metal-poor stars extending to [Fe/H]$= -2.0$ and a number of stars as metal-rich as [Fe/H]$= -0.4$.

The Carina dwarf is another example of dSph galaxy with a prominent intermediate-age population. With two main bursts occurring 5 and 11 Gyr ago, and a small population of stars on a younger main sequence (up to 0.6 Gyr ago), Carina is the best example of an episodic star formation history (see Monelli *et al.* 2003; and refs. therein). This galaxy also provides the most dramatic example of an "age-metallicity degeneracy", i.e. the balance of competing and superposed effects of a younger age and an increasing metallicity on the CMD distribution of red giant stars. Yet its very narrow red giant branch, seemingly contrasting the multiple star formation episodes, does not imply a small range in [Fe/H]. A quantitative analysis of the CMD of Carina based on synthetic diagrams shows that the narrowness of the RGB is indeed consistent with the combined effects of the star formation and chemical enrichment history (Rizzi *et al.* 2003). The presence of a wide range in metallicity is confirmed by spectroscopy (Koch *et al.* 2004).

Figure 1 (right panel) shows the color-magnitude diagram of Sculptor. Sculptor is an "old" dwarf spheroidal galaxy lacking a significant star formation in the last 5 Gyr. It shows a particularly strong radial variation in the HB morphology index, which is accompanied by a spatial gradient in the RGB stars in the outer regions (Harbeck et al. 2001; Tolstoy *et al.* 2004; Babusiaux et al. 2005). The composite nature of the RGB (in particular the existence of a substantial metal-poor component) suggested by previous studies is not confirmed by recent photometry (Babusiaux *et al.* 2005). The presence of these gradients in both the HB morphology and the RGB implies that a significant metallicity gradient must be present (Rizzi *et al.* 2004). Using spectroscopic estimates of [Fe/H] for RGB stars, Tolstoy *et al.* (2004) indeed find two components among red giant stars, with the metal-rich component more centrally concentrated than the metal-poor stars.

Similar CMD morphology gradients are seen in Sextans dSph (Harbeck et al. 2001; Lee *et al.* 2003). Our own synthetic CMD analysis suggests that in this case the stellar population gradient is driven by both metallicity and age variations. A hint for the origin of the changes in the stellar populations within Sextans dSph may also come from a comparison with internal kinematics. Indeed, Kleyna et al. (2004) associate a cold, inner stellar component having velocity dispersion close to zero with the change in the stellar populations near the center.

Other recent studies of old, metal-poor dwarf spheroidals include Draco, Ursa Minor, and Cetus (Aparicio *et al.* 2001; Carrera et al. 2002; Wyse *et al.* 2002; Sarajedini et al. 2002). In Draco and Ursa Minor, the metallicity distribution is narrow yet with a measurable spread and slightly different shapes indicating different star formation histories (Bellazzini *et al.* 2002).

2.2. *An extended star formation in old dwarf spheroidals ?*

One interesting result of wide-field observations is that star formation may have continued at a low rate up to a recent (a few Gyr) epoch even in predominantly old dwarf spheroidals. In the CMD of Sextans, Ursa Minor, and Sculptor (see Fig. 1), a "plume" of stars is obviously present above the main-sequence turnoff. If those stars are main sequence stars, their presence implies a modest rate of star formation up to ~ 2 Gyr ago, after the bulk of stars was formed at old epochs. However, the possibility cannot be ruled out that these stars are "blue stragglers" like those found in globular clusters, which are likely to be the evolved products of mass transfer in old binary systems (see Lee et al. 2003). The recent studies of Ursa Minor and Sextans conclude that the "blue plume" is more likely to be made of old GC-like "blue stragglers" rather than intermediate-age normal main-sequence stars (Carrera et al. 2002; Lee *et al.* 2003). A similar interpretation is proposed for the "blue plume" stars in Sculptor (Rizzi *et al.* 2004), while a different conclusion is reached by Aparicio *et al.* (2001) for Draco, where an analysis of the red clump and subgiant branch stars supports the existence of an intermediate age population, up to 2–3 Gyr ago. These different interpretations obviously affect our picture of early star formation in old dwarf spheroidals and their relation to the reionization era (see Grebel & Gallagher 2004).

2.3. *Evolved stellar populations*

The intermediate-mass stars on the extended asymptotic giant branch (AGB), in particular the carbon-rich (C) stars, are important tracers of intermediate-age populations in dwarf galaxies (for a review of AGB stars in Local Group galaxies, see Groenewegen 2004). Because of their high luminosities and distinctive spectral features, C stars are relatively easy to identify on the basis of narrow-band and near-infrared colors. Carbon star surveys have been conducted by different groups (e.g., Kerschbaum *et al.* 2004; Battinelli & Demers 2005; Harbeck *et al.* 2004). The formation of C stars is both a function of the star formation history and metallicity. Because of this dependence on metallicity, not only is the ratio between carbon-rich and oxygen-rich cool giants, C/M, globally a function of galaxy metallicity (see, e.g., Battinelli & Demers 2005, and references therein) but also it can be used to map *local* abundance variations across galaxies. New models including revised effects of variable opacity in carbon-rich stars are now able to better reproduce the observed distribution of luminous AGB stars in near-infrared CMDs (Marigo 2002).

Since even a few luminous AGB stars can significantly contribute to the integrated light of galaxies, studying the behavior of AGB stars in nearby dwarfs represents an important step towards a successful intepretation of the spectra of distant dwarf galaxies. Evolutionary population synthesis models focusing on intermediate-age populations and their contribution to the integrated light have been presented by Mouhcine & Lançon (2003).

The cool AGB as well as the RGB stars are best observed in the near-infrared, where bolometric corrections to intrinsic luminosities and effective temperatures are smaller, making interpretation of the observations on the basis of stellar evolution models more reliable. The optical-near infrared colors of RGB stars are more sensitive to metallicity than optical (or infrared) colors alone, in particular for old, metal-poor systems. This alleviates the age-metallicity degeneracy and allows an improved estimate of both mean metallicities and metallicity distributions (Babusiaux *et al.* 2005; Gullieuszik *et al.*, this conference). If we exclude the Sagittarius dSph, which is close enough to be studied with the 2MASS database (e.g., Cole 2001), the sensitivity and small size of near-infrared instrumentation have limited the number of studies to very few dwarf spheroidals (Menzies *et al.* 2002; Pietrzyński *et al.* 2003; Babusiaux et al. 2005). Ongoing wide-area studies of dSph galaxies by our group as well as other teams aim at filling this gap, especially using the new wide-field instruments.

2.4. *Variable stars as stellar population tracers*

Pulsating variable stars have been studied in dSph galaxies for many years (Mateo 1998). Besides providing independent distance estimates to the galaxies, variable stars can trace the presence and properties of different stellar generations. In particular, RR Lyrae variables originate from the oldest stars ($\gtrsim$ 10 Gyr), so that an old stellar population can be detected even when the data are too shallow or confusion-limited to reach the main sequence turnoff of low-mass stars. Measurements of RR Lyrae metallicity based on their pulsational properties (period and amplitude of the light curve) help us to reconstruct the metal enrichment history of dwarf spheroidals. In addition, the radial distribution of variable stars in different classes can be used to map the gradients in their parent stellar populations (e.g., Gallart *et al.* 2004).

Several recent studies have addressed the properties of variable stars in dSph galaxies (e.g., Leo I: Held *et al.* 2001; Fornax: Bersier & Wood 2002, Poretti *et al.* 2005, in prep.; Carina: Dall'Ora *et al.* 2003; Draco: Bonanos *et al.* 2004; Phoenix: Gallart *et al.* 2004; and references therein). The RR Lyrae in most dSph galaxies appear to be intermediate between those in Oosterhoff type I and type II globular clusters (e.g., Dall'Ora *et al.* 2003), a characteristic shared by the globular clusters belonging to Fornax dSph (Mackey & Gilmore 2003).

Other classes of variable stars bear information on the stellar content of local dwarfs. An interesting class is Anomalous Cepheids, which are found numerous in dSph galaxies, where their number per unit luminosity seems to be correlated to the galaxy metallicity and luminosity (e.g., Pritzl *et al.* 2005). That Anomalous Cepheids are related to the short-period Classical Cepheids found in (dwarf) irregular galaxies is still a matter of debate (Dolphin 2002; Baldacci *et al.* 2004; Marconi et al. 2004; Gallart *et al.* 2004; Pritzl et al. 2005). As such, the Anomalous Cepheid would trace the intermediate-age populations rather than the old component.

2.5. *Modeling the star formation histories*

Presently, the main challenge is to interpret observational facts in terms of galaxy evolution and understand the chemical enrichment and star formation histories of dSph galaxies (see, e.g., Gallagher & Grebel 2004). Disentangling the age and metallicity variations of the stellar populations is a complex task that models based on evolutionary tracks and synthetic color-magnitude diagrams have recently undertaken (Hernandez *et al.* 2000; Ikuta & Arimoto 2002; Dolphin 2002; Rizzi *et al.* 2004; Pont et al. 2004). While in many cases the models assume a metal enrichment law, some chemical evolution models have started to shed some on the interplay of star formation and chemical evolution of dwarf spheroidal galaxies (e.g., Ikuta & Arimoto 2002; Carigi *et al.* 2002; Lanfranchi & Matteucci 2004). For example, in the Lanfranchi & Matteucci (2004) models, the metallicity distributions of stars in dSph galaxies are predicted using the star formation histories derived from the CMDs together with physical constraints such as the mass and gas content, and the observed abundance patterns from high-resolution spectroscopy (e.g., Tolstoy et al. 2003; Shetrone 2004). The overall properties of dwarf spheroidals imply very low star formation rates and a high efficiency of galactic winds in order to reproduce the observed gas fraction and abundance ratios, a conclusion on which the different models mostly agree.

Environmental factors, such as interaction with the Galaxy and M 31, may also influence the evolution of dSph galaxies. In the case of the Milky Way satellites, our understanding of the relationship between history of star formation and *orbital motions* can benefit from recent analyses of proper motions. For Fornax, the work of Dinescu *et al.* (2004) suggests that the termination of star formation about 200 Myr ago indicated by color-magnitude diagrams coincides with the time when Fornax crossed the Magellanic Plane. However, in the case of Carina, the history of star formation is not explained by passages through perigalacticon and the Galactic disk – the interval between multiple passages being too short to explain the main star formation episode about 6-7 Gyr ago (Piatek et al. 2003). Thus, while interaction with the Milky Way is suggested to be able to trigger or stop star formation (by gas removal), it is certainly not the only driving factor of their star formation histories.

I wish to thank Ivo Saviane, Yazan Momany, Luca Rizzi, Marco Gullieuszik, and Gianpaolo Bertelli, my collaborators on stellar population research, and the organizers of this enjoyable meeting. The search for variable stars in local dwarfs is done in collaboration with teams led by G. Clementini, E. Poretti, H. Smith, and M. Catelan.

References

Aparicio, A., Carrera, R. & Martínez-Delgado, D. 2001, *AJ* 122, 2524

Babusiaux, C., Gilmore, G. & Irwin, M. 2005, *MNRAS*, 331

Baldacci, L., Clementini, G., Held, E.V., Marconi, M. & Rizzi, L. 2004, in JENAM 2003 Minisymposium on "Asteroseismology and Stellar Evolution", *Communications in Asteroseismology* 145, 32

Battinelli, P. & Demers, S. 2005, *A&A* 434, 657

Bellazzini, M., Ferraro, F.R., Origlia, L., Pancino, E., Monaco, L. & Oliva, E. 2002, *AJ* 124, 3222

Bersier, D. & Wood, P.R. 2002, *AJ* 123, 840

Bonanos, A.Z., Stanek, K.Z., Szentgyorgyi, A.H., Sasselov, D.D. & Bakos, G.Á. 2004, *AJ* 127, 861

Carigi, L., Hernandez, X. & Gilmore, G. 2002, *MNRAS* 334, 117

Carrera, R., Aparicio, A., Martínez-Delgado, D. & Alonso-García, J. 2002, *AJ* 123, 3199

Cole, A.A. 2001, *ApJL* 559, L17

Da Costa, G. S. 1998, in "Stellar astrophysics for the Local Group: VIII Canary Islands Winter School of Astrophysics", eds. A. Aparicio, A. Herrero, & F. Sanchez, Cambridge, New York : Cambridge University Press, p.351

Dall'Ora, M., *et al.* 2003, *AJ* 126, 197

Dinescu, D.I., Keeney, B.A., Majewski, S.R., & Girard, T.M. 2004, *AJ* 128, 687

Dolphin, A.E. 2002, *MNRAS* 332, 91

Gallagher, J.S. & Grebel, E.K. 2004, in "Origin and Evolution of the Elements', Carnegie Observatories Centennial Symposia, Cambridge University Press, eds. A. McWilliam and M. Rauch

Gallart, C., Aparicio, A., Freedman, W.L., Madore, B.F., Martínez-Delgado, D. & Stetson, P.B. 2004, *AJ* 127, 1486

Grebel, E.K. 2000. in "Star formation from the small to the large scale", ESLAB Symp. 33, eds. F. Favata, A. Kaas, & A. Wilson, The Netherlands: European Space Agency, p.87

Grebel, E.K. & Gallagher, J.S. 2004, *ApJL* 610, L89

Groenewegen M. 2004, in ESO Workshop "Planetary Nebulae beyond the Milky Way" (arXiv:astro-ph/0407282)

Harbeck, D., *et al.* 2001, *AJ* 122, 3092

Harbeck, D., Gallagher, J.S. & Grebel, E.K. 2004, *AJ* 127, 2711

Held, E.V., Saviane, I., Momany, Y. & Carraro, G. 2000, *ApJL* 530, L85

Held, E.V., Clementini, G., Rizzi, L., Momany, Y., Saviane, I. & Di Fabrizio, L. 2001, *ApJL* 562, L39

Hernandez, X., Gilmore, G. & Valls-Gabaud, D. 2000, *MNRAS* 317, 831

Ikuta, C. & Arimoto, N. 2002, *A&A* 391, 55

Kerschbaum, F., Nowotny, W., Olofsson, H. & Schwarz, H.E. 2004, *A&A* 427, 613

Kleyna, J.T., Wilkinson, M.I., Evans, N.W., & Gilmore, G. 2004, *MNRAS* 354, L66

Koch, A., Grebel, E.K., Harbeck, D., Wilkinson, M.I., Kleyna, J.T., Gilmore, G.F., Wyse, R.F.G. & Evans, N.W. 2004, in "Chemical Abundances and Mixing in Stars", eds. L. Pasquini & S. Randich, in press (arXiv:astro-ph/0410435)

Lanfranchi, G.A. & Matteucci, F. 2004, *MNRAS* 351, 1338

Lee, M.G., *et al.* 2003, *AJ* 126, 2840

Mackey, A.D. & Gilmore, G.F. 2003, *MNRAS* 345, 747

Marconi, M., Fiorentino, G. & Caputo, F. 2004, *A&A* 417, 1101

Marigo, P. 2002, *A&A* 387, 507

Mateo, M.L. 1998, *ARA&A* 36, 435

Menzies, J., Feast, M., Tanabé T., Whitelock, P. & Nakada, Y. 2002, *MNRAS* 335, 923

Monelli, M., *et al.* 2003, *AJ* 126, 218

Mouhcine, M. & Lançon, A. 2003, *MNRAS* 338, 572

Piatek, S., Pryor, C., Olszewski, E.W., Harris, H.C., Mateo, M., Minniti, D. & Tinney, C.G. 2003, *AJ* 126, 2346

Pietrzyński, G., Gieren, W. & Udalski, A. 2003, *AJ* 125, 2494

Pont, F., Zinn, R., Gallart, C., Hardy, E. & Winnick, R. 2004, *AJ* 127, 840

Pritzl, B.J., Armandroff, T.E., Jacoby, G.H., & Da Costa, G.S. 2005, *AJ* 129, 2232

Rizzi, L., Held, E.V., Bertelli, G. & Saviane, I. 2003, *ApJL* 589, L85

Rizzi, L., Held, E.V., Bertelli, G. & Saviane, I. 2004, *MmSAI* 75, 110

Sarajedini, A., *et al.* 2002, *ApJ* 567, 915

Saviane, I., Held, E.V. & Bertelli, G. 2000, *A&A* 355, 56

Shetrone M.D. 2004, in "Origin and Evolution of the Elements', Carnegie Observatories Centennial Symposia, Cambridge University Press, eds. A. McWilliam & M. Rauch, p. 220

Tolstoy, E. 2003, in "A Decade of Hubble Space Telescope Science", eds. M. Livio, K. Noll, & M. Stiavelli STScI Symp. Series Vol. 14, Cambridge, UK: Cambridge University Press, p. 128

Tolstoy, E., Venn, K.A., Shetrone, M., Primas, F., Hill, V., Kaufer, A. & Szeifert, T. 2003, *AJ* 125, 707

Tolstoy, E., *et al.* 2004, *ApJL* 617, L119

van den Bergh, S. 2000, "The galaxies of the Local Group", Cambridge Astrophysics Series vol. 35 (Cambridge, UK: Cambridge University Press)

Wyse, R.F.G., Gilmore, G., Houdashelt, M.L., Feltzing, S., Hebb, L., Gallagher, J.S. & Smecker-Hane, T.A. 2002, *NewA* 7, 395

Near-Field Cosmology with Dwarf Elliptical Galaxies
Proceedings IAU Colloquium No. 198, 2005
H. Jerjen and B. Binggeli, eds.

doi:10.1017/S1743921305003418

Modeling of star formation histories of dwarf ellipticals

N. Arimoto

National Astronomical Observatory of Japan, Mitaka, Tokyo 181-8588, Japan
email: arimoto@optik.mtk.nao.ac.jp

Abstract. Cosmological numerical simulations suggest that dSphs and dEs formed hierarchically via assembling much smaller stellar/gas clumps. Star formation had occurred intensively during the building up phase that is shorter than ~1 Gyr. Dark matter assembled hierarchically, while gas collapsed dissipatively; as a result the metallicity gradient was built up and stars at the galaxy centre tend to have larger rotational velocity and smaller velocity dispersion. There are increasing evidences, however, that suggest rather long-lasted star formation in dwarf galaxies with much smaller SFRs, with which cosmological simulations need to be reconciled.

Keywords. galaxies: abundances, galaxies: dwarf, galaxies: elliptical and lenticular, cD, galaxies: evolution, galaxies: formation, (galaxies:) Local Group, galaxies: stellar content

---

1. Introduction

Star formation histories of dwarf elliptical galaxies (dEs) are unknown. Classical scenarios suggested that a dE should have experienced an early intensive burst of star formation, which followed quickly by a galactic wind induced by supernovae explosions (eg, Yoshii & Arimoto (1988)). Thus, in short, star formation in a dE is intensive and short. As a result, a dE is composed of old and metal-poor stars whose chemical abundances are similar to the yield patterns of type II supernovae (SNeII). There is an increasing evidence, however, suggesting that dEs might have experienced long lasting star formation with moderately low star formation rate (SFR). For example, metal-poor RGB stars ($[Fe/H] \leqslant -2$) of dwarf spheroidals (dSphs) in the Local Group have much lower $[\alpha/Fe]$ ratios compared to the equivalents of Galactic halo (eg, Shetrone *et al.* (2001)). Spectroscopic analyses of M32, a compact dE in the Local Group, have shown that this galaxy has young/intermediate age stellar populations at least at the galaxy centre (eg, Rose *et al.* (2005)). dSphs in the Local Group often show narrow RGBs in spite of the fact that RGB stars show a wide spread in metallicities, which implies that metal-rich stars are much younger than metal-poor ones (Ikuta & Arimoto (2002)), ie, star formation lasted much longer than expected by a classical monolithic collapse scenario.

Are dwarf galaxies building blocks of more massive ones? Recent Subaru (Suprime-Cam) imaging of Leo A, a dwarf irregular galaxy, has shown that this very small galaxy has an old and metal-poor stellar halo (Vansevicius *et al.* (2004)). This might indicate that even the smallest dwarf galaxies had been formed hierarchically.

Indeed, latest cosmological numerical simulations of dSph/dE galaxies are successful in reproducing chemical and kinematical properties of dSph galaxies by building them up hierarchically, but these models always predict that stars formed intensively at the beginning of galaxy evolution and stopped well before the onset of type Ia supernovae (SNeIa), which apparently contradicts recently piling up evidences for long-lasting star formation.

2. Chemical Enrichment in dSph Galaxies

The abundance patterns of the Draco, Sextans, and Ursa Minor dSphs all suggest that the chemical enrichment occurred with SFRs much lower than normal spiral galaxies (Ikuta & Arimoto (2002)). Figure 1 reproduces the observed stellar abundances of the Draco, Sextans, and Ursa Minor dSphs in the $[Mg/Fe]-[Fe/H]$ diagram (Shetrone *et al.* (2001)). Despite the narrow widths of the RGB, the iron abundances exhibit a large dispersion ($-3 \leqslant [Fe/H] \leqslant -1.4$), and the abundance ratios of the α-elements (e.g., Mg) to iron are near or below solar. The abundance ratio $[Mg/Fe]$ starts to decrease at $[Fe/H] \simeq -2$. On the other hand, $[Mg/Fe]$ in the solar neighbourhood decreases with $[Fe/H]$ at around $[Fe/H] \geqslant -1$, which is generally interpreted as the onset of SNeIa explosions (e.g., Matteucci & Greggio (1986)). Supposing that the decline of $[Mg/Fe]$ in the dSphs is also caused by SNeIa, the star formation should have continued longer than the lifetime (typically $1-2$ Gyr) of progenitors of SNeIa and the SFRs should be much lower than that of the solar neighbourhood. A time-delay model (Matteucci & Brocato (1990)) of the SNeIa precisely predicted that the low SFR results in lower $[Mg/Fe]$ ratios relative to stars in the Galaxy. Ikuta & Arimoto (2002) have shown that the observed abundance patterns of these dSphs can best be reproduced by chemical evolution models in which SFR, initial mass function (IMF), and time duration of initial star formation are in the range of $\epsilon = 0.01 - 0.005$ Gyr^{-1} (cf, $\epsilon_{\odot} = 0.2$ Gyr^{-1} in the solar neighbourhood), the Salpeter IMF, and $\Delta T_{\rm SF} = 3.9 - 6.5$ Gyrs, respectively. This star formation history can reproduce morphologies of the observed CMDs as well, such as narrow RGBs and red HBs and succeeds in solving the second parameter problem of the Draco dSph. Hence, both of the abundance patterns and the CMD morphologies can be explained by the star formation histories characterised by the low SFR and the long duration of the star formation period.

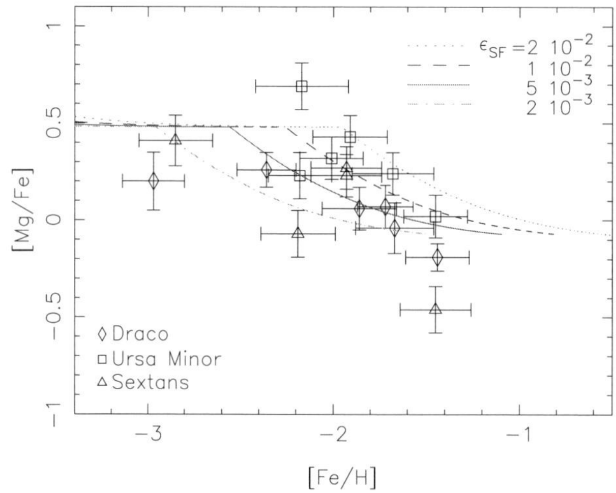

Figure 1. Theoretical abundance patterns together with the observations (Shetrone *et al.* (2001)) for stars in the Draco, Sextans, and Ursa Minor dSphs. The meaning of the marks is written on the panel. The same IMF (the Salpeter IMF: x=1.35) with upper ($m_{\rm u} = 60\ M_{\odot}$) and lower ($m_l = 0.1\ M_{\odot}$) mass limits are adopted in all the model.

3. Subaru/HDS Observation of dSph Ursa Minor

Ikuta & Arimoto (2002) have implicitely assumed that all dSphs should have almost identical star formation history which, however, might not be true judging from variety of CMD morphologies. Keeping this in our mind, we have started a campaign of Subaru/HDS observation of Ursa Minor dSph and have so far obtained high-resolution optical region spectra of 11 RGBs (Sadakane *et al.* (2004); Sadakane *et al.* (2005)). The mean stellar metallicities of 11 RGBs is $\langle [Fe/H] \rangle = -2.0 \pm 0.3$ and many Ursa Minor RGBs show significantly low $[Na/Fe] \leqslant -0.5$ abundances, which is contrary to Galactic GC stars. As was expected, $[Mg/Fe]$, $[Si/Fe]$, $[Ca/Fe]$, and $[Ti/Fe]$ are systematically lower than the Galactic counterparts, starting to decrease at around $[Fe/H] \simeq -2$. $[Mn/Fe]$ shows a mirrow images of $[\alpha/Fe]$, increasing at $[Fe/H] \simeq -2$. All these strongly suggest a significant contribution of SNeIa at such early stage of chemical enrichment of Ursa Minor dSph. In one star COS 82 ([Fe/H] $= -1.5$), however, we have found a large excess of heavy neutron-capture elements with a general abundance pattern similar to the scaled solar system r-process abundance curve. This is a bit puzzling, because it means there is no hint of an s-process (i.e., AGB stars) contribution, even at [Fe/H] $= -1.5$, suggesting a peculiar nucleosynthetic history of the Ursa Minor dSph galaxy (Sadakane *et al.* (2004)).

4. CMD Morphologies of dSph/dE/dIrr Galaxies

CMD morphologies of dSphe/dEs clearly indicate that star formation had lasted 3.5 – 8.0 Gyrs in almost all dSph/dE galaxies. We have studied the star formation histories of a sample of dwarf galaxies in the Local Group, via the analysis and modeling, with the mean of an evolutionary stellar population synthesis, of their CMDs (Ikuta, Arimoto, & Jablonka (2005)). We have built a uniform set of CMDs derived from images taken with the *Hubble Space Telescope*. Our sample is composed of Andromeda I, Andromeda II, Andromeda III, Fornax, GR8, IC10, IC1613, Leo I, Leo II, Leo A, M32, NGC147, NGC185, NGC205, Pegasus, Phoenix, Sextans A, and WLM. We could distinguish several groups, based on their CMD morphology. The width of the RGB determines two different classes of objects, referenced as the Class I and II. Class I galaxies have RGBs narrower than the Class II galaxies. Based on the fraction of young (<1 Gyr) or intermediate age (1 – 5 Gyr) stellar populations, the Class I and Class II are further divided into three different sub-classes from (a) to (c). By means of our CM diagram simulator, we show that the star formation histories of the Local Group dwarf galaxies can be simply described by two main parameters, the SFR and the length of the star formation. Galaxies grouped in the Class I have similar SFRs, regardless of their morphology. The variation of length of their star formation period explains their difference. This suggests a common origin for the early and late type dwarf galaxies. We find a correlation between the duration of the star formation in a given dwarf galaxy and its distance to M31 or the Milky Way; star formation lasted longer in galaxies at larger distance. This suggests that environmental effects play an important role in the evolution of dwarf galaxies and that strong tidal forces might well transform late-type dwarf galaxies into early-type ones. The Class II galaxies are characterised by SFRs stronger than what is necessary to reproduce the properties of the Class I objects (see figure captions of Figs.2–3). We find a possible correlation between the mass of the galaxies and their SFRs, ie, more massive galaxies tend to have much higher SFRs. This correlation might be the origin of the observed luminosity-metallicity relation in the Local Group dwarf galaxies.

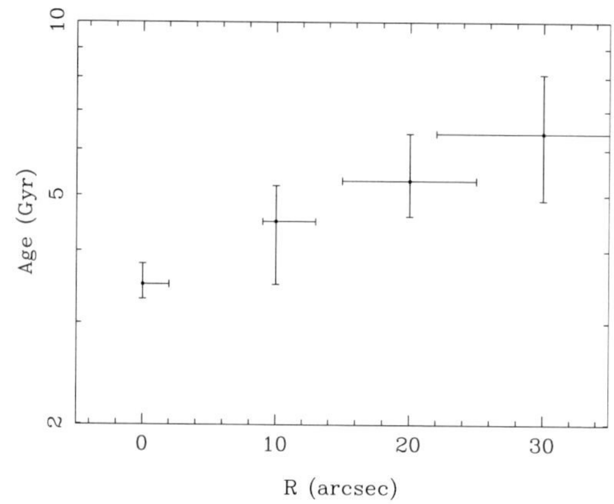

Figure 4. Spectroscopic age gradient of M32. Ages are derived from the robust age indicator H_{γ_σ} defined by Vazdekis & Arimoto (1999).

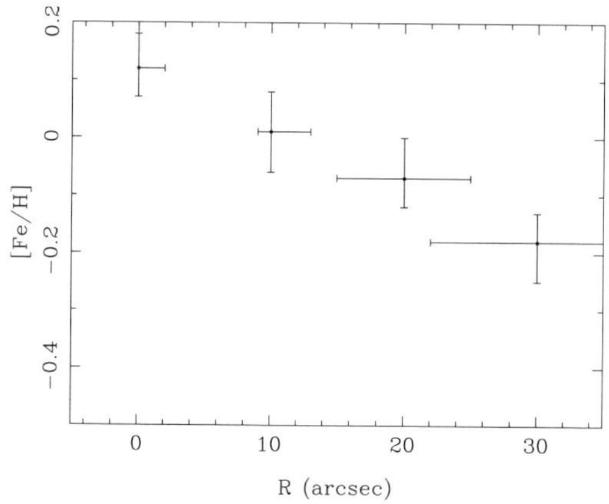

Figure 5. Spectroscopic metallicity gradient of M32. $[Fe/H]$ are derived from Fe3 index. Contrary to Fe3 feature, Mg_2 index does not show any strong gradient, demonstrating very little gradient in $[Mg/H]$ abundance.

of the H I envelope (Sternberg, McKee & Wolfire (2002)), $a = 7.5' - 8.0'$; and (5) a sky background zone where we derived a number density of contaminants to the RGB stars, $a = 8.0' - 12.0'$. From the discovery of an old stellar halo and a sharp stellar edge as well as stellar and gaseous content, Leo A is found to resemble massive disc galaxies. This implies that galaxies of very low stellar mass are also able to develop complex structures.

Another evidence for complex early evolution of dSph galaxies is recently discovered by Tolstoy *et al.* (2004) who found evidence for the presence of two distinct ancient stellar components (both $\geqslant$ 10 Gyr old) in the Sculptor dSph by using the ESO Wide Field Imager in conjunction with the VLT/FLAMES spectrograph. We find that two components are discernible in the spatial distribution of HB stars in our imaging and in the $[Fe/H]$ and $v_{\rm hel}$ distributions for our large sample of spectroscopic measurements. Sculpter dSph is composed of a "metal-poor" component ($[Fe/H] < -1.7$) and a "metal-rich" component ($[Fe/H] > -1.7$). The metal-poor stars are more spatially extended than the metal-rich stars, and they also appear to be kinematically distinct. Therefore, both Leo A (dIrr) and Sculpter (dSph) indicate that even these simplest of galaxies appear to have had a surprisingly complex early evolution.

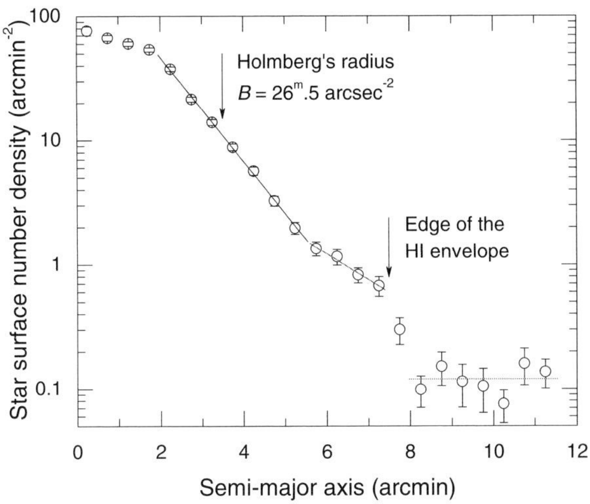

Figure 6. Radial profile of the RGB star surface number density in Leo A.

7. Modeling Star Formation History of Dwarf Ellipticals

To explain complex structure of the Sculpter dSph, we have performed cosmological simulations (Kawata & Arimoto (2005)). Basic features of our model are described by Kawata & Gibson (2003). In the simulation, dSph galaxies formed via hierarchical clustering, but stars formed from cold gas. Star formation started at $z = 20$ and terminated by $z = 4.8$, thus lasted only $\sim$1.0 Gyr, too short to expect significant contribution of SNeIa in chemical enrichment. Stars formed at the galaxy centre tend to form from metal-rich infall gas, which builds up the metallicity gradients. Infalling gas has larger rotational velocity and small velocity dispersion due to dissipative processes. The simulations can explain 1) the metallicity gradient and 2) the kinematic behaviours of stars at different distance from the galaxy centre, but it clearly fails to reproduce the observed abundance pattern of $[\alpha/Fe]$ vs $[Fe/H]$ which is now well established for most of dSph galaxies so far observed, requiring a significant improvement in the modeling SFH of dEs.

8. Conclusions

Modeling of star formation histories of dwarf ellipticals is not yet satisfactory. Cosmological numerical simulations can reproduce the complex features of chemical and kinematical properties of dSph galaxies, but fail to explain why SNeIa had contributed so sigificantly in a very early stage of chemical enrichment.

Acknowledgements

This work was financially supported in part by a Grant-in-Aid for the Scientific Research (No.16540223) by the Japanese Ministry of Education, Culture, Sports and Science.

References

Ikuta, C., Arimoto, N., & Jablonka, P. 2005, *in preparation*
Ikuta, C. & Arimoto, N. 2002, *A&A* 391, 55
Kawata, D. & Arimoto, N. 2005, *in preparation*
Kawata, D. & Gibson, B.K. 2003, *MNRAS* 346, 135
Mateo, M. 1998, *ARA&A* 36, 435
Matteucci, F. & Brocato, E. 1990, *ApJ* 365, 539
Matteucci, F. & Greggio, L. 1986, *A&A* 154, 279
Rose, J.A., Arimoto, N., Caldwell, N., Schiavon, R., Vazdekis, A. & Yamada, Y. 2005, *AJ* 129, 712
Sadakane, K., Arimoto, N., Aoki, W., Ohkubo, M., Ohnishi, K. & Tajitsu, A. 2005, *in preparation*
Sadakane, K., Arimoto, N., Ikuta, C., Aoki, W., Jablonka, P. & Tajitsu, A. 2004, *PASJ* 56, 1041
Shetrone, M.D., Côtè, P. & Sargent, W.L.W. 2001, *ApJ* 548, 592
Sternberg, A., McKee, Ch. & Wolfire, M.G. 2002, *ApJS* 143, 419
Tolstoy, E., *et al.* 2004, *ApJL* 617, L119
Vansevicius, V., Arimoto, N., Hasegawa, T. *et al.* 2004, *ApJL* 611, L93
Vazdekis, A. & Arimoto, N. 1999, *ApJ* 525, 144
Young, L.M. & Lo, K.Y. 1996, *ApJ* 462, 203
Yoshii, Y. & Arimoto, N. 1987, *A&A* 188, 13

Discussion

LIN: The total heavy element content in dSph is comparable to the output of a few supernovae. Unless there is a significant preferential loss of supernovae ejecta, dSph may be over polluted by self contamination. However the formation of several generation of stars, with a metallicity dispersion, requires gas retention. How do you balance heavy element loss, gas retention, and protracted star formation process in your model?

ARIMOTO: Perhaps dSphs were more massive at the beginning, but lost substantial amount of mass later.

FERGUSON: What is the explanation for the different widths of the RGB in your class Ia and IIa models? Is the narrow RGB in the Ia models simply a consequence of the wider distribution of age contradicting the color-spread that comes from metallicity? Or does the different charactor of winds in the Ia and IIa models also playing a role?

ARIMOTO: It is due to the age-metallicity degeneracy effect.

LISKER: In your sample of Virgo dEs, do you have nucleated and non-nucleated dEs, and if so, do you see differences in their spectral indices?

ARIMOTO: No, we do not see any difference.

SILVA: I have two comments: 1) Integrated light (features) is always luminosity-weighted. So, the mass driving the feature may be quite small. Note that due to age-metallicity degeneracy, this issue can affect "age" features and "metal" features. 2) When considering the larger cosmological context, remember that it is now clear that galaxies, supermassive black holes, and large scale structure are forming faster in the real universe than in current generator λCDM based hierarchical merger simulations. In short, in the real universe, the clock is running fatser than in the simulated universe.

Near-Field Cosmology with Dwarf Elliptical Galaxies
Proceedings IAU Colloquium No. 198, 2005
H. Jerjen and B. Binggeli, eds.

doi:10.1017/S174392130500342X

The star formation history of the Fornax dSph galaxy using the new synthetic color–magnitude diagram code IAC-STAR

Carme Gallart,[1] Antonio Aparicio,[1] R. Zinn,[2] R. Buonanno,[3] E. Hardy[4] and G. Marconi[5]

[1]Instituto de Astrofísica de Canarias, 38200 La Laguna, Tenerife, Spain
email: carme@iac.es, antapaj@iac.es

[2]Dep. of Astronomy, Yale University, P.O. Box 208101, New Haven, CT 06520-8101, USA,
email: zinn@astro.yale.edu

[3] Università di Roma Tor Vergata, via della Ricerca Scientifica 1, 00133 Rome, Italy,
email: buonanno@mporzio.astro.it

[4] NRAO, Santiago de Chile, email: ehardy@nrao.edu

[5] ESO, Alonso de Cordova 3107, Vitacura, Santiago de Chile, email: gmarconi@eso.org

Abstract. We present VLT-FORS1 color–magnitude diagrams (CMDs) reaching the oldest main-sequence turnoffs for three fields at two galactocentric distances in the Fornax dSph galaxy. We derive the star formation history (SFH) for each field through comparison with synthetic CMDs and show that the main epoch of star formation has extended to more recent epochs toward the center of the galaxy. A small burst of star formation about 2 Gyr ago is a robust feature in the SFH of all fields. It might be related to the shell structure found in Fornax by Coleman *et al.* (2004), which in turn has been interpreted as the relic of a merger event that occurred in Fornax about 2 Gyr ago.

1. Introduction

The Fornax dwarf spheroidal (dSph) galaxy is - in competition with the Sagittarius dwarf - the most luminous and massive dSph satellite in the Milky Way. Both contain their own globular cluster system. In addition, except for Leo I and Leo II, Fornax is the most distant satellite of the Milky Way. Early hints that Fornax contained a substantial population of intermediate-age stars were provided by the discovery of a large number of C stars by Aaronson & Mould (1980). The first evidence of bright main-sequence stars was provided by Buonanno *et al.* (1985). The intermediate-age and young (down to 100 Myr) population of Fornax, and its stellar population gradients were shown clearly for the first time by Stetson, Hesser, & Smecker-Hane (1998) with a color–magnitude diagram (CMD) of 100 000 stars on a $\frac{1}{3}$ deg^2 field. A deeper HST-WFPC2 CMD reaching the oldest main-sequence turnoffs in a small field near the center of the galaxy was presented by Buonanno et al. (1999). This study indicated that star formation began in the Fornax field $\simeq$ 12 Gyr ago and continued to at least 0.5 Gyr ago. Apparently separated subgiant branches provided tantalizing evidence that star formation occurred in discrete bursts.

In this paper, we present VLT-FORS1 CMDs as deep as the HST one obtained by Buonanno *et al.* (1999) but covering a much larger area, and sampling the Fornax stellar populations at two galactocentric radius. Preliminary results on the detailed star formation history (SFH) in these fields, obtained through comparison with synthetic CMDs, are discussed.

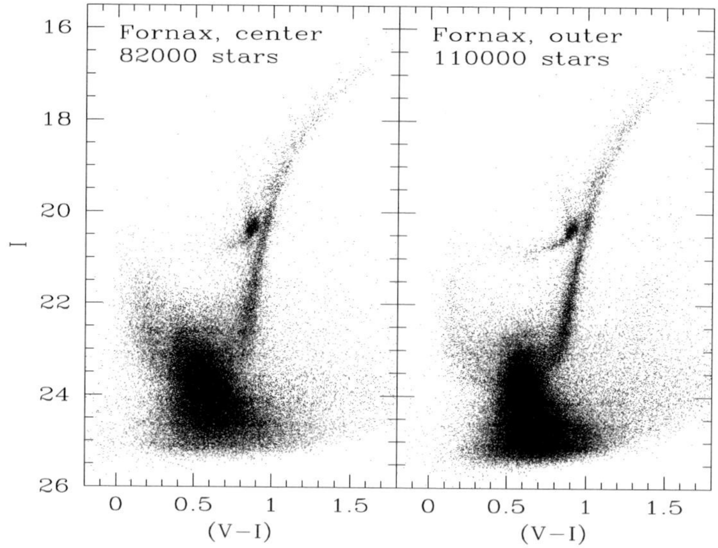

Figure 1. VLT CMDs of Fornax for the central and the outer fields discussed in the text.

2. The Data

We obtained old main-sequence turnoff photometry for three fields at two galactocentric distances in Fornax using FORS1 at the VLT. The observations were performed in service mode in July 2000. The seeing was typically 0.6″, but in some of the frames, with exposure times as long as 750 s, the measured seeing was as good as 0.4″. In the central field containing globular cluster Fornax#4, a total of 3650 s in V and 4700 s in I were accumulated. The other two fields were situated at about 11 arcmin north of the center of the galaxy, and total integration times were 1850 s in V and 4600 s in I. The locations of these off-center fields were chosen taking into account the change in surface brightness across Fornax and selecting fields that would contain similar numbers of stars in the central field and in the other two fields combined.

Figure 1 shows the corresponding CMDs. Note that their depth is similar to that of the HST-WFPC2 CMD by Buonanno *et al.* (1999), but that the number of stars is much larger, thus providing a better representation of the SFH that is less affected by small number statistics.

3. The synthetic CMD code: IAC-STAR

We are performing the analysis of the Fornax CMDs in terms of the SFH and its gradients using the new codes IAC-STAR (available on the Web at http://iac-star.iac.es; Aparicio & Gallart 2004) and IAC-POP (Hidalgo & Aparicio 2005), using a similar method as described in Gallart *et al.* (1999).

IAC-STAR allows the computation of synthetic CMDs for any combination of the star formation rate, SFR(t), chemical enrichment, $Z(t)$, initial mass function (IMF), and frequency and mass distribution of binary stars. Composite stellar populations are calculated on a star-by-star basis, by computing the luminosity, effective temperature, and gravity of each star by direct interpolation in the metallicity and age grid of a library of stellar evolution tracks. Visual and infrared magnitudes are also provided for each star after applying bolometric corrections. The version currently available offers the choice of the stellar evolution libraries of Bertelli *et al.* (1994), Girardi *et al.* (2000), and Pietrinferni *et al.* (2004), as well as various bolometric corrections libraries.

To compare model and observations, the synthetic stars are distributed in an array of "partial" models containing stars within small intervals of age and metallicity, and a set of boxes is defined in the CMD. An array, M_i^j, containing the number of stars from partial model S_i populating box j is computed. The same operation is made in the observational CMD, producing a vector O^j containing the number of observed stars in box j. With this information, the distribution of stars in the defined boxes can be calculated for any SFH as a linear combination of the M_i^j: $M^j = A\sum_i \alpha_i M_i^j$, where A is a scaling constant. The SFH best matching the distribution, O^j, of the observational CMD can be found using a merit function such as χ^2 or, better, χ^2_γ (Mighell 1999), which provides the best solution and a test of its goodness. Finally, the corresponding SFH can be written as $S = A\sum_i \alpha_i S_i$. The best-fitting SFH has been calculated using the code IAC-POP, which uses a genetic algorithm (Charbonneau 1995) to optimize the minimization of χ^2_γ.

4. The star formation history and its gradients

In this section, we present preliminary results of our derivation of the Fornax SFH. The grid of synthetic CMDs used combined the following parameters and functions:

i) Ten $Z(t)$ producing a color distribution in the model CMDs compatible with the observed one; **ii)** four values of the binary fraction, $f = 0.1, 0.3, 0.6, 0.9$, assuming a minimum mass ratio $q = 0.6$; and **iii)** three parameterizations of the IMF, as in Gallart et al. (1999), with a common slope $m^{-1.3}$ for $0.08 \leqslant m \leqslant 0.5$, as in Kroupa, Tout, & Gilmore (1993) for the disk of the Milky Way, and three different slopes for the upper mass interval bracketing those of Kroupa *et al.*

The Bertelli *et al.* (1994; Padova hereinafter) and Pietrinferni *et al.* (2004; Teramo) stellar evolution models were used to create synthetic CMDs with the grid of ingredients listed above. The comparison of these synthetic CMDs with the observed CMD, after the observational errors were simulated in them following the procedure described in Gallart *et al.* (1999), allowed us to identify a number of $Z(t)$ for which we can find an SFR(t) that successfully reproduces the number of stars across the observed CMD. It was not possible to discriminate between the different parameterizations of the IMF. While strictly speaking no unique solution could be found, the accepted solutions define a narrow range of $Z(t)$ and SFR(t) combinations. The $Z(t)$ range is compatible with the age–metallicity relation derived by Pont *et al.* (2004) from CaII triplet spectroscopy. In Figure 2, examples of these solutions are displayed.

The main difference between the SFR(t) in the central and outer fields, which appears independently of the stellar evolution model set used, is that the main epoch of star formation ends about 5 Gyr ago in the outer field, while it extends up to 3 Gyr ago in the central field. Low-intensity star formation, more prominent in the center of Fornax, has continued till just a few hundred Myr ago. In the case of the Padova models, the existence of a small burst about 1.5 Gyr ago is a robust and ubiquitous feature that appears in basically all accepted solutions. This feature disappears, or rather, gets shifted to older ages, in the solutions with the Teramo models. This is a consequence of the known differences between these two model sets, as discussed in Gallart, Zoccali, & Aparicio (2005). In particular, at intermediate ages, the Teramo models assign an older age to stars in the same position of the CMD than the Padova models. For example, a $Z = 0.001$, 2 Gyr old Teramo isochrone matches almost perfectly a 1 Gyr old Padova isochrone. The age of the young burst of star formation, as inferred with the Padova models, is in very good agreement with the age of the stellar population that dominates the shell structure

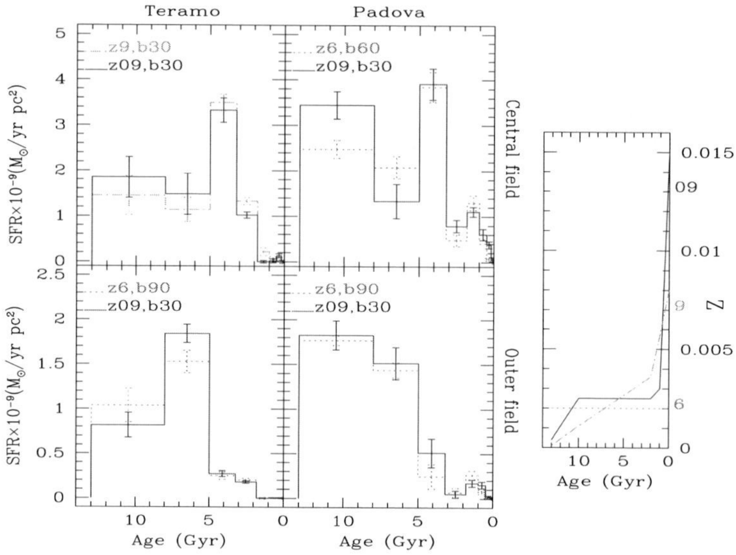

Figure 2. SFHs derived for the two Fornax fields (see text for details), for both the Teramo and Padova models. The labels in each panel indicate the assumed metallicity law (the three possibilities are displayed on the right panel, with their corresponding code indicated), and the percentage of binary stars.

found in Fornax by Coleman *et al.*, (2004), which was in fact dated using the same models. These authors interpret this shell structure possibly as the remnant of a merger with Fornax of a smaller, gas-rich system, about 2 Gyr ago. Since we find the increased SFR at approximately the same age in all three fields sampled, this may indicate that the stars incorporated into Fornax by this smaller system are not confined to the shell structure but have mixed substantially over a large volume of the galaxy.

Acknowledgements

This research is part of a Joint Project between Universidad de Chile and Yale University, partially funded by the Fundación Andes. CG acknowledges partial support from the Spanish Ministry of Science and Technology (Plan Nacional de Investigación Científica, Desarrollo e Investigación Tecnológica, AYA2004-06343) and from the European Structural Funds.

References

Aaronson, M. & Mould, J. 1980, *ApJ* 240, 804

Aparicio, A. & Gallart, C. 2004, *AJ* 128, 1465

Bertelli, G., Bressan, A., Chiosi, C., Fagotto, F. & Nasi, E. 1994, *A&AS* 106, 275

Buonanno, R., Corsi, C.E., Fusi Pecci, F., Hardy, E. & Zinn, R. 1985, *A&A* 152, 65

Buonanno, R., Corsi, C.E., Castellani, M., Marconi, G., Fusi Pecci, F. & Zinn, R. 1999, *AJ* 118, 1671

Charbonneau, P. 1995, *ApJS* 101, 309

Coleman, M., Da Costa, G.S., Bland-Hawthorn, J., Martínez-Delgado, D., Freeman, K.C. & Malin, D. 2004, *AJ* 127, 832

Gallart, C., Freedman, W.L., Aparicio, A., Bertelli, G. & Chiosi, C. 1999, *AJ* 118, 2245

Gallart, C. Zoccali, M. & Aparicio, A. 2005, *ARAA* 43, in press

Girardi, L., Bressan, A., Bertelli, G. & Chiosi, C. 2000. *A&AS* 141, 371

Hidalgo, S. & Aparicio, A. 2005, *in preparation*
Kroupa, P., Tout, C.A. & Gilmore, G. 1993, *MNRAS* 262, 545
Mighell, K.J. 1999. *ApJ* 518, 380
Pietrinferni, A., Cassisi, S., Salaris, M. & Castelli, F. 2004, *ApJ* 612, 168
Stetson, P.B., Hesser, J.E. & Smecker-Hane, T.A. 1998, *PASP* 110, 533

Discussion

MATEO: Do you see different spatial distributions as a function of age? I know the field is small, but the young clump may inject stars in a highly asymmetric manner.

GALLART: It is a good idea. But I haven't actually analyzed this yet. However, note that the "excess" young population seems to be found everywhere we look, even in the off-center field, which is situated opposite to the shell structure with respect to the center of the galaxy.

DA COSTA: Can you briefly summarize the differences between the Padova and Teramo evolutionary models? A factor of two difference in age at 1 Gyr is somewhat embarrassing.

GALLART: The Teramo models have updated most physics input to very recent determinations (equation of state, nuclear reactions, etc.). In this age range the overshooting treatment is also very relevant. It is very difficult to isolate which are the key ingredients that produce this remarkable difference. Pietrinferni et al. (2004) briefly discuss the possible origin of the discrepancy and conclude that only a fraction of it is related to differences in the overshooting treatment, the remainder having to be ascribed to the different input physics. But they didn't isolate the key ingredient. I think it would be very important, in any case, to fine-tune the overshooting treatment, through comparison with observations, specifically for each stellar evolution library.

Near-Field Cosmology with Dwarf Elliptical Galaxies
Proceedings IAU Colloquium No. 198, 2005
H. Jerjen and B. Binggeli, eds.

doi:10.1017/S1743921305003431

Carbon stars in the M 31 dwarf spheroidals: evolutionary implications

D. Harbeck[1], J.S. Gallagher[2], E.K. Grebel[3] and P. Guhathakurta[4]

[1]UC Berkeley, Space Sciences Laboratory, 7 Gauss Way, Berkeley, CA 94720, USA

[2]University of Wisconsin, Madison, 475 N. Charter Street, Madison, WI 53706, USA

[3] Astronomical Institute of the University of Basel, Venusstr. 7, 4102 Binningen, Switzerland

[4] UCO/Lick Observatory, UC Santa Cruz, 1156 High Street, Santa Cruz, CA 95064, USA

Abstract. We present results of a photometric survey for carbon stars in the M 31 dwarf spheroidal galaxies. From the low frequency of carbon stars in these galaxies compared to their Milky Way cousins we conclude that they consist of predominantly ancient stellar populations. However, the complexity and variety of the metallicity distributions in the M 31 dwarf spheroidals suggest that they still experienced an extended epoch of star formation.

Keywords. Local Group, dwarf galaxies, stellar populations

1. Introduction

While similar from a distance, the two massive spirals of the Local Group (LG), the Milky Way (MW) and M 31, appear different when looked at more closely. For example, the halo of M 31 is more metal-rich than that of the MW, suggesting different merger histories of the two systems; M 31 might have experienced a major merger some 7 Gyr ago. In this paper we explore if the differences of the two massive spirals are also reflected in their dwarf galaxy satellite systems. Of particular interest are the dwarf spheroidal (dSph) galaxies, which might be the representation of unused building blocks in hierarchical cold dark matter scenarios of galaxy formation. Their small total and baryonic masses make their star formation history and chemical evolution most susceptible to external regulating mechanisms, such as the onset of reionization or local ram pressure stripping (summarized, e.g., in Grebel *et al.* 2003). Any local differences between the M 31's and MW's halos might also be unveiled by different episodes of star formation in their satellite galaxies.

The proximity of the MW's dSph galaxies allows a detailed study of their star formation histories based on deep color magnitude diagrams that reach the oldest main sequence turn-off, and, more recently, massive spectroscopic surveys allow one to paint a detailed picture of their chemical evolution. Such an approach is not practical for the M 31 companions, where a reasonable commitment of HST time only allows to obtain photometry to the horizontal branch level. To examine the star formation history of the M 31 companions, in particular their content of intermediate-age stellar populations, we depend on indirect tracers of stellar ages such as luminous AGB stars. For the sparse dSphs galaxies, simple star-counts in the luminosity range of AGB stars are unreliable due to foreground star contamination. However, surveys for carbon (C) stars, AGB stars that have undergone deep mixing, would be less confused by foreground stars due to the distinct spectral features of such stars. C-stars can be readily detected from four-color photometry, by using two broad band filters and two narrow band filters centered on the near infrared CN and TiO bands (Cook, Aaronson, & Norris 1986). Here we report on

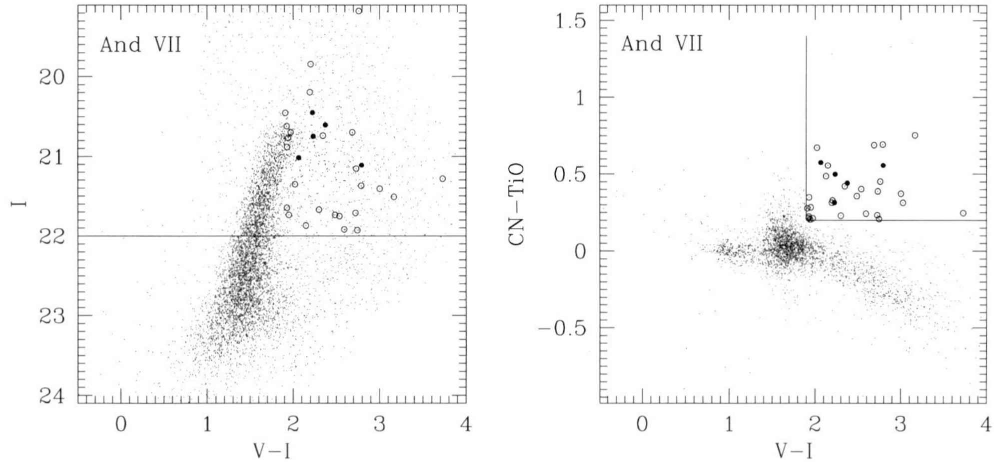

Figure 1. Color-magnitude diagram (left) and two-color diagram (right) of And VII as an example of the survey technique. Open circles show misidentifications due to bright star stray light or cosmic rays. Bona-fide C-stars are drawn with filled circles. Stars above the tip of the red giant branch are considered as C-stars, while less luminous stars are most likely CH stars, evolved dwarf carbon stars.

our survey for C-stars as a tracer of 1 to 10 Gyr old stellar populations in the M 31 dSphs And III, And V, And VI, And VII, and AND IX with the WIYN 3.5m telescope (details of the survey are described in Harbeck *et al.* 2004, 2005).

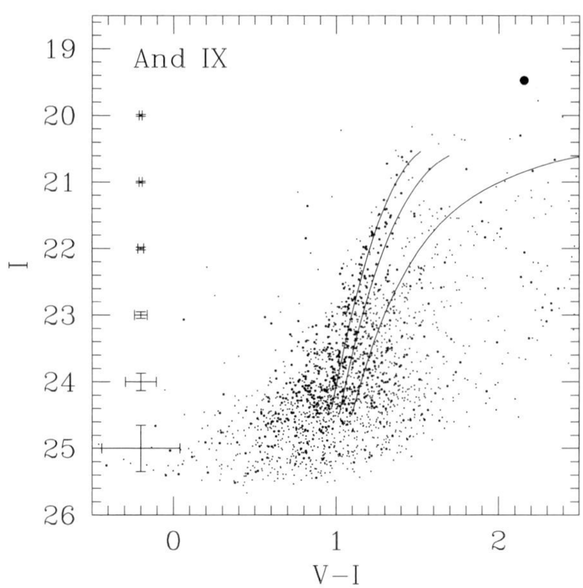

Figure 2. CMD of And IX, based on observations with the WIYN telescope under excellent (0.5″) seeing conditions. Stars within the central arc-minute are plotted with larger dots. We overlay globular cluster fiducials (Da Costa & Armandroff 1990) with metallicities of -2.26, -1.62, -0.76 dex. And IX seems to be very metal poor, which is expected given its low baryonic mass. The more metal-rich red giant branch of the M 31 halo is visible as well.

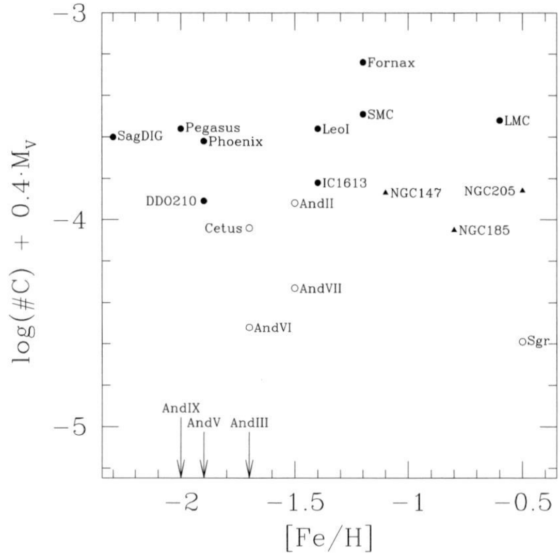

Figure 3. The number of carbon stars – corrected by the host galaxy's luminosity – is compared among Local Group dwarf galaxies. Dwarf galaxies with recent star formation are plotted with filled circles, those with predominantly ancient stellar populations with open circles; triangles refer to dE galaxies. Data for And II taken from Kerschbaum *et al.* (2004).

2. WIYN survey for Carbon stars

The M 31 dSphs galaxies where observed with the WIYN 3.5 meter telescope in Fall 2003 and Fall 2004. Exposure times were 500 s each in V and I filters and 1500 s each in the narrow band TiO and CN filters. The seeing was always better than $0.8''$ and as good as $0.5''$. We present a sample color magnitude diagram (CMD) and two color diagram (TCD) of our survey in Fig. 1, where we show the results for And VII. From the color cut in the TCD diagram we select Carbon star candidates that are cross referenced with the original images. False candidates due to bright saturated stars or cosmic rays are rejected. Stars more luminous than the tip of the RGB are regarded as genuine C-stars, while fainter C-stars are probably evolved dwarf carbon ("CH") stars. To summarize our survey results, we find: no C-stars in And III, And V, and And IX; one C-star in And VI; three C-stars in And VII. There is one C-star candidate in the line of sight towards And IX (Fig. 2), however, given the proximity of And IX to M 31 this candidate might equally belong to M 31's outer disk.

Fig. 3 compares the number of C stars normalized by the luminosity (i.e., baryonic mass) of their host galaxy of Local Group dwarf galaxies. We concentrate on those galaxies where surveys are believed to be complete (see Harbeck *et al.* 2004 for a description of the sample). Two conclusions can be drawn from this graph: (i) compared to the MW dSph galaxies, the And companions appear under-abundant in C-stars, i.e., there are no Fornax-type dSphs among the Andromeda dSphs. It appears that the M 31 dSphs contain predominantly ancient (i.e., older than 10 Gyr) stellar populations. The M 31 dSph show no imprint of M 31's major merging event. (ii) In particular with And II, there seems to be sufficient overlap in the C-star frequency in the M 31 and MW dSphs to conclude that their nature is not substantially different.

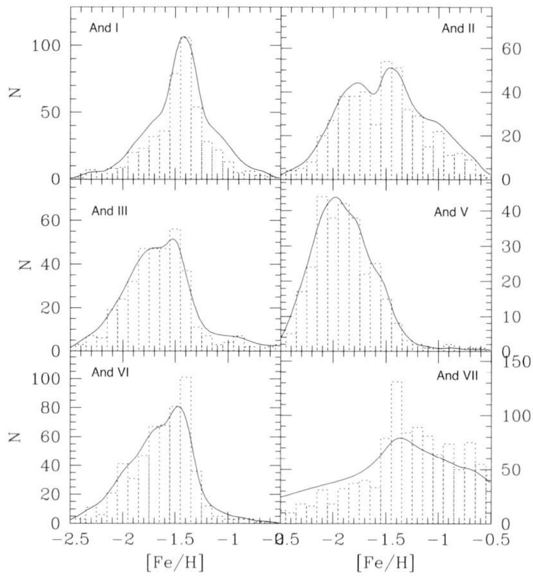

Figure 4. Metallicity distribution of M 31 dSphs galaxies based on RGB star colors (derived from HST photometry, see Harbeck *et al.* 2001 for details). The distribution of And VII is dominated by photometric errors. The variety and broadness of the metallicity distributions suggests very different, partly complex star formation histories of these galaxies.

3. Ancient, but not trivial

The under-abundance of C-stars in the M 31 dSphs suggests a fast decline of star formation activities in these galaxies compared to their Milky Way cousins. However, this does not imply simple enrichment histories. In Fig. 4 we plot the stellar metallicity distributions of M 31 dSphs derived from HST photometry. These metallicity distributions are calculated by comparing colors of red giant branch stars to globular cluster fiducials (Da Costa & Armandroff 1990). We are assuming predominantly old stellar populations in the dSphs (which we have seen to be justified for most M 31 dSphs - maybe with the exception of And II). The variety of different metallicity distributions among the M 31 dSphs is apparent, emphasizing that each galaxy is an individual case. To achieve a broad metallicity distribution as it is seen most of the M 31 dSphs an extended (e.g., at least 4 Gyr) period of star formation is required (Ikuta & Arimoto 2002).

4. Conclusions

C-stars offer a possibility to derive a rough picture of a galaxy's intermediate-age star formation history where deep photometry is not available; at least a classification into galaxies dominated by old or intermediate stellar populations is possible. The apparent under-abundance of C-stars in the M 31 dSphs suggests a faster decline of star formation activity compared to the MW dSphs. The M 31 dSphs' stellar populations are both older and more metal-poor than the M 31 halo; we thus conclude that they are indeed an primordial legacy and are not tidal dwarfs produced during a later major merger event. And II and And VII, with a moderate content of C-stars and broad metallicity might most closely resemble Galactic dSphs with extended epochs of star formation.

Acknowledgements

DH gratefully acknowledges support by an IAU travel grant.

References

Cook, K.H., Aaronson, M. & Norris, J. 1986, *ApJ* 305, 634
Da Costa, G.S. & Armandroff, T.E. 1990, *AJ* 100, 162
Grebel, E.K., Gallagher, J.S. & Harbeck, D. 2003, *AJ* 125, 1926
Harbeck, D. *et al.* 2001, *AJ* 122, 3092
Harbeck, D., Gallagher, J.S. & Grebel, E.K. 2004, *AJ* 127, 2711
Harbeck, D., Gallagher, J.S., Grebel, E.K., Koch, A. & Zucker, D.B. 2005, *ApJ* 623, 159
Ikuta, C. & Arimoto, N. 2002, *A&A* 391, 55
Kerschbaum, F., Nowotny, W., Olofsson, H. & Schwarz, H.E. 2004, *A&A* 427, 613

Discussion

FERGUSON: If you assume all the stars were the same age, how young could you push the age without violating your carbon star constraint?

HARBECK: C-stars (and the lack of them) constrain atellar populations with ages from 10 Gyr to 1 Gyr. Stellar populations younger than 1 Gyr can be excluded from existing CMDs. What happens within the 1-10 Gyr range cannot be constrained by just counting C-stars.

CONSELICE: Could the higher number of Carbon stars in the Milky Way be the result of contamination, as the field of view needed to observe the Milky Way system is larger?

HARBECK: At the luminosity of MW dSph Carbon stars, potential contaminators could only be dwarf Carbon (dC) stars. These, however, are very rare and are unlikely to be a significant contaminator. As dC stars are faint, they would be nearby and could be identified by their proper motions.

Near-Field Cosmology with dwarf Elliptical Galaxies
Proceedings IAU Colloquium No. 198, 2005
H. Jerjen and B. Binggeli, eds.

doi:10.1017/S1743921305003443

The stellar populations of dSph galaxies in nearby groups

G.S. Da Costa

Research School of Astronomy & Astrophysics, The Australian National University
email: gdc@mso.anu.edu.au

Abstract. In this contribution initial results are presented from a program to study in detail the stellar populations of dwarf Spheriodal (dSph) galaxies in three nearby groups. The long-term aim of the program is to assess the influence of environment in governing the evolution of these low-luminosity systems. Specific results described here include the detection and measurement of intermediate-age upper-AGB populations in dSphs in the M81 and Cen A groups, and the discovery that four of the five low-luminosity early-type dwarfs in the low density Sculptor group contain modest amounts of neutral hydrogen gas.

Keywords. galaxies:dwarf, galaxies:stellar content, galaxies: evolution, stars: AGB

1. Introduction

We have known for more than two decades that the dwarf Spheroidal (dSph) companion galaxies to the Milky Way show a variety of star formation histories. These range from essentially a single old stellar population (e.g. Ursa Minor, Olszewski & Aaronson 1985, Carrera *et al.* 2002) through to the complex star formation histories of dSphs such as Carina (e.g. Monelli *et al.* 2003), Fornax (e.g. Saviane *et al.* 2000) and Leo I (e.g. Gallart *et al.* 1999), which contain stars as young as ~1 Gyr, or perhaps even younger. The derivation of these star formation histories is now based on deep colour-magnitude (c-m) diagrams that reach below the main sequence turnoff for the oldest populations, but it is important to recall that the first evidence for the existence of intermediate-age (i.e. age between ~1 and 10 Gyr) populations in dSphs came from the observation that some systems contain upper-AGB stars (cf. Aaronson & Mould 1980). Upper-AGB stars are stars with sufficient mass that they evolve to luminosities well above the tip of the red giant branch in the c-m diagram. Provided the underlying population is relatively metal-poor ($<[Fe/H]> \leqslant -1.0$, approximately), the presence of upper-AGB stars, which are frequently seen as carbon stars, in a stellar population is an unambiguous signature of the existence of an intermediate-age population, and the luminosity of the brightest upper-AGB stars is a measure of the age of the youngest intermediate-age population present.

With advent of HST it has also become possible to determine c-m diagrams that reach below the level of the horizontal branch for the dSph† companions to M31 (e.g. Da Costa *et al.* 2002), and thus gain some first-order information on their star formation histories. All the systems studied contain RR Lyrae variables, indicating that they contain at least some stars with ages comparable to that of the Milky Way globular clusters (this is also true of the Milky Way dSph companions). Nevertheless, the ubiquitous red horizontal branches in the c-m diagrams argue for extended epochs of star formation and "mean ages" younger than that of the Galactic globular clusters (much like the Galactic dSph Leo II – see Mighell & Rich 1996). However, the M31 dSph companions appear to lack the

† Throughout this contribution the term 'dSph' is synomous with 'low luminosity dE'.

substantial intermediate-age upper-AGB populations seen in many of the Milky Way's dSph companions. Confirmed populations of upper-AGB stars are known only in some of the M31 dSph companions, such as And II and And VII (Cas) (see Harbeck *et al.* 2004 and Harbeck, these proceedings), but the luminosities of these stars suggest ages of perhaps 6–9 Gyr, in contrast to the younger (and more luminous) stars seen in Carina, Leo I and Fornax (cf. Da Costa *et al.* 2000).

Within the Local Group there are a further four known low-luminosity early-type dwarf galaxies. These are the transition-type (dIrr/dE) systems Phoenix and LGS 3, that are outlying companions of the Milky Way and M31, respectively, and the relatively isolated dSph systems Tucana and Cetus. Both Phoenix and LGS 3 have modest populations of young stars, contain small amounts of HI gas, and clearly have had star formation from the distant past to the present (cf. Holtzman *et al.* 2000, Miller *et al.* 2001). The somewhat limited data for Cetus shows that this dSph has a red horizontal branch morphology (Sarajedini *et al.* 2002), and that it lacks any obvious upper-AGB population (Harbek *et al.* 2004). It is apparently broadly analogous to the M31 dSph companions. The isolated dSph Tucana, however, shows a relatively strong blue horizontal branch morphology in its c-m diagram (cf. Da Costa 1998); indeed other than the Galactic dSph companion Ursa Minor, the horizontal branch morphology of Tucana is the bluest known of the Local Group dSphs, suggesting a relatively brief star formation epoch in the distant past. A recent determination of the radial velocity of Tucana (Tolstoy, personal communication) has removed any possibility of an association between this dSph and an HI cloud that lies nearby on the sky (cf. Osterloo *et al.* 1996). Given this result, the Oosterloo *et al.* (1996) upper limit of $M_{HI}/L_B \approx 10^{-2}$ in solar units indicates that the Tucana dSph is similar to the Galactic dSphs in lacking neutral gas.

Do we understand what causes this variety of star formation histories in early-type low-luminosity galaxies? In general the answer is 'no' but it is clear from the existing results that 'environment', represented by, for example, proximity to a luminous galaxy, and the type of that galaxy, plays a role in influencing the star formation histories of these systems. Van den Bergh (1994) noted that for the Galaxy's dSph companions, there is a tendency for the systems with larger intermediate-age component to lie at larger galactocentric distances. The same trend may also apply for the M31 dSphs, as the two systems with likely intermediate-age populations, And II and And VII (Cas), are among the most distant from M31. Further, the difference in relative frequency of intermediate-age populations between the Galactic and M31 dSph systems may have its origin in the different 'parent' galaxy types: M31 has a much larger bulge than the Galaxy.

Clearly if we wish to understand the role of environment in influencing the evolution of these low luminosity systems, we need to study the stellar populations of dSphs in groups beyond the Local Group. Fortunately, the nearest groups provide a variety of different environments. The Sculptor group is a loose aggregation of galaxies ranging in distance from ~1.5 to ~4 Mpc. It contains half a dozen or so early-type low-luminosity systems though the group contains mainly late-type galaxies. In contrast, the M81 and Cen A groups are relatively dense, compact groups that lie at distances of ~3.5 to ~4 Mpc. Both groups contain at least a dozen or more low-luminosity early-type galaxies. In all three groups the dSphs cover a range of internal (absolute magnitude, size, etc) and external (distance from nearest large galaxy, local galaxy density) properties.

Together with my collaborators I am involved in a number of programs aimed at studying in detail the stellar populations of the dSphs in these groups. We are using HST ACS/WFC and existing 'snapshot' data to determine distances and metal abundances, Gemini-North NIRI and VLT ISAAC near-infrared images to study upper-AGB populations, and, for the two southern groups, Australia Telescope Compact Array and Parkes radiotelescope observations to determine, or place limits on, neutral gas contents. In this contribution I present initial results for some of the dSphs in these groups.

2. M81 Group

The first investigation of the stellar populations of dSph galaxies in this group was that of Caldwell *et al.* (1998), who used multi-orbit HST/WFPC2 V and I imaging to determine c-m diagrams for two M81 group dSphs, F8D1 and BK 5N. The latter system has $M_V \approx -11.3$ and lies approximately 70kpc from M81 and ~30kpc from NGC 3077 (in projection). The Caldwell *et al.* (1998) c-m diagram clearly shows that an upper-AGB population is present in this galaxy, indicating on-going star formation. Caldwell *et al.* (1998) estimate the upper-AGB termination luminosity as $M_{bol} \approx -4.3$, which suggests that stars formed in this dSph until ~8 Gyr ago. F8D1, on the other hand, is a luminous ($M_V \approx -14.2$) low surface brightness dSph that lies ~120kpc from M81 and approximately 35kpc from the spiral galaxy NGC 2976 (in projection). The c-m diagram shows a strong upper-AGB population, which remains even when allowance is made for the likely presence of 47 Tuc-like long period variables, i.e. stars that lie above the red giant branch tip, but which are not of intermediate-age. Such stars can occur whenever the mean metallicity of a system exceeds [Fe/H] ≈ -1.0 dex. Caldwell *et al.* (1998) indicate that the most luminous of the upper-AGB stars in F8D1 have $M_{bol} \approx -5.0$. This dSph has also been observed in the near-infrared using NIRI on the Gemini-N telescope. The observations (see Da Costa 2004 for a full discussion) yield very similar results to those of Caldwell *et al.* (1998): the most luminous stars again have $M_{bol} \approx -5.0$ and the luminosity functions for the upper-AGB stars calculated from the I-band and K-band data agree well (Da Costa 2004). Comparisons with LMC and SMC cluster data, and with near-IR data for the Galactic dSph companion Leo I (cf. Menzies *et al.* 2002), suggest that F8D1 contains stars as young as 1–2 Gyr.

Gemini-N NIRI imaging has now also been analysed for an additional M81 group dSph, known as kkh57 (Da Costa *et al.*, in preparation). This dSph is a small, low luminosity ($M_V \approx -10.9$) system, that lies in a rather isolated location in the group, approximately 370kpc in projection from M81. The c-m diagram for kkh57 clearly shows the presence of upper-AGB stars, indicating on-going star formation in this dSph as well. The brightest of the upper-AGB stars have $M_{bol} \approx -4.3$ to -4.6, somewhat more luminous than for BK 5N yet fainter than for F8D1. This would appear to be prima facie evidence that the diversity of star formation histories seen among the Local Group dSphs is also present in the dSphs of the M81 group.

3. Cen A Group

In this group we have recently obtained VLT ISAAC near-IR imaging of 14 dwarf systems (see Rejkuba *et al.*, these proceedings). Analysis of this dataset continues but I present here initial results for AM1343-452 (kk217), a dSph with absolute visual magnitude $M_V \approx -11.5$ that lies approximately 290 kpc in projection from Cen A. Karachentsev *et al.* (2002) classify it as a Cen A companion. The left panel of Fig. 1 shows an optical c-m diagram derived from the HST snapshot data for this galaxy. *The stars plotted are only those that are measured on both the HST and the VLT ISAAC data.* The tip of the red giant branch is clearly visible at $I \approx 23.9 \pm 0.1$. Overplotted are standard globular cluster giant branches which suggest a mean metallicity for the dSph of $<$[Fe/H]$> \approx -1.55 \pm 0.3$ dex. The field contamination is substantial, given the comparatively low

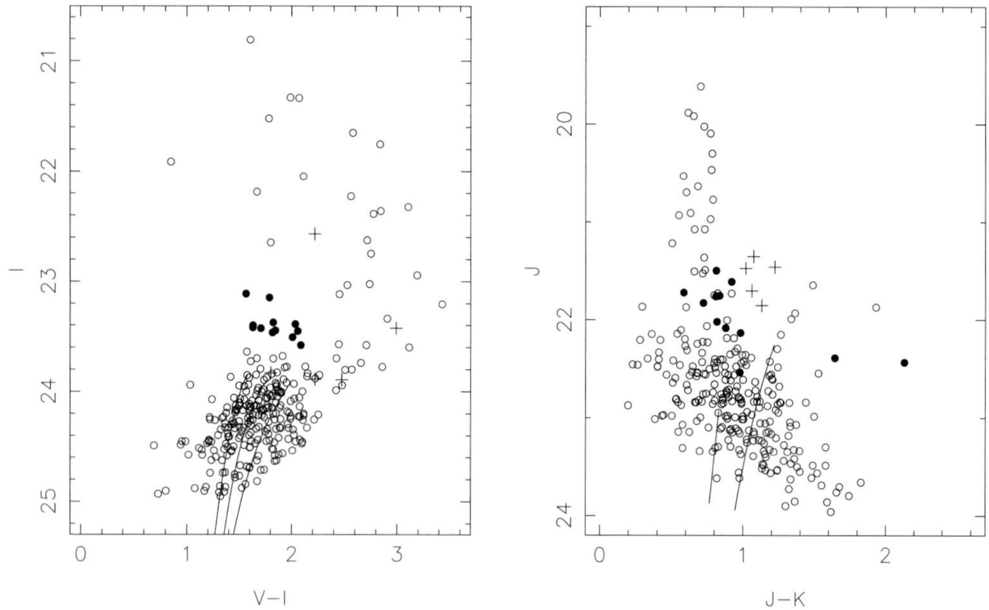

Figure 1. Colour-magnitude diagrams for the Cen A group dSph AM1343-452 (kk217). The left panel shows V, I photometry derived from HST snapshot images. Shown also are the red giant branches (RGBs) of the Galactic globular clusters M15 ([Fe/H]=–2.12), M2 (–1.62) and NGC 1851 (–1.22) shifted to the distance modulus and reddening of the dSph. The right panel shows J, K photometry derived from VLT ISAAC images together with the RGBs for the globular clusters M92 (–2.28) and 47 Tuc (–0.76). In both panels the stars plotted are those measured on both sets of data. A number of potential upper-AGB stars are plotted as filled symbols, for those selected in the left panel, and as plus-signs, for those selected in the right panel. The same symbol is used to show the location of these stars in the corresponding diagram.

galactic latitude, but nevertheless there is some indication of a population of stars that lie above the red giant branch tip. A sample of such potential upper-AGB stars, selected using the c-m diagram of BK 5N of Caldwell *et al.* (1998) as a guide, are plotted as filled symbols in this panel.

The right panel of Fig. 1 shows the near-IR colour magnitude diagram derived from the ISAAC data. Plotted in this panel are the red giant branches of the Galactic globular clusters M92 and 47 Tuc, using the same AM1343-452 reddening and distance modulus as for the left panel. In this c-m diagram most of the Galactic field stars have colours of $J - K \approx 0.7$, and so form a vertical sequence in the c-m diagram. Five possible upper-AGB stars are plotted as plus symbols in this panel: the left panel shows that these stars also fall at or above the RGB tip in the optical c-m diagram. Similarly, while some of the upper-AGB candidates selected in the left panel fall in the field star sequence in the near-IR c-m diagram, the majority are above the RGB tip and two have quite red near-IR colours. These two stars may well be carbon stars.

Regardless of the actual status of individual stars, these data suggest that AM1343-452 does contain a population of upper-AGB stars. This interpretation confirmed by an analysis of the field-subtracted J-band luminosity function (see Rejkuba *et al.*, these proceedings), which reveals a clear excess of stars brighter than the RGB tip. The most luminous of these stars have M_{bol} values in the range of –4.5 to –4.8, corresponding to ages of a few Gyr. Clearly there has been on-going star formation in this particular Cen A group dSph.

While the complete data set remains to be analysed in detail, it appears that a sizeable fraction (perhaps as much as 50%) of the Cen A group dwarfs observed in the near-IR show evidence for upper-AGB populations, and thus on-going star formation. Nevertheless, there are clearly some dSphs in our sample that apparently lack such populations. Further modelling of the image completeness as a function of magnitude and colour is required to be certain of this result, but it is likely that some of the dSphs in this group lack intermediate-age populations. These would be dSphs in which essentially all the stars are older than 8–10 Gyr, making them Cen A group analogues to Local Group systems such as Sculptor and And I. When the full analysis is complete it will be interesting to see where these particular dwarfs lie in the group, as compared to the locations of the dSphs with intermediate-age populations.

4. Sculptor Group

In this group there are five low-luminosity early-type galaxies with confirmed group membership. Four of these five have been classified by Jerjen *et al.* (2000) on the basis of their morphology: three are classed as transition types while the fourth is classed as a dwarf elliptical. The fifth system is classified as a dSph by Karachentsev *et al.* (2000). We do not as yet have any near-IR imaging data for these systems, though HST snapshot imaging is available for all five. These data, together with existing ground-based optical imaging data for ESO 540–032, reveal the presence of blue stars in the central regions of at least two of the dwarfs (cf. Karachentsev *et al.* 2000, Jerjen & Rejkuba 2001), suggesting that they have had modest levels of relatively recent star formation.

We have observed all five systems with the Parkes radio telescope to determine, or place limits on, their neutral hydrogen contents. The results are described in more detail in Bouchard *et al.* (these preceedings) but in summary, four of the five systems, including all three classified as transition types, and including both of the systems with central blue stars, were detected in HI. The M_{HI}/L_B values range from 0.08 to 0.19 in solar units for the detections, comparable to those for the Local Group transition systems LGS 3 and Phoenix, while the 3σ upper limit for the non-detection corresponds to $M_{HI}/L_B <$ 0.08 in solar units. For three of these four systems, knowledge of the HI radial velocity has allowed the generation of HI maps from data available in the Australia Telescope Compact Array archive. These maps show that for two systems, the center of the HI distribution is offset from that of the optical image, while in the third, the HI, which is apparently quite extended, peaks near the center of the optical image (Bouchard *et al.*, in preparation). Again these situations are reminiscent of that for the Local Group transition objects Phoenix and LGS 3, but considerably more information than is currently available is required before we can attempt to connect the gas dynamics with the recent star formation histories (cf. Gallart *et al.* 2001).

5. Conclusions

Although the sample of objects analysed so far is not large, it does seem clear that in the M81 and Cen A groups a substantial fraction of the dSphs have had on-going star formation, and that like the Local Group, there is considerable diversity in the extent and duration of this phenomenon. On the other hand, in the Scl group, which is a less dense environment, the low-luminosity early-type galaxies are predominantly of the transition type, possessing modest amounts of neutral gas and showing indications of relatively recent star formation. This would appear to be a direct observational confirmation of the premise of Grebel *et al.* (2003), who suggested that "transition-type dwarfs ... should replace dSphs in isolated locations where stripping is ineffective".

the total surface density in the vicinity of the feature to be 1.3 ± 0.1 times that of the remainder of F4. That is, at least one quarter of the stars in the overdense region are part of the feature, while the remainder form the background population. We removed the background population and, assuming the feature lies at the same distance modulus as Fornax [$(m - M) = 20.70$; Mateo 1998], then it has an integrated absolute magnitude of $M_V = -6.7 \pm 0.2$. Additionally, two 'lobe' structures are visible in the contour plot. Similar to the two shells, these lie on the Fornax minor axis.

4. Analysis

A full description of the analysis techniques are presented in C05. We applied two techniques to measure the level of stellar clustering beyond the tidal radius. The first technique resulted in a function which described the probability of measuring a given density in the extra-tidal region. This provided a significance limit on the north-western shell and the two lobed structures on the minor axis. Secondly, we applied the two-point angular correlation function, which is commonly used to measure the level of galaxy clustering in cosmological studies.

From the probability density function, it was found that the extra-tidal region consists of three populations with distinguishable densities. The north-western shell was a $\sim 20\sigma_{\rm f}$ result, where $\sigma_{\rm f}$ is the standard deviation in the density of the field population. The density of the two lobes was measured to be separated from that of the field population by $4.5\sigma_{\rm f}$, implying they are not due to random fluctuations in the field star distribution. Assuming they are at the same distance as Fornax, with a similar stellar population, they measure $\sim$8 kpc in length with an integrated absolute visual magnitude of $M_V \sim -9$. The integrated luminosity of these structures imply that the merged companion was at least as bright as a small dwarf galaxy such as Ursa Minor or Draco.

5. Discussion

The detected structures provide a strong argument for a recent merger in the Fornax system. However, another possibility is that the two large lobes of material extending to either side of the dSph represent tidal tails. They qualitatively match simulations of tidal tails (Helmi & White 2001; Mayer *et al.* 2001) which are caused by interaction with the Milky Way. These models also indicate that material leaves the satellite in bursts, which may produce the observed shell-like structure.

However, clear tidal tails have only been observed in a single dSph satellite of the Galaxy; the Sagittarius dSph. The Sagittarius orbit brings it within approximately 15 kpc from the Galactic centre with an orbital eccentricity of $e \approx 0.75$ (Law, Johnston, & Majewski 2004), and thus Sgr experiences substantial tidal forces. In contrast, Fornax is located at a distance of 140 kpc (Mateo 1998) and proper motion measurements indicate it is on a polar orbit with a small eccentricity ($e = 0.27$) and is currently at perigalacticon (Dinescu *et al.* 2004). Moreover, Fornax contains a mass larger than all the other Galactic dSphs combined (excluding Sagittarius), which further precludes the possibility of stars escaping from this relatively deep potential well. Thus, given the relatively large mass, distance and orbital parameters of Fornax, we do not expect to observe such clear evidence of Milky Way interaction.

Further, the inner shell described in C04 demonstrated a clear signature; it was dominated by stellar population with an age of 2 Gyr, significantly different to the mixed stellar population observed in Fornax (Stetson *et al.* 1998). If the observed substructure

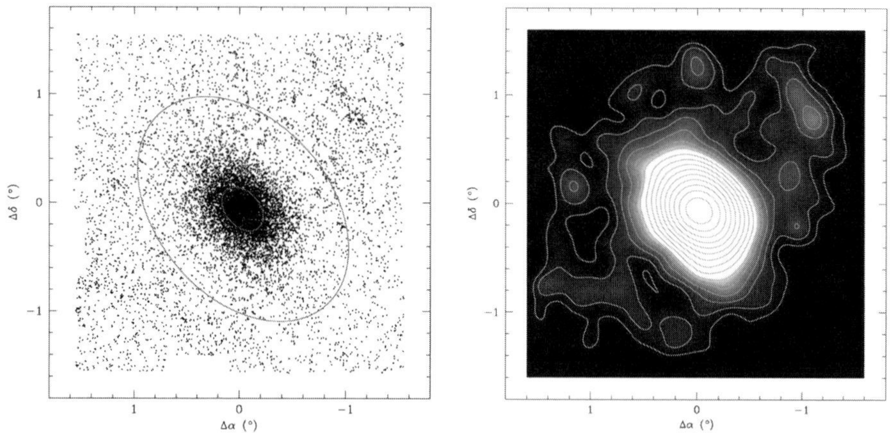

Figure 2. *Left panel:* Distribution of RGB-selected stars on the sky. The inner and outer ellipses are the core and tidal radii respectively (Mateo 1998). *Right panel:* Distribution of Fornax RGB stars where each star has been convolved with a Gaussian of radius 40″. The contours are logarithmically spaced, and are fitted with a smoothing length of 3.0′. The first contour represents a stellar density $3.3\sigma_{\mathrm{field}}$ above the field star population.

is due to the gravitational influence of the Galaxy, then it would contain the same stellar population as the remainder of the system.

The high density structure located 1.3° NW of Fornax is clearly shell-like in appearance. In the case of shell structure, the lobes would represent those portions of shell for which we do not see a limb brightening effect; in essence, they lie in front of, or behind, the Fornax system. These appear remarkably similar to the simulations of interacting dark halos presented by Knebe *et al.* (2005). To further investigate the shell possibility, we will obtain deep photometry of the shell region. If the stellar population is similar to the Fornax system, this would indicate the substructures are tidal tails. However, if the stars in the shell display a marked age difference from those in Fornax, this would support the possibility of shell structure in a dSph.

References

Coleman, M., Da Costa, G.S., Bland-Hawthorn, J., Martínez-Delgado, D., Freeman, K.C. & Malin, D. 2004, *AJ* 127, 832 (C04)

Coleman, M.G., Da Costa, G.S., Bland-Hawthorn, J. & Freeman, K.C. 2005, *AJ* 129, 1443 (C05)

Dinescu, D.I., Keeney, B.A., Majewski, S.R. & Girard, T.M. 2004, *AJ* 128, 687

Helmi, A.& White, S.D.M. 2001, *MNRAS* 323, 529

Hernquist, L.& Quinn, P.J. 1988, *ApJ* 331, 682

Hernquist, L.& Quinn, P.J. 1989, *ApJ* 342, 1

Knebe, A., Gill, S.P.D., Kawata, D. & Gibson, B.K. 2005, *MNRAS* 357, L35

Law, D.R., Johnston, K.V. & Majewski, S.R. 2005, *ApJ* 619, 807

Malin, D.F. 1978, *Nature* 276, 591

Malin, D.F. & Carter, D. 1980, *Nature* 285, 643

Malin, D.F. & Carter, D. 1983, *ApJ* 274, 534

Mateo, M. 1998, *ARA&A* 36, 435

Mayer, L., Governato, F., Colpi, M., Moore, B., Quinn, T., Wadsley, J., Stadel, J. & Lake, G. 2001, *ApJ* 559, 754

Stetson, P. B., Hesser, J. E. & Smecker-Hane, T. A. 1998, *PASP* 110, 533

Discussion

SKILLMAN: Why do you favour a two-lobe interpretation over a spherical halo?

COLEMAN: The two observed shells support a merger scenario where the companion approached on a radial vector along the minor axis. A two-lobe structure is a natural consequence of such a merger, however further observations may reveal a spherical shell system.

CONSELICE: Can you rule out the possibility that Fornax is being tidally stripped, which could explain these observations?

COLEMAN: Not conclusively. However, the large mass, distance and relatively gentle orbit of Fornax compared to the other Galactic dSphs make such tidal structures highly improbable.

VADUVESCU: I could not understand how you subtracted the foreground stars.

COLEMAN: Briefly, using Monte Carlo simulations using stars in a shell around the galaxy.

Near-Field Cosmology with Dwarf Elliptical Galaxies
Proceedings IAU Colloquium No. 198, 2005
H. Jerjen and B. Binggeli, eds.

doi:10.1017/S1743921305003467

A near infrared view of Local Group dwarf galaxies

M. Gullieuszik[1,2], E. V. Held[1], Y. Momany[2], I. Saviane[3], L. Rizzi[4] and S. Ortolani[2].

[1]Osservatorio Astronomico di Padova, INAF, vicolo dell'Osservatorio 5, I-35122 Padova, Italy

[2]Dip. di Astronomia, Università di Padova, vicolo dell'Osservatorio 2, I-35122 Padova, Italy

[3]European Southern Observatory, Casilla 19001, Santiago 19, Chile

[4]Institute for Astronomy, University of Hawaii, 2680 Woodlawn Drive, Honolulu, HI 96822

Abstract. We present the results of near-infrared imaging of the dwarf spheroidal galaxies Fornax and Leo I as part of a photometric survey of stellar populations in Local Group dwarf galaxies. Wide-field observations in the J, H, and K_s bands have been obtained with the SOFI camera at the ESO NTT at La Silla, Chile. The aim of this project is to study the evolution of Local Group dwarf spheroidal galaxies, with special regard to star formation at intermediate epochs. The near-infrared data, together with optical catalogs from the Padova Local Group wide-field survey, provide a very large color baseline ideal to study the properties of red giant branch (RGB) and asymptotic giant branch (AGB) stars.

Keywords. Local Group, galaxies: dwarf, galaxies: individual (Fornax , Leo I), stars: AGB and post-AGB, infrared: galaxies.

1. Near-infrared color-magnitude diagrams

The infrared color-magnitude diagrams (CMD) of Fornax and Leo I are presented in Fig. 1 (upper panel). The CMDs show the extended AGBs of these two galaxies, characterized by prominent intermediate-age stellar populations, along with literature identification of carbon stars and other AGB stars. We have outlined some of the regions empirically defined by Nikolaev & Weinberg (2000) to classify the evolved star content of the LMC from 2MASS photometry: upper RGB stars, oxygen-rich, carbon-rich, and dusty AGB stars. We note in particular the presence of dust-enshrouded AGB stars in both Leo I (see also Menzies *et al.* 2002) and Fornax.

2. Distance and metallicity

In the case of Fornax dSph, our photometry is deep enough to reach the "red clump" of core He burning stars above $1M_{\odot}$. The K_s magnitude of the red clump is one of the best distance indicators for resolved galaxies, since it is less affected from reddening and age dependence than optical bands. We adopted the Alves (2000) calibration and the prescriptions of Salaris & Girardi (2002) to account for the actual star formation rate and chemical evolution. The resulting distance is $(m-M)_0 = 20.74$, in excellent agreement with the I-band RGB tip distance (Saviane *et al.* 2000).

The wide color baseline of optical-IR photometry is extremely useful to constrain the metallicity distribution function (MDF) of stars in galaxies. An MDF-related observable has been constructed by interpolating the location of red giant stars in the M_{K_s}, $(V-K_s)$ CMD against the fiducial RGB lines of Galactic globular clusters (Valenti *et al.* 2004).

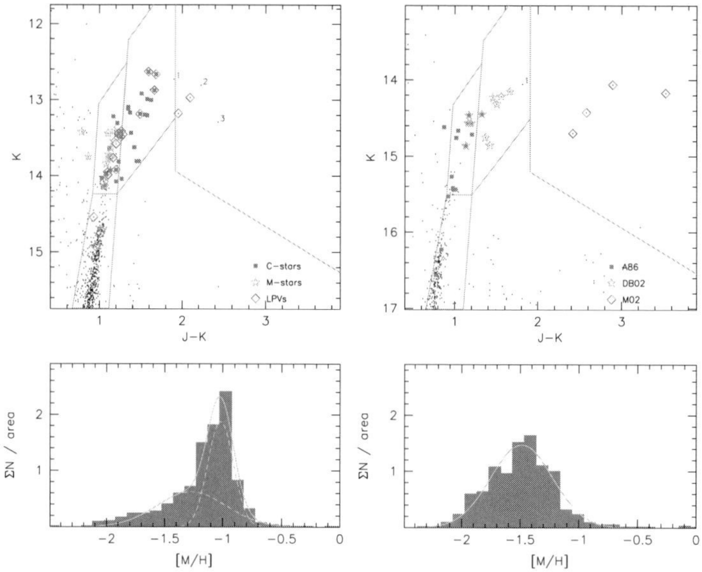

Figure 1. *Upper Left:* The near-IR CMD of Fornax along with literature classification of C-stars, M-giants, and Long Period variable stars (Stetson *et al.* 1998 and refs. therein; Bersier & Wood 2002, Westerlund *et al.* 1991). *Upper Right:* The CMD of Leo I dwarf spheroidal. C-stars were identified by Azzopardi *et al.* (1986), Demers & Battinelli (2002), and Menzies *et al.* (2002) using spectroscopy (A86), narrow-band (DB02) or infrared (M02) photometry. *Lower Left:* The observed MDF of Fornax. *Lower Right:* The observed MDF of Leo I.

The MDFs of Fornax and Leo I are shown in Fig. 1 (lower panels). This distribution exactly represents the distribution of stellar metallicities only in the case of old stellar populations. In the general case of extended star formation and younger stellar populations, as in Fornax and Leo I, these observed distribution functions still represent an important observable for the chemical evolution models to confront with. The observational MDF derived for Fornax is well modeled by the sum of two Gaussian functions. The principal component is centered at $[M/H] = -1.02$, with a tail of metal-poor stars. The MDF of Leo I is more like a single wide Gaussian centered at $[M/H] = -1.56$. Using theoretical isochrones to estimate the effects of age and metallicity on $(V - K_s)$ colors, we obtain $[M/H] \sim -0.9$ and $[M/H] \sim -1.2$ as our preliminary estimates of the age-corrected mean metallicities of Fornax and Leo I.

References

Alves, D.R. 2000, *ApJ* 539, 732
Azzopardi, M., Lequeux, J. & Westerlund, B.E. 1986, *A&A* 161, 232
Bersier, D. & Wood, P.R. 2002, *AJ* 123, 840
Demers, S. & Battinelli, P. 2002, *AJ* 123, 238
Menzies, J., Feast, M., Tanabé, T., Whitelock, P. & Nakada, Y. 2002, *MNRAS* 335, 923
Nikolaev, S. & Weinberg, M.D. 2000, *ApJ* 542, 804
Salaris, M. & Girardi, L. 2002, *MNRAS* 337, 332
Saviane, I., Held, E.V. & Bertelli, G. 2000, *ApJ* 355, 56
Stetson, P.B., Hesser, J.E. & Smecker-Hane, T.A. 1998, *PASP* 110, 533
Valenti, E., Ferraro, F.R. & Origlia, L. 2004, *MNRAS* 351, 1204
Westerlund, B.E., Edvardsson, B. & Lundgren, K. 1987, *A&A* 178, 41

Near-Field Cosmology with Dwarf Elliptical Galaxies
Proceedings IAU Colloquium No. 198, 2005
H. Jerjen and B. Binggeli, eds.

doi:10.1017/S1743921305003479

Near IR imaging of Centaurus group dwarf elliptical galaxies

M. Rejkuba[1], H. Jerjen[2], G. Da Costa[2], B. Binggeli[3] and M. Zoccali[4]

[1]European Southern Observatory, Karl-Schwarzschild-Strasse 2, D-85748 Garching, Germany
email: mrejkuba@eso.org

[2]Research School of Astronomy and Astrophysics, Australian National University, Mount Stromlo Observatory, Cotter Road, Weston ACT 2611, Australia

[3]Astronomical Institute of the Univ. of Basel, Venusstrasse 7, CH-4102 Binningen, Switzerland

[4]P. Univ. Católica de Chile, Astronomy & Astrophysics Dept., Casilla 306, Santiago 22, Chile

Abstract. Detailed studies of dwarf galaxies in the Local Group have revealed an amazing diversity in their star formation histories. Young and intermediate-age stellar populations are observed in some gas-poor dwarf ellipticals, like Fornax and Carina, where naively one would have expected to find preferentially old stellar populations. There are evidences that the environment might drive the evolution of these galaxies.

In order to explore stellar populations in dwarf galaxies in an environment denser than that of the Local Group we have observed 14 dwarf elliptical galaxies in Centaurus group in near-IR J_s and K_s bands. The intrinsic properties of these galaxies, the metallicities and ages of their stellar populations and the presence or absence of bright, intermediate-age AGB stars will be correlated with extrinsic properties like the local galaxy density. Here we summarize our observations, present the sample and show near-IR color-magnitude diagrams and luminosity functions for ESO269-066 and AM1343-452 dwarf galaxies.

Keywords. galaxies: dwarf, galaxies: fundamental parameters, galaxies: photometry, galaxies: stellar content

1. The Data

Deep J_s and K_s band images for 14 Centaurus Group dwarf galaxies were obtained with the short wavelength arm of ISAAC near-IR array at Antu (UT1) Very Large Telescope (VLT) at ESO Paranal Observatory. Field of view of the short wavelength arm of ISAAC is $2.'5 \times 2.'5$ and the detector is Hawaii Rockwell array with a pixel scale of $0.''148$. The list of all the targets is given in the Table 1. The observations were carried out in service mode between April and September 2004. Each galaxy was observed once in J_s with a total exposure of 35 min, and twice in K_s band with a total exposure of 72 min. Seeing was typically between $0.''4 - 0.''6$ and only for some images up to $0.''8$. The good seeing and high sensitivity of ISAAC allowed us to resolve well upper red giant branch stars in 7 galaxies, while for the other targets we could resolve only few brightest stars.

The images were reduced using IRAF. For all the targets PSF fitting photometry was done using the suite of DAOPHOT and ALLFRAME programmes (Stetson 1994). The photometry was then tied to 2MASS photometric system by matching all the point sources from 2MASS observed in our fields with our J_s and K_s band detections.

Table 1. List of the 14 dwarf galaxies observed with ISAAC. Columns 4 and 5 report distance moduli (DM) from the literature, and the methods used to derive them: TRGB=Tip of the red giant branch, SBF=surface brightness fluctuation, MEM=average distance to the group whose membership is based on the morphology. The last column lists references to the literature where these distances were originally published (see the list of references for details).

Galaxy	RA_{J2000}	DEC_{2000}	Type	DM	method	Reference
AM1320-230	13:23:29.8	−23:23:35.0	dE	28	MEM	JBF00
AM1339-445	13:42:05.8	−45:12:20.9	dE	27.77, 27.87	TRGB, SBF	K+02, JFB00
AM1343-452	13:46:17.8	−45:41:04.9	dE	27.92, 27.99	TRGB, SBF	K+02, JFB00
Cen8	13:22:56.2	−33:34:19.9	Im/dE	28	MEM	JBF00
CenA-dE1	13:12:47.7	−41:50:03.0	dE	28	MEM	JBF00
CenA-dE2	13:21:31.9	−31:53:12.0	dE/Im	28	MEM	JBF00
CenA-dE3	13:46:00.7	−36:19:45.0	dE	28	MEM	JBF00
CenA-dE4	13:46:39.6	−29:58:42.9	dE	28	MEM	JBF00
CenA-dE5	14:30:04.9	−33:28:45.0	dE	28	MEM	JBF00
ESO219-010	12:56:09.6	−50:08:38.0	dE,N	28.34	SBF	JFB00
ESO269-066	13:13:07.9	−44:53:21.9	dE,N:	27.75	SBF	JFB00
ESO384-016	13:57:01.2	−35:19:59.0	dS0/Im	27.85	SBF	JFB00
SGC1319.1-4216	13:22:00.4	−42:32:08.0	dE	28	MEM	JBF00
UGCA365	13:36:30.7	−29:14:11.0	dIrr/dE	28	MEM	JBF00

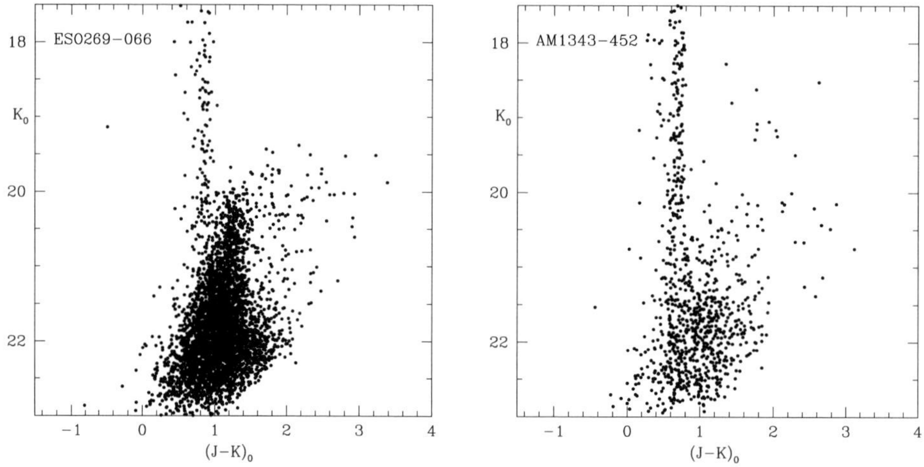

Figure 1. Color-magnitude diagrams of ESO269-066 and AM1343-452.

2. Near-IR color-magnitude diagrams and luminosity functions

Matching the J_s and K_s band photometry color-magnitude diagrams (CMDs) were produced for all the stars in each galaxy field. Near-IR color-magnitude diagrams of two well resolved dwarf galaxies ESO269-066 and AM1343-452 are presented in Fig. 1, showing well populated upper RGB, and possibly some evidence of intermediate-age AGB stars. Due to rather low galactic latitude of our targets ($b \sim 20°$) there is significant galactic foreground contamination. Most of the Galactic stars have colors between $(J-K)_0 = 0.65 - 0.85$ and can be easily seen as the vertical sequence at these colors.

Assuming the distance modulus from the literature (see Tab. 1), we can determine the average metallicity of the RGB stars from the mean $(J-K)_0$ color at $M_K = -5.5$ (Valenti *et al.* 2004). It is $[Fe/H] = -0.65 \pm 0.15$ for ESO269-066, and -1.1 ± 0.3 dex

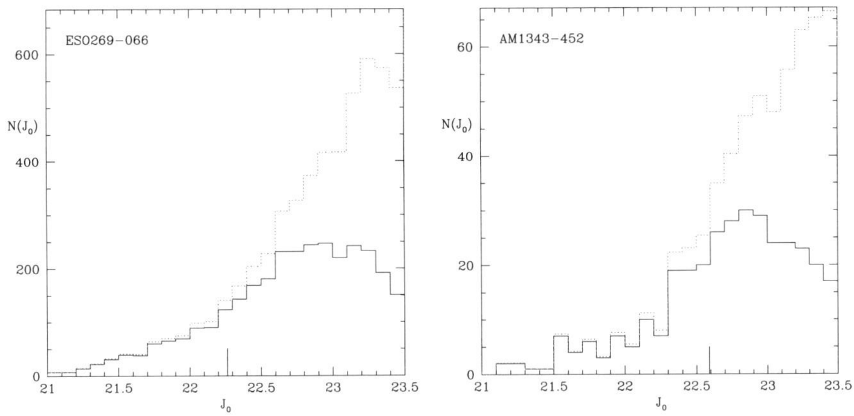

Figure 2. J band LFs of ESO269-066 and AM1343-452. Solid histogram shows the observed LF with foreground contamination statistically subtracted. The dotted histogram is used for the completeness corrected LF. The short vertical line indicates the expected RGB tip break, assuming the distance modulus from Table 1 and metallicity determined from the RGB color at $M_K = -5.5$ (Valenti *et al.* 2004).

for AM1343-452. The larger uncertainty for the latter is due to larger distance modulus and consequently larger photometric errors and incompleteness.

The radial density profiles were used to determine the extent of each galaxy and the area where foreground stars dominated. These outer regions were used to statistically subtract the foreground stars. The subtraction is good for AM1343-452, while it works less well for ESO269-066, because of the larger radial extent of that galaxy. The J band luminosity functions (LFs) of ESO269-066 and AM1343-452, with foreground contamination statistically subtracted, are presented in Fig. 2. The solid histogram is used for the observed, and the dotted line for the completeness corrected LF. The short vertical line indicates the expected magnitude of the RGB tip break, assuming the distance from the literature and the average metallicity for the RGB population.

The future work will include more detailed investigation of the intrinsic properties of the galaxies in our sample which will then be correlated with extrinsic properties like the local galaxy density in order to investigate the role of environment on their evolution. For more discussion of the environmental influences on the star formation histories of dwarf galaxies in groups the interested reader is referred to contributions of Da Costa and Karachentsev in this volume.

Acknowledgements

The participation of MR at the IAUC 198 was partially supported by an IAU Grant. The presented data were obtained at ESO Paranal Observatory within the observing programme 073.B-0131. HJ's and GDaC's work is supported in part by ARC Discovery Grant DP0343156.

References

Jerjen, H., Binggeli, B. & Freeman, K.C., 2000, *AJ* 119, 593 (JBF00)
Jerjen, H., Freeman, K.C. & Binggeli, B., 2000, *AJ* 119, 116 (JFB00)
Karachentsev, I.D., *et al.*, 2002, *A&A* 385, 21 (K+02)
Stetson, P.B., 1994, *PASP* 106, 250
Valenti, E., Ferraro, F.R. & Origlia, L., 2004, *MNRAS* 351, 1204

Near-Field Cosmology with Dwarf Elliptical Galaxies
Proceedings IAU Colloquium No. 198, 2005
H. Jerjen and B. Binggeli, eds.

doi:10.1017/S1743921305003480

Kinematics of dSph and dE galaxies

Mario Mateo

Department of Astronomy; University of Michigan; Ann Arbor, MI 48109-1042, USA

Abstract. We have entered a new era in the observational study of the kinematics of dwarf spheroidal and dwarf elliptical galaxies. Large telescopes with multi-object spectrographs can now obtain radial velocities of hundreds if not thousands of individual stars in the nearer dSph systems, while new sensitive longslit spectrometers can measure the kinematics of dE galaxies from their integrated light to unprecedented faint surface brightness levels. I review where we have been, where we are, and where we are going in the kinematic studies of these dwarfs systems.

---

1. Introduction

The attendees to IAUC 198 and possibly many of the readers of this review need little motivation to justify the study of dwarf Spheroidal (dSph) and dwarf Elliptical (dE) galaxies. These systems have proven to be closely related to the earliest dark matter (DM) halos to have ever formed: they are both high-density and small, pushing them to the earliest formation epochs in any hierarchical cosmology (Press and Schechter 1974; Navarro, Frenk and White 1997). We see direct evidence that dwarfs are merging into large, local structures (Ibata *et al.* 1994), and that this process may be ubiquitous at the present epoch (Shang *et al.* 1998; Ferguson *et al.* 2002). Because dwarfs live in such an extreme gravitational state – their stars reside in extremely broad, and hence, low-acceleration potentials – they are natural tests of non-standard gravity models (Gerhard and Spergel; Łokas 2001, 2002). These small galaxies can be used to address some of the biggest issues in astrophysics; their kinematics play an important role in this regard.

But it was not always this way. It is well known that Marc Aaronson almost single-handedly started the field of dSph/dE with his seminal 1983 paper. Though Aaronson clearly wanted to use this paper to launch investigations to the sorts of issues I summarized above, everyone else met it with far more prosaic questions. How could one claim to derive *any* useful information on the dynamical state of a pressure-supported system with just three (!) stars? How could one possibly use Carbon stars – some known to be variables – as kinematic tracers at the few km/s level? And what about binary stars? These might contribute significantly to the inferred dispersions of 8-10 km/s. Many wondered how one could possibly obtain 1 km/s precision for such very faint stars ($V \sim 18!$).

New telescopes and instruments have rejuvenated the field of dSph/dE kinematics, providing the opportunity to obtain large samples of stars in the nearer systems, and making it possible to carry out integrated-light spectroscopy at high enough resolution to study dE systems beyond the Local Group. My own efforts are built around a new instrument – the Michigan/MIKE Fiber system – on the Magellan telescope. With MMFS we can obtain spectra at a resolution of 15,000 to 19,000 of up to 256 targets simultaneously. We observe the region from 5120–5200 Å which contains strong Mg features and many weaker (mostly Fe) lines. In 2–3 hours we obtain sufficient S/N to $V \sim 21$ to measure velocities of late-type giants to a precision of 1–2 km/s. We have spectra now of over 2500 stars in six galaxies; by comparison, all of the kinematic results summarized in Mateo 1998 were based on observations of about 450 stars.

The early concerns – sample size, stellar atmospheric and binary motions, and the intrinsic velocity precision of the observations – consumed the early discussions of dSph kinematics through much of the 1980's. *They remain important issues,* and it is important to not ignore these more prosaic questions. Nonetheless, as these problems become more settled, we can turn to the much more interesting questions that dSph/dE kinematics address. Do the kinematics of these dwarfs arise solely from internal effects? If external effects matter, what are these processes and do they operate on all dSph to a similar degree? Finally, are the basic assumptions we bring to the analysis of these systems secure? To make for a more constrained discussion here, I will address four issues below that relate to these more general questions.

2. Does mass follow light in dSph/dE galaxies?

Pryor and Kormendy (1990) first tried to address this question using only data on the central velocity dispersions of two dSph satellites of the Milky Way. Not surprisingly, it was impossible to rule out or confirm extended DM halos from these data; what was clear was that dispersion *profiles* were required. Figure 1 is a plot of dispersion profiles for *five* nearby systems based on observations of 1400 stars (Mateo *et al.* 2005, Walker *et al.* 2005). Dispersion profiles have also been produced for some of these galaxies by others, including Sextans (Kleyna *et al.* 2004), Draco (Kleyna *et al.* 2001), Sculptor (Tolstoy *et al.* 2004), and Leo I (Sohn *et al.* 2005). In Sculptor, the dispersion profiles appear to depend on metallicity, a potentially very exciting result (see below). Dispersion and rotation profiles are now also being produced for dE galaxies as distant as the Virgo and Fornax clusters (van Zee *et al.* 2004; Geha *et al.* 2002, 2003; De Rijcke *et al.* 2001, 2004). Apart from the very closest dEs (e.g. NGC 205, Geha 2004), this work remains limited to integrated spectroscopy of galaxies with surface brightnesses considerably higher than non-nucleated dSph galaxies. Nonetheless, the results as they stand suggest a plausible (but not yet proven) kinematic connection between dSph galaxies (at the low surface brightness, low mass, and small size end of the relations) and dE's (see Geha *et al.* 2003).

Among the dE galaxies, what is immediately evident is the sheer variety in the radial kinematic profiles of these galaxies. In contrast, the situation for dSph galaxies is somewhat less clear; some studies claim to see cold cores (Sextans; Kleyna *et al.* 2004), with other profiles falling at large radii (Ursa Minor and Draco; Wilkinson *et al.* 2004). The flat profiles in Figure 1 for some of the same galaxies suggest that the purported variety of the dSph kinematics may be due to small-sample effects. While none of the lowest-luminosity dSph systems exhibit dynamically significant rotation, some dE galaxies some minor-axis rotation, while others – for no recognizable reason (e.g. Geha *et al.* 2003) – show no rotation at all. For these galaxies their shapes appear to be maintained either fully by internal pressure support or fully by rotation.

But what of our question above? In every case measured where the kinematic data extend well outside the core radii of dSph and (a few) dE systems, dispersion profiles appear to strongly favor mass profiles, $M(R)$, that extend beyond the limits of the visible components of the galaxies (e.g Wang *et al.* 2005). There is also growing indications that the kinematics become progressively more radially anisotropic, at least in some dSph galaxies (Wilkinson *et al.* 2002), as stars approach the classical tidal radii. Walker *et al.* (2005) have specifically addressed this issue in Fornax. Under the assumptions of spherical symmetry and an isotropic velocity distribution, they find that mass *must* be considerably more extended than visible matter in that galaxy. Early results for the other dSph systems in Figure 1 confirm this conclusion under the same assumptions. Future

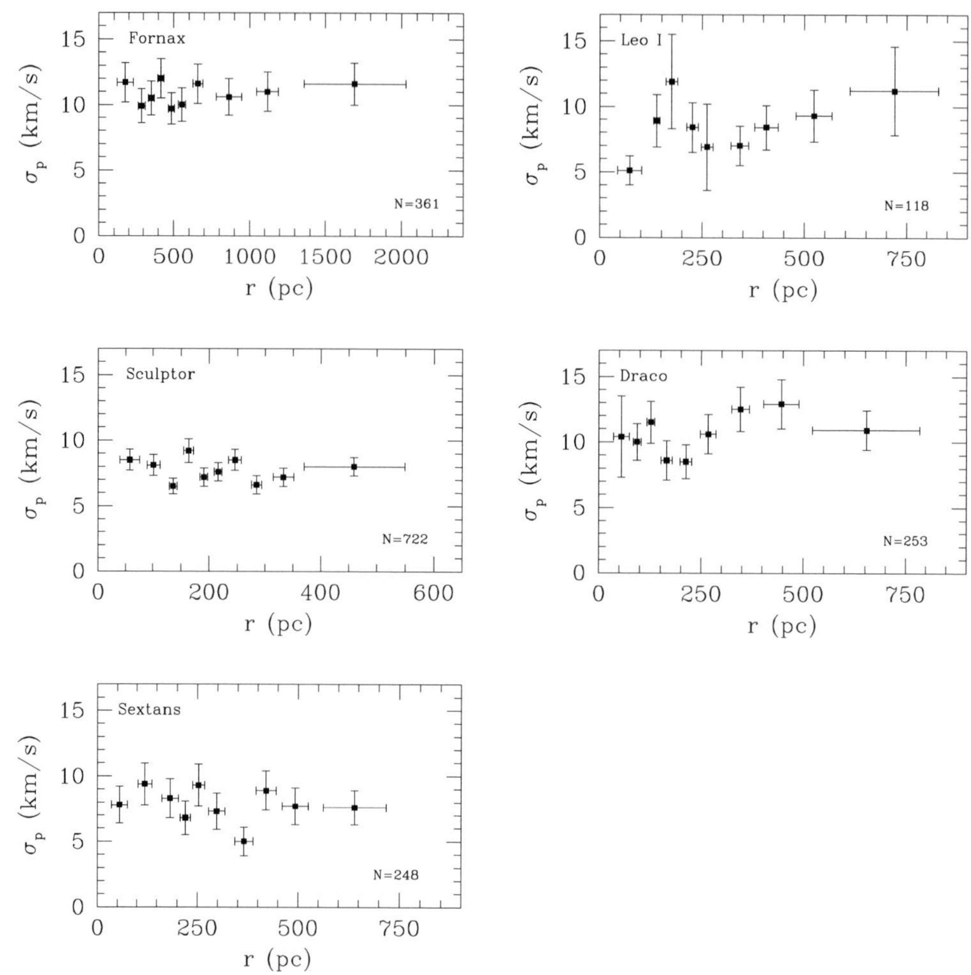

Figure 1. Binned radial velocity dispersion profiles for five of the Milky Way's dSph satellite companions. The data summarized here were taken with the Michigan/MIKE Fiber System at the Magellan Observatory (For, Scl, and Sex) or with Hectochelle at the MMT (Leo I and Dra).

work will clearly revolve around the validity of these assumptions in real systems and how to drop them when they are not justified (see Wilkinson *et al.* 2002).

3. Substructure in dE and dSph Galaxies

Despite the low luminosities and seemingly simple structures of dwarf ellipsoidal systems, these galaxies frequently display internal substructure. Geha *et al.* (2002; 2003) produced some extraordinary maps that illustrate the nature and variety of substructures present in dE galaxies. Some show extended halos, some bright cores, some inner disks and some miniature spiral patterns. The variety is astounding, but no clear correlation of these structures with external (location relative to neighboring systems), or internal properties (luminosity, kinematics, surface brightness) has been established.

Most dSph galaxies reveal star-formation histories that are decidedly non-monotonic (Mateo 1998), and which often exhibit complicated variations with location within individual systems. The latter is usually seen in the form of a core-halo age or color gradient (Harbeck *et al.* 2001; Tolstoy *et al.* 2004), but highly asymmetric structures such as clumps and possibly shells (Coleman *et al.* 2004, Coleman and Da Costa 2005) are also seen. These stellar population variations have long fueled speculation that dSph galaxies in particular may have evolved from low-surface brightness dIrr systems (Mayer *et al.* 2001a,b). The behavior of the core of NGC 205 is particularly instructive. Here, tidal effects seem to have driven gas towards the galaxy center where star formation occurs within a kinematically cold core (McConnachie *et al.* 2004; Geha 2004; Butler and Martínez-Delgado 2005). Since much of the substructure seen in other dE systems tends to focus near the galaxy cores (Geha *et al.* 2002, 2003), and so similar tidal influences may be the ultimate cause.

The shells described by Coleman *et al.* (2004. 2005) are more surprising, as they may indicate that some dSph galaxies have undergone mergers with yet smaller systems in the recent past. These authors note that the substructure in this particular galaxy has a different age distribution than the galaxy as a whole (see also Olszewski *et al.* 2005) and that the purported merger has produced a halo-like structure aligned with, presumably, the plane along which the merged system entered Fornax (which happens to be nearly perpendicular to the major axis of Fornax).

Kleyna *et al.* (2003) illustrate the use of substructure as a dynamical probe by modeling the kinematically distinct clump offset from the core of the Ursa Minor dSph galaxy. Their observations confirm the system to be a cold-substructure within the overall potential of UMi. Their models demonstrate such a structure can only survive if the central potential of UMi arises from a near-constant density central DM halo, one with an effective core radius much larger than that of the visible material. It remains to be seen how substructure in other dwarfs can be exploited to constrain the underlying potentials of these systems, but this lesson in UMi illustrates that the kinematics of substructure can be most revealing. Direct kinematic evidence of substructure has also been detected in some of the more luminous dE systems in the form of strong core-halo kinematic discontinuities (Geha *et al.* 2003; De Rijcke *et al.* 2001, 2004).

4. The Role of Tides

There is simply no question now that tides affect the structures and kinematics of dwarf galaxies. We see strong tidal signatures in systems ranging from the Magellanic Clouds to fully disrupted streams orbiting the Milky Way and nearby galaxies. The crucial question is 'to what extent to tides dominate the kinematics of dwarfs?' Much of the motivation behind the kinematic surveys of dSph and dE systems is to constrain the properties of their dark-matter halos. But if tides dominate, such analyses are at worst invalid if they rely on static models such as King or NFW profiles, and at best greatly complicated by the need to include the role of tides. As illustrated by the many high-resolution n-body simulations presented at this meeting, tides have conceivably have affected *all* of the dwarf satellites of our Milky Way to some extent.

Closer to the Milky Way the distinction between internal kinematics and tidal effects becomes more problematic. At this meeting Evans et al. argued that Draco shows a distinct drop in its dispersion profile at large radii, a hint that tides do not necessarily dominate that galaxy's kinematics. An independent measurement of this profile (Mateo *et al.* 2005; see Figure 1) does not confirm this result. In the case of Ursa Minor, another dSph plausibly affected strongly by tides, Palma *et al.* (2003) and Gómez-Flechoso and

Martínez-Delgado (2003) suggests that UMi merges into a stream-like outer structure aligned with the galaxy's projected space-motion. The fact that both Draco and UMi lie close to an 'exclusion boundary' of 60-70 kpc within which dSph galaxies are rarely seen surrounding L_* galaxies provides some empirical hint that perhaps these systems may be showing incipient tidal effects.

My goal here is not to try to settle this particular debate, but to emphasize that to better account for tides we need (a) better constraints on the form of the Milky Way potential, and (b) better estimates of the orbits of the dwarf systems. The Sgr Stream is a powerful tool to address the first point, though there is still disagreement about the basic structure of the Galactic Halo from this system (Johnston *et al.* 2005; Helmi 2004). Ground-based and HST measurements of the space motions of dSph systems have begun to reveal their orbital parameters, but there is at present only modest agreement between individual measurements (e.g. Piatek *et al.* 2002, 2003; Dinescu *et al.* 2004). Nonetheless, continued observations with existing techniques as well as space-based interferometric proper motions (SIM and GAIA) of these galaxies are sure to reveal their orbital parameters in the relatively near future.

The certain existence of tides should make it clear that the concept of a 'tidal radius' is at best ill-defined. From an observational standpoint, this is typically measured from the photometric profiles and fits to King or King-like models. The extreme sensitivity of any tidal radius, or, by implication of 'tidal breaks', to the background should now be very clear (consider the interesting history of Fornax: Hodge and Smith 1974; Eskridge 1988a,b; Irwin and Hatzidimitriou 1995; Walcher *et al.* 2003; Coleman and Da Costa 2005). Moreover, it should be clear that a true King tidal radius is a fiction for any galaxy in which the stars interact internally at any level, and which are bound by a tidal field. This suggests, at the very least, that the utility of spherically symmetric, isotropic dynamical models may be nearing its end (Wilkenson *et al.* 2002; Kleyna *et al.* 2001, 2002 explore anisotropic models in some detail). In this respect, tides are likely to elongate systems *and* cause the velocity distribution to become more radial near the tidal limits of the systems (Piatek and Pryor 1995; Oh *et al.* 1995).

A possible corollary of this discussion is that tides may *dominate* the kinematics of dSph systems in particular. So much so, in fact, that dark matter may not be required at all to explain their structures or kinematic properties (Klessen and Zhao 2002; Fleck and Kuhn 2003). This interpretation seems far too strong to me. Specific predictions of large line-of-sight extensions in individual systems are not borne out by observations (Klessen and Zhao 2002; Klessen et al. 2003). Indeed, what's more surprising perhaps is how *subtle* tidal effects seem to be – that is, until systems are fully disrupted! Tides do not appear to eradicate the need for dark matter halos in dwarf systems, but it is clear they certainly do complicate the kinematic models we require to fully understand these halos. More detailed models that investigate the kinematic behavior of realistic systems in realistic tidal fields would be most helpful.

5. Nagging Problems

The ubiquity of DM halos inferred from the kinematics of dwarf systems (tides notwithstanding) poses an interesting dilemma: namely, if systems can be essentially dominated by DM, why must DM halos – especially the smallest ones – have *any* baryons associated with them? This is simply a restatement of the long-standing 'missing halo' problem of CDM models. For the Milky Way, this implies that each visible dSph system is simply the visible tip of a DM iceberg consisting of dozens of other dark halos that lack baryons (see also Kormendy and Freeman 2004 for a different angle on this problem). As with many aspects of CDM, what you do not see is supposed to not bother you. But, for me, it does. If small DM halos are so common, do they ever interact with their visible brethren? If so, what do such interactions do to the kinematics of the systems we can directly study? Why do visible dSph systems outside of 50-70 kpc from the Galaxy center appear to be reasonably undisturbed, while inside this radius we see only streams? If there is truly a disconnect between DM and baryons, shouldn't we see a few very massive – and tidally stable – halos with just enough baryons to be detectable located close to the Milky Way?

The subtlety of substructure seems also to be a problem. Why do so many dSph and dE galaxies appear to be well-fit by King profiles (or, if you prefer, exponential or Sersic profiles)? These are the baryons, after all; if the DM halos are much more extended so that the cores – where the baryons supposedly reside – are essentially near-constant density harmonic potentials, substructures should be long-lived (Klypin *et al.* 2003). One might expect the baryons to be distributed in quite chaotic ways within these harmonic cores. Perhaps tidal effects such as those described Mayer *et al.* 2001 have forced the baryons into a much more restricted family of structural models near the centers of the halo potentials. From the perspective of the DM cores, the homogeneity of the dispersion profiles (Figure 1) raises the inverse question: do the baryons affect the dynamics at all? Also, why are the DM halos all so similar in central density and extent (again, as implied by Figure 1)? CDM models predict a range in these properties, but the DM halos of local dwarfs do not seem to exhibit this range.

These sorts of questions have made dwarf systems a testing ground for non-standard formulations of gravity. MoND (Modified Newtonian Dynamics, Milgrom 1983a,b; Bekenstein 2004) represents the most common form of altered gravity applied to astronomical kinematics. The fact is, MoND fits dSph and dE kinematics just as adequately as standard gravity (e.g. Lokas 2001, 2002). But some of the issues raised above, notably tides and the non-spherical structures and anisotropies that tides induce, remain free parameters that allow both CDM and MoND models to fit these data reasonably well. The hope is that as the datasets increase in size and as the models become more refined, a critical test to use dwarf galaxy kinematics to distinguish MoND and DM models may emerge. To this end, it seems to me that it would be interesting to attempt fully-MoNDian n-body simulations of individual dwarf systems in a (MoNDian, naturally) Milky Way potential. The lack of apparent precession in the orbit of the Sgr Stream (Helmi 2004; Johnston *et al.* 2005) has been cited as evidence against a severely oblate potential (as a disk-like system with no DM must be), but detailed models may provide some surprises.

6. Postscript from Les Diablerets

The issues above reflect the contents of my talk at Les Diablerets. Here are some related thoughts – in highly abbreviated form – that developed as I spoke with and listened to various colleagues at the conference:

• Though the kinematic study of both dSph and dE galaxies is now entering a new observational era, we must remain wary of the same problems that people worried about early in this game: precision, binaries and atmospheric jitter. The multiple groups are attacking the problem (for both dSph and dE galaxies) must carry out critical comparisons of their kinematic datasets; for this to work, all groups need to publish their data as fully and quickly as practical.

• For the dE galaxies, longslit spectra are still the norm (however, see Geha 2004); Two-dimensional mapping of the kinematics of these galaxies will require clever use of integral field units and Fabry-Perot instruments.

Solanes, J.-M., *et al.* 2001, *ApJ* 548, 97
Tully, R.B. & Pierce, M.J. 2000, *ApJ* 533, 744
van Zee, L. 2001, *AJ* 121, 2003
van Zee, L., Skillman, E.D. & Haynes, M.P. 2004, *AJ* 128, 121

Discussion

CONSELICE:: The fact that the dEs in the center of Virgo do not tend to be the rotating systems, is it possible that dynamic effects are halting the rotation, or is another dE formation scenario necessary?

VAN ZEE:: We expect tidal stirring to also change the kinematics of some dEs. One thing to keep in mind, however, is that the "non-rotating" dEs have only been observed along the major axis; if they are tumbling, or if we are looking at face-on systems ($\epsilon < 0.1$), we would not expect to see a velocity gradient with the data in hand.

GALLAGHER:: Have you considered the possibility that a wider range of galaxy types (somewhat more massive than Irrs) could feed into the present day dEs? For example, a galaxy falling in at an earlier epoch and being stripped would have a much lower surface brightness (more extended stellar pop and fewer stars), and so we might, e.g., get M33-types into rotating dwarfs?

VAN ZEE:: There are two observations that indicate that the progenitor population is *not* bright galaxies that have been disrupted. First, the rotation curve data is inconsistent with the velocity widths expected for giant spirals. Second, the observed metallicities are low, similar to the abundances measured in dwarf irregular galaxies at similar luminosity.

Near-Field Cosmology with Dwarf Elliptical Galaxies
Proceedings IAU Colloquium No. 198, 2005
H. Jerjen and B. Binggeli, eds.

doi:10.1017/S1743921305003510

Internal kinematics and stellar populations of dE galaxies: clues to their formation/evolution

Philippe Prugniel[1,2], Igor Chilingarian[1,3], Olga Sil'chenko[3] and Victor Afanasiev[4]

[1]CRAL Observatoire de Lyon, France
[2]GEPI Observatoire de Paris-Meudon, France
[3]Sternberg Astronomical Institute of MSU, Russia
[4]Special Astrophysical Observatory of RAS, Russia

Abstract. What is the origin of the numerous population of diffuse elliptical galaxies (dE)? These galaxies formed their stars several billion years ago and lost their gas. Though the stellar winds resulting from star formation and the interactions with the environment undoubtedly play a role, their respective role and details of this evolution are still debated.

In this presentation we will review the first 3D spectroscopic observations of a handful of dE galaxies. These data reveal complex kinematical structures, with embedded discs and counter rotating cores, and they open extremely promising perspectives for studying the history of the stellar population throughout these various features.

The presence of disks, which was already known from detailed image analysis, their complex kinematics, and the new constraints on the stellar population enforce the hypothesis of the evolutionary connection between dEs and disk galaxies.

Keywords. methods: data analysis, galaxies: dwarf, kinematics and dynamics, stellar content

1. Introduction

Though diffuse elliptical galaxies represent majority of the galaxy populations in dense regions of the nearby Universe like rich clusters, their origin and evolution remain still a matter of debate. Two main scenarios proposed so far are (a) early hierarchical collapse with a feedback of star formation (Dekel & Silk 1986, Nagashima & Yoshii 2004), and (b) environmental evolution: ram pressure stripping of disc galaxies in clusters (Mori & Burket 2000) and groups (Marcolini *et al.* 2003), or gravitational harassment (Moore *et al.* 1998). The predictions of the popular now CDM cosmological simulations for the number and intrinsic structures of the nearby dwarf galaxies does not agree with the real situation, as well as the multi-burst star formation histories of dSph of the Local Group are quite unexpected in the frame of the hierarchical concept (Koch *et al.* 2004). So observations remain the only valid way to find a clue to their formation and evolution.

The dynamical structure of the dEs is not clear too. From a sample of 35 dEs, only a few are not consistent with rotational flattening (prolate or triaxial) (Prugniel & Simien 2003). However, there are non-rotating ellipsoidal dwarfs like tidally stressed objects (like NGC205), or some other non-rotating flat objects, like IC 794, probably anisotropic. Fine structures, embedded bars and discs sometimes with spiral patterns, have been found in some dEs (Jerjen *et al.* 2000, Barazza *et al.* 2002). They may argue for a hypothesis of common origin of dEs and dIrrs.

By using the benefits of the integral-field spectroscopy providing two-dimensional distributions of the kinematical parameters, as well as of absorption-line indices, we aim to

Name	Date	seeing	t_{exp}	S/N_{cent}	σ_{cent}
IC 3468	2004/Mar/20	1.5"	2.5h	20	30
IC 3653	2004/May/24	1.4"	2h	35	70
NGC 770	2004/Oct/07	2"	2h	50	100

Table 1. Parameters of observations

search for (1) kinematical counterparts of the morphological fine structures revealed by the surface photometry, (2) related subpopulations in the stellar content.

2. Observations, data reduction and analysis

In 2004 we started to observe dE galaxies in the Virgo cluster with the MPFS integral field unit (IFU) spectrograph at the Russian 6-m telescope. The technique of 3D spectroscopy allows to obtain spatially resolved distribution of kinematical characteristics (radial velocity and velocity dispersion fields) independently for gas and stars, and stellar population parameters (for example, maps of line-strength indices) in a single exposure.

Observations of the galaxies have been performed with the MPFS IFU spectrograph on the 6m telescope of SAO RAS during three observing runs in 2004. The parameters of the observations are presented in Tab. 1

Data processing and analysis are discussed by Chilingarian *et al.* (this volume). We refer this paper for explanations of adaptive binning technique and fitting procedures.

3. Results

Here we present the results for three diffuse galaxies: IC 3468 and IC 3653 in the Virgo cluster, and NGC 770 in the NGC 772 group. We are modeling the simplest star formation history containing one burst.

3.1. *IC 3468*

Rather low S/N ratio doesn't allow very precise measurements. The results shown are obtained for tessellations with target S/N=10. No young subpopulation is detected. For a main burst with $t = 5$ Gyr with low metallicity ([m/H] $= -0.4$) we see some metallicity gradient (from -0.3 in the centre till -0.55 in the outskirts). Velocity dispersion map shows stripe with low values (near the limit of the measurements, about 20 km/s) aligned along the major axis of the isophotes. Radial velocity field is quite irregular and asymmetric, though showing some kind of rotation with the amplitude of 10 km/s.

3.2. *IC 3653*

Good S/N ratio allows to build precise maps of the parameters till $7''$ from the centre of the galaxy. Disc rotation with the amplitude of about 40 km/s is seen in the radial velocity field with the strongest gradient in the inner region, that is quite unusual for objects of such type. Velocity dispersion shows gradient from 70 km/s in the centre till 40 km/s in the outer regions. The age map is almost flat with a mean age of 7 Gyr, all details seen are not statistically significant except young values in the very centre. Metallicity distribution shows shallow gradient from $[m/H] = 0$ in the centre to $[m/H] = -0.2$ at the outskirts with a small metal-rich nuclear region ($[m/H] = 0.13$). There is a marginal detection of younger population $t \sim 1$ Gyr in the centre which may contain a mass fraction of 1 to 3 percent. The maps are computed for the target S/N values of 25.

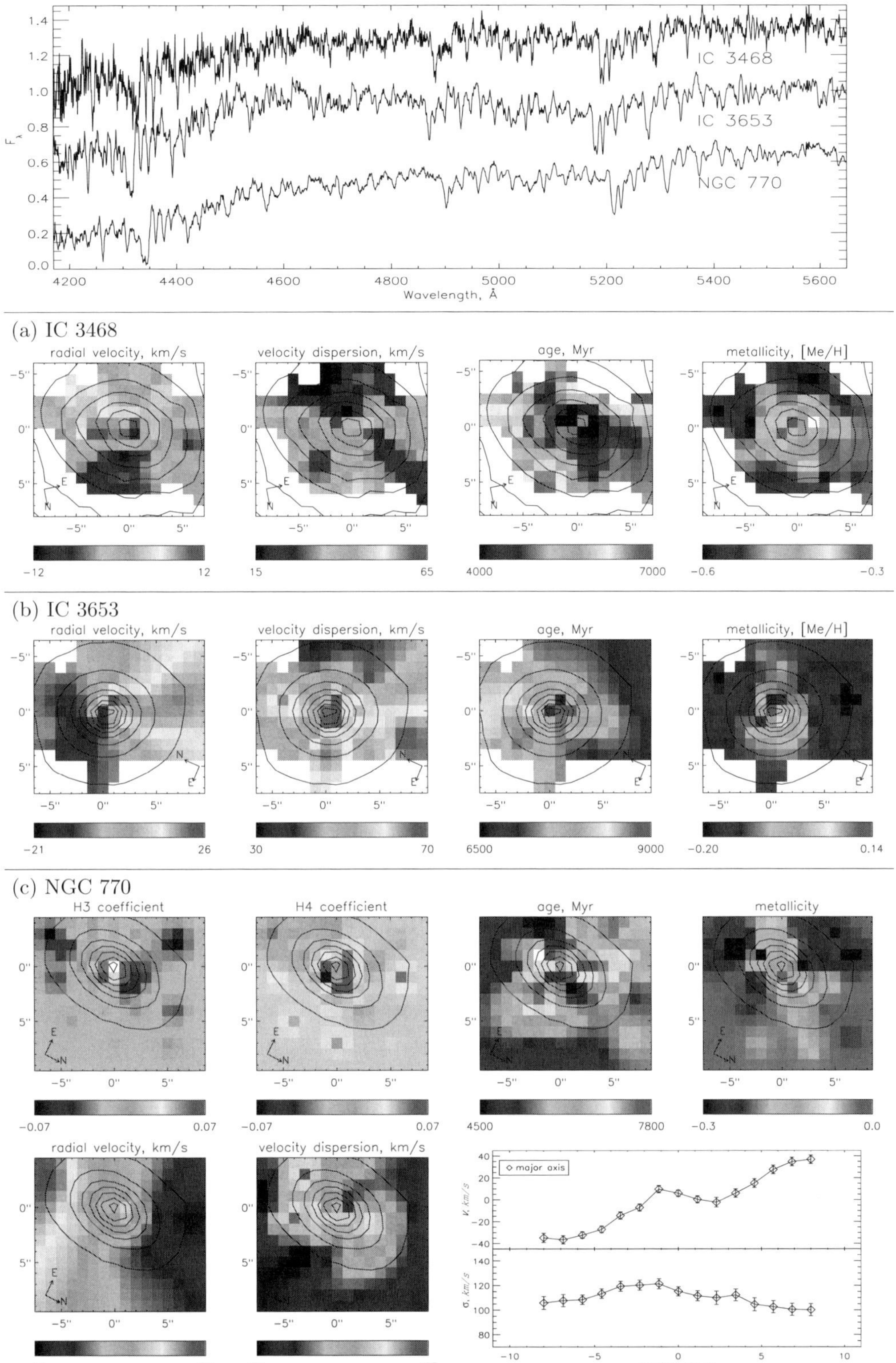

Figure 1. On the upper figure spectra of the central regions of three galaxies are shown. (a) and (b) Results for IC 3468 and IC 3653: 2D fields of radial velocity, velocity dispersion, age and metallicity in the single burst SFH model. (c) Results for NGC 770: the same as previous, besides the maps of H3 and H4 Gauss-Hermite coefficients, and kinematical profiles along major axis are shown.

3.3. *NGC 770*

The galaxy experiences strong tidal interaction with the giant spiral NGC 772. Its luminosity ($M_B = -18.4$) and velocity dispersion (110-120 km/s) place it between giants and dwarfs. Very good S/N ratio allows precise measurements of all the parameters (for example, $\Delta v \sim 1$ km/s). Good sampling of LOSVD due to high velocity dispersion allows to measure H3 and H4 Gauss-Hermit coefficients. We see impressive kinematically decoupled core in this object. This kinematical decoupling has been recently noticed by Geha *et al.* 2005. There are evidences for considering it to be a counter-rotating young metal-rich highly inclined stellar disc aligned almost with the major axis of the galaxy: (1) velocity dispersion map shows larger values on the "switches" of the velocity, there is a stripe with lower values orthogonal to the structures in age/metallicity distributions; (2) map of H3 coefficient shows regions with positive and negative values before and after "switch" of the velocity; (3) age and metallicity distributions clearly show the decoupled structure aligned almost along the major axis having younger age ($t \sim 4.5$ Gyr vs 7 Gyr) and higher metallicity ($[m/H] \sim -0.05$ vs -0.2) than the surrounding spheroid. The galaxy probably had a partially dissipative merger event, that induced a burst of star formation some 4.5 Gyr ago producing the central stellar structure that we see now as a counter rotating subpopulation with the high metallicity.

4. Conclusions

Kinematical features found in the dE galaxies strengthen the connection between dEs and dIrrs. Inner discs are probably the remnants of the stellar discs existing before gas removal.

Whatever the gas-removal mechanism is (winds or ram pressure stripping) the signatures would be the same. These systems probably did not undergo strong relaxation which would have erased the structures and population gradients. Most dEs should have the same origin, i.e. the same type of progenitors, but they had a different history, i.e. interplay between feedback, ram pressure stripping, and tidal effects. The study of the properties of the kinematical and population substructures will help to understand this history.

Acknowledgements

We would like to thank International Astronomical Union providing financial support for attending IAU Colloquium 198, 6-m telescope time allocation committee, INTAS foundation providing PhD fellowship for I.C., EGIDE and CNRS for additional support.

References

Barazza, F.D., Binggeli, B. & Jerjen, H. 2002, *A&A* 391, 823
Dekel, A. & Silk, J. 1986, *ApJ* 303, 39
Franx, M. & Illingworth, J. 1988, *ApJ* 327, L55
Geha, M., Guhathakurta, P. & van de Marel, R.P. 2005, *AJ* in press, astro-ph/0503370
Jerjen, H., Kalnajs, A. & Binggeli, B. 2000, *A&A* 358, 845
Koch, A. *et al.* 2004, *ANS* 325, 44
Le Borgne D. *et al.* 2004, *A&A* 425, 881
Marcolini, A., Brighenti, F. & D'Ercole, A. 2003, *MNRAS* 345, 1329
Moore B., Lake G. & Katz N. 1998, *ApJ* 495, 139
Mori M. & Burkert A. 2000, *ApJ* 538, 559
Nagashima, M. & Yoshii, Y. 2004, *ApJ*, 610, 23
Prugniel, P. & Simien, F. 2003, *Ap&SS*, 284, 603

Near-Field Cosmology with Dwarf Elliptical Galaxies
Proceedings IAU Colloquium No. 198, 2005
H. Jerjen and B. Binggeli, eds.

doi:10.1017/S1743921305003522

On the origin of dwarf spheroidal galaxies. Clues from cosmological simulations

Massimo Ricotti[1] and Nicholay Y. Gnedin[2]

[1]Dept. of Astronomy, University of Maryland, College Park, MD 20742-2421, USA
email: ricotti@astro.umd.edu

[2]Dept. of Astronomy, University of Colorado, Boulder, CO 80309, USA
email: gnedin@casa.colorado.edu

Abstract. We compare the properties of dwarf galaxies in the Local Group with the simulated galaxies formed before reionization in a cosmological simulation of unprecedented spatial and mass resolution including radiative feedback effects. We find that a subset of the Local Group dwarfs, the dwarf Spheroidals, are remarkably similar to the simulated dwarf galaxies in all their properties already before reionization. Simulated and observed dwarf Spheroidal galaxies not only have similar properties but also follow the same scaling relations. Based on this similarity and on the observed ages of their dominant stellar populations we propose the hypothesis that Local Group dwarfs form in two different ways: (i) most dwarf Spheroidals are pristine fossils of the pre-reionization era and (ii) dwarf irregulars are more massive galaxies that formed most of their stars later, after reionization. There is also a group of "polluted fossils" with properties that are intermediate between these two main groups. We predict the existence of many more dwarf Spheroidals, fainter and with lower surface brightness than the observed population.

Keywords. early universe, Local Group, galaxies: dwarf, formation, high-redshift, individual (UMi, Draco, Antlia, Sculptor, Phoenix, Sextans, Cetus, Tucana, KKR25 AndI-III, AndV-VI)

1. Introduction

The punchline in this talk is to provide evidences in support of the primordial origin of dwarf Spheroidals (dSphs). I will present a comparison between the abundant observational data on Local Group dwarfs and the results of detailed cosmological simulation of formation of dwarf galaxies in the pre-reionization era.

Almost all Local Group dwarf Spheroidals exhibit a prominent old population and a fraction show a decline of their star formation rates about 10 Gyr ago. In the standard Λ-CDM theory stellar reionization took place 12.5 Gyr ago but current observational techniques for measuring absolute stellar ages are not able to differentiate between 10 and 12.5 Gyr (*e.g.*, Krauss & Chaboyer 2003). Although the observed drop of star formation rate is consistent with reionization at redshift $z \sim 6$, as noted by Grebel & Gallagher 2004, the metallicity spreads observed in the old populations may imply a star formation history protracted over about 2 Gyr, to redshift $z \sim 3$. In this talk I critically discuss this important issue. I will show that our simulations reproduce the observed metallicity spreads without a protracted period of star formation. The metallicity spreads observed in our simulated dwarfs can be understood in terms of hierarchical accretion of sub-haloes containing stars with different metallicities.

The key difference between our proposal for the origin of dSphs and the tidal hypotheses (Stoehr *et al.* 2002, Kravtsov *et al.* 2004) is whether most of the stars in a given dwarf galaxy formed in *halos of mass* $\lesssim 10^8\ M_\odot$ before reionization ($z > 6$, 12.5–13 Gyr ago) or in *more massive halos* after reionization, during the epoch of formation of the Milky Way and Andromeda ($z \sim 2$–3, 10–11 Gyr ago). We find that most dSphs are fossils of galaxies with mass $M_{\rm dm} < 10^8$ M$_\odot$ that formed before reionization, while dIrr formed later in more massive halos. The original contribution of our study is that for the first time we have been able to simulate the first population of galaxies including the relevant feedback processes that are known to self-limit their formation. Before our work (Ricotti, Gnedin, & Shull 2002a, 2002b) negative feedback effects (namely the rapid photo-dissociation of H_2) were thought to suppress the formation of all galaxies with mass smaller than 10^8 M$_\odot$. We instead found that these galaxies do form and, at the epoch of their formation, they closely resemble dSph galaxies. This second result is the one I would like to emphasize in the present talk. Our simulations also show that reionization feedback is not the dominant mechanism that suppresses star formation in most small mass halos. Negative feedbacks such as photo-heating by the stars inside each galaxy and inefficient cooling produce the observed properties of dSphs well before reionization of the IGM is complete. Reionization does not affect substantially their properties because they have already lost most of their gas due to internal ionizing sources. After reionization we expect that the formation of small mass galaxies will stop and their stellar populations will evolve only passively. Only the few most massive dwarfs in our simulation retained some of their gas and may continue to form stars after reionization.

2. Simulation

The simulation we use in this paper has been thoroughly described in Ricotti *et al.* (2002a, 2002b) as run "256L1p3". Here we just remind the reader that the simulation included 256^3 dark matter particles, an equal number of baryonic cells, and more than 700,000 stellar particles. The simulation includes most of the relevant physics, including time-dependent spatially-variable radiative transfer, detailed radiative transfer in Lyman-Werner bands, non-equilibrium ionization balance, etc. In addition, here we also include the effect of reionization, see Ricotti and Gnedin (2005) for details. The limitation of our simulation is that it stops at redshift $z = 8.3$. Clearly, it is a long way to go from $z = 8.3$ to the present day. Therefore, in the rest of this talk, we are assuming that the stellar component of our galaxies does not change after $z = 8.3$ except due to passive stellar evolution. This hypothesis, obviously, does not apply to *all* observed dwarfs, but it may apply to *some* of them. We estimated that about 5-10% of the galaxies present in our simulation at $z = 8.3$ may have survived almost unspoiled until today. In order to compute the luminosities of the simulated galaxies at $z = 0$, we adopt a fiducial ratio of $M_*(z = 8.3)/L_V(z = 0) = 3$. However, our results are not immensely sensitive to that number - changing it to 2 or 10 would not change our conclusions.

3. Results

Clearly, not all dwarfs are true fossils of the pre-reionization era, since many of them had recent episodes of star formation. But is there a subset of Local Group dwarfs that formed before reionization? We can attempt to answer this question by comparing our simulated dwarfs as they would appear at $z = 0$ with the real observed galaxies. Figure 1(left) presents such a comparison for the basic structural parameters: the luminosity L_V, the core radius r_c, and the central luminosity density I_c, defined as $I_c = \Sigma_V/2r_c$, where Σ_V is the surface brightness. From the top panel of that figure it is clear that the simulated galaxies, shown by the gray squares without error-bars, do not exceed about $10^7 L_\odot$. Moreover, it appears that the observed galaxies (shown by squares with error-bars, open circles, and filled circles - the different symbols will be explained later in the text) also fall into two groups - those with luminosities on the order of $10^7 L_\odot$ and below,

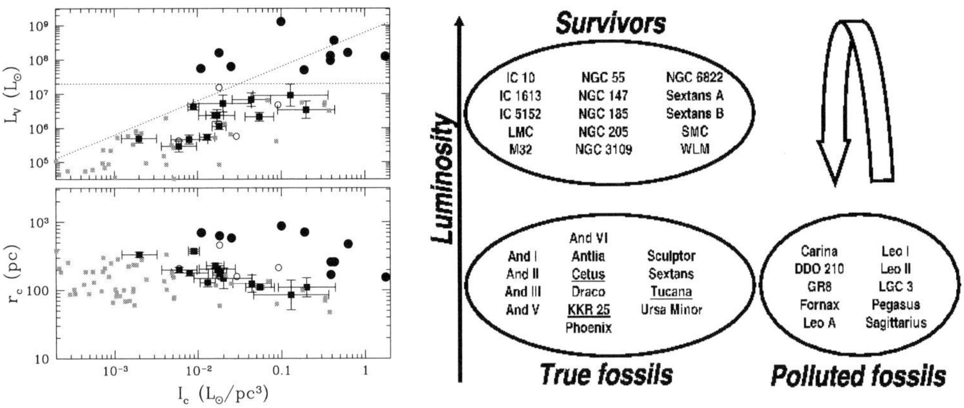

Figure 1. *(Left)* The V-band luminosity (top) and the core radius (bottom) vs the central luminosity density for the Local Group dwarfs (squares with error-bars, open circles, and filled circles - the different symbols are explained in the text) and for the simulated galaxies passively evolved to $z = 0$ (gray squares). Two dotted lines show two possible boundaries between the "fossils" and "survivors". The data is from Mateo (1998). We have complemented the data using updated distances and photometry from van den Berg (2000) and metallicities and metallicity spreads from Grebel *et al.* (2003). *(Right)* A tentative separation of all Local Group dwarfs into three groups: "survivors", that formed via the tidal scenario (mostly dIrr), "true fossils", that formed via the reionization scenario, and "polluted fossils", that went through both stages. In our tentative classification most dSphs (14) are true fossils: five are M31 satellites, six are Milky Way satellites and three (the underlined names) are isolated from both M31 and the Milky Way.

and those with luminosities in excess of $5 \times 10^7 L_\odot$. In addition, these two groups are also clearly separated in $r_c - I_c$ plane. The low luminosity dwarfs appear to trace the same distribution as the simulated galaxies, having smaller core radii as the central luminosity density increases. This observation can be understood in terms of the less efficient cooling of the gas in dark matter halos of smaller mass.

We can now see how our separation in these two groups (based only on the luminosity) compare with the available data on the star formation history of observed dwarfs. Hereafter we will adopt the star formation histories compiled by Grebel & Gallagher (2004). In order to schematically separate the observed dwarfs in groups according to their star formation histories, we call those dwarfs that formed few (say, less than 30%) stars after reionization "true fossils", and those that did form a substantial amount of stars "polluted fossils". The various symbols in Figure 1(left) and the following ones correspond to these classes: "true fossils" are shown with filled squares with error bars, "polluted fossils" are shown with open circles. Galaxies with dominant younger stellar populations (they mostly are dIrrs) are shown with larger filled circles and we call them "survivors". All the galaxies in our simulation have almost identical properties to the "true fossils" and most "polluted fossils". All these galaxies have a prominent old stellar population as expected. Instead all the galaxies with a prominent young population (filled circles) are part of a distinct population that is not present in our simulation. Figure 1(right) presents a tentative separation of the Local Group dwarfs into these three types.

Figure 2(left) compares the surface brightness and the core radii of simulated and observed galaxies as a function of the luminosity. The simulation reproduces very well the structural properties and scaling relations of the galaxies that we classified as "true fossils". The lines show, for comparison, the scaling relations derived by Kormendy & Freeman (2004) for more luminous late-type galaxies (Sc-Im) with luminosities in the

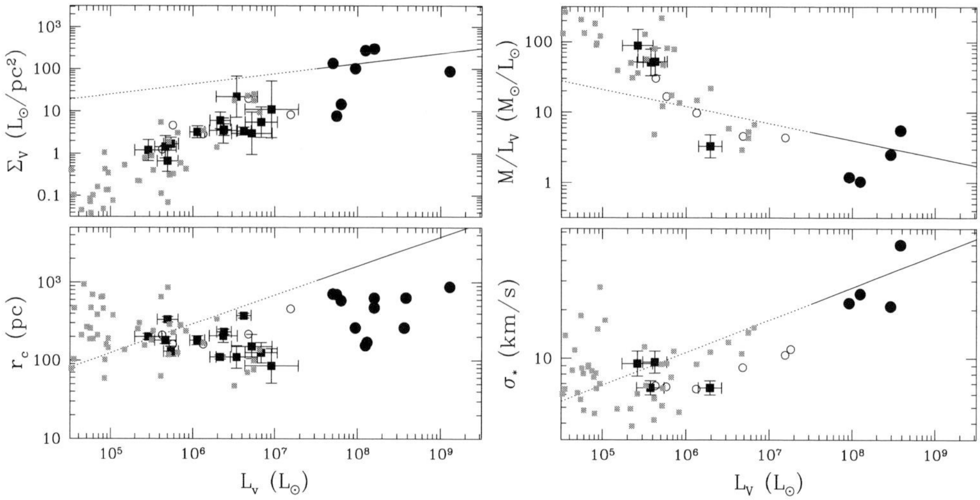

Figure 2. *(Left)* Surface brightness (top) and core radius (bottom) as a function of the V-band luminosity for the simulated and observed dwarfs. *(Right)* Total mass-to-light ratio (top) and stellar velocity dispersion (bottom) for the simulated and observed dwarfs. Symbol markings and the meaning of the lines are the same as in Fig. 1.For comparisons we show (solid lines) the scaling relationships for the more luminous Sc-Im galaxies ($10^8 L_\odot \lesssim L_B \lesssim 10^{11} L_\odot$) derived by Kormendy & Freeman 2004.

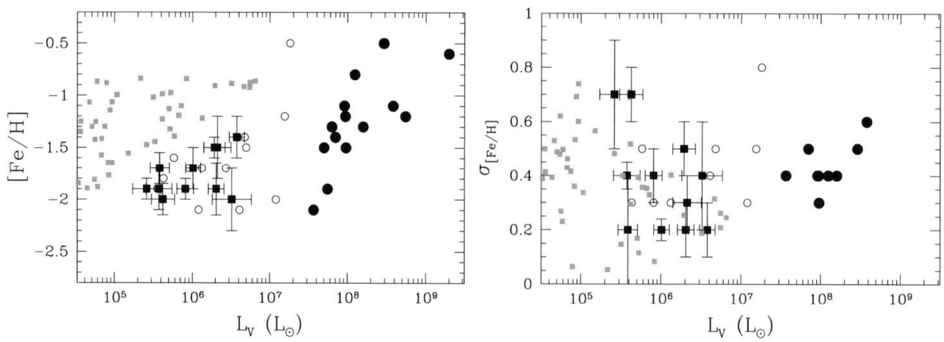

Figure 3. Stellar metallicity (left) and the metallicity spreads (right) as a function of stellar luminosity. We adopt solar abundances and yield y=$\rho_Z/\rho_* = 1\ Z_\odot$. Symbol markings and the meaning of the lines are the same as in Fig. 1. The spreads $\sigma_{[Fe/H]}$ is defined as the variance of the number weighted metallicity distribution of the "stellar particles" in each dwarf.

range $10^8 L_\odot \lesssim L_B \lesssim 10^{11} L_\odot$. Figure 2(right) compares the dynamical state of these galaxies: the stellar velocity dispersion and the total mass-to-light ratio. The latter quantity for the simulated galaxies is computed in the same way as for the observed ones ($M/L = 136\sigma_*^2 r_c/L$, see Mateo 1998). Within the errors the $\sigma - L_v$ relation seems to be the same power law for all the galaxy types, but "true fossils" obey a steeper relationship between the mass-to-light ratio and luminosity than dIrr and Sc-Im galaxies.

Figure 3 gives yet another comparison between the simulation and the data: the metallicity (left) and the metallicity spreads (right) as a function of stellar luminosity. We find that the typical metallicity variance in each dwarf is about 0.5 dex, in good agreement with observations. The agreement between stellar metallicities in the simulation and the data is somewhat worse. At this point we are not worried by this disagreement as our simulation is not sophisticated enough to predict the stellar metallicities to within a factor of two. The accretion of stars from the hierarchical distribution of sub-haloes partially explains the large metallicity spread that we find in our simulation. Thus we conclude that the large metallicity spread observed in the old stellar population of dSphs is not a definite signature of star formation extended over 2–4 Gyrs as argued by Grebel & Gallagher (2004). Neither the argument based on the long timescale (*e.g.*, 1–4 Gyr) needed to have a substantial iron production by Type Ia SN is a strong argument in favour of prolonged star formation. The typical time scale for iron enrichment by Type Ia SN is uncertain and very likely is dependent on the mode of star formation. Matteucci & Recchi (2001) estimated that this time scale varies from 40–50 Myr for an instantaneous starburst to 4–5 Gyr for a disk of a spiral galaxy like the Milky Way.

4. Discussion

In this paper we have proposed that dwarf galaxies of the Local Group (and, by implication, all other dwarf galaxies in the universe) formed in three different evolutionary paths: i) *"true fossils"* formed most of their stars in the pre-reionization era, and have had little (say, less than 30%) star formation since then; ii) *"polluted fossils"* started as true fossils, but had substantial episode(s) of subsequent star formation as they continued accreting mass and were tidally shocked during the formation of the parent galaxy halo; iii) finally *"survivors"* started forming stars mostly after reionization. This conclusion is motivated by considering star formation histories of Local Group dwarfs and by comparing their properties with properties of simulated galaxies that formed all their stars before cosmological reionization. The fact that our simulated galaxies are remarkably similar to the subset of the Local Groups dwarfs that we identified as "true fossils" in Figs 1–3 renders support to our conclusions. This subset includes about all known dSphs. It is worth noting that the structural properties of the galaxies that we call "true fossils" are the result of feedback processes in action prior to reionization. The effect of reionization is simply to preserve those properties by suppressing the star formation rate in halos with masses that remain smaller than the filtering mass in the IGM.

To conclude, we would like to emphasize another exciting possibility. From the results of our simulations, we feel quite confident in predicting the existence of fainter and lower surface brightness galaxies than the one currently discovered (see the gray squares at the extreme left corners in Figs. 1–3). Two promising candidates for such a class of dwarf galaxies were recently discovered in M31 and near the Milky Way (Zucker *et al.* 2004, Willman *et al.* 2005).

As we mentioned above, it is mostly likely that the majority of reionization fossils will merge into larger objects by $z = 0$, and only a small fraction of them will survive. Thus, we can introduce a "survival probability" S (which can be a function of other parameters, in particular the luminosity) of a reionization fossil to remain intact by the present time. Figure 4 shows our predictions for the abundance of these "ultra-dwarfs" in the Local Group. The solid line in Fig. 4 refers to the case in which the survival probability of reionization fossils is kept constant at 3% – in this case the observations may be missing about half of the satellites with luminosities in the range $2 \times 10^5\,\mathrm{L}_\odot \lesssim L_V \lesssim 6 \times 10^5\,\mathrm{L}_\odot$. Alternatively, we can assume that the observations are complete in this luminosity range, in which case the survival probability must be a weak function of luminosity. Fitting the logarithmic dependence (as an example) to the abundance of observed dwarfs in the three lowest luminosity bins ($2 \times 10^5\,\mathrm{L}_\odot \lesssim L_V \lesssim 6 \times 10^6\,\mathrm{L}_\odot$), we find $S(L_v) = 3\%(L_v/2 \times 10^6\,\mathrm{L}_\odot)^{1/3}$. This case is shown with the dashed line in Fig. 4. As one can see, a modest change in the survival probability makes a large change in the predicted numbers of dwarf galaxies in the $L_V \lesssim 10^5\,\mathrm{L}_\odot$ luminosity range. Although, the detection of such

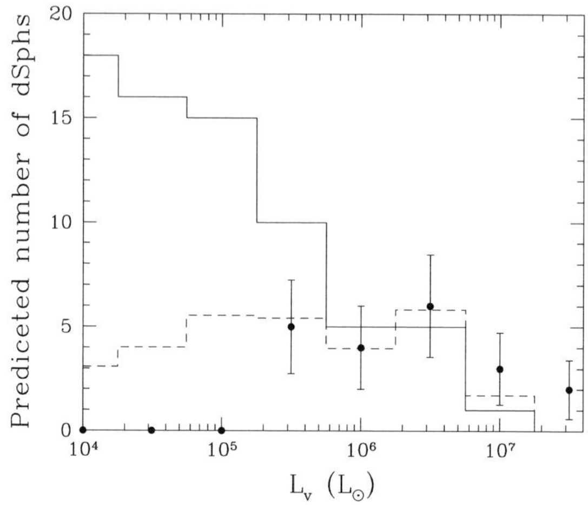

Figure 4. Prediction of the number of faint dSphs per logarithmic luminosity bin in the Local Group (histograms). The points with error bars (Poisson errors) show the number of observed dwarf galaxies in the Local Group. The solid line refers to the case in which the survival probability of reionization fossils is constant as a function of luminosity, while the dashed line shows an example of a case in which the survival probability is mildly luminosity dependent (see text for details).

objects in the Milky Way is very difficult if not impossible using photometric techniques (due to contamination from foreground stars), in the future it might become possible to search for these objects using spectroscopic techniques by measuring the radial velocities of many halo stars.

References

Grebel, E.K., Gallagher, J.S. & Harbeck, D. 2003, *AJ* 125, 1926
Grebel, E.K. & Gallagher, J.S. 2004, *ApJL* 610, L89
Kormendy, J. & Freeman, K.C. 2004, *IAU Symposium* 220, 377
Krauss, L.M. & Chaboyer, B. 2003, *Science* 299, 65
Kravtsov, A.V., Gnedin, O.Y. & Klypin, *A&A* 2004, *ApJ* 609, 482
Mateo, M. 1998, *ARAA* 36, 435
Matteucci, F. & Recchi, S. 2001, *ApJ* 558, 351
Ricotti, M., Gnedin, N.Y. & Shull, J.M. 2002a, *ApJ* 575, 33
Ricotti, M., Gnedin, N.Y. & Shull, J.M. 2002b, *ApJ* 575, 49
Ricotti, M. & Gnedin, N.Y. 2005, *ApJ* 629,
Stoehr, F., White, S.D.M., Tormen, G. & Springel, V. 2002, *MNRAS* 335, L84
van den Bergh, S. 2000, *PASP* 112, 529
Willman, B., *et al.* 2005, *ApJL* submitted (astro-ph/0503552)
Zucker *et al.* 2004, *ApJ* 612, L121

Discussion

MATEO: How do I interpret the number of galaxies at very low luminosities in your plots? That is, can I just scale by the number of known galaxies in the regions of your plots where the data and simulations overlap?

RICOTTI: Yes, that is the way to do it. There might be second order corrections due to a possible dependence of the survival rate on the mass (or luminosity) of the primordial dwarfs.

ZAGGIA: Can you say anything about rotation in dwarfs?

RICOTTI: The rotation velocity is negligible with respect to the velocity dispersion of the stars.

READ: What is the origin of the anisotropy (radial) in the outermost point of the stars which is not present in the dark matter?

RICOTTI: I do not know. Perhaps it is a signature of accretion of stellar subunits. Remember that most dark matter halos are dark. Stars are formed in a small fraction of DM halos.

GALLAGHER: Have you modeled abundance ratios of elements that might distinguish the short starburst and extended star formation models, for example the r-process or α-elements relative to iron?

RICOTTI: No.

Near-Field Cosmology with Dwarf Elliptical Galaxies
Proceedings IAU Colloquium No. 198, 2005
H. Jerjen and B. Binggeli, eds.

doi:10.1017/S1743921305003534

Andromeda and the seven dwarfs

Alan McConnachie†[1], Mike Irwin[1], Scott Chapman[2], Rodrigo Ibata[3], Annette Ferguson[4], Geraint Lewis[5], and Nial Tanvir[6]

[1]Institute of Astronomy, Madingley Road, Cambridge, CB3 0HA, U.K.
email: alan@ast.cam.ac.uk

[2] California Institute of Technology, Pasadena, CA, 91125

[3]Observatoire de Strasbourg, 11, rue de l'Universite, F-67000, Strasbourg, France

[4]Institute for Astronomy, University of Edinburgh, Royal Observatory, Blackford Hill, Edinburgh, EH9 3HJ, U.K.

[5]Institute of Astronomy, School of Physics, A29, University of Sydney, NSW 2006, Australia

[6]Physical Sciences, Univ. of Hertfordshire, Hatfield, AL10 9AB, U.K.

Abstract. The satellite system of M31 is the next closest satellite system to our own, and is well within reach of current generations of ground-based instrumentation. We have conducted a survey of this system with the Isaac Newton Telescope Wide Field Camera, particularly focussing on the seven dwarf spheroidal companions, and derived homogeneous, fundamental information for them. This reveals the presence of severe tidal distortion and multiple structural components in some systems. In addition, a comparison of the M31 dwarf spheroidals to their Galactic counterparts shows that the M31 dwarf spheroidals have scale radii generally twice as large as the Galactic dwarfs, suggesting that the tidal field of M31 may be substantially weaker than that for the Galaxy. A Keck/DEIMOS study of the newly discovered satellite And IX suggests that this galaxy has an anomalously large metallicity for its luminosity, and shows it to be strongly dark matter dominated.

Keywords. (galaxies:) Local Group, galaxies: dwarf, galaxies: stellar content, galaxies: structure, galaxies: kinematics and dynamics

---

1. Introduction

Most of our detailed knowledge on the structure of dwarf spheroidal (dSph) galaxies comes from observations of the nine dSphs which make up part of the Galactic satellite system – Draco, Ursa Minor, Carina, Fornax, Sculptor, Sextans, Sagittarius, Leo I and Leo II. For these, it is possible to reach below the main sequence turn-off with modest instrumentation and conduct a detailed analysis of their stellar populations. Irwin, & Hatzidimitriou (1995) were the first to analyse the structure of these systems by mapping their resolved star counts, providing much fundamental structural information. Generally, they found that the stellar structure of all the dSphs was well described by a single component King model or an exponential profile. A generic feature was an excess of stars at large radii, beyond the best-fitting tidal radius. These stars have generally been interpreted as a sign of tidal disturbance from interaction with the gravitational potential of the Milky Way, although other interpretations are possible. Choi, Guhathakurta & Johnston (2002) have shown that a break in the radial profile – which may lead to an excess of stars observed in this way - is a natural consequence of tidal perturbation.

† New address from October 2005: Department of Physics and Astronomy, University of Victoria, 3800 Finnerty Road, Victoria, BC V8P 1A1, Canada

More recent studies have shown that the structure of dSphs may be more complex than that which has previously been assumed. Harbeck, *et al.* (2001) conducted a systematic and homogeneous study of population gradients in several Local Group dSphs, by looking for spatial variations in the colour of the red giant branch and the horizontal branch. They find evidence for gradients in many of the systems analysed. Tolstoy, *et al.* (2004) have since analysed the global population of stars in Sculptor from photometry and kinematic information and showed that it has a spatially, chemically and kinematically distinct second component, in the form of a more centrally concentrated core. Additionally, Kleyna, *et al.* (2004) have shown that the Sextans dSph has a kinematically cold core. They postulate that this is a disintegrating star cluster, whose orbit has decayed by dynamical friction to the centre of Sextans. This is similar to Ursa Minor, which shows a distinct concentration of stars offset from the geometric center of this galaxy (Irwin, & Hatzidimitriou (1995), Kleyna, *et al.* (2003)) and also shows evidence of kinematic substructure at the location of the secondary over-density.

Given these findings in the Galactic dSph system, it is timely to look closely at the next closest dSph system to our own, that of M31. Approximately 16 satellites make up this subsystem, of which seven – Andromeda I, II, III, V, VI, VII, and IX – are classified as dSphs. In comparison to the dSphs which orbit the Galaxy, relatively little is known in detail about the stellar populations of the Andromeda dSphs. Deep HST fields have been obtained for Andromeda I, II and III which extend below the horizontal branch (Da Costa *et al.* (1996), Da Costa *et al.* (2000), Da Costa *et al.* (2002)). Horizontal branch analysis shows that each has had an extended epoch of star formation and variations in their horizontal branch morphology suggest that their star formation histories have been notably different. In addition, Andromeda I is observed to display a gradient in its horizontal branch morphology, such that their are more blue horizontal branch stars located at larger radius from the center of the dwarf (Da Costa *et al.* (1996)). Some evidence for AGB components have also been seen in these systems (most recently by Kerschbaum, *et al.* (2004) and Harbeck, *et al.* (2004)), although a strong intermediate age component similar to some of the Galactic dSphs is generally lacking.

We have obtained Johnston V and Gunn i photometry for the majority of the members of the M31 subgroup using the Isaac Newton Telescope Wide Field Camera (INT WFC). Figure 1 shows two typical dSph colour magnitude diagrams (CMDs) created from our data for And I & II. This photometry is deep enough to observe the top few magnitudes of the red giant branch in each system. We have used this to conduct a homogeneous study of these satellites and derive fundamental information for them, including distances, metallicities, and structural parameters for the dSphs. Detailed descriptions of these studies are to be found in McConnachie, *et al.* (2004), McConnachie, *et al.* (2005), McConnachie & Irwin (2005a), McConnachie & Irwin (2005b). The main M31 survey is detailed in Ferguson, *et al.* (2002). In addition, we have conducted a spectroscopic analysis of the newest dSph, And IX, using the Keck/DEIMOS spectrograph (Chapman, *et al.* (2005)). Here, we present a broad overview of this study and some of the main results.

2. Distances and metallicities

The tip of the red giant branch (TRGB) is an ideal standard candle for older, more metal poor systems. Physically, this stage in stellar evolution represents the point of the core helium (He) flash of low mass stars. During this phase of stellar evolution the temperature of the degenerate, quasi-isothermal He core is only dependent upon the properties of the thin hydrogen burning shell around it. This, in turn, varies only very slightly with chemical abundance and surrounding mass. The bolometric luminosity of

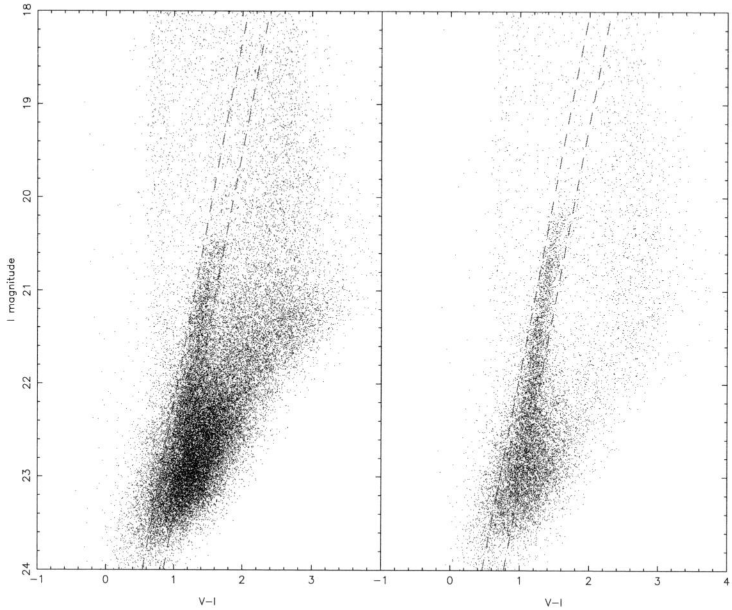

Figure 1. CMDs for And I (left panel) and And II (right panel) demonstrating the quality of the INT WFC photometry. The red giant branch of each system is clearly visible. The additional red feature in the And I CMD is due to the M31 giant stellar stream discovered as part of our main M31 survey.

the TRGB is thus of roughly constant intrinsic brightness. The absolute I band magnitude of the TRGB (M_I^{TRGB}) is also found to be nearly constant for these stars as the stellar spectrum peaks near this wavelength range and absorption effects in the stellar atmosphere are minimal. For more metal rich stars, molecular absorption causes M_I^{TRGB} to dim. Barker, Sarajedini & Harris (2004) have recently conducted a study of the reliability of the TRGB as a distance indicator for stellar populations with a wide variety of star formation histories and find that it is reliable for any system which is not dominated by a metal-rich population ([Fe/H]> -0.3) or a population with a substantial young ($\lesssim 1.7$ Gyrs) component. This makes it ideal for determining distances to Local Group galaxies.

The TRGB is clearly visible in Figure 1 for both And I & II. In McConnachie, *et al.* (2004), we devise an algorithm in order to measure this point, based upon a generalised least-squares fitting procedure. This allows the distances to be derived for each of the 17 galaxies for which we have INT WFC data. In order to account for the slight metallicity dependency of M_I^{TRGB}, we also self-consistently derive the median metallicity of each system based upon the colour of its red giant branch. The net result is a large, homogeneous set of distances and metallicities for the Local Group galaxies M31, M33, And I, II, III, V, VI, VII, IX, NGC 205, 185, 147, Pegasus, WLM, LGS3 and Aquarius. The results are listed in Table 1 and 2 of McConnachie, *et al.* (2005), and are ideal for studies of Local Group galaxy properties. As an example, Figure 2 is a projection of the M31 satellite system in M31-centric spherical coordinates, derived from these estimates. The positions of these satellites around M31 are much better constrained than previously, as systematic errors have been minimised and each distance is accurate to $\sim$5 %.

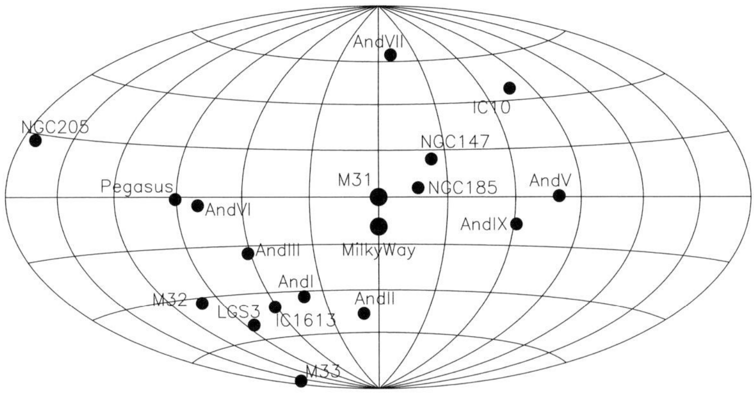

Figure 2. An Aitoff projection of the satellite galaxies of M31 in a native coordinate system. l is measured around the disk of M31 (where the Milky Way is defined to lie at $l = 0$) and b is measured from the disk of M31. The position of each satellite around M31 is better constrained than previous studies due to the homogeneous set of distances used.

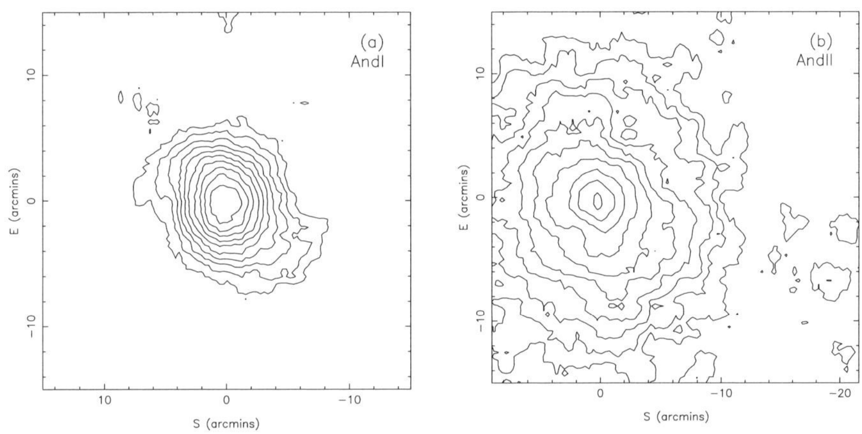

Figure 3. Isopleth maps for the And I & II (left and right panels respectively) based upon resolved star counts. The outermost contours are $2\,\sigma$ above the background. And I shows clear evidence of severe tidal disruption, whereas And II is tidally unperturbed. This galaxy, however, shows compelling evidence for a secondary 'core' component (Figure 4).

3. Structural properties of the M31 dSphs

Figure 3 shows contour maps of the red giant branch spatial distribution for And I (left panel) and And II (right panel). The outer contour is $2\,\sigma$ above the background in both plots. Both of these galaxies are relatively large, and have distinct morphologies.

And I is a very non-spheroidal dSph, and displays a distinct and very well-defined S-shape. This is an unambiguous indicator of strong tidal disruption, and it is possible that And I will have a low surface brightness tidal stream associated with it. In addition, And I may have a relatively low density dark matter halo surrounding it, as this would otherwise act to shield the stellar component from severe tidal disruption. Several other M31 dSphs also show tentative evidence for tidal disturbance, but none as convincing as And I.

And II, on the other hand, is nearly perfectly circular in projection, and displays no evidence of tidal effects. However, examination of its isopleths shows that the central few contours appear more concentrated that the outer contours. Figure 4 shows the surface

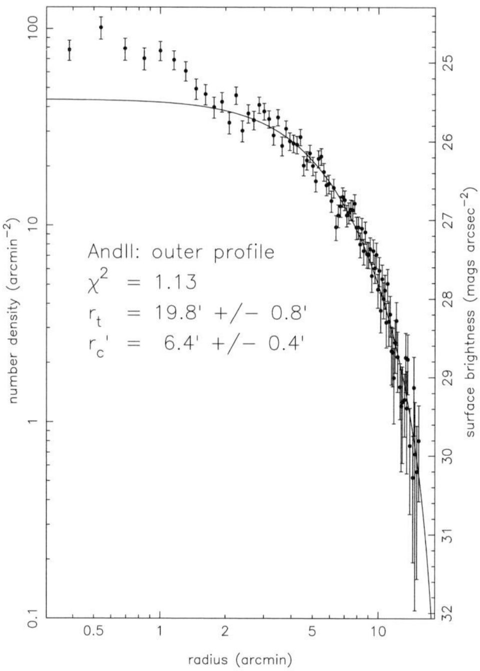

Figure 4. The best fit King profile to the surface brightness profile of Andromeda II, ignoring the inner $2'$. The formal value of the fit is very good, and the divergence of the King profile to the data at small radii suggests the presence of a secondary core component in this dSph.

brightness profile of And II, derived from the stars counts. The solid line is a King profile fit to the outer regions of the profile, and clearly demonstrates that there is a break in the profile and a large excess of stars at small radii. This 'core' is reminiscent of the sort of structure seen by Tolstoy, *et al.* (2004), and provides compelling evidence that And II is also a two-component dSph. Deeper global photometry extending below the horizontal branch is required.

4. The Local Group dSph populations compared

The central surface brightnesses, integrated magnitudes, half light radii and scale radii of the best fitting King, exponential and Plummer profiles are derived for And I, II, III, V, VI, VII and Cetus in McConnachie & Irwin (2005a). Coupled with the distance information to each of these systems, this allows us to compare the structural characteristics of the M31 dSph population to their Galactic counterparts.

As an example of the sort of comparison that is possible, Figure 5 shows the half-light radii (left panel) and tidal radii of the best-fitting King profile (right panel) as a function of integrated luminosity for the M31 dSphs (triangles), Galactic dSphs (squares) and the isolated Cetus dSph (diamond). For a given absolute magnitude, the M31 dSphs are generally at least twice as extended as the Galactic dSphs. This is true both in terms of the half-light radius and also in terms of the tidal radii. It may be that this reflects environmental differences. For example, if the tidal field of M31 is significantly less than for the galaxy, then we would expect the M31 population of dSphs to have generally larger tidal radii. It is not so clear, however, how tidal effects could significantly affect the half-light radii to produce the trend that we observe.

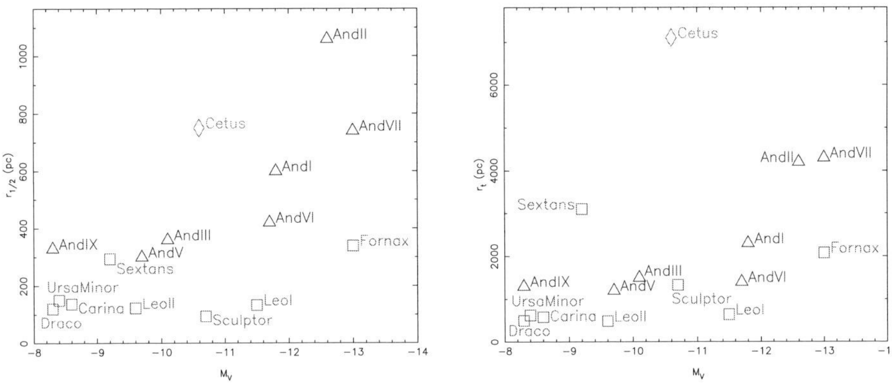

Figure 5. Half-light radii (left panel) and tidal radii (right panel) for the Local Group dSphs. Satellites of M31 are shown as triangles, satellites of the Galaxy are shown as squares, and the isolated dSph in Cetus is shown as a diamond. For a given absolute magnitude, the scale radii of the M31 dSphs are generally at least twice as large as their Galactic counterparts. This could reflect environmental differences between the populations.

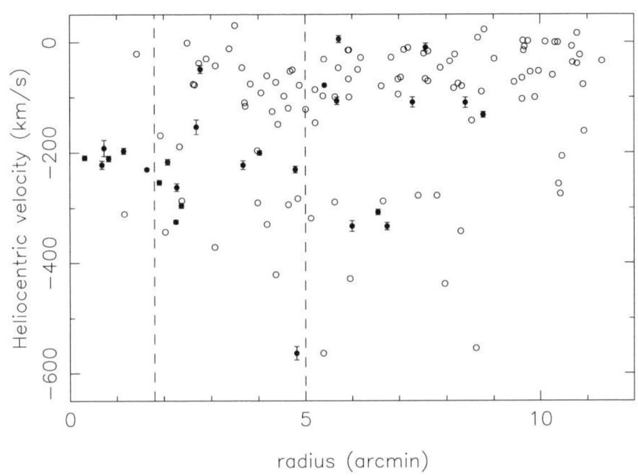

Figure 6. Radial velocities of all the star's in our Keck/DEIMOS field centred on And IX. Black points represent stars which lie on the And IX red giant branch. The vertical dashed lines are 1.8 and 5 arcmins from the center of M31. Due to the close proximity of And IX to M31, there is significant contamination in this field. However, within 1.8 arcmins of the center there is a clear kinematic grouping of stars with the correct colour to belong to And IX. We identify these as And IX red giant branch stars.

5. Kinematics of the faintest dSph, Andromeda IX

The discovery of And IX has only recently been reported by Zucker, *et al.* (2004) and a detailed study of its stellar content is not yet available, although it appears as if And IX consists only of an old stellar population, with no evidence of young or intermediate-age components. As part of our group's ongoing kinematic survey of M31 using the DEep Imaging Multi-Object Spectrograph on the Keck II telescope, we have conducted a kinematic study of this object to constrain its dynamical evolution and dark matter content (Chapman, *et al.* (2005).

Figure 6 shows the measured radial velocities of all the stars in our Keck/DEIMOS field centered on And IX. The black points represent stars which additionally have colours and magnitudes which place them on the red giant branch of And IX in a CMD. And IX is sufficiently close to M31 that there is significant contamination from M31 in our DEIMOS field. However, by applying a tight spatial cut then we can identify a group of stars within

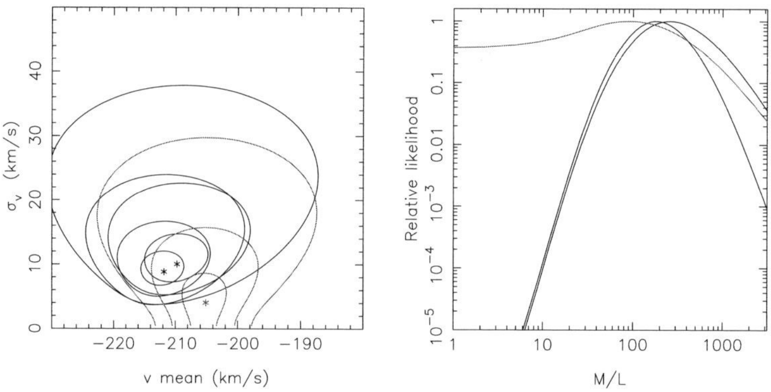

Figure 7. Left panel: Maximum likelyhood contours for the radial velocity and velocity dispersion of And IX. The dark line corresponds to the analysis using all six stars within 1.8 arcmins highlighted in Figure 6. The lighter line is the same analysis but neglecting the star whose velocity is furthest from the mean of this group. Right panel: Relative probability for the mass-to-light ratio of And IX. Dark and light lines are the same as in the left panel.

1.8 arcmins of the center of And IX which are tightly correlated in radial velocity and which have the correct colours to be And IX red giant branch stars. We identify these six stars as And IX members and conduct a maximum likelyhood analysis of these stars to obtain the radial velocity and dispersion of And IX. The corresponding maximum likelyhood contours are shown as the darkest lines in the left panel of Figure 7 and imply a radial velocity for And IX of $v_r = 210 \pm 5\,\mathrm{km\,s^{-1}}$, with a velocity dispersion of $\sigma = 10^{+5}_{-3}\,\mathrm{km\,s^{-1}}$. The darkest line in the right panel of Figure 7 is the relative probability for the mass-to-light ratio assuming And IX is not very tidally disturbed. We find that the mass-to-light ratio of And IX is > 27 at the 99 % confidence level.

Of the six stars on which our analysis is performed, it is very unlikely that any of them belong instead to M31. If, however, the star whose velocity is furthest from the mean in our grouping is actually from M31, then our results change dramatically – the velocity dispersion of And IX would then be consistent with zero within the $1\,\sigma$ errors. This is represented by the lighter lines in Figure 7. Nevertheless, this is unlikely.

By stacking the six spectra, we are able to measure the equivalent width of the Calcium triplet and imply a metallicity. This corresponds to [Fe/H] = −1.5 dex. This is approximately 0.5 dex higher than would be expected from the faint luminosity of And IX. It may be that the chemical enrichment processes in this faintest of galaxies differs from many other dSph galaxies in the Local Group.

6. Summary

Our INT WFC photometric survey of M31's satellites has allowed the derivation of a large number of fundamental parameters for this subsystem, including distances and metallicities, and reveals the presence of tidal effects and multiple components in several of the dSphs. In addition, important difference appear to exist in the scale sizes of the M31 and Galactic dSphs, which may reflect environmental differences. Our Keck/DEIMOS survey of And IX suggest that And IX is a highly dark matter dominated satellite, and that the metallicity is anomalously large for such a faint system, perhaps suggesting that And IX is an atypical dSph.

Acknowledgements

AM would like to thank the organisers for a very enjoyable and exciting meeting, and the IAU for providing financial assistance.

References

Barker, M.K., Sarajedini, A. & Harris, J. 2004 *ApJ* 606, 869

Bellazzini, M., Ferraro, F.R. & Pancino, E. 2001 *ApJ* 556, 635

Chapman, S., Ibata, R., Lewis, G., Ferguson, A., Irwin, M. & McConnachie, A.W. & Tanvir, N. 2005 *ApJ* submitted

Da Costa, G.S. & Armandroff, T.E. 1990 *AJ* 100, 162

Da Costa, G.S., Armandroff, T.E., Caldwell, N. & Seitzer, P. 1996 *AJ* 112, 2576

Da Costa, G.S., Armandroff, T.E. & Caldwell, N. 2000 *AJ* 119, 705

Da Costa, G.S., Armandroff, T.E. & Caldwell, N. 2002 *AJ* 124, 332

Choi, P.I., Guhathakurta, P. & Johnston, K.V. 2002 *AJ* 124, 310

Ferguson, A., Irwin, M.J., Ibata, R.A., Lewis, G.F. & Tanvir, N. 2002 *AJ* 124, 1452

Harbeck, D., Gallagher, J.S. & Grebel, E.K. 2004 *AJ* 127, 2711

Harbeck, D., Grebel, E.K., Holtzman, J., Guhathakurta, P., Brandner, W., Geisler, D., Sarajedini, A., Dolphin, A., Hurley-Keller, D. & Mateo, M. 2001 *AJ* 122, 3092

Irwin, M. & Hatzidimitriou, D. 1995 *MNRAS* 277, 1354

Kerschbaum, F., Nowotny, W., Olofsson, H. & Schwarz, H.E. 2004 *A&A* 427, 613

Kleyna, J.T., Wilkinson, M.I., Evans, N.W. & Gilmore, G. 2004 *MNRAS* 354, 543

Kleyna, J.T., Wilkinson, M.I., Gilmore, G. & Evans, N.W. 2003 *ApJL* 588, L21

McConnachie, A.W., Irwin, M.J., Ferguson, A., Ibata, R.A., Lewis, G.F. & Tanvir, N. 2004 *MNRAS* 350, 243

McConnachie, A.W. & Irwin, M.J. 2005 *MNRAS* submitted

McConnachie, A.W. & Irwin, M.J. 2005 *MNRAS* submitted

McConnachie, A.W., Irwin, M.J., Ferguson, A., Ibata, R.A., Lewis, G.F. & Tanvir, N. 2005 *MNRAS* 356, 979

Tolstoy, E., Irwin, M.J., Helmi, A., Battaglia, G., Jablonka, P., Hill, V., Venn, K.A., Shetrone, M.D., Letarte, B., Cole, A.A., Primas, F., Francois, P., Arimoto, N., Sadakane, K., Kaufer, A., Szeifert, T. & Abel, T. 2005 *ApJL* 617, L119

Zucker, D.B, *et al.* 2004 *ApJL* 612, L121

Discussion

CALDWELL: The excess light in the center of And III is intriguing, but I worry that the presence of foreground stars may have made your measurements more difficult there. Does your finding hold true even in areas not contaminated by these stars?

MCCONNACHIE: Yes, although foreground stars do complicate the analysis, it is possible to account for them in the analysis procedure and so there presence does not affect our conclusions.

BAILIN: How robust are the And IX results to the exclusion of any of the 6 stars from the spectrum?

MCCONNACHIE: If we exclude the most discrepant star of the 6 then the results change dramatically – the velocity dispersion is consistent with zero. However, it is very improbable that this star is an interloper.

MATEO: How far is And IX from the center of M31 in angular and linear units?

MCCONNACHIE: Andromeda IX is approximately 1.5 to 2 degrees from the center of M31, and has a 3D separation from M31 of about 40 kpc.

Near-Field Cosmology with Dwarf Elliptical Galaxies
Proceedings IAU Colloquium No. 198, 2005
H. Jerjen and B. Binggeli, eds.

doi:10.1017/S1743921305003546

A case for extended dark matter halos in dwarf spheroidal galaxies

Sergey Mashchenko, H.M.P. Couchman and Alison Sills

Department of Physics and Astronomy, McMaster University, Hamilton, ON, L8S 4M1, Canada
email: syam,couchman,asills@physics.mcmaster.ca

Abstract. We explore the recently proposed idea that the Galactic dwarf spheroidal galaxies are significantly (by 2 orders of magnitude) more massive than the conventional mass estimates of $\sim 10^7$ $M_\odot$. In the larger mass case, the observed distribution of stars in these galaxies should have been entirely shaped by internal processes (formation and dynamic relaxation of stars in the potential of the dark matter halo), and not by the Galactic tidal field. We carried out numerical n-body simulations aimed at testing this scenario. Observed properties of three Galactic dwarf spheroidal galaxies were found to be consistent with our model. From our analysis, these dwarfs appear to be massive enough to alleviate the "missing satellites" problem of cold dark matter cosmologies.

Keywords. dark matter, early universe, galaxies: dwarf, galaxies: formation, methods: n-body simulations

1. Introduction

Cosmological cold dark matter (CDM) simulations predict that in a Milky Way sized halo there should be 1–2 orders of magnitude more subhalos than the observed number of Milky Way satellites (Moore *et al.* 1999). It has been suggested (Stoehr *et al.* 2002, Hayashi *et al.* 2003) that the "missing satellites" problem can be alleviated if for some reason only the most massive subhalos managed to form stars.

There are also astrophysical reasons to believe that the Galactic dwarf spheroidals (dSphs) can be much more massive than the conventional estimates of $\sim 10^7$ $M_\odot$. Odenkirchen *et al.* (2001) used a multi-color filtering technique to derive a high quality surface brightness map of Draco. The lack of tidal features in the outermost isodensity contours implied that Draco is more dark matter (DM) dominated in its outskirts than at its center, with a virtually unconstrained upper limit on total mass. Also, there is a contradiction between complex star formation histories and apparent absence of interstellar medium (ISM) in Galactic dSphs. Burkert & Ruiz-Lapuente (1997) and Mashchenko *et al.* (2004) proposed two different scenarios to explain this puzzle. Both models assume that the ISM is still there, but the gas is fully ionized [either by supernovae Ia (Burkert & Ruiz-Lapuente 1997) or by external ionizing radiation (Mashchenko *et al.* 2004)], rendering it virtually undetectable. The common requirement is that the DM halos should be massive ($\gtrsim 10^9$ $M_\odot$) to keep the ionized gas gravitationally bound.

If dSphs are massive enough, the distribution of stars in these galaxies (including "extratidal stars" in their outskirts) should have been entirely shaped by internal processes, and not by the Galactic tidal field. We test this idea using a simple model for formation and dynamic relaxation of stars in dSphs with extended DM halos. The results of simulations are compared with the observed properties of four Galactic dSphs: Draco, Sculptor, Carina, and Fornax.

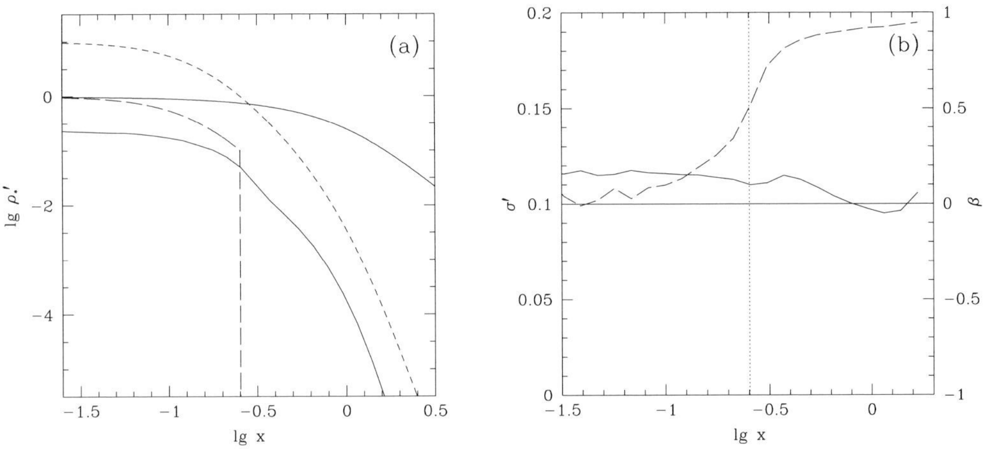

Figure 1. Radial profiles for Burkert model with $c_g' = 0.16$, $\rho_{g,0}' = 10$, $\xi = 0.1$, and $\lambda = 0.1$. (a) Density profiles for DM (*thin solid line*), initial gas distribution (*short-dashed line*), initial stellar distribution (*long-dashed line*), and relaxed stellar cluster (*thick solid line*). (b) Total velocity dispersion σ' (*thick solid line*) and anisotropy parameter β (*long-dashed line*) profiles for the relaxed stellar cluster. Vertical dotted line marks the initial stellar cluster radius. Horisontal solid line corresponds to $\beta = 0$.

2. Model

The main assumptions in our model are as follows. Initially, gas is in hydrostatic equilibrium in the static potential of the DM halo (short-dashed line in Figure 1a). We take into account self-gravity of the gas, which is assumed to be isothermal. In a DM halo with a universal density profile, the distribution of gas depends only on two free parameters: the central gas density $\rho_{g,0}'$ (in units of the DM halo scaling density) and the sound speed in the gas c_g' (in units of the halo scaling velocity). We consider two types of DM halos: cuspy Navarro, Frenk, & White (1997; NFW) halos and flat-cored Burkert (1995) halos. We assume that all stars in the galaxy formed in a single starburst, with a constant efficiency ξ out to a radius r_λ where the gas density drops down to $\lambda\rho_{g,0}'$. (Here ξ and λ are free parameters.) No stars are formed beyond r_λ (see long-dashed line in Figure 1a). The newly born stars have one-dimensional velocity dispersion equal to the sound speed in the gas c_g and isotropic distribution of velocity vectors. We assume that the leftover gas is instantaneously removed by stellar feedback. The stellar cluster is not in equilibrium initially, and will expand before reaching a relaxed state (solid line in Figure 1a). We study this process using n-body simulations.

In the relaxed stellar cluster, stellar orbits are almost isotropic at the center and strongly radially anisotropic in the outskirts, beyond the initial stellar radius r_λ, whereas the total velocity dispersion does not change much with radius (see Figure 1b). In projection, such a galaxy would appear to have a flat central line-of-sight velocity dispersion profile and kinematically cold outskirts, similarly to what is observed in some dSphs (Kleyna *et al.* 2004, Wilkinson *et al.* 2004).

To simulate the relaxation of stars in a static potential of the DM halo, we used a parallel n-body code GADGET (Springel *et al.* 2001). We ran around 700 models for different values of the four initial parameters $\rho_{g,0}'$, c_g', ξ, and λ – for both NFW and Burkert DM halo profiles. Stars were represented by 10^4 collisionless equal mass stellar particles.

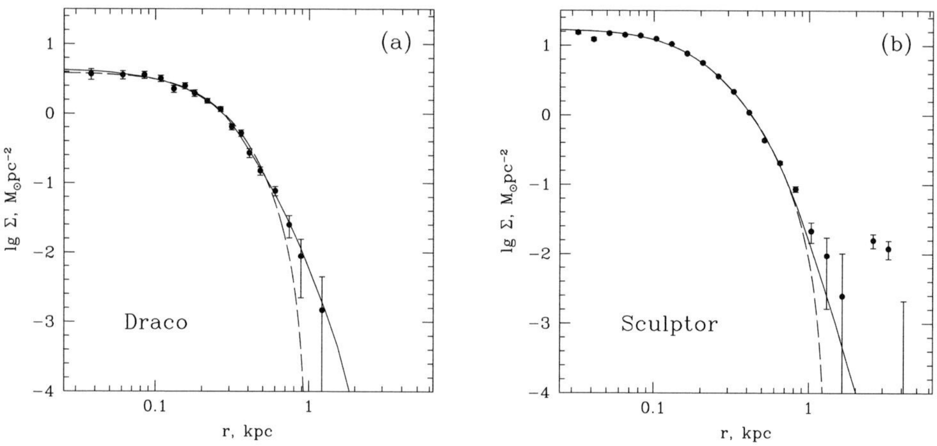

Figure 2. Surface density profiles for (a) Draco and (b) Sculptor. The observed (*circles with errorbars*), our best fitting model (*solid lines*), and the best fitting theoretical King model (*dashed lines*) profiles are shown.

We used three criteria to choose models which are a good match to observed dSphs. (1) The model stellar surface density profile should be a better match to the observed one than the best fitting theoretical King model profile: $\chi^2_{\rm model} \lesssim \chi^2_{\rm King}$. (2) The model central line-of-sight velocity dispersion value should be equal to the observed one within the observational uncertainties. (3) DM halo comoving number density in the concordance ΛCDM cosmology should be similar to or larger than the comoving number density of Local Group dSphs. (DM halo parameters are inferred after fitting the model surface density profile to the observed one.)

3. Results

We used star density profiles for Draco (two datasets), Sculptor, Carina, and Fornax from Odenkirchen *et al.* (2001), Walcher *et al.* (2003), and Wilkinson *et al.* (2004). Typically, we had ∼5–10 "good models" (which would satisfy all the three selection criteria from the previous section) per galaxy. The only exception was Fornax, where not a single model would satisfy the selection criteria.

In Figure 2 we show the best fitting models for Draco and Sculptor (Carina's fit is similar to that of Sculptor). For comparison, we also show the best fitting theoretical King models. As one can see, our models describe the central parts of the galaxies (out to a radius where the surface density becomes 2–2.5 orders of magnitude lower than at the center) very well and are virtually identical to the King models. In the outer parts of the galaxies, our models significantly improve upon the King models by having less steep surface brightness profiles. In our models, the "extratidal stars" (stars located beyond the nominal King tidal radius) are gravitationally bound to dSphs and are a part of their extended stellar halos.

Our "good models" have the following parameters. Their virial masses (the masses they had when they were formed) are $\sim 10^9$ M$_\odot$ (up to 2×10^{10} M$_\odot$), which are large enough to alleviate the "missing satellites" problem of CDM cosmologies. They tend to have shallow inner DM density profiles (with the slope $\gamma = -0.5 \ldots 0$). The gas temperature is $\sim 10^4$ K. The star formation efficiency is $\sim 10\%$. The galaxy formation redshifts are $\sim 2 \ldots 6$ (lookback time $\sim$10–13 Gyr), which is consistent with the age of the oldest stars in these galaxies.

4. Conclusions

Our simple star formation model for dSphs fits very well the observed properties of Draco, Sculptor, and Carina. The inferred large DM halo masses are consistent with the idea that Galactic dSphs are hosted by the most massive subhalos predicted to populate the Milky Way halo by cosmological ΛCDM simulations.

Acknowledgements

The n-body simulations reported in this paper were carried out at the Canadian Institute for Theoretical Astrophysics.

References

Burkert, A. 1995, *ApJ* (Letters) 447, L25
Burkert, A. & Ruiz-Lapuente, P. 1997, *ApJ* 480, 297
Hayashi, E., Navarro, J.F., Taylor, J.E., Stadel, J. & Quinn, T. 2003, *ApJ* 584, 541
Kleyna, J.T., Wilkinson, M.I., Evans, N.W. & Gilmore, G. 2004, *MNRAS* 354, L66
Mashchenko, S., Carignan, C. & Bouchard, A. 2004, *MNRAS* 352, 168
Moore, B., Ghigna, S., Governato, F., Lake, G., Quinn, T., Stadel, J. & Tozzi, P. 1999, *ApJ* (Letters) 524, L19
Odenkirchen, M., *et al.* 2001, *AJ* 122, 2538
Springel, V., Yoshida, N. & White, S.D.M. 2001, *New Astronomy* 6, 79
Stoehr F., White S.D.M., Tormen G. & Springel V. 2002, *MNRAS* 335, L84
Walcher, C.J., Fried, J.W., Burkert, A. & Klessen, R.S. 2003, *A&A* 406, 847
Wilkinson, M.I., Kleyna, J.T., Evans, N.W., Gilmore, G.F., Irwin, M.J. & Grebel, E. K. 2004, *ApJ* (Letters), 611, L21

Discussion

READ: A live halo should be important. If anisotropy is introduced through mass loss to the stars then it should also be introduced to the dark matter. Also it is important to model the contraction (often *not* adiabatic) of the dark matter in response to the initial gas collapse. See Read & Gilmore 2205 (MNRAS).

MASHCHENKO: I agree that this can be an important effect. We plan to test it by resimulating some of our best fitting models with live DM halos.

HENSLER: If you relax your strongly constraining assumption of instantanious gas removal, your results (stellar velocity anisotropy, density distribution, etc.) should change significantly. Did you study this effect?

MASHCHENKO: No, we didn't study this effect. I do not think that it will change significantly our results. E.g., even in the instantanious gas removal case the stellar cluster does not expand much (globally), so the relaxed stellar surface brightness should not depend much on the way we remove gas. The anisotropy in the outskirts will change (it will become less severe), so the outer l.o.s. velosity dispersion profile will also change. Fortunately, this data was not used in our fitting.

DRINKWATER: A few years ago I heard Bill Harris speaking about extratidal light in globular clusters. Can your type of model be applied to globulars?

MASHCHENKO: Yes, if globular clusters had a primordial origin with DM halos (à la Peebles). Recently we published two papers dealing with this question (Mashchenko & Sills 2005a,b). In these papers we actually had even more consistent simulations, as our DM halos were live (non-static).

Near-Field Cosmology with Dwarf Elliptical Galaxies
Proceedings IAU Colloquium No. 198, 2005
H. Jerjen and B. Binggeli, eds.

doi:10.1017/S1743921305003558

The Canis Major dwarf galaxy as the progenitor of the Monoceros tidal stream

D. Martínez-Delgado[1]†, J. Peñarrubia[2], D.I. Dinescu[3], D.J. Butler[2] and H.W. Rix[2]

[1]Instituto de Astrofísica de Andalucía (CSIC), Spain. [2] Max-Planck-Institut-fur Astronomie, Germany. [3] Astronomy Department, Yale University, USA.

Abstract. The Sloan Digital Sky Survey has recently discovered a coherent ring of stars at low galactic latitude that is believed to be the tidal stream of a merging dwarf galaxy in the Galactic plane (named the Monoceros tidal stream). The existence and location of the core of its progenitor galaxy is still controversial. The best candidate is the Canis Major dwarf galaxy, a distinct overdensity of red stars discovered in the 2MASS survey, but also interpreted as the signature of the Galactic warp viewed in projection. In this paper, we report a variety of new observational evidence that supports the notion that CMa is the remnant of a partially disrupted core of a dwarf satellite. The comparison of the orbit derived from our theoretical model for the parent galaxy of this ring-like structure with an accurate determination of CMa orbit leads to the conclusion that this satellite is the best candidate for the progenitor of the Monoceros tidal stream

Keywords. Galaxy:formation, Galaxy:structure, galaxies:dwarf

1. Introduction

The Sloan Digital Sky Survey has recently discovered a coherent ring-like structure at low galactic latitude spanning about 100 degree in the sky and surrounding the Galactic disk at Galactocentric distances from ∼15 kpc to ∼20 kpc (Newberg *et al.* 2002; Yanny *et al.* 2003). The tremendous observational and theoretical effort to understand its origin have proven that the structural characteristic and kinematics of this stellar ring are consistent with the properties expected for the tidal stream of a merging dwarf galaxy in the Galactic plane (named the Monoceros tidal stream). If this ring is a tidal tail feature, it must have had a parent galaxy, which may or may not be completely disrupted by now.

Unlike the Sagittarius tidal stream, the Monoceros stream has been detected prior to locating the main body of its progenitor galaxy. The best available candidate is the Canis Major (CMa) dwarf galaxy, a strong elliptically shaped stellar overdensity of red giant stars discovered by Martin *et al.* (2004) from an analysis of the 2MASS survey. As an alternative interpretation, Momany *et al.* (2004) suggested that this overdensity is only the signature of the Galactic warp. In this paper, we address the controversy on the origin of this stellar system and its relation with the Monoceros stream by answering the following questions: Can we constrain the position of the progenitor galaxy with N-body simulations from the distribution and kinematics of Monoceros tidal debris? Is the CMa over-density the remmnat of the core of a dwarf galaxy? And if it is, is CMa the progenitor of the Monoceros tidal stream?

† email: ddelgado@iaa.es

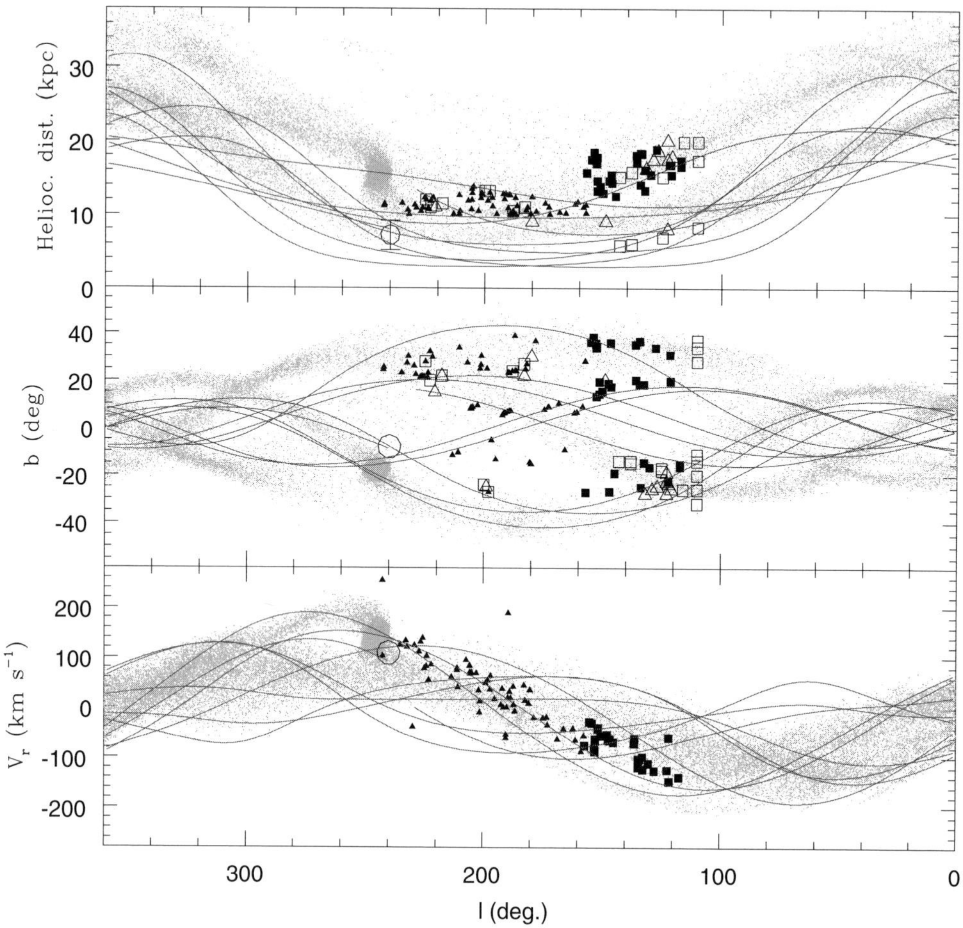

Figure 1. Comparison of the best fitting N-body simulation of the Monoceros tidal stream (small dots;see Peñarrubia *et al.* 2005 and references therein for a complete description of the Monoceros stream detections) with the orbit of the CMa dwarf derived from its absolute proper motion (solid line) by Dinescu *et al.* (2005)

2. Constraining the progenitor position with theoretical model of the Monoceros tidal stream

Peñarrubia *et al.* (2005) have attempted to determine the present position of the Monoceros stream progenitor through N-body simulations. These authors constrain the motion and mass of a possible progenitor by fitting the available Monoceros stream detections to thousands of orbits of dwarf galaxies with different masses. This method has been proven to provide powerful constrains on the eccentricity and orbital inclination of the possible progenitor, finding that the best solutions are for a satellite galaxy moving on a low eccentric (e = 0.10 ± 0.05), low orbital inclination (i = 25 ± 5 deg) prograde orbit (Figure 1). Owing to the small area of the sky where the Monoceros stream has been detected, the solutions are degenerated for some free parameters of the model: Namely, the axis-ratio of the Milky Way halo, the mass and the present location of the stream progenitor cannot be sufficiently constrained. Focusing on the last point, the model reproduces the geometrical and kinematical distribution of debris if the progenitor's remnants are

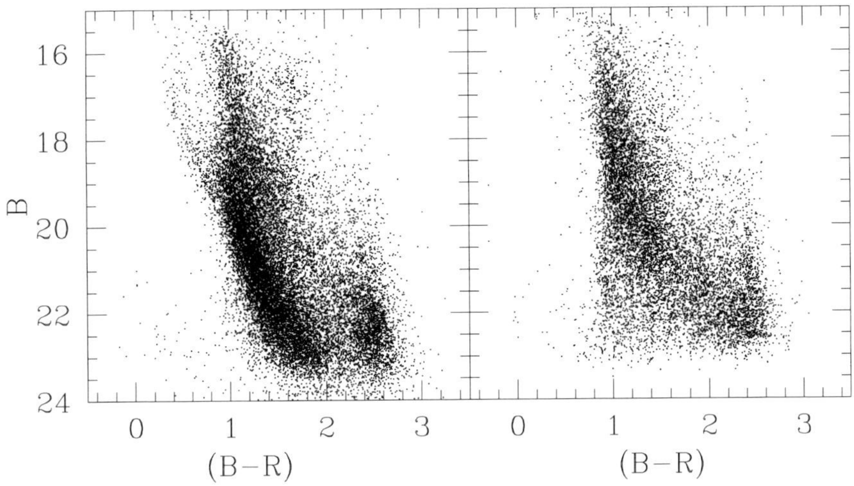

Figure 2. (*left panel*) Color-magnitude diagram of the center of the Canis Major overdensity ((l,b) = (240,−8)) (see Martínez-Delgado *et al.* 2005 for a detailed description); (*right panel*) Color-magnitude diagram of a control field situated at (l,b)=(240,+8), showing the distribution of Galactic background/foreground stars and an absence of the dwarf galaxy features.

located within $260 > l > 120$ at heliocentric distance of ∼14 kpc. That solution can be futher constrained with future detections in a wider range of longitudinal directions. The CMa system is located at $l = 240$, which enters in the range obtained from N-body simulations, although at a closer distance (d_{helioc} ∼8 kpc). The close distance of CMa appears to indicate that, if this system is the progenitor of the Monoceros stream, there are parts of the tidal stream at closer heliocentric distances that have not yet been detected.

3. The nature of the Canis Major dwarf

Fig 2a shows the color-magnitude diagram (CMD) of the putative center of the CMa over-density discovered by Martin *et al.* (2004). The most remarkable feature is a narrow main-sequence (MS), with a high constrast with respect to the thin/thick/halo background contamination, of which distribution is better observed in the CMD of a control field (Fig. 2b). The CMD morphology in Fig. 1a is in agreement with the expected for a stellar system composed by a mean metal-rich stellar population and that has undergone at least two distinct epochs of star formation (the last one only 1–2 Gyr ago).

From the MS feature, we derive a line-of sight depth for this system of 0.9 ± 0.3 kpc, consistent with the interpretation of a remnant of a partially disrupted dwarf satellite (Martínez-Delgado *et al.* 2005). This limited line-of-sight depth is also very difficult to reconciled with the hypothesis of a warped Galactic outer disk viewed in projection, as proposed by Momany *et al.* (2004) (see also Sec. 3). The derived surface brightness ($\mu_{V,0} = 23.3 \pm 0.1$ mag) and absolute magnitude ($M_V = -14.5 \pm 0.1$ mag) of CMa places it in the category of dwarf galaxy in the known size-luminosity relation followed by dwarf galaxies (see Pasquali *et al.* 2005). Additional evidence on the dwarf nature of CMa from its orbital motion can be found in Sec. 4. At a distance of 8 kpc, CMa is the closest dwarf galaxy known.

4. The orbit of the Canis Major dwarf

The question about whether the CMa dwarf could be the progenitor of the Monoceros tidal stream cannot be answered without a better constrain on the motion of this posible satellite. Recent absolute proper motion measurements of a sample of *bona-fide* CMa star members (Dinescu *et al.* 2005) provide an accurate determination of the CMa orbit. The orbit has a pericenter of 10 ± 0.9 kpc and an apocenter of 14 ± 0.2 kpc. The orbit inclination is 15 ± 3 deg and the eccentricity is 0.14 ± 0.04. Currently, CMa is at its apocenter, and it should undergo tidal disruption.

The orbital motion of the CMa also provides important clues on the controversy about the origin of this stellar system (dwarf galaxy *versus* Galactic pertubation). While the CMa orbit is not very dissimilar from orbits of thick disk stars, the Monoceros stream stars reach a maximum distance from the Galactic plane of ~2 kpc, that is larger than the thick disk scale height ($\leqslant$ 1 kpc). In addition, the derived W velocity component (W=-49 $\pm$ 15 km/s) shows significant (3σ) motion perpendicular to the the disk and its negative value is inconsistent (7σ) with the expected motion (e.g., Drimmel, Smart & Lattanzi 2000) of the warp at this Galactic location (see Dinescu *et al.* 2005).

The derived orbit parameters are fairly similar to those predicted in the theoretical model by Peñarrubia *et al.* (2005; see Sec. 3) for the Monoceros stream's progenitor (within 1σ). Fig. 1 shows a comparison between the best fitting N-body model (small dots) of the Monoceros stream and the orbit of CMa derived from its proper motion (solid line). This agreement supports the argument that CMa is the best candidate for the parent galaxy of the Monoceros stream.

References

Dinescu, D.I., Martínez-Delgado, D., Girard, T.M., Peñarrubia, J., Rix, H.W., Butler, D.J. & van Altena, W.F. 2005, *ApJ* Letters submitted

Martin, N.F., Ibata, R.A., Bellazzini, M., Irwin, M.J., Lewis, G.F. & Dehnen, W. 2004, *MNRAS* 348,12

Drimmel, R., Smart, R.L. & Lattanzi, M.G. 2000, *A&A* 354 ,67

Martínez-Delgado, D., Butler, D.J., Rix, H.W., Franco, Y.I., Peñarrubia, J., Alfaro, E.J. & Dinescu, D.I. 2005, *ApJ* (Letters) in press (astro-ph/0410611)

Momany, Y., Zaggia, S.R., Bonifacio, P., Piotto, G., De Anfeli, F., Bedin, L.R. & Carraro, G. 2004, *A&A* 421, L29

Newberg, H.J. *et al.* 2002, *ApJ* 569, 245

Pasquali, A., Larsen, S., Ferreras, I., Gnedin, O.Y., Malhotra, S., Rhoads, J.E., Pirzkal, N. & Walsh, J.R. 2005, *AJ* 129, 148

Peñarrubia, J., Martínez-Delgado, D., Rix, H.W. *et al.* 2005, *ApJ* (in press) (astro-ph/0410448)

Yanny,B. *et al.* 2003, *ApJ* 588, 824

Discussion

MATEO: a)Was the Galactic potential in the model static?; b) How long did the model run?; c)What is the total mass of the contents of the stream?

MARTÍNEZ-DELGADO: a) In our simulations, the Galaxy reacts to the presence of the Monoceros stream, but our Galaxy model does not implement a cosmological evolution; b) The available observational data can be reproduce within the last 3 Gyr of a possible progenitor orbit; c) The initial mass was $6 \times 10^8 M_\odot$ and the satellite lost half of the mass at the end of the simulation.

Near-Field Cosmology with Dwarf Elliptical Galaxies
Proceedings IAU Colloquium No. 198, 2005
H. Jerjen and B. Binggeli, eds.

doi:10.1017/S174392130500356X

The Sagittarius dwarf mass-to-light ratio

S. Zaggia[1], P. Bonifacio[1], M. Bellazzini[2], E. Caffau[1], F. Ferraro[3], G. Marconi[4], L. Monaco[1,4], S. Monai[1] and L. Sbordone[4]

[1]INAF – Osservatorio Astronomico di Trieste, Trieste
email: zaggia@ts.astro.it

[2]INAF – Osservatorio Astronomico di Bologna; [3]Dipartimento di Astronomia, Università di Bologna; [4]European Southern Observatory, Santiago

Abstract. We report on the use of high-resolution spectra to obtain a detailed description of the Sagittarius dwarf spheroidal internal dynamics, its Mass and Mass to Light ratio (M/L). Our direct measure of the central velocity dispersion of SGR give $\sigma_{SGR} = 8.1 \pm 0.4$ km/s which translates in a total mass estimate of $M_{SGR} = 1.6 \times 10^8 M_\odot$ and corresponding $(M/L)_{SGR} = 9.1(M/L)_\odot$. We also report on a possible detection of rotation in the core of SGR.

Keywords. dwarf galaxies – kinematics – mass

1. Introduction

The Sagittarius (SGR) dwarf galaxy is actually the nearest companion of the Milky Way undergoing a tidal disruption. Its vicinity represent a unique opportunity to look in detail a possible Searle & Zinn (1978) galactic fragment formed in the outer halo, later captured by the inner Galaxy and actually under its final stages of disruption: in the hierarchical galaxy formation scenarios this kind of galaxies may represent the basic building blocks of larger galaxies.

A very important aspect of the SGR structure is represented by its internal kinematics and the kinematics of its stellar tails. While for this later aspect the group of Majewski (Law *et al.* 2005) has recently performed a very complete observational-theoretical study of the SGR dynamical history using the kinematics of the tails, no direct measurements yet exist on the dynamics of the core of SGR and of its relation with the M54 globular cluster. The only kinematic information on the body of SGR comes from the work of Ibata *et al.* (1997), who used radial velocity measurements from low resolution spectra claiming an $(M/L)_{SGR}$ of as much as $\simeq$50, implying the presence of a big dark halo. This estimate has been revised by Majewski *et al.* (2003) using an updated total luminosity for the SGR body proposing an $(M/L)_{SGR} \simeq 25$. The recent work of Law *et al.* (2005) on the modeling of the stellar tidal tails, set the following constraints on the total mass and mass to light ratio: $M_{SGR} = 2 \div 5 \times 10^8 M_\odot$ and $(M/L)_{SGR} = 14 \div 36(M/L)_\odot$.

In this paper we present the first direct measurement of the central velocity dispersion and Mass estimate based on high-resolution spectra for a large sample of stars in the core of SGR.

2. Observations

The data presented here are part of the Guaranteed Time Observations (GTO) awarded to the *Ital-FLAMES* consortium (see Cacciari 2003) who collaborated with ESO at the development, construction and installation of the VLT FLAMES facility (Pasquini *et al.*

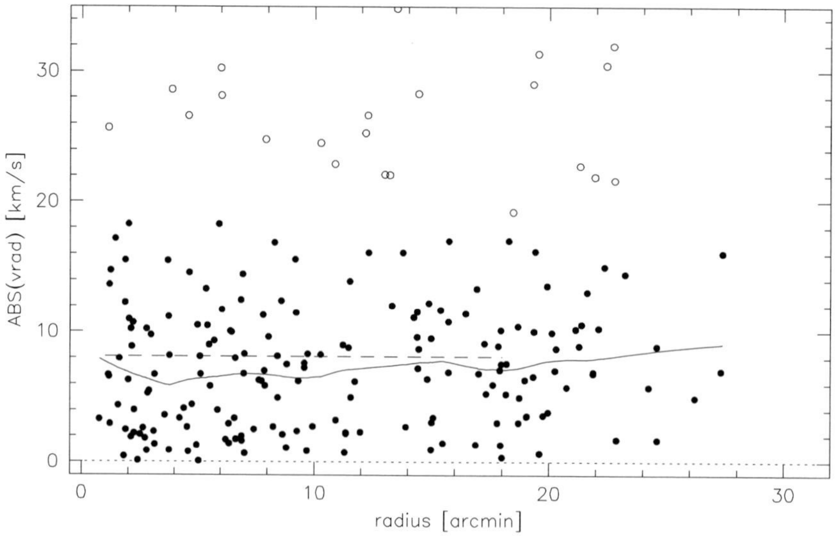

Figure 1. Radial velocity dispersion profile in the core of the SGR galaxy. The single radial velocities are folded around the mean: filled dots SGR members, open dots non-members; the continuous line is the LOWESS estimate of the velocity dispersion profile; the dashed line is the velocity dispersion of the whole sample of objects.

2000). FLAMES is installed at the Nasmyth A focus of the VLT/UT2 telescope and is composed by a fiber positioner, OzPoz, which feeds the dedicated medium-high resolution GIRAFFE (resolving power $\sim$20000) and the UVES (resolving power $\sim$40000) spectrographs. FLAMES has a multiplex capability of 132 single fibers with GIRAFFE (in the MEDUSA mode) and 8 fibers with UVES, over a large field of view of $\simeq$28 arcmin in diameter. We aimed at obtaining a S/N ratio of the spectra of better than $\simeq$20 in two different GIRAFFE setups: HR09 centered at 515 nm and HR14 at 650 nm.

For the present program it was used 1 GTO night in May 2003 in order to observe a total of 280 stars with GIRAFFE, both in the SGR field and in the Globular Cluster M54. One year later, in July 2004, we obtained additional observing time for a total of 5 hours spread on two adjacent fields connected to the 2003 fields. In this case we observed a total of 230 stars. The two samples, 2003 and 2004, totaled 510 stars of which 189 turned out to be radial velocity members of SGR.

The 4 observed FLAMES fields cover: the center of the cluster M54 (supposed to be coincident with the center of SGR, Monaco *et al.*, 2000; Majewski *et al.* 2003), and 3 fields at 12 arcmin East, West and North of the center. Being FLAMES a "blind" instrument (it has no pre-imaging capabilities) it needs a very high precision astrometric catalog for the input targets. The selection of the targets were performed on the Monaco *et al.* (2002) V, I photometric and astrometric catalog (based on EIS Pre-FLAMES images Zaggia *et al.* 2001) which contains nearly half-million objects over 1 square degree around the center of SGR. We selected objects with magnitude in the range $16 \div 17.5$ in a diagonal strip of the (V$-$I) color which cross the RGBs of M54 and SGR in order to search for any possible metallicity present.

The reduction of the GIRAFFE multi-spectra have been performed using the FLAMES data reduction pipeline developed by the Geneva Observatory Blecha *et al.* (2001) while UVES spectra have been reduced with a modified UVES pipeline. The radial velocities have been extracted using a traditional cross correlation technique. We also obtained the

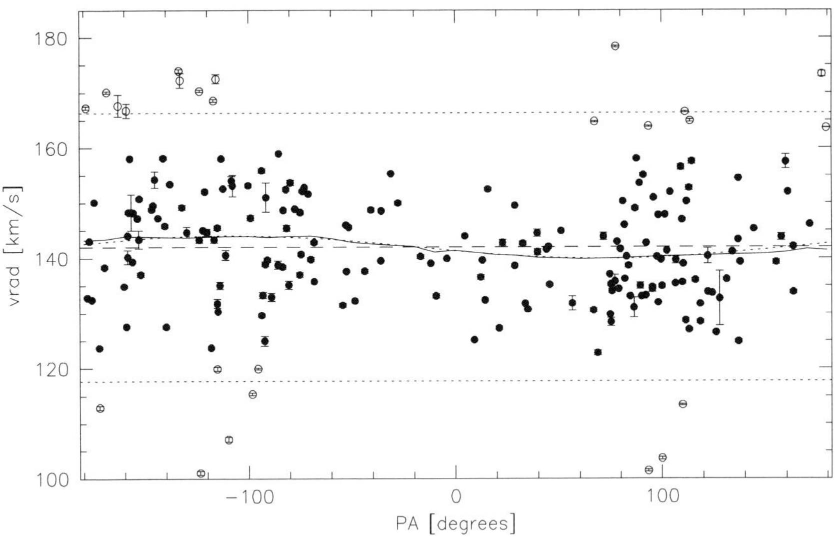

Figure 2. Radial velocity rotation curve. Filled dots are SGR members while open dots are non-members. The continuous line shows the running mean of radial velocity along the position angle calculated in strip of 10°. The dotted line a sinusoidal fit.

metallicity of each object using an automatic program properly suited for this kind of spectra (Bonifacio and Caffau 2003).

3. SGR Central Dynamics

The derived Mass and $(M/L)_{SGR}$ depends on the structural parameters of the galaxy and on its central radial velocity dispersion. Majewski *et al.* (2003) from a complete analysis of SGR M-stars with 2MASS obtained new values for the core radius, tidal radius and total luminosity of the galaxy and used them to estimate a better value of the Mass and M/L (compared to the estimate of Ibata *et al.* (1997)) with the standard King method. The value obtained is M/L= 25 which is based on an *inferred* central velocity dispersion of $\simeq$11.4 km/s from the Ibata *et al.* (1997) field "f7" in the SGR central regions. With a different approach the tidal tails modeling of Law *et al.* (2005) constrains on the total mass in the regime $2 \div 5 \times 10^8 M_\odot$ and the mass to light ratio in $14 \div 36 (M/L)_\odot$.

Our high precision radial velocity measurements permitted us to perform the first analysis of the central radial velocity dispersion of the SGR galaxy. Radial velocity measurements have a mean error of 0.33 km/s with 97% of objects with errors <0.65 km/s. Membership selection was performed in a multi-parametric way. Objects have been classified as dwarfs or giants on the base of the comparison with appropriate templates and only giants were kept. Second, a 3 sigma clipping in radial velocity has been applied and, finally, we used the metallicity measurements to isolate and remove the M54 globular cluster stars.

The radial velocity dispersion profile of the galaxy is presented in Figure 1 where the folded (around the mean) radial velocities of the single objects are shown. The velocity dispersion of the whole sample of 189 stars is $\sigma_{SGR} = 8.1 \pm 0.4$ km/s. The radial profile do not seems to show strong internal gradients as shown after a smoothing with the LOWESS estimator (continuous line). The velocity dispersion value is significantly lower than the

Ibata *et al.* (1997) value of field "f7" and imply a total mass of $M_{SGR} = 1.6 \times 10^8 M_\odot$ and a corresponding $(M/L)_{SGR} = 9.1(M/L)_\odot$ calculated following the King methodology and SGR morphological parameters of Majewski *et al.* (2003).

This low value of M_{SGR} seems to be compatible with the expected range of masses implied by the central velocity dispersion from the modeling of the tails of SGR of Law *et al.* (2005), but is incompatible with the masses constrained by the measured velocity dispersion of the SGR tails: in this case the "lower" bound of the acceptable mass of SGR is $2.5 \times 10^8 M_\odot$, i.e. ~50% larger than our measured value. Our Mass value seems to be compatible with the "Model II" of Helmi & White (2001) having an initial mass of 1.7×10^9 $M_\odot$ that, when evolved for a Hubble time, reproduce all data available for SGR.

We finally show in Figure 2 a possible detection of rotation in the core of SGR. In the figure the radial velocity shows a clear sinusoidal trend with Position Angle. A fitted sin curve to the running mean (in step of 10°) of radial velocity members give a $v_{rot} = 2.1 \pm 0.4$ km/s at a PA of maximum rotation of $\mathrm{PA}_{rot} = -100° \pm 15°$. This value is larger than the expected trend in radial velocity due to projection effects or to the orbit of SGR (see Ibata *et al.* 1997). The possible rotation is mis-aligned with the Major Axis of SGR by ≃30 and could possibly be the signature of a decoupled core. A better understanding of this aspect could only be obtained with a bigger sample of objects at larger radial distances.

References

Blecha, P. *et al.* 2001, *FLAMES Users Workshop*
Cacciari, C. 2003, *Mem. Sat Suppl.* 3, 86
Bonifacio, P. & Caffau, E. 2003, *A&A* 399, 1183
Helmi, A. & White, S., 2001, *MNRAS* 307, 495
Ibata, R. *et al.* 1997, *AJ* 113, 634
Law, D.R., Johnston, K.V. & Majewski, S.R. 2005, ApJ, 619, 807
Monaco, L. *et al.* 2002, *ApJ* 578, L47
Majewski, S.R., Skrutskie, M.F., Weinberg, M.D. & Ostheimer, J.C. 2003, *ApJ* 599, 1082
Pasquini, L., *et al.* 2000, *SPIE* 4008, 129
Searle, L. & Zinn, R. 1978, *ApJ* 225, 357
Zaggia, S. *et al.* 2001, *The Messenger* 105, 25

Discussion

ZHAO: Does the new M/L ≈10 of Sgr bring down the total mass of Sgr or bring up the total luminosity?

ZAGGIA: It brings down the total mass of Sgr. The luminosity I used is the one of Majewski *et al.* (2003, astro-ph/0304198).

CALDWELL: Doesn't the difference in mean metallicity between M54 and the core of Sgr argue that M54 is not the (a) nucleus?

ZAGGIA: Monaco *et al.* (2004) clearly demonstrated from a morphological point of view that M54 is not the nucleus of Sgr. This finding is also confirmed by the metallicity difference.

Near-Field Cosmology with Dwarf Elliptical Galaxies
Proceedings IAU Colloquium No. 198, 2005
H. Jerjen and B. Binggeli, eds.

doi:10.1017/S1743921305003571

3D Spectroscopic studies of dE galaxies

Igor Chilingarian[1,2], Philippe Prugniel[2,3], Olga Sil'chenko[3] and Victor Afanasiev[4]

[1]Sternberg Astronomical Institute of MSU, Russia
[2]CRAL Observatoire de Lyon, France
[3]GEPI Observatoire de Paris-Meudon, France
[4]Special Astrophysical Observatory of RAS, Russia

Abstract. We present first studies of dE galaxies by means of 3D spectroscopy, and address questions concerning data processing and analysis techniques used to extract internal kinematics and stellar population properties from integrated light spectra.

Keywords. methods: data analysis, galaxies: dwarf, kinematics and dynamics, stellar content

1. Introduction

Diffuse elliptical galaxies is a dominant population in the nearby Universe, but their origin and evolution remain controversial. Two main scenarios are (a) early collapse with a feedback of star formation, and (b) environmental effects: ram pressure stripping of disc galaxies in clusters and groups, gravitational harassment. To choose the scenario we propose to search for kinematical counterparts of the morphological fine structures found in some dEs (Jerjen *et al.* 2000) and subpopulations in the stellar content.

In 2004 we started the project of observing dEs with the MPFS integral field unit (IFU) spectrograph at the Russian 6-m telescope. 3D spectroscopy is a rapidly developing direction in a modern astrophysical experiment. Based on the ideas of Courtes described in the sixties, integral field spectroscopy has 17 years of successful scientific exploitation since first realization ("Tiger" at CFHT). This technique allows to obtain spatially resolved distribution of kinematics (radial velocity and velocity dispersion fields) and stellar population parameters (for example maps of line-strength indices).

Here we address technical questions concerning data reduction and analysis. Scientific aspects, presentation of the results for three dEs (IC3468, IC3653, and NGC770) are given by Prugniel *et al.* (this volume).

2. Data Processing and Analysis

IFU data require some new methods dealing with them besides specific process of data reduction. One of those techniques is so-called adaptive binning approach. Basically the idea is to achieve equal S/N ratio over the field of view by binning the data spatially: as lower signal is as larger bins should be. We use the implementation of Voronoi 2D binning procedure (Cappellari & Copin 2003). Specifying different target S/N ratio we obtain various tessellations to extract of different physical parameters of galaxies.

For the spectrum in each bin we apply direct template fitting procedure: the template spectrum (or the superposition of several ones) is fitted into the observational data using the parameterized LOSVD. We use high-resolution synthetic spectra computed with the new PEGASE.HR evolutionary synthesis code (Le Borgne *et al.* 2004). For a moment, these models exist only for solar abundance ratios of α/Fe and Mg/Fe. This puts some limitations on the upper limit of mass of galaxies being modeled, and allows to study

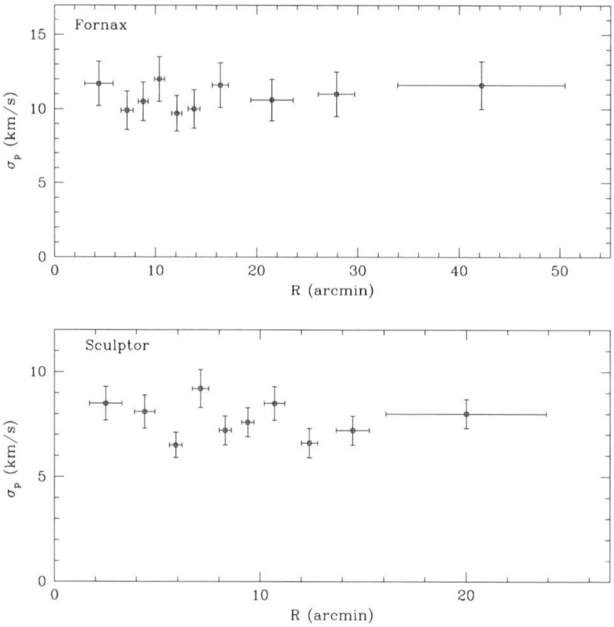

Figure 1. Projected radial velocity dispersion profiles for Fornax and Sculptor. There are 35 stars per bin in the Fornax profile, 72 stars per bin in the Sculptor profile.

A New Nonparametric Mass Estimation Technique

Wang *et al.* (2005) introduce a non-parametric method for estimating mass distributions. Implicit are assumptions of dynamic equilibrium, velocity isotropy, and spherical symmetry. Mass is not assumed to follow light, and no stellar distribution function is adopted. One operates directly on photometric data to estimate the deprojected stellar density profile, and then on the radial velocity data to estimate the mass profile. Trials performed on simulated data drawn from a Plummer distribution indicate that samples containing >1000 stars are sufficient to estimate $M(r)$ to within a factor of two. Applying this method to the Fornax and Sculptor samples gives $M(r < 1500\text{pc}) \sim 10^8 M_{sun}$ for Fornax, and $M(r < 400\text{pc}) \sim 10^7 M_{sun}$ for Sculptor, which has only been observed out to a few core radii. Considered with the measured luminosities of these galaxies, these values correspond to lower limits on the M/L of ~7 for Fornax and ~4.5 for Sculptor.

Acknowledgements

I would like to thank my collaborators, Mario Mateo, Edward Olszewski, Michael Woodroofe, and Xiao Wang for outstanding guidance and support.

References

Mateo, M., Olszewski, E., Welch, D.L., Fischer, P. & Kunkel, W. 1991, *AJ* 102, 914

Walker, M.G., Mateo, M., Olszewski, E.W., Wang, X. & Woodroofe, M. *in prep.*

Wang, X., Woodroofe, M., Mateo, M. & Walker, M.G. 2005, *ApJ accepted*

Wilkinson, M.I., Kleyna, J.T., Evans, N.W., Gilmore, G.F., Irwin, M.J. & Grebel, E.K. 2004, *ApJ* 611L, 21

Near-Field Cosmology with Dwarf Elliptical Galaxies
Proceedings IAU Colloquium No. 198, 2005
H. Jerjen and B. Binggeli, eds.

doi:10.1017/S1743921305003595

Chemodynamical evolution of dwarf elliptical galaxies

Gerhard Hensler

Institut für Astronomie, Universitäts-Sternwarte Wien,
Türkenschanzstr. 17, A–1180 Vienna, Austria
email: hensler@astro.univie.ac.at

Abstract. The growing number of observational details questions the standard picture of the evolution of dwarf ellipticals and argues for a complex evolution dominateded by environmental effects. Here we compile some standard and most recent observational issues and discuss the results of simple chemo-dynamical models in comparison with observations. The necessity of further model developments are justified and prospected.

Keywords. galaxies: dwarf, galaxies: ellipticals, galaxies: evolution

1. Introduction

Dwarf elliptical galaxies (dEs) are the most common galaxies in the local universe (see e.g. Ferguson & Binggeli (1994) for review). They are frequently cited as examples of "stellar fossil" systems in which the bulk of their star formation (SF) occurred in the past. Most known dEs are located in regions with high galaxy densities, and dEs are the most numerous type of galaxy in nearby galaxy clusters, including Coma (Thompson & Gregory 1993, Secker, Haris, & Plummer 1997), Fornax (Caldwell 1987, Rakos *et al.* 2001), Virgo (e.g., Binggeli, Sandage, & Tammann 1985), and recently Perseus (Conselice, Gallagher, & Wyse 2003a). Cluster dEs are usually almost free of interstellar gas and contain few young stars (Bothun *et al.* 1985).

In trying to understand the dE galaxies, we must also consider several regularities in their structures. A positive correlation is known since years to exist in dEs between optical surface brightness and luminosity (e.g., Binggeli *et al.* 1984, Wirth & Gallagher 1984, Kormendy 1985) and between luminosity and stellar velocity dispersion. The latter parameter also correlates with metallicity (e.g., Dekel & Silk 1986, Petersen & Caldwell 1993). Furthermore dE galaxies often have flattened profiles but are mostly kinematically supported by their stellar velocity dispersions rather than by rotation (Bender, Paquet, & Nieto 1991).

The combination of low gas mass fractions and moderate-to-low stellar metallicities in dE and the related dwarf spheroidal (dSph) galaxies, the least luminous and least massive dEs, is a key feature of this class (Grebel, Gallagher, & Harbeck 2003). In galaxies where gas is depleted by SF stellar abundances are predicted to be near the solar value. The lower abundances of stars in dEs (about 0.1 of solar or less; Gallagher & Wyse 1994, Cellone & Forte 1996, Han *et al.* 1997) suggest that extensive gas loss occurred during their evolution and SF ceased due to a lack of raw materials rather than exhaustion of the gas supply through SF. Galactic winds are therefore a hallmark of modern models for dE galaxies, starting from the basic consideration by Larson (1974) and continued with the studies by Dekel & Silk (1986) and Vader (1986) and they are commonly assumed to have cleaned out dE galaxies soon after their formation. MacLow & Ferrara (1999) have, however, demonstrated that gas exhaustion by means of galactic winds even in low-mass systems requires a dark-to-baryonic matter ratio much smaller than attributed to DGs in the classical picture (see e.g. Mateo 1998).

A variety of observations are now available which bring this standard picture for the evolution of dE galaxies into question. Recent HI studies of Virgo cluster dEs (Conselice, Gallagher, & Wyse 2003b) and also those of the Fornax cluster (see e.g. Michielsen *et al.* 2004) have unveiled that a small but significant fraction of them contains gas, has experienced recent SF, and can be argued from internal kinematics and cluster distribution data to form an infalling class of different types of gas-rich galaxies in the state of morphological transformation. Recent findings of a significant fraction of rotationally supported dEs in the Virgo cluster (van Zee, Skillman, & Haynes 2004) also supports the possibility of morphological transformation from dwarf irregular galaxies to dEs thru gas exhaustion. In addition, ultra-compact DGs have been detected and classified as a new type of cluster dEs that differ by their intrinsic structure and brightness (Phillipps *et al.* 2001). Nevertheless, if the 'threshing' scenario holds to produce this type of DG they stem from normal dEs by transformation in cluster environments (De Propris, Phillips, & Drinwater 2005). Another plausible mechanism to produce them from tidal tail compact star clusters has been elaborated (see e.g. Kroupa, this conference).

The situation in Local Group dEs is similarly complicated, with NGC 185 and NGC 205 both being well known to have cool interstellar gas and trace young stellar populations, while NGC 147 contains neither an ISM nor young stars (Young & Lo 1997). Furthermore, high-quality color-magnitude diagrams for nearby dSph and dE galaxies often reveal substantial intermediate age stellar populations (Han *et al.* 1997, Martínez-Delgado & Aparicio 1998, Hurley-Keller, Mateo, Nemec 1998, Stetson, Hesser, & Smecker-Hane 1998, Ikuta & Arimoto 2002, Tolstoy *et al.* 2003) which means that small spheroidal galaxies did not form all their stars in single events near the time of giant galaxy formation. This agrees with the findings of large internal metallicity dispersions in local dSphs from high-resolution stellar spectra (Shetrone, Cote, & Sargent 2001, Tolstoy *et al.* 2003). A second and possibly related line of evidence has come from deep surveys of faint galaxies. The frequent presence of compact, narrow emission line galaxies in moderate redshift galaxy samples (Koo *et al.* 1995, Guzman *et al.* 1996) leads naturally to the suggestion that such galaxies are small systems experiencing major starburst events perhaps 5–8 Gyr before the present, and likely predecessors of the current dE systems (Babul & Rees 1992, Babul & Ferguson 1996).

A further intriguing fact that requires explanation within the framework of evolutionary scenarios is that in the Carina dSph and in a few others almost no metal enrichment is perceivible from stellar spectra taken with UVES (Tolstoy *et al.* 2003) although SF has occurred in several distinct epochs with clear interrupts. More extremely, the $[\alpha/H]$ values in Carina and Sculptur drop significantly below zero.

Another interesting question concerns the structural discontinuity between dEs and giant Es (Kormendy 1985). This is e.g. still confirmed by De Rijcke *et al.* (2005) but questioned by Graham (see article in this volume).

All these new results have changed our impression that dEs can be easily understood from their structure and represent a simple and homogeneous evolutionary scenario. On the contrary, they gave reason for this conference and will do for a lot more in the future because they provide strong motivations for a reexamination of theoretical models for the evolution of these small spheroidal galaxies.

2. Evolutionary Models of dEs

A variety of evolutionary models for dEs have been performed by several groups. They are focussing on different aims and, by this, are constrained accordingly to the addressed questions.

For isolated dEs e.g. numerical models have been presented by several authors (Spaans & Norman 1997, Mori, Yoshii, & Nomoto 1999, Lia, Carraro, & Salucci 2000, Carraro, Chiosi, & Girardi 2001, Pasetto, Chiosi, & Carraro 2003, Hensler, Theis, & Gallagher 2004). Since dEs show clear concentrations towards galaxy cluster centers, another important perspective is the effect of the cluster gas environment. Hydrodynamical simulations of a moving gas-rich dwarf galaxy through cluster gas (IGM) in order to study the ram-pressure stripping (RPS) with particular emphasis on dEs has been undertaken by another group of authors (Murakami & Babul 1999, Mori & Burkert 2000, Marcolini, Brighenti, & D'Ercole 2003). All those models agree that for low-mass systems (around 10^9 $M_\odot$ and below) the whole gas component can be stripped efficiently by the a normal IGM. Arecent systematic study (Roediger & Hensler 2005) of the RPS phenomenon has shown that even in the outskirts of galaxy clusters infalling galaxies suffer gas stripping. Mayer *et al.* (see this conference) successfully cleaned off the Milky Way dSphs by this RPS effect acting by the hot galactic halo gas but applied a yet implausibly high gas density. Nevertheless, it is still unsolved to what extent dense cool gas can survive a RPS epoch due to the time-limited momentum transfer from the wind and whether the interstellar clouds are even compressed and stabilized instead of being swept-up and/or evaporated. This same uncertainty holds for galactic winds (MacLow & Ferrara 1999, Murakami & Babul 1999, Marcolini, Brighenti, & D'Ercole 2004). The gas expansion by winds facilitates the RPS effect because of lowering its gas binding energy.

In present cosmological scenarios small galaxies ought to be formed as the oldest objects if cold dark matter (CDM) fluctuations harbor sufficient baryonic substrate. Lateron larger galaxies are assembled by hierarchical accumulation of smaller units. Since the gas must cool and, subsequently, collapse so that those condensations led to enhanced SF such scenarios would expect a very early epoch of dwarf galaxy (DG) formation and their early appearance in the universe. This, however, stands in conflict with the above-mentioned observations that they seem to have formed at around $z = 1$ and that even the dSphs with old stellar populations show evidence for SF extending over several Gyr. This obvious inconsistency is not yet resolved, but could be due to the impact of the intergalactic UV radiation field that slowed the evolution of DG formation (e.g., Kepner, Babul, & Spergel 1997, Barkana & Loeb 1999, Grebel & Gallagher 2004).

There is, however, at present still a lack of global models that account self-consistently for all the different processes in scope to be affecting the evolution of dEs. This means that calculations must combine a large spatial range with a high complexity of spatially resolved processes and with different timescales, a yet unfeasible task.

Progress, however, also requires a detailed and appropriate despription of the coexisting gas phases, their mutual interactions and plasmaphysical processes, and of the stellar component, its formation, death and energetic coupling to the multi-phase interstellar medium (ISM). Such basic recipes are combined in the chemo-dynamical prescription of galaxy evolution (see reviews by Hensler 2000, Hensler 2003 and also Samland, Hensler, & Theis 1997).

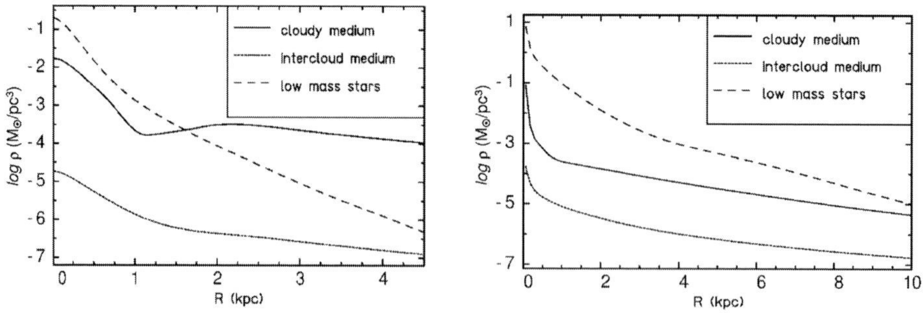

Figure 1. Radial density distribution of cloudy and intercloud medium as well as low-mass stars after an age of 10 Gyr of the 10^9 $M_\odot$, 1σ model (*left*) and the 10^{10} $M_\odot$, 3σ model (*right*) (from HTG).

3. Chemodynamical models

In a series of papers we have applied the chemodynamical treatment to various galaxy types. These models range from 1d to 3d Since chemo-dynamical simulations are numerically highly complicated and not always numerically stable because of the various competing timescales and large discontinuities it took years from the principle formulation until the first set of models. And even these could only be constructed one-dimensionally like for non-rotating massive galaxies (Theis *et al.* 1992), for dwarf ellipticals (Hensler, Theis, & Gallagher 2004, hereafter: HTG) and for the vertical settling of the galactic disk (Burkert, Truran, & Hensler 1992). The next step to 2d models was rough. At first again we started with more massive disk galaxies (Samland & Hensler 1996) and in particular the MWG (Samland, Hensler, & Theis 1997) and moved to dwarf irregular galaxies (dIrrs) where in a series of papers various aspects have been illuminated (Hensler, Rieschick, & Köppen 1999, Hensler & Rieschick 2002, Rieschick & Hensler 2000, Rieschick & Hensler 2004). In this review I wish to refer to recently published simple 1d chemo-dynamical models of dEs by us (HTG). The motivation of these studies was to extent the models of massive non-rotating galaxies, plausibly giant Es, to the low-mass regime.

Since our aim here is to consider isolated entities as a first chemo-dynamical approach to DG evolution, i.e. systems unperturbed by external effects and without inherently implied absolute timescales, on the one hand, our model evolution can be considered to start after the delayed recombination of the embedded gas. Moreover, the existence of a DM halo is expected to accelerate the collapse and to lead to enhanced SF (Mori, Yoshii, & Nomoto 1999). If the local dSph system is produced in a merger event of our Milky Way, an attractive scenario in order to understand the pecularities of this system, they should by free of DM.

HTG have presented 1d models of diffuse dEs in the mass range from 10^9 to 10^{10} $M_\odot$ that collapse from 1σ and 3σ cosmological perturbations consisting only of baryonic matter. These highly idealized models of dE galaxies fit already qualitatively quite well to several key observations and are briefly summarized here (For further information and discussion of the results see HTG.):

1) These models experience rapid compression of their initial gas content and undergo violent initial starbursts. As anticipated by earlier models, such as Dekel & Silk (1986), this leads to extensive gas loss and a re-expansion of the remaining stellar system. However, due to the inclusion of a multi-phase representation of the ISM in our models, gas ejection is not complete and some of the model galaxies can sustain SF for extended

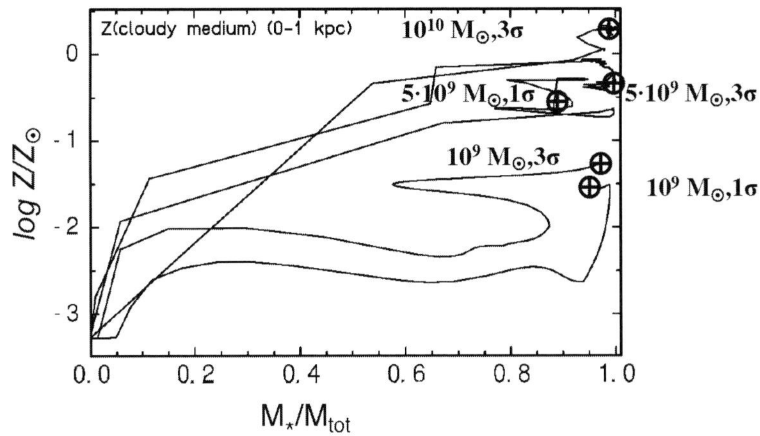

Figure 2. Evolutionary tracks of the total metallicity Z of the cloudy medium with the stellar mass fraction for the inner 1 kpc radius after 10 Gyr of the 5 models (labelled) presented in HTG. The final state is marked by the plus.

time periods (>5 Gyr) after the initial collapse. The initial behaviour is the same in the presence of a DM halo, while the dominance of the DM potential on larger scale canmore easily capture the expanding gas.
2) They predict that dE galaxies will contain a range of stellar ages and stellar metallicities, and that dEs do not have to be gas-free systems, what is consistent with some observations.
3) A 10^9 $M_\odot$ galaxy suffers a dominant single SF event within the first 200 Myrs in which 10% of the gas are converted into stars but 80% gas are lost instantaneously, driven by SNeII. The key issue is that the stellar component dominates the central range of 1 kpc by 90% of the included mass and drops exponentially, while the remaining gas extends radially over more than 5 kpc with almost constant density (see Fig. 1). Such gaseous halo is ideally exposed to tidal stripping as well as RPS.
4) In the mass range of $5{\times}10^9$ $M_\odot$ the chemo-dynamical models separate in their evolution: the 1σ model ceases to form stars after a single initial peak as a consequence of the lower gas density where self-regulation acts efficiently. More gas remains available than in the 3σ model that consumes most of its gas in an initial rapid starburst but only some gas remains gravitationally bound and supports oscillating SF on a 300 Myr timescale. These details of the SF history would be difficult to observe in present-day DGs but the model allows to understand a SF history stretch over billion years even with low gas content but SF fluctuations.
5) A 1σ initial density fluctuation for $10^{10}M_\odot$ starts with an initial collapse at a low-level SF followed by a violent initial starburst. A peak SF rate of about 100 $M_\odot$ yr^{-1} is sustained for about 60 Myr and declines below 10 $M_\odot$ yr^{-1} 200 Myr after the beginning of the starburst. This active phase is followed by a 3 Gyr plateau of SF oscillations between 0.01 and 1 $M_\odot$ yr^{-1}. At an age of almost 6 Gyr the system is seriously depleted in CM gas. Thus a key feature of these models is that the bulk of the stars in dE galaxies should have ages of $\geqslant$8 Gyr for galaxies which formed 10 Gyr ago. The radial density distribution of the long-living low-mass stars (fig. 1) traces the dissipational evolution of the collapse and the radial brightness has not developed a single exponential profile.
6) The stellar metallicities increase with higher masses as is observed (see Fig. 2);

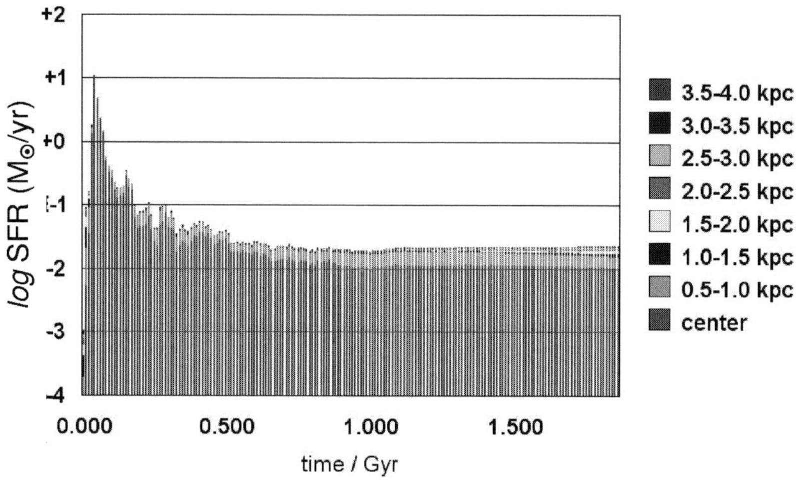

Figure 3. Star-formation history of an almost non-rotating ($\lambda \approx 0.001$) 2d 10^9 $M_\odot$ chemodynamical model for different radial zones. While the darker much extended plot region belongs to the very central 0.5 kpc the next annulus between 0.5-1.0 kpc contributes only almost 10%. Significantly, the star-fromation rate is actively self-regulated after already 0.5 Gyrs

Some problems and caveats are discussed in this paper (HTG). These include:
The models predict more cool gas than is observed in dE galaxies.
The modelled cores are very compact with implicitly high central surface brightnesses.
While the models make some intermediate-age stars, most stars are made in less than 1 Gyr after formation.
The very simple nature of these 1d chemo-dynamical models overestimates the momentum transfer between the gas phases and leads to an overly efficient gas removal.
No environment could be included that would lead to an extended period of inflow delayed due to the ionization of the IGM and feedback from SF within the DG.
No external pressure on the galaxy from an IGM that could hamper the gas expansion.

More complex models are required to bring the qualitative agreement with observations into the quantitative regime.

Therefore the chemo-dynamical treatment was soonly extended to 2d mainly to explore the evolution of dwarf irregular galaxies but can also be applied to slowly rotating dEs. Already a first rough comparison with a 2d chemo-dynamical model of 10^9 $M_\odot$ even with DM halo of additional 10^{10} $M_\odot$ demonstrates how self-regulated and smoothly the evolution proceeds (Fig. 3). After a very short (100 Myrs) initial burst due to the collapse of the gas after recombination, the central 1 kpc region cooks at a moderate SF rate of 10^{-2} $M_\odot$/yr while the annulus between 0.5 and 1 kpc reaches an almost constant SF rate of almost three times the center according to its larger area. At radii larger than 1.5 kpc the SF becomes negligible.

4. Conclusions

One further observation can basically demonstrate the complexity of the dE evolution we are faced with and that should be treatible in models: Sub-solar [α/Fe] abundance ratios are derived by Hill *et al.* for many stars of the Sculptur dSphs with VLT/FLAMES (Tolstoy 2005; see also talk by Eline Tolstoy, this conference) set a clear sign that SNII gas from an earlier SF epoch was lost from the galaxy so that the ISM capable for the most recent SF event consisted predominantly of elements released on longer timescales, i.e. from intermediate-mass stars that replenished the evacuated gas. The recurrent SF episode argues in favour of additional gas infall.

The aims to simulate these coupled evolutionary processes and their effects on both, internally and externally, are highly challenging tasks to computations. Although 1d chemodynamics have their limitations by the lack of spatial degrees of freedom for the gas and stellar flows, the multi-phase description and the interaction network (Hensler 2003) are the substantial requirements for any reliable chemo-dynamical modelling of galaxy evolution. While a grid-based treatment reaches its computational limits at 2d or low spatial resolution in 3d, the new strategy aims to treat also the *Smooth Particle Hydrodynamics* (SPH) chemo-dynamically. Developments from the single-gas phase SPH programs are under way (e.g. Semelin & Combes 2002, Berczik *et al.* 2003, Harfst, Theis, & Hensler 2004, Harfst, Theis, & Hensler 2005).

Acknowledgements

I would like to thank Bruno Binggeli and Helmut Jerjen for the organisation of this excellent and most interesting IAU Coll. The participation on this conference was suppted by a travel grant from the University of Vienna.

References

Babul, A. & Ferguson, H.C. 1996, *ApJ* 458, 100
Babul, A. & Rees, M.J. 1992, *MNRAS* 255, 346
Barkana, R. & Loeb, A. 1999, *ApJ* 523, 54
Bender, R., Paquet, A. & Nieto, J.-L. 1991, *A&A* 246, 349
Berczik, P., Hensler, G., Theis, C. & Spurzem, R. 2003, in *The Evolution of Galaxies. – III. From simple approaches to self-consistent models*, eds. G. Hensler *et al.*, *Ap. Space Sci.* 284, 865
Binggeli, B., Sandage, A. & Tarenghi, M. 1984, *AJ* 89, 64
Binggeli, B., Sandage, A. & Tammann, G. 1985, *AJ* 90, 1681
Binggeli, B. & Cameron, L.M. 1991, *A&A* 252, 27
Bothun, G.D., Mould, J., Wirth, A. & Caldwell, N. 1985, *AJ* 90, 697
Burkert, A., Truran, J.S.W. & Hensler, G. 1992, *ApJ* 391, 651
Caldwell, N. 1987, AJ, 94, 1116
Carraro, G., Chiosi, C., Girardi, L. & Lia, C. 2001, *MNRAS* 327, 69
Cellone, S.A. & Forte, J.C. 1996, *ApJ* 461, 176
Conselice, C.J., Gallagher, J.S.III. & Wyse, R.F.G. 2003a, *AJ* 125, 66
Conselice, C.J., Gallagher, J.S.III. & Wyse, R.F.G. 2003b, *ApJ* 591, 167
Dekel, A. & Silk, J. 1986, *ApJ* 303, 39
De Propris, R., Phillips, S., Drinkwater, M.J., *et al.* 2005, *ApJ* in press (astro-ph 0503377)
De Rijcke, S., Michielsen, D., Dejonghe, H., *et al.* 2005, *A&A* in press
Ferguson, H.C. & Binggeli, B. 1994, *A&AR* 6, 67
Friaca, A.C.S. & Terlevich, R. 1998, *MNRAS* 298, 399
Gallagher, J.S. & Wyse, R.F.G. 1994, *PASP* 106, 1225
Gallagher, J.S., Tolstoy, E., Dohm-Palmer, R.C., *et al.* 1998, *AJ* 115, 1869
Grebel E.K., 1997, *Rev. Mod. Astron.* 10, 29
Grebel E.K., 1999, in *The Stellar Content of the Local Group*, IAU Symp. 192, eds. P. Whitelock & R. Cannon (ASP Conf. Ser.), p. 17
Grebel, E. & Guhathkurta, P. 1999, *ApJ* 511, 101
Grebel, E.K. & Gallagher, J.S. 2004, *ApJ* 610, 89
Grebel, E.K., Gallagher, J.S. & Harbeck, D.E. 2003, *AJ* 125, 1926
Guzman, R., Koo, D.C., Faber, S.M., *et al.* 1996, *ApJ* 460, 5
Han, M., Hoessel, J.G., Gallagher, J.S., III., *et al.* 1997, *AJ* 113, 1001
Harfst, S., Theis, C. & Hensler, G. 2004, *PASA* 21, 228
Harfst, S., Theis, C. & Hensler, G. 2005, *A&A* submitted

Hensler, G. 2000, *Star, Gas and Dust in Galaxies: Exploring their Links*, eds. D. Alloin *et al.*, *ASP Conf. Ser.* 221, 1

Hensler, G. 2003, in *CNO in the Universe*, eds. C. Charbonnel *et al.*, *ASP Conf.Ser.* 304, 371

Hensler, G. & Rieschick A. 2002, *Modes of Star Formation*, eds. E. Grebel & W. Brandner, *ASP Conf. Ser.* 285, 341

Hensler, G., Rieschick A. & Köppen, J. 1999, *The Evolution of galaxies on Cosmological Timescales*, eds. J. Beckman & T.J. Mahoney, *ASP Conf. Ser.* 187, 214

Hensler, G., Theis, C. & Gallagher, J.S. 2004, *A&A* 426, 25 (HTG)

Hurley-Keller, D., Mateo, M. & Nemec, J. 1998, *AJ* 115, 1840

Ikuta, C. & Arimoto, N., 2002, *A&A* 391, 55

Kawata, D. & Gibson, B. 2003, *MNRAS* 340, 908

Kepner, J.V., Babul, A. & Spergel, D.N. 1997, *ApJ* 487, 61

Köppen, J. & Hensler, G. 2005, *A&A* 434, 531

Köppen, J., Theis, Ch. & Hensler, G. 1995, *A&A* 296, 99

Köppen, J., Theis, Ch. & Hensler, G. 1998, *A&A* 328, 121

Koo, D.C., Guzman, R., Faber, S.M., *et al.* 1995, *ApJ* 440, L49

Kormendy, J. 1985, *ApJ* 295, 73

Larson, R.B. 1974, *MNRAS* 169, 229

Lia, C., Carraro, G. & Salucci, P. 2000, *A&A* 360, 76

MacLow, M.-M. & Ferrara, A. 1999, *ApJ* 513, 142

Marcolini, A., Brighenti, F. & D'Ercole, A. 2003 *MNRAS* 345, 1329

Marcolini, A., Brighenti, F. & D'Ercole, A. 2004 *MNRAS* 352, 363

Martínez-Delgado, D. & Aparicio, A. 1998, *AJ* 115, 1462

Mateo, M. 1998, *ARA&A* 36, 435

Michielsen, D., De Rijcke, S., Zeilinger, W.W., *et al.* 2004 *MNRAS* 353, 1293

Mori, M. & Burkert, A., 2000, *ApJ* 538, 559

Mori, M., Yoshii, Y. & Nomoto, K. 1999, *ApJ* 511, 585

Murakami, I. & Babul, A. 1999, *MNRAS* 309, 161

Pasetto, S., Chiosi, C. & Carraro, G. 2003, *A&A* 405, 931

Petersen, R.C. & Caldwell, N. 1993, *AJ* 105, 1411

Phillipps, S., Drinkwater, M. J., Gregg, M. D. & Jones, J. B. 2001, *ApJ* 560, 201

Rakos, K., Schombert, J., Maitzen, H. M., *et al.* 2001, *AJ* 121, 1974

Rieschick, A. & Hensler, G. 2000, *Cosmic Evolution and Galaxy Formation: Structure, Interactions, and Feedback*, eds. J. Franco *et al.*, San Francisco, *ASP Conf. Ser.* 215, 130

Rieschick, A. & Hensler, G. 2004, *A&A* submitted

Roediger, E. & Hensler, G., 2005, *A&A* 433, 875

Samland, M., Hensler, G. & Theis, Ch. 1997, *ApJ* 476, 277

Secker, J., Harris, W.E. & Plummer, J.D. 1997, *PASP* 109, 1377

Semelin, B. & Combes, F. 2002, *A&A* 388, 826

Shetrone, M.D., Cote, P. & Sargent, W.L.W. 2001, *ApJ* 548, 592

Spaans M. & Norman, C.A. 1997, *ApJ* 483, 87

Stetson, P.B., Hesser, J.E. & Smecker-Hane, T.A. 1998, *PASP* 110, 533

Theis, Ch., Burkert, A. & Hensler, G. 1992, *A&A* 265, 465 (TBH)

Thompsen, L.A. & Gregory, S.A. 1993, *AJ* 106, 2197

Tolstoy, E. 2005, Proceedings of the ESO/Arcetri-workshop on *Chemical Abundances and Mixing in Stars*, in press (astro-ph/0412067)

Tolstoy, E. & Venn, K. 2004, in *Highlights in Astronomy* Vol. 13, IAU 2003, ed. O. Engvold, in press

Tolstoy, E., Venn, K.A., Shetrone, M.D., *et al.* 2003, *AJ* 125, 707

Vader, J.P. 1986, *ApJ* 305, 669

van Zee, L., Skillman, E.D. & Haynes, M.P. 2004, *AJ* 128, 121

Wirth, A. & Gallagher, J.S., III. 1984, *ApJ* 282, 85

Young, L.M. & Lo K.Y. 1996, *ApJ* 462, 203

Young, L.M. & Lo K.Y. 1997, *ApJ* 476, 127

Discussion

KROUPA: It is to be noted that the star-formarion timescale for your low-mass models (no DM) is about 2-3 Gyrs which is what we see in the low-mass dSph satellites.

HENSLER: Wonderful indeed, isn't it? Yes, the typical timescales of star-formation relevant processes are all in the same range for an individual galaxy and governed by the gravitational energy density.

READ: How do you decide what fraction of supernova energy to pump into the gas and to what extent does this affect your results?

HENSLER: As we could show in a few papers (see e.g. Köppen, Theis, & Hensler 1995,1998, Samland, Hensler, & Theis 1997) and in PhD theses, the network of self-regulation processes makes the models almost unaffected by the applied supernova energy within one order of magnitude. We imply 10^{51} ergs per SN Ia and II, 72% as thermal, the rest as kinetic energy.

SAVIANE: Do you predict metallicity gradients? What is their amount?

HENSLER: As I mentioned in the 1d models the metallicity gradient is positive for the least-massive models and changes sign towards higher masses and initial densities.

Near-Field Cosmology with Dwarf Elliptical Galaxies
Proceedings IAU Colloquium No. 198, 2005
H. Jerjen and B. Binggeli, eds.

doi:10.1017/S1743921305003601

The chemical composition of Local Group dwarf spheroidals

Eline Tolstoy

Kapteyn Institute, University of Groningen, Postbus 800, 9700AV Groningen, the Netherlands

Abstract. I will review the progress of VLT spectroscopy of large numbers of individual stars in nearby dwarf spheroidal galaxies. This spectroscopy has allowed us to obtain detailed insights into the chemical and dynamical properties of the resolved stellar population in these nearby systems.

1. Introduction

The dwarf galaxies being considered here are the lowest luminosity (and mass) galaxies that been have found. It is perhaps not a surprise that there is a lower limit in mass for an object to be able to form more than one generation of stars which will be related to the limit below which one supernova will completely destroy a galaxy. Numerical and analytic models tell us that the limit must be around a few $\times 10^6 M_\odot$ (e.g., Ferrara & Tolstoy 2000). Globular clusters, for example, typically have much lower masses than this limit.

It is among dwarf galaxies, irregular (dI) and spheroidal (dSph), that we find the lowest mass objects. For example the Sculptor dSph galaxy which apparently has a present day mass of $6\times10^6 M_\odot$, or the dI galaxy DDO 210 which has a present day mass of $5.4\times10^6 M_\odot$ (Mateo 1998). The only difference between the lowest mass dSphs and dIs seems to be the presence of gas and current star formation in dIs. It has already been noted that dSphs predominantly lie close to our galaxy (<250kpc away), and dIs predominantly further away (>400 kpc away). This suggests that the proximity of dSph to our Galaxy played a role in the removal of gas from these systems. Although the range of properties found in dSph and dI galaxies does not allow a straight forward explanation; particularly not the large variations in star formation histories and chemical evolution paths that have now been observed in these systems.

Here I am going to discuss some recent observations of individual stars in nearby dSph galaxies, predominantly from the large survey being undertaken by DART (Dwarf Abundances and Radial-velocities Team) to measure abundances and velocities for several hundred individual stars in a sample of three nearby dSph galaxies: Sculptor, Fornax and Sextans. We have used the VLT/FLAMES facility to obtain more than 100 spectra per 25arcmin diameter field of view across several positions in each galaxy out to the tidal radius (see Figure 1). We have used the low resolution setting to obtain Ca II triplet metallicity estimates as well as accurate radial velocity measurements (Tolstoy *et al.* 2004; Battaglia *et al.*, in prep) over a large area in each galaxy, and we have also used the high resolution settings to concentrate on about 100 stars in the central field of each galaxy – where we are able to obtain accurate abundances for a range of interesting elements such as Mg, Ca, O, Ti, Na, Eu to name a few (Hill *et al.*, in prep; Letarte *et al.*, in prep).

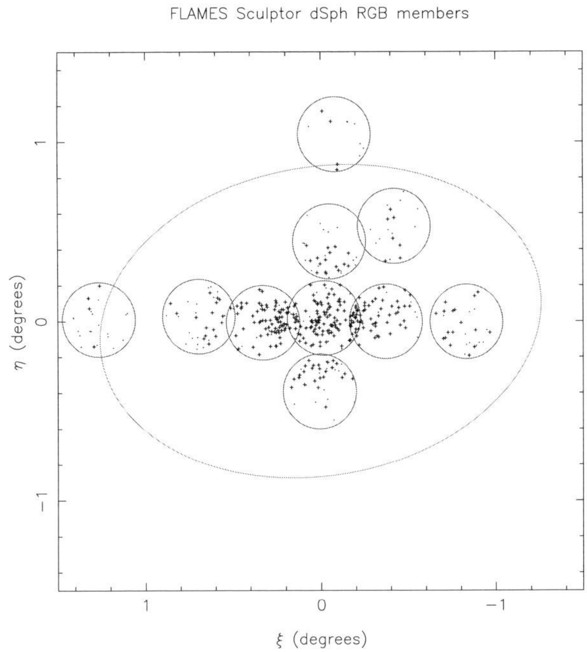

Figure 1. The distribution of candidate members RGB stars (shown as small crosses) and velocity non-members (dots) in the Sculptor dSph galaxy chosen from velocities measured in VLT/FLAMES spectra. The ellipse is the tidal radius of Sculptor, determined by Irwin & Hatzidimitriou (1995). The circles represent the 10 individual FLAMES pointings made by DART in this galaxy.

2. How do we understand the evolution of galaxies

When we have resolved the individual stars in a galaxy and accurately measured their colours and magnitudes down to the oldest main sequence turn-offs ($M_V \sim +3.5$) in a Colour-Magnitude Diagram we can in principle come to detailed conclusions about how the star formation rate has varied with time all the way back to the first star formation in the galaxy (e.g., Tolstoy & Saha 1996; Gallart *et al.* 1999; Hurley-Keller *et al.* 1999). This approach is the most accurate for intermediate age populations, but for stars older than about 10 Gyr the time resolution gets quite poor (and the stars are getting very faint), and it becomes hard to distinguish a 12 Gyr old star from a 10 Gyr old star. Here it becomes useful to consider the Horizontal Branch stars ($M_V \sim 0.$) which are the bright He-burning phase of low mass stars >10 Gyr old. The ratio of red to blue horizontal branch stars tells us about the age and metallicity variation (Lee *et al.* 2001) of the oldest stars in the galaxy (e.g., for Sculptor dSph, see Figs 2 & 3).

The Red Giant Branch (RGB, $-3 < M_V < 0.$) contains stars with ages >1 Gyr old, back to the oldest stars in the galaxy. In the majority of RGB stars it is believed that the atmosphere of the star remains an unpolluted sample of the interstellar medium out of which it was formed. This means we have small pockets of interstellar medium of different ages (enriched by different numbers of processes) conveniently covering the nuclear burning core of stars. This hot stellar core provides a useful bright background source to be absorbed in the stellar atmosphere allowing very detailed studies of the elemental abundances in these ancient gas samples. Thus, a spectroscopic analysis of the variation of the abundances of different elements seen in absorption in atmospheres of

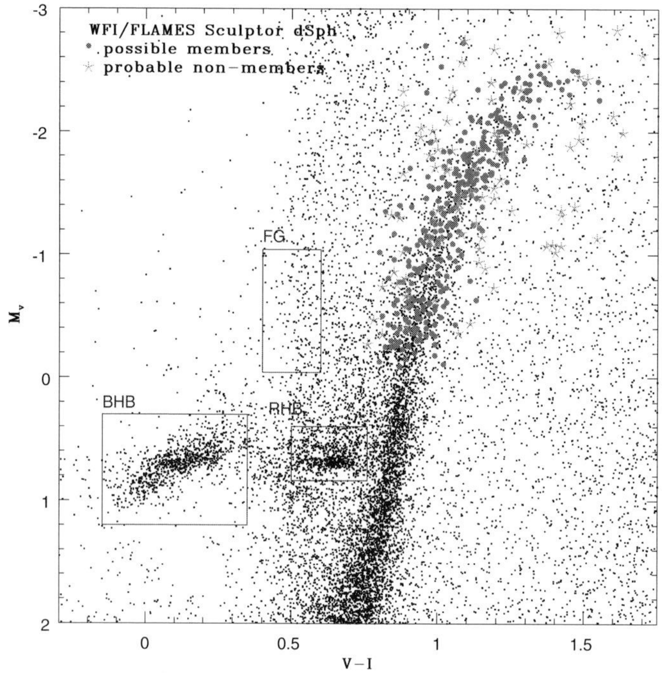

Figure 2. The Colour-Magnitude Diagram for the WFI coverage of Scl shown in Figure 1. The circle and star symbols are the Red Giant Branch stars which were observed with VLT/FLAMES and for which we have accurate v_{hel} and [Fe/H] measurements (S/N>10). The potential members of Scl are shown as circles and non-members shown as stars. Also shown are the regions used to define the Blue Horizontal branch (BHB), the red horizontal branch (RHB) and foreground comparison (FG) populations.

different age stars allows us to trace the detailed chemical enrichment history of a galaxy with time.

The age-metallicity degeneracy on the RGB (graphically illustrated by Cole *et al.* 2005; their Figure 8) leads to serious problems using photometric colour alone to disentangle age and metallicity. This uncertainty is the leading reason for quite different star formation histories to be inferred from the same Colour-Magnitude Diagrams. For example, in the case of the Carina dSph three groups determined three different star formation histories from two different data sets (Hurley-Keller *et al.* 1998; Hernandez *et al.* 2000; Dolphin 2002), some of the (significant) differences can be traced to different assumptions about the chemical evolution of Carina. However if similar chemical enrichment histories are assumed then the consistency between models by different groups is significantly improved (e.g., Skillman *et al.* 2003).

The observed magnitude and colour (e.g., M_V, V−I) of a star combined with a measured metallicity, [Fe/H], allows us to effectively remove the age-metallicity degeneracy and determine the age of an RGB star from an isochrone. However the uncertainties in this age increase with *decreasing* metallicity, and often the photometric errors on the colour measurements are of the same order of magnitude as the entire age range of the isochrones from the oldest (13 Gyr) to the youngest (2 Gyr). This is before even considering other effects such as the variations in the Galactic reddening, α-element abundance, distance, metallicity measurement error or errors in the interpolation of the models.

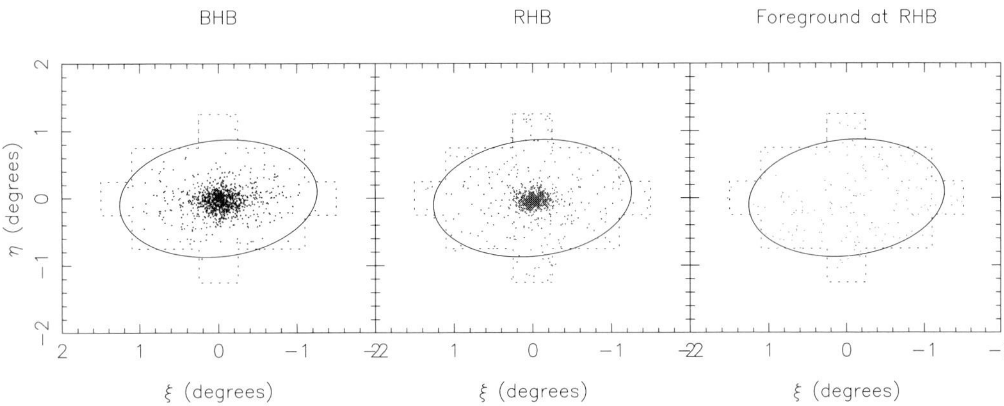

Figure 3. The distribution of Horizontal Branch stars from ESO/WFI imaging of the Scl dSph. The three panels show the different spatial distributions of blue horizontal branch stars (BHB), and red horizontal branch stars (RHB) as selected from the M_v, V−I Colour-Magnitude Diagram (CMD) shown in Figure 2. Also shown, to illustrate the foreground contamination in the RHB distribution, are a CMD-selected sample of foreground stars to match the RHB contamination density (see Figure 2). The ellipse is the tidal radius of Scl as defined by Irwin & Hatzidimitriou (1995).

The interpretation of all these stellar measurements rely upon stellar evolution models which are primarily based upon our understanding of Galactic globular cluster stars and we must make the assumption that this knowledge is generally applicable to stars in dwarf galaxies even though the initial abundance ratios are different, and certainly the stellar density is very different.

3. High Resolution Spectroscopy: Detailed Abundance Analysis

The best way to measure the chemical evolution of a galaxy is to determine the relative abundances of different chemical elements. This has been called chemical tagging by Freeman & Bland-Hawthorn (2002). Different elements are created in different circumstances and if we are able to determine the abundance of elements known to be created in a particular set of physical conditions, we can assess the importance and frequency of these conditions during the history of star formation in a galaxy. This clearly can give insights into the star formation rates at different times in a galaxy's history and the environments in which most of the stars were formed.

For example, the abundance of light elements (e.g., O, Na, Mg, Al) are considered to be tracers of "deep mixing" patterns which are found only in globular cluster environments, which gives a limit to the number of dissolved globular clusters which can exist in a stellar population. These typical (Galactic) globular cluster abundance patterns have also been recently found in the globular clusters of Fornax dSph (Letarte *et al.* 2005), but not (so far) in the field star populations (Shetrone *et al.* 2003).

The creation of α-elements (e.g., O, Mg, Si, Ca, Ti) occurs predominately in Supernovae type II explosions, i.e. the explosion of massive stars a few $10^6 - 10^7$ yrs after their formation. The abundance of the different α-elements is quite sensitive to the mass of the SNII progenitor (e.g., Woosley & Weaver 1995) so the α/Fe ratio traces the mass function of the stars which contributed to the creation of the α elements, and ratios of different α elements themselves can put limits on the highest mass star which has enriched a galaxy and also the typical mass range (e.g., McWilliam 1997).

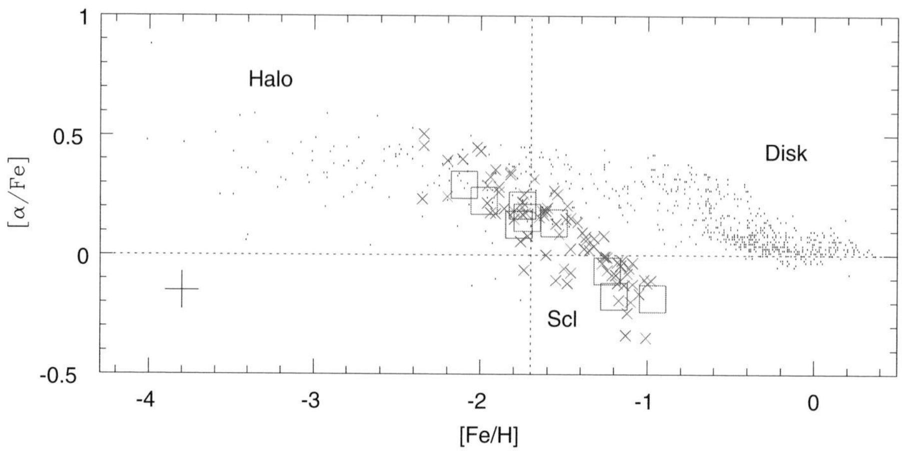

Figure 4. The α-abundance (average of Ca, Mg and Ti) for stars in our Galaxy compared to those in Scl. The VLT/FLAMES high resolution measurements of 92 members in the central field are shown as crosses (from Hill *et al.*, in prep). The Galactic stars come from standard literature sources (see Venn *et al.* 2004 for references). The 10 open squares are UVES measurements of individual stars in Scl (Shetrone *et al.* 2003; Geisler *et al.* 2005).

Heavy Elements ($Z > 30$) are a mix of r- and s- process elements. That is to say elements which were produced by rapid or slow neutron capture which tells us about the environment in which enrichment occurred. Rapid capture (r-process) is assumed to occur in high energy circumstances, such as supernovae explosions. For example Eu is considered to be an element produced almost exclusively by the r-process. The slow capture (s-process) is thought to be created by more quiescent processes such as the stellar winds common in AGB type stars of intermediate age (and mass). Typical elements which are thought to be created by the s-process are Ba and La. The ratio between r- and s-process element abundances gives an indication of the relative importance of these different enrichment processes during the history of star formation in a galaxy. Dwarf spheroidal galaxies are generally found to have strong s-process enhancement. This may indicate that supernovae products are typically lost to a shallow potential well, or that the slow star formation rate means that massive stars are not very common (Tolstoy *et al.* 2003; Venn *et al.* 2004).

With an 8m class telescope it is possible to measure the abundance of these elements in dSph galaxies in a few hours of integration time with a high-resolution high-throughput spectrograph, e.g. KECK/HIRES (Shetrone *et al.* 2001); VLT/UVES (Tolstoy *et al.* 2003; Shetrone *et al.* 2003; Geisler *et al.* 2005; Letarte *et al.* 2005, in prep). But it has been quite slow and laborious to obtain a statistically meaningful sample of stars in dSph galaxies which have fairly complicated star formation histories. This complex star forming history means that a sizeable sample of stars of different ages is required to trace the full range enrichment processes. The real break-through comes with VLT/FLAMES (and other fibre systems, such as MIKE on Magellan). In one shot spectra can be obtained for over 100 stars in one 25arcmin field of view. The sensitivity and the field of view are a very nice match to the Galactic dSph galaxies.

The results published to date still lack good statistics to make a detailed comparison between our Milky Way and individual dSph, although attempts have been made (e.g, Tolstoy *et al.* 2003; Venn *et al.* 2004). From these results, and the early results from our FLAMES surveys (e.g., Hill *et al.*, in prep; Battaglia *et al.*, in prep) a match is not found between the detailed abundance properties of stellar populations found in dSph galaxies and any (significant) component of our Galaxy. There remains a discrepancy between the metallicity ([Fe/H]) distribution of the stars randomly selected in a dSph (e.g., Scl: Tolstoy *et al.* 2004; also true for Fornax and Sextans, Battaglia *et al.* in prep) and samples of stars in our Milky Way. This is also true when more detailed comparisons are made, for example considering how [α/Fe] varies with [Fe/H] (Hill *et al.* 2004, in prep; see Figure 4).

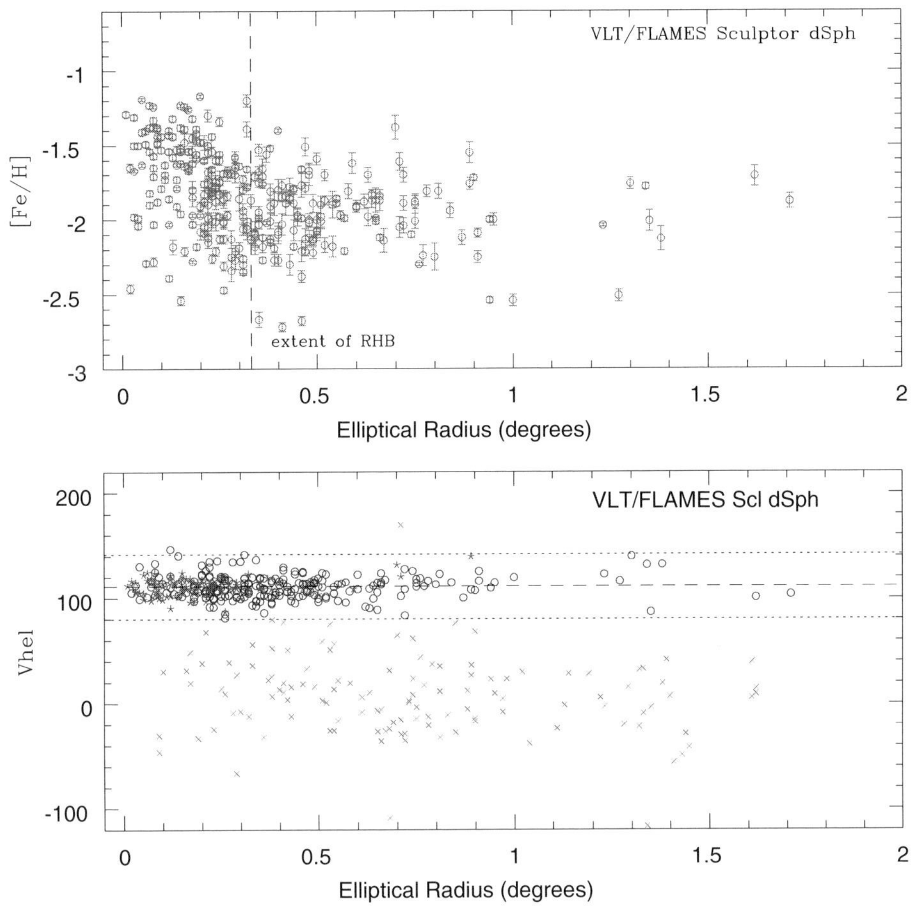

Figure 5. The upper panel shows the VLT/FLAMES spectroscopic measurements of [Fe/H] for the 321 potential RGB velocity members of Scl (and S/N>10). We see a clear trend of metallicity with radius. The lower panel shows v_{hel} as a function of elliptical radius for all stars satisfying S/N > 10. Likely Scl members are clearly seen clustered around the systemic velocity of 110 km/s. The 321 stars which are potential members are plotted as stars ([Fe/H] > -1.7) and circles ([Fe/H] < -1.7), while the 128 crosses are assumed to be non-members.

4. Low Resolution Spectroscopy: Metallicities

The ideal is to be able to obtain high resolution spectra for individual stars in nearby dSph over a large wavelength range, and make a detailed analysis of a range of different elements. However, this is quite time consuming both in telescope time (at the distance of dSph) and in analysis. One of the most simple ways to get an estimate of the metallicity of RGB stars is with the Ca II triplet (e.g., Cole *et al.* 2004). This is a basic metallicity indicator requiring only low or intermediate spectral resolution, based on three lines around 8500Åwhich have been empirically calibrated from observations of stars in globular clusters stars with high resolution abundances (e.g., Armandroff & Da Costa 1991).

With the Ca II triplet we obtain an overview of the global metallicity range of the RGB stars in a dwarf galaxy. In the case of the DART project these measurements are complementary to the high resolution observations made in the centre of each dSph. In the low resolution larger area survey we can assess how representative our detailed study is of the stellar population of the whole galaxy.

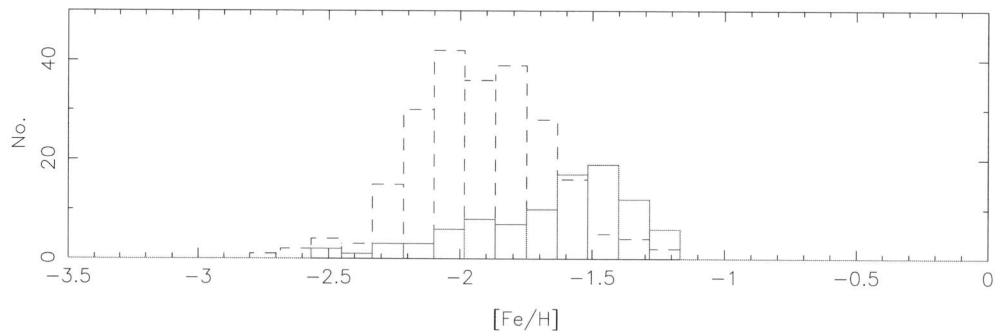

Figure 6. Here is the histogram distribution of [Fe/H] for Scl dSph: the 94 stars within the central, r < 0.2 degree region (solid line); and the 227 stars beyond r > 0.2 degrees (dashed line).

Our first VLT/FLAMES results, based upon Ca II triplet, clearly show that Scl dSph contains two distinct stellar components with different spatial, kinematic and abundance properties (Tolstoy *et al.* 2004, see Figure 5). There appears to be a metal-rich, $-0.9 >$ [Fe/H] > -1.7, and a metal-poor, $-1.7 >$ [Fe/H] > -2.8 component. The metal-rich component is more centrally concentrated than the metal poor, and on average appears to have a lower velocity dispersion, $\sigma_{metal-rich} = 7 \pm 1$ km/s, whereas $\sigma_{metal-poor} = 11 \pm 1$ km/s.

The full abundance analysis of the FLAMES HR data (Hill *et al.* in prep) will provide more details of the chemical enrichment history of Scl. This will enable us to distinguish between two episodes of star formation or more continuous star formation, manifested as a gradient in velocity dispersion and metallicity from the centre of the galaxy.

It is clear from the histogram of [Fe/H] measurements in Scl dSph (see Figure 6) that this distribution lacks a low metallicity tail, in fact the lowest metallicity star in our sample of more than 300 stars is [Fe/H] $= -2.7$. We find a similar lack of low metallicity stars in Fornax and Sextans. Although it is difficult to make an accurate comparison with Galactic samples, where the completeness of different samples is hard to quantify, there appears to be a significantly different distribution between dSph and the (metal-poor) halo of the Milky Way.

5. Conclusions

There are indications that the presence of two populations is a common feature of dSph galaxies. Our preliminary analysis of Horizontal Branch stars, v_{hel} and [Fe/H] measurements in the other galaxies in our sample (Fornax and Sextans dSph; Battaglia *et al.*, in prep) also show similar characteristics to Scl, especially in the most metal poor component. Pure radial velocity studies (Wilkinson *et al.* 2004; Kleyna *et al.* 2004) have also considered the possibility that kinematically distinct components exist in Ursa Minor, Draco and Sextans dSph galaxies.

What mechanism could create two ancient stellar components in a small dwarf spheroidal galaxy? A simple possibility is that the formation of these dSph galaxies began with an initial burst of star formation, resulting in a stellar population with a mean [Fe/H] $\leqslant -2$. Subsequent supernovae explosions from this initial episode could have been sufficient to cause gas (and metal) loss such that star formation was inhibited until the remaining gas could sink deeper into the centre (e.g., Mori *et al.* 2002). Thus the subsequent generation(s) of stars would inhabit a region closer to the centre of the galaxy, and have a higher average metallicity and different kinematics. Another possible cause is external influences, such as minor mergers, accretion of additional gas or the kinematic stirring by our Galaxy. It might also be that events surrounding the epoch of reionisation influenced the evolution of these small galaxies (e.g., Skillman *et al.* 2003) and resulted in the stripping or photo-evaporation of the outer layers of gas in the dSph, meaning that subsequent more metal enhanced star formation occurred only in the central regions.

Acknowledgements

I am grateful for support from a fellowship of the Royal Netherlands Academy of Arts and Sciences, and the exceptional collaborators that make up DART, especially Mike Irwin, Giuseppina Battaglia, Amina Helmi, Andrew Cole, Vanessa Hill, Kim Venn, Bruno Letarte, Pascale Jablonka, Matthew Shetrone & Andreas Kaufer.

References

Armandroff, T. & Da Costa, G.S. 1991, *AJ* 101, 1329
Cole A.A., Smecker-Hane T.A., Tolstoy E. *et al.* 2004, *MNRAS* 347, 367
Cole A.A., Tolstoy E., Gallagher J.S. & Smecker-Hane T.A. 2005, *AJ* 129, 1465
Dolphin A. 2002, *MNRAS* 332, 91
Ferrara A. & Tolstoy E. 2000, *MNRAS* 313, 291
Freeman K. & Bland-Hawthorn J. 2002, *ARA&A* 40, 487
Gallart C., Freedman W.L., Aparicio A., Bertelli G. & Chiosi C. 1999, *AJ* 118, 2245
Geisler, D., Smith, V., Wallerstein, G., Gonzalez, G. & Charbonnel, C. 2005, *AJ* 129, 1428
Hernandez X., Gilmore, G. & Valls-Gabaud, D. 2000, *MNRAS* 317, 8831
Hurley-Keller D., Mateo M., & Nemec J. 1998, *AJ* 115, 1840
Hurley-Keller D., Mateo M., & Grebel E.K. 1999, *ApJL* 523, 25
Irwin M. & Hatzidimitriou D. 1995, *MNRAS* 277, 1354
Kleyna J.T., Wilkinson M.I., Evans N.W. & Gilmore G. 2004, *MNRAS* 354, L66
Lee Y.-W. *et al.* 2001 in: T. von Hippel *et al.* (eds.) *Astrophysical Ages and Time Scales*, ASP Conference Series, vol. 245, p. 343
Mateo M. 1998, *ARA&A* 36, 435
McWilliam A. 1997, *ARA&A* 35, 503
Mori M., Ferrara A. & Madau P. 2002, *ApJ* 571, 40
Shetrone M.D., Cote P. & Sargent W.L.W. 2001, *ApJ* 548, 592
Shetrone M.D., Venn K.A., Tolstoy E., Primas F., Hill V. & Kaufer A. 2003, *AJ* 125, 684
Skillman E.D., Tolstoy E., Cole A.A., Dolphin A.E. *et al.* 2003, *ApJ* 596, 253
Tolstoy E. & Saha A. 1996, *ApJ* 462, 672
Tolstoy E., Venn K.A., Shetrone M.D., Primas F., Hill V. *et al.* 2003, *AJ* 125, 707
Tolstoy E., Irwin M.J., Helmi A., Battaglia G., Jablonka P., Hill V. *et al.* 2004, *ApJL* 617, 119
Venn K., Irwin M., Shetrone M.D., Tout C.A., Hill V. & Tolstoy E. 2004, *AJ* 128, 1177
Wilkinson M., Kleyna J.T., Evans N.W., Gilmore G. *et al.* 2004, *ApJL* 611, 21

Near-Field Cosmology with Dwarf Elliptical Galaxies
Proceedings IAU Colloquium No. 198, 2005
H. Jerjen and B. Binggeli, eds.

doi:10.1017/S1743921305003613

Galactic chemical abundance distributions in a ΛCDM universe

Andreea S. Font[1], Kathryn V. Johnston[1], James S. Bullock[2] and Brant Robertson[3]

[1]Van Vleck Observatory, Wesleyan University Middletown, CT , 06459, USA
[2]Department of Physics & Astronomy, University of California, Irvine, CA 92687, USA
[3]Harvard-Smithsonian Center for Astrophysics, 60 Garden Street, Cambridge, MA 02138, USA

Abstract. Observations suggest systematic differences between chemical abundances of stars in satellite galaxies and those in the Milky Way halo. Specifically, for the same [Fe/H] values, stars in surviving satellite galaxies display significantly lower [α/Fe] ratios than stars in the stellar halo.

Here we investigate whether the observed differences can be explained in the framework of hierarchical structure formation. We model the chemical enrichment of a typical Milky Way galaxy in a ΛCDM Universe using, in combination, *i)* a semi-analytical code and numerical simulations that model the accretion and disruption of halo substructure and *ii)* a chemical evolution model that takes into account each satellite's star formation, metal enrichment and stellar feedback. Our results suggest that the observed chemical abundance patterns are a natural outcome in the process of hierarchical assembly of the Galaxy. We find that the stellar halo is built up from satellite galaxies accreted early on (more than 8–9 Gyr ago) and enriched in α-elements produced in Type II supernovae (average [α/Fe] values between 0.2–0.5). In contrast, satellites which survive today were typically accreted late (within the last 4–5 Gyr) and had at the time of accretion nearly solar [α/Fe] values as a result of the longer contribution of Type Ia supernovae.

1. Introduction

Recent observations have revealed interesting chemical abundance patterns in the Milky Way halo. Specifically, for the same [Fe/H] values, stars in neighboring satellite galaxies display [α/Fe] ratios significantly lower (by about 0.1–0.2 dex) than stars in the stellar halo (Shetrone *et al.* 2001, Shetrone *et al.* 2003, Venn *et al.* 2004). If ΛCDM is indeed a valid model for how galaxies form and evolve, then it has to explain the discrepancy in chemical abundances of different Galactic components. Following the chemical enrichment of the Galaxy in the full cosmological context is non trivial. In order to tackle this problem one has to have a model that incorporates a realistic merger history of a Milky Way-type galaxy, a detailed follow-up of the kinematics of substructure inside the galactic potential well and a chemical evolution model that includes an accurate treatment of stellar yields, the IMF and of winds from massive and intermediate mass stars. We present here a blending of two theoretical methods, one of constructing Milky Way-like stellar halos from accreting substructure (Bullock & Johnston 2004, Johnston & Bullock, in prep.) and the other of modeling the chemical evolution of individual satellite galaxies by taking into account both the inflow and outflow of matter (Robertson *et al.* 2005). The simulated galaxy has properties that match a series of observational constraints for the stellar halo (eg. total luminosity, gas content) and for its infalling satellites as based on observations of Local Group dwarfs (eg. the stellar mass – circular velocity of dark halo relation, stellar mass - metallicity relation).

We describe below the results for a typical Milky Way halo. A detailed analysis of a larger sample of simulated halos will be presented in Johnston & Bullock (in prep.) and Font *et al.*(in prep.).

2. Results

Figure 1 shows the mass accretion history of one of our simulated stellar halos binned in time intervals of 0.2 Gyr. The top panel shows the accretion of the entire halo, and the bottom panel that of the inner $R < 20$ kpc region (here the radial distance R is centered on the position of the Sun, $R_{\odot} = 8.5$ kpc). The filled histograms represent the stellar material still gravitationally bound at present time ($t = 0$), whereas the empty histograms represent the entire accreted stellar mass. The figure shows that the stellar halo assembles early on. For example, ~80% of the inner halo (which is the counterpart of the "local halo" typically accessible in observations) is already in place by look-back time ~9 Gyr. The material still bound today (i.e. in surviving satellites) has been accreted more recently, generally within the past few Gyr. Not surprisingly, none of the surviving satellites are found in the inner region of the halo where tidal forces are strongest.

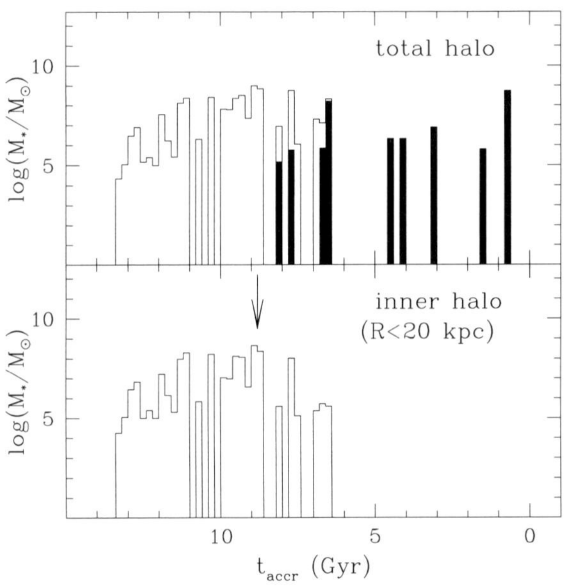

Figure 1. Mass accretion history of the simulated galaxy halo binned in time intervals of 0.2 Gyr. Filled histograms represent the stellar material still gravitationally bound at present time ($t = 0$), and empty histograms represent the entire accreted stellar mass. Top panel corresponds to the entire halo and bottom panel to the inner $R < 20$ kpc halo. The arrow in the bottom panel denotes the look-back time when ~80% of the inner halo has assembled.

These results are typical of all simulated halos and suggest that present day satellite galaxies of the Milky Way are likely to have been accreted recently. This implies that they also had more time available for sustaining star formation and hence further their chemical enrichment (our model assumes that most of star formation in satellite galaxies occurs prior to their accretion onto the main halo, after which their gas is stripped by ram pressure and thus star formation is shut off; the result is however largely independent of this assumption, since the majority of satellites are fully disrupted within 1–2 orbits after accretion).

The chemical evolution of infalling substructure is investigated in Figure 2, which shows the [α/Fe] versus [Fe/H] ratios of all baryonic satellites (values are averages weighted by the contribution of different stellar populations). Here [α/Fe] represents the average of [O/Fe] and [Mg/Fe] abundances. The left panel shows the abundance ratios plotted with different symbols function of their time of accretion. In the right panel we plot the same quantities, this time the different symbols denoting whether satellites are fully disrupted or still containing some bound material at $t = 0$. This figure shows that those satellites that enter the main potential well later tend to have higher [Fe/H] and lower [α/Fe]. This is because Fe content increases with time, with the most significant contribution coming from Type Ia supernova activity which sets on after first ~1–2 Gyr (Larson 1974). In contrast, α-elements being produced very early on in Type II supernovae have abundances that remain relatively constant in time (hence the decrease in [α/Fe] with time). Surviving satellites have the lowest [α/Fe] ratios and, on average, these values are 0.1–0.2 dex lower than that of the stellar halo (that is, of the material originating in the disrupted satellites) – in agreement with the observational results.

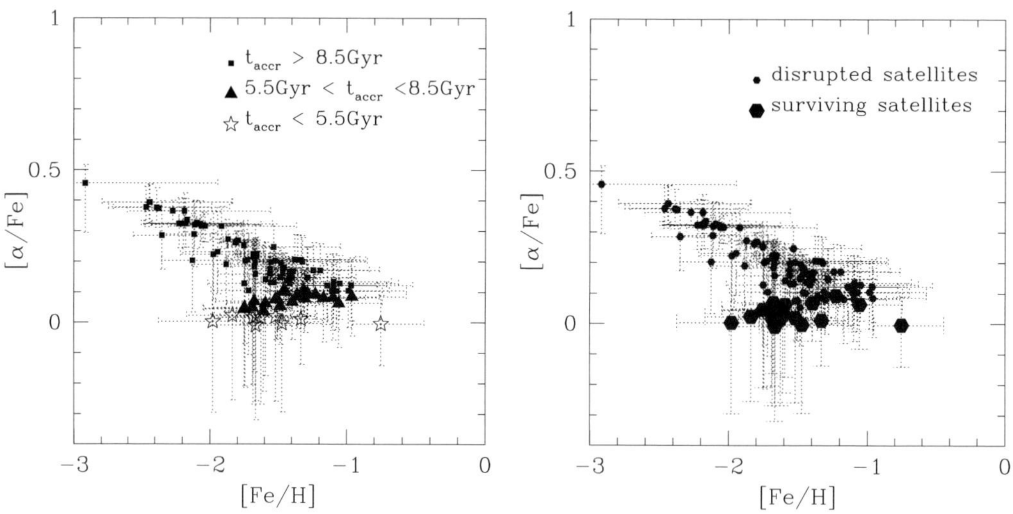

Figure 2. [α/Fe] versus [Fe/H] for all baryonic satellites accreted onto the main halo of the Milky Way -type galaxy. *Left:* Satellites are plotted with different symbols function of their look-back time of accretion: squares denote satellites accreted at $t_{accr} > 8.5$ Gyr, triangles those accreted $5.5 < t_{accr} < 8.5$ Gyr, and stars with $t_{accr} < 5.5$ Gyr. *Right:* Small and large symbols denote whether satellites are fully disrupted or still containing some bound material at present time ($t = 0$), respectively. Errorbars in both panels represent 10% and 90% of the weighted average values.

3. Conclusions

We have shown that the observed differences between chemical abundances of surviving satellites and halo can be naturally understood in the framework of the ΛCDM model. Thus satellites surviving at present time were generally accreted late and had more time available for chemical enrichment (notably Fe enrichment by Type Ia supernovae). In contrast, satellites accreted early (which make up, by mass, most of the halo) were typically more deficient in Fe than their later counterparts, hence had higher [α/Fe] values when accreted.

Acknowledgements

A.F and K.V.J's contributions were supported through NASA grant NAG5-9064 and NSF CAREER award AST-0133617.

References

Bullock, J.S. & Johnston, K.V. 2004, in *Satellites and Tidal Streams*, ed. F. Prada, D. Martinez-Delgado & T. Mahoney (San Francisco: ASP)
Larson, R.B. 1974, *MNRAS* 169, 229
Robertson, B. , Bullock, J. S., Font, A.S., Johnston, K.V. & Hernquist, L. 2005, *ApJ* submitted
Shetrone, M.D., Côté, P. & Sargent, W.L.W. 2001, *ApJ* 548, 592
Shetrone, M.D., Venn, K.A., Tolstoy, E., Primas, F., Hill, V. & Kaufer, A. 2003, *AJ* 125, 684
Venn, K.A., Irwin, M., Shetrone, M.D., Tout, C.A., Hill, V. & Tolstoy, E. 2004, *AJ* 128, 1177

Discussion

DEKEL: What is the origin of the difference in metallicity between the stellar halo and the surviving satellites? How does the initial satellite mass translate to the final difference? Aren't gradients within the initial satellites relevant?

FONT: In terms of the [Fe/H] distribution, the stellar halo and surviving satellites look about the same. The difference arises in the [α/Fe] ratios: surviving satellites tend to have lower [α/Fe] values than the local stellar halo. Among the satellites which become disrupted at present time, those with high initial mass are expected to contribute the most to the make-up of the halo. However, the stellar halo has also a non-negligible contribution from the low-mass, early accreted satellites, which tend to have higher [α/Fe] ratios. These satellites make most of the difference in [α/Fe] patterns of the stellar halo. Initial metallicity gradients are important and they will be preserved in the halo.

Near-Field Cosmology with Dwarf Elliptical Galaxies
Proceedings IAU Colloquium No. 198, 2005
H. Jerjen and B. Binggeli, eds.

doi:10.1017/S1743921305003625

IMF variations in dwarf galaxies

C. Weidner and P. Kroupa

Observatory of the University of Bonn, Auf dem Hügel 71, D-53121 Bonn, Germany
email: cweidner, pavel@astro.uni-bonn,de

Abstract. Dwarf galaxies tend to have a large spread in metal abundances and overall lower metallicites than large Spiral or Elliptical ones. Here we propose that if all stars are formed from an invariant canonical IMF within each star cluster which are distributed according to an embedded cluster mass function (ECMF), then the integrated stellar IMF over the whole galaxy (the integrated galaxial IMF, IGIMF) must be steeper than the canonical IMFs within the clusters, and additionally they must depended on the stellar mass of the galaxy. In particular, dwarf galaxies and low-surface brightness galaxies show steeper galaxial IMFs which vary strongly with the star-formation rate. The resulting number of Supernovae II per low-mass star, and the chemical enrichment history of these galaxies therefore vary substantially depending on the galaxy assembly history.

Keywords. galaxies: abundances, galaxies: dwarf, galaxies: luminosity function, mass function, galaxies: star clusters, galaxies: stellar content

1. The stellar and the (embedded) cluster IMF

Stars form in embedded clusters with a few to many million stars. Inside these clusters they appear to form following a universal initial mass function (IMF) with a Salpeter power-law slope ($\alpha_3 = 2.35$, Salpeter 1955, Kroupa 2002) above $1\,M_\odot$, $\xi(m) \propto m^{-\alpha_i}$, with $\alpha_0 = 0.30$ for $0.01 \leqslant m/M_\odot \leqslant 0.08$, $\alpha_1 = 1.30$ for $0.08 \leqslant m/M_\odot \leqslant 0.50$, $\alpha_2 = 2.30$ for $0.50 \leqslant m/M_\odot \leqslant 1.00$ and $\alpha_3 = 2.35$ for $1.00 \leqslant m/M_\odot$. This has been found to be the case for a wide range of different conditions in the MW, LMC and SMC and other galaxies.

Star clusters also seem to be distributed according to a (universal?) power-law embedded cluster mass function (ECMF), $\xi_{\rm ecl} \propto M_{\rm ecl}^{-\beta}$, where $M_{\rm ecl}$ is the total mass in stars. We assume a single-slope power-law ECMF with $\beta = 2.35$ between $5\ M_\odot$ and $10^6\ M_\odot$.

2. The integrated galaxial initial mass function (IGIMF)

In our model all stars are born in clusters following a universal IMF but also clusters are formed from a universal ECMF. An important constrain is that the mass of a cluster limits the mass of the most massive star in it (Fig. 1a). The ECMF is connected to the star-formation rate of a galaxy via the following equation (Weidner, Kroupa & Larsen 2004, Fig. 1b):

$$\log(M_{\rm ecl,max}) = \log(k_{\rm ML}) + (0.75 \cdot \log {\rm SFR}) + 6.77,$$

where $k_{\rm ML}$ is the mass-to-light ratio, typically 0.0144 for young (< 6 Myr) clusters.
So the mass function for all stars born in all clusters, which we call the integrated galaxial initial mass function (IGIMF), becomes the following integral (Kroupa & Weidner 2003),

$$\xi_{\rm IGIMF}(m,t) = \int_{M_{\rm ecl,\,min}}^{M_{\rm ecl,\,max}(SFR(t))} \xi(m \leqslant m_{\rm max})\, \xi_{\rm ecl}(M_{\rm ecl})\, dM_{\rm ecl}.$$

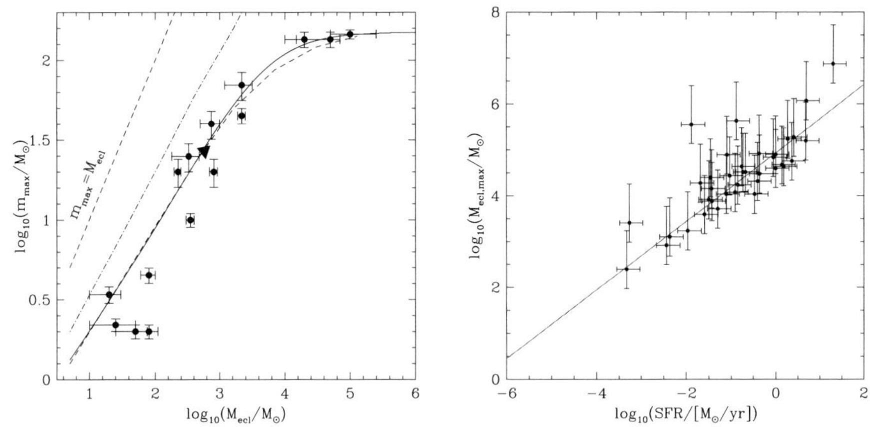

Figure 1. Left panel: The thick solid line shows the dependence of the maximum stellar mass on the cluster mass for $\alpha_3 = 2.35$ from the semi-analytical model (Weidner & Kroupa 2004). The dashed line which follows close the semi-analytical result shows the mean maximum stellar mass for Monte-Carlo experiments when a fundamental upper mass limit of $m_{\max *} = 150\ M_\odot$ is invoked and the stars are sorted and then added until they reach the cluster mass $M_{\rm ecl}$. The straight dashed line marks Monte-Carlo results without such a limit and for fully random sampling. The dots with error bars are observed clusters, while the triangle is a result from a star-formation simulation with an SPH code by Bonnell, Bate & Vine (2003). Right panel: The solid line shows the star-formation-rate–maximal-cluster-mass relation from Weidner, Kroupa & Larsen (2004) while the dots with the error bars are the observations.

As visible in Fig. 2a,b the resulting IGIMFs are always steeper than the input canonical stellar IMF.

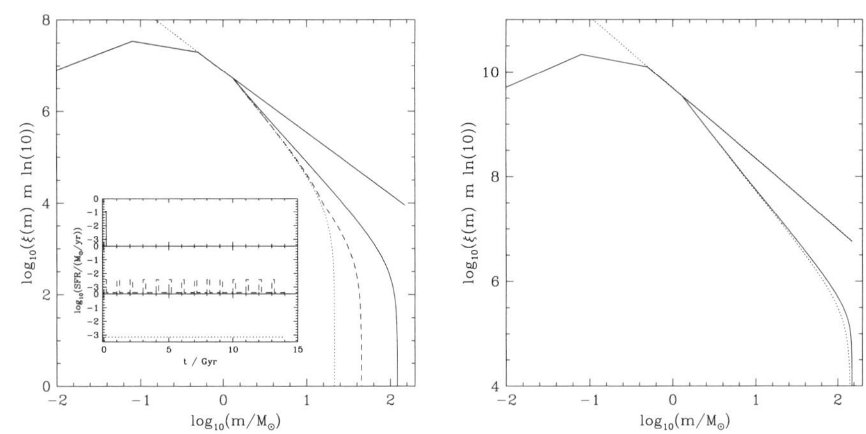

Figure 2. Left panel: Integrated galaxial mass functions for a galaxy with a stellar mass of $10^7 M_\odot$ (dE, dSph, dIrr, LSB) with different star-formation histories (solid curved line: single initial-burst of 100 Myr, dashed line: periodic SF of 100 My every 900 Myr and dotted line: constant SFR over 14 Gyr, respectively, from top to bottom) shown in the small box. The input IMF slope above 1 $M_\odot$ is $\alpha_3 = 2.35$, while the ECMF slope β is 2.35. Right panel: The same for a galaxy with $10^{10}\ M_\odot$ stellar mass. Both taken from Weidner & Kroupa (2005).

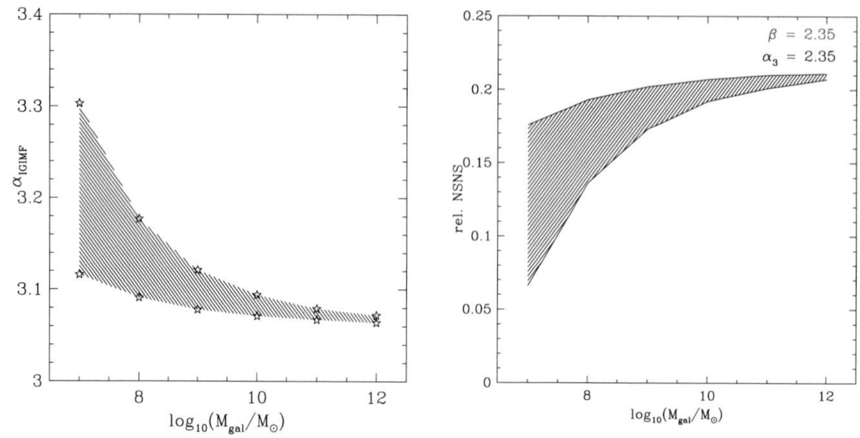

Figure 3. Left panel: IGIMF slopes above $\sim 1.3\,M_\odot$ in dependence of the stellar galaxy mass for $\alpha_3 = 2.35$ and $\beta = 2.35$. The lower limit of the shaded region results from single initial-burst models while the upper limit is from continuous models. The IGIMF thus becomes steeper with reducing stellar mass, thus resulting in systematic differences in the chemical evolution of different galaxy types. Also in the low-mass regime a larger scatter in chemical properties is to be expected. Right panel: The number of SNII per star relative to the same number for a constant canonical IMF as a function of galaxy mass for $\alpha_3 = 2.35$ and $\beta = 2.35$. The upper limit of the shaded region results from single burst models while the lower limit is deduced from continuous star-formation models. While for galaxies with large stellar masses the value is about 20% of the corresponding canonical IMF value it drops significantilly for galaxies with lower stellar mass. Both panels are from Weidner & Kroupa (2005).

3. Results & Conclusions

- The IGIMFs must be steeper (Fig. 3a) than the stellar IMF and vary with galaxy type (Fig. 2).
- The number of SNII per star, the chemical enrichment histories and mass-to-light ratios calculated with an invariant Salpeter IMF may not be correct for any galaxy.
- The number of supernovae per star (Fig. 3b) is possibly significantly lower over cosmological times than for an invariant canonical IMF.
- Irrespective of how old a galaxy is it will always appear less chemically evolved than a more massive equally-old galaxy as a result of the steeper IGIMF (Fig. 4).
- The scatter in chemical properties must increase with decreasing galaxy mass.

Acknowledgements

This work has been funded by DFG grant KR1635/3.

References

Bonnell I.A., Bate M.R. & Vine S.G. 2003 *MNRAS* 343, 413

Kroupa P. 2002, *Science* 295, 82

Kroupa P. & Weidner C. 2003, *ApJ* 598, 1076

Salpeter E.E. 1955, *ApJ* 121, 161

Tremonti C.A., Heckman T.M., Kauffmann G., Brinchmann J., Charlot S., White S.D.M., Seibert M., Peng E.W., Schlegel D.J., Uomoto A., Fukugita M. & Brinkmann J. 2004, *ApJ* 613, 898

Weidner C. & Kroupa P. 2004, *MNRAS* 348, 187

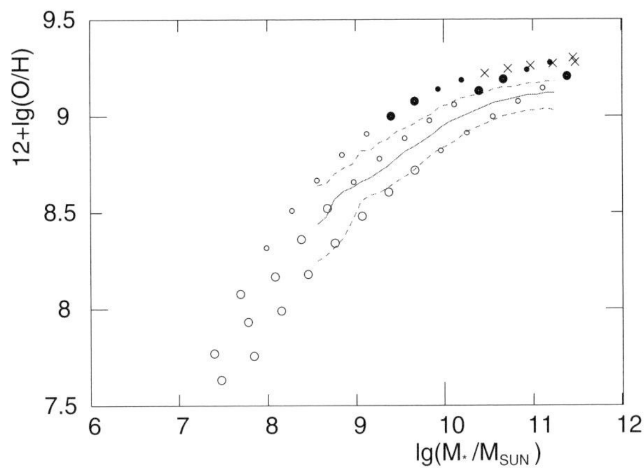

Figure 4. The relation between oxygen abundance and stellar mass of galaxies. Dashed lines and the solid line indicate the 16, 50, and 84 percentiles of the observed metallicity distribution by Tremonti *et al.* (2004). The symbols are results from closed-box models and indicate the present-day gas fraction: large circle > 0.5, small circle 0.5...0.25, large filled circle 0.25...0.15, small filled circle 0.15...0.1, and crosses < 0.1.

Weidner C. & Kroupa P. 2005, *ApJ* in press (astro-ph/0502525)
Weidner C., Kroupa P. & Larsen S.S. 2004, *MNRAS* 350, 1503

Discussion

GALLAGHER: Can you indirectly test your model by predicting where observable massive stars, such as WR stars, will become rare due to IMF variations?

WEIDNER: Yes, one has to normalise the model on the number of observed unevolved stars and then calculate the expected number of WR and massive stars using stellar evolution models.

READ: How does your IMF for dwarf galaxies compare with direct observations e.g. Wyse *et al.* for the UMi dSph galaxy?

WEIDNER: The Wyse *et al.* results for masses above 1 $M_\odot$ are also based on indirect constrains (abundances) and still subject to large uncertainties (e.g. yields). Other authors (e.g. Tolstoy et al. 2004, AJ, 125, 707) find for Sculptor, Fornax, Carina and Leo I dSph indications for an non-Salpeter IMF. We are in the progress of looking into the chemical evolution consequences of our model.

HENSLER: The IGIMF slope for constant SF in e.g. dIrrs is much steeper then Salpeter. Does one significantly underestimate the past SFR if one applies a Salpeter slope with M/L to observed luminosities? To what extent?

WEIDNER: There is a systematic error and we are looking into it.

Near-Field Cosmology with Dwarf Elliptical Galaxies
Proceedings IAU Colloquium No. 198, 2005
H. Jerjen and B. Binggeli, eds.

doi:10.1017/S1743921305003637

The chemical evolution of subpopulations in the Carina dwarf spheroidal

A. Koch[1], M. Wilkinson[2], E.K. Grebel[1], D. Harbeck[3], J. Kleyna[4], R. Wyse[5], G. Gilmore[2] and W. Evans[2]

[1]Astronomical Institute of the University of Basel, Venusstr. 7, 4102 Binningen, Switzerland
email:koch@astro.unibas.ch

[2]Institute of Astronomy, Cambridge University, Madingley Road, Cambridge CB3 0HA, UK

[3]University of California, Space Sciences Laboratory, 7 Gauss Way, Berkeley, CA 940720

[4]Institute for Astronomy, University of Hawaii, 260 Woodlawn Drive, Honolulu, HI 96822

[5]The John Hopkins University, 3701 San Martin Drive, Baltimore, MD 21218

Abstract. We present the observed metallicity distribution for 437 red giants in the Carina dwarf spheroidal (dSph) galaxy that was derived from FLAMES medium-resolution spectroscopy of the near-infrared Ca triplet (CaT). We find a mean [Fe/H] of −1.7 dex and a full spread in metallicities of at least one dex. Most of this width is due to the occurrence of three populations at different [Fe/H], which are likely to be related to the three star forming (SF) episodes that governed Carina's evolution and also correlate with peaks in the age-distribution. Both the lack of any correlation between stellar colour and metallicity and a comparison of the metallicities with chemical evolution models, which suggest a major role of infalling gas, contribute to the detailed understanding of Carina's evolutionary history. Moreover, we find a mild dependence of metallicity on radial location, reflected in a shift of the distribution's peak towards the metal poor end in the outer regions.

Keywords. galaxies: abundances, galaxies: evolution, galaxies: dwarf

1. Stellar populations in Carina

Photometric analyses of the faint Carina dSph have disclosed the presence of various stellar populations (e.g., Monelli *et al.* 2003), exhibiting a prominent old (>11 Gyr), intermediate-age (5–6 Gyr), and a young (3 Gyr) population. This implies that Carina must have undergone several SF episodes with at least three significant pulses. In apparent defiance of the resulting spread in ages, the galaxy's colour-magnitude diagram (CMD) features a remarkably narrow red giant branch (RGB). A reason for that can be a counteracting spread in metallicities, where metal rich, young stars have colours comparable to the older, more metal poor ones. Such an undesired age-metallicity degeneracy can be resolved if independent, accurate [Fe/H] measurements are available so that stellar ages as the remaining parameter can be obtained from isochrones. Moreover, the overall shape and possible spatial variations of the metallicity distribution functions (MDFs) themselves provide vital ingredients for analysing Carina's unusual SF history.

With the aim of studying the chemical *and* kinematical properties of Carina's separate populations we launched an ESO Large Programme. In this vein, spectroscopic observations of five fields in Carina were performed, comprising 1257 red giants out to the tidal radius, thus providing a wide spatial coverage and good number statistics. Our observations were performed during 23 nights with the multi-object spectrograph FLAMES at the VLT in medium-resolution mode (grating LR 8), centered at the near-infrared CaT around 860 nm. For details on the data reduction and calibration see Koch *et al.*,

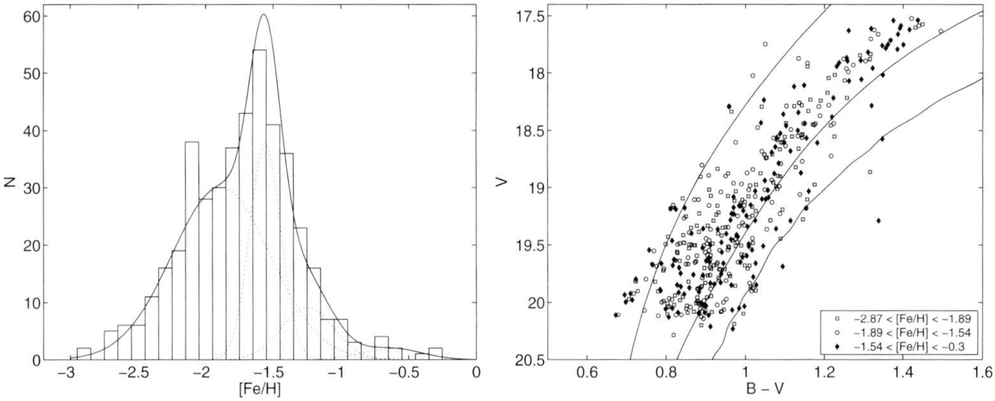

Figure 1. *Left panel*: MDF for our sample of red giants in Carina. Also indicated are the populations as revealed by a KMM-test (dotted lines) and the sum of these populations (solid). The *right panel* illustrates the distribution of metallicities across the RGB, where different symbols refer to three ranges of [Fe/H]. Solid lines indicate globular cluster fiducials from Sarajedini & Layden (1997): M 15 ([Fe/H] = −2.17), NGC 6752 (−1.54) and NGC 1851 (−1.29).

in preparation. Typical signal-to-noise (S/N) ratios for the spectra lie between 25 and 150 pixel^{-1}. Rejection of foreground stars could be achieved by iterative error-weighted Gaussian fits of the radial velocity distribution upon convergence, which yields a distinctive peak at 224 km s^{-1} with a respective dispersion of 7.5 km s^{-1}. This procedure (cutting at $3\,\sigma$), after additional rejection of stars with too low a S/N to measure equivalent widths, left 437 radial velocity members to be analysed. Finally, metallicities were derived from the reduced equivalent width W′ (Rutledge *et al.* 1997, Cole *et al.* 2004) of the CaT, which we calibrated against a common reference dataset of galactic globular clusters, obtained during the same observational runs.

2. Metallicity distributions

The resulting MDF for our 437 red giants is displayed in the left panel of Fig. 1. Its formal mean lies at an [Fe/H] of -1.72 ± 0.01 dex, and the dispersion is $\sigma = 0.37$ dex, whereas the distribution implies that the full spread in [Fe/H] is at least one dex. Considering this wide spread and the measurement uncertainties, this is well in agreement with the spectroscopic study of Smecker-Hane *et al.* (1999), who found a value of -1.99 ± 0.08 dex from a low number of stars, and also with the estimate of Rizzi *et al.* (2003) of -1.91 ± 0.18 dex as inferred from their CMDs after correction of age effects. The apparent presence of a secondary peak in the MDF at lower metallicities and the occurrence of three SF episodes in Carina led us to perform a KMM-test for multiple populations (cf. left panel of Fig. 1). As a result, four populations are preferred against the hypothesis of only two or three peaks and their significance versus a one-population model lies at a confidence of 98.1%. The major part of stars (∼67%) in our sample has peak metallicities at −1.89 dex, whereas 21% of the targets can be assigned to a population around −1.56 dex. The remaining 12% form the more metal rich tail composed of peaks at −1.2 and −0.6 dex.

2.1. *Absence of colour trends*

A puzzling feature of Carina's CMD is the RGB, which is, in spite of the wide age range covered by the galaxy's stellar populations, remarkably narrow. However, the ever so wide observed spread in metallicities is able to explain this narrowness in terms of a counteraction of age versus metallicity in the sense that metal rich, young stars have colours

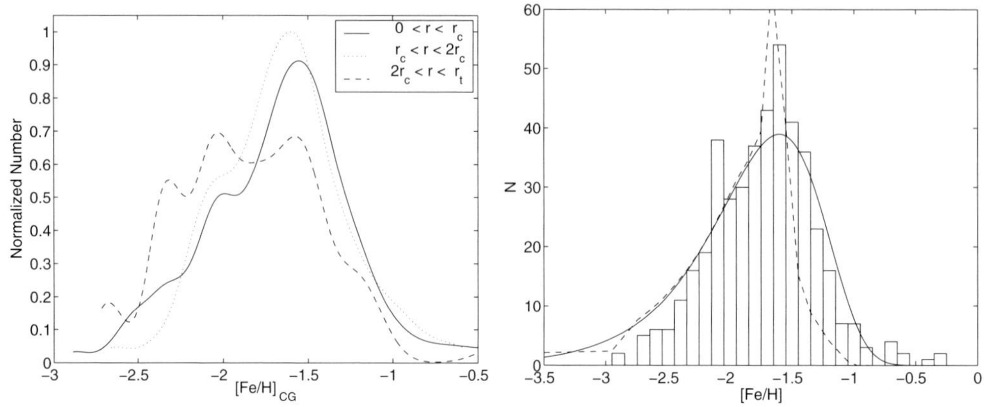

Figure 2. *Left:* Generalized MDFs after convolution with an error distribution for different radial bins in Carina. The *right* panel are overplots of the best-fit curve from a closed box model (solid line) and of a model from Lanfranchi & Matteucci (2004) on the observed MDF.

comparable to the older, more metal poor ones (Smecker-Hane *et al.* 1999). This in fact confirms that a complex mixture of stellar populations as in Carina does not necessarily contradict the RGB's narrowness. As the right panel of Fig. 1 implies, there is no apparent trend of [Fe/H] with the colour of the red giants – different metallicity subgroups are equally widely spread across the whole width of the RGB. Thus producing metal rich and metal poor giants at the same location in the CMD requires the more metal poor stars to be also older and, vice versa, that the metal rich stars at the respective locus tend to have younger ages. Hence there seems to be an indication of an age-metallicity relation of some kind in Carina, although this is not immediately obvious from the CMD.

2.2. *Presence of radial trends*

The different stellar populations in Carina have often photometrically been shown to take different spatial distributions, where the intermediate age red clump stars are clearly concentrated towards the center compared to the old HB (see Harbeck *et al.* 2001). Such a radial trend seems to be a common feature among the dSphs of the Local Group. In order to assess whether such a gradient is also reflected in our spectroscopic metallicity sample, we plot in Fig. 2 (left panel) the error-weighted MDFs for three radial bins, each with a width of approximately one core radius (8.′8). The distribution from the central region shows evidence of the different populations, reflected in the occurrence of peaks at several metallicities. When proceeding outwards to the middle annulus, the peak of the entire MDF is shifted by ca. 0.1-0.2 dex towards the metal poor end, whereas there is an increasing contribution from the secondary peak at ~ -2.1 dex. Finally, the outermost bin shows an increasing tendency towards the metal poor tail of the distribution, with a similar contribution from all the different peaks. In addition to this spatial separation of stellar populations in terms of their chemical poperties, there is also evidence of kinematic substructure, see Wilkinson – this proceedings.

2.3. *Models of chemical evolution*

Overplotted on the observed MDF in Fig. 2 (right) is the curve for a best-fit simple closed box model. This model is able to roughly reproduce the metal poor peak and the metal rich tail of the distribution. Likewise, a more sophisticated model from Lanfranchi & Matteucci (2004), which incorporates the infall and outflow of material from the galaxy and is characterised by a remarkably low SF efficiency and a high wind rate, predicts the

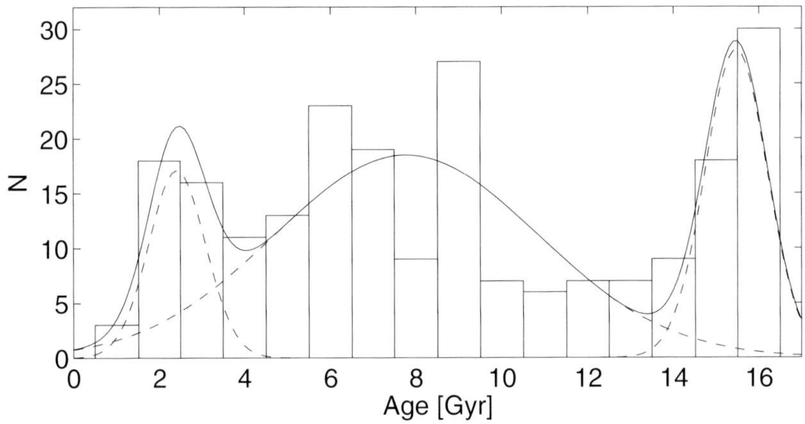

Figure 3. Preliminary age distribution of stars in Carina, together with separate fits of the underlying populations from a KMM-test.

metal poor peak and the slow decline towards the metal poor end of the distribution†. However, all such models tend to overpredict the number of extremely metal poor stars, which we do not see in the observations. This is in fact comparable to the (solar) G-dwarf problem. The general failure of a simple closed-box assumption points to the fact that Carina cannot be taken as a closed system and it must be rather considered as likely that the star forming bursts were triggered by the infall of (pre-enriched) gas.

2.4. *Stellar ages*

With the accurate [Fe/H] measurements on hand stellar ages could be estimated from isochrone fits to each star's location in the CMD. The *preliminary* distribution shown in Fig. 3 is drawn from a set of Y^2-isochrones with no α-enhancement. Again, there is strong evidence of at least three underlying populations, where a KMM-test reveals a young population at 2.5 Gyr (with 13% of all stars), a rather old population at ca. 15 Gyr (23%) and the major part (64%) of the giants at intermediate ages around 7.8 Gyr. Both the young and old population appear to be rather narrow, with a formal width of 0.8 Gyr, whereas the dominant intermediate age population spans a range of at least 3 Gyr.

3. Conclusions

The analysis of a large number of spectroscopic metallicities in the Local Group dSph Carina has revealed the presence of detailed substructure in this particular galaxy, which is most likely related to the occurrence of extended SF bursts. Both the MDF itself and also the age distribution show three distinct populations, whereas the radial variation of metallicities and comparison with chemical evolutionary models argue in favour of a dominant role of infalling gas onto the galaxy. Sophistication of the isochrone grids and completion of the ongoing analysis of multi-object high-resolution spectroscopy will then deepen the detailed understanding of Carina's evolutionary history.

† Note, however, that the latter model was not fit to our data, but we rather overplotted the curve from literature, which the authors labelled their "best-fit model".

References

Cole, A. *et al.* 2004, *MNRAS* 347, 367
Lanfranchi, G. & Matteucci, F. 2004, *MNRAS* 351, 1338
Monelli, M. *et al.* 2003, *AJ* 126, 218
Rizzi, L., Held, E.V., Bertelli, G. & Saviane, I. 2003, *ApJ* 589, L85
Rutledge, G.A., Hesser, J.E. & Stetson, P.B. 1997, *PASP* 109, 907
Sarajedini, A & Layden, A. 2003, *AJ* 113, 264
Smecker-Hane, T.A. *et al.* 1999, *ASP* 192, 159

Discussion

ZAGGIA: What is the impact of your radial velocity errors on the chemical abundance calculations?

KOCH: Our median measurement error is 1.2 km s^{-1}. Any possible inclusion of one or two remaining non-members will *not* alter the conclusions from the metallicity distributions.

Near-Field Cosmology with Dwarf Elliptical Galaxies
Proceedings IAU Colloquium No. 198, 2005
H. Jerjen and B. Binggeli, eds.

doi:10.1017/S1743921305003649

The Origins of dEs

Evan D. Skillman

Astronomy Department, University of Minnesota, 116 Church St. SE, MN 55455, USA
email: skillman@astro.umn.edu

Abstract. With the new discoveries that some dEs are rotationally supported and that dIs may show a large variety in star formation histories, the perceived relationships between these two families of galaxies are changing. There are at least three viable channels for the origin of dwarf elliptical galaxies with strong observational evidence that support their reality. I will discuss the observational evidence for each of these channels and the likely physical processes which determine each channel.

Keywords. galaxies: clusters: general, galaxies: dwarf, galaxies: formation, galaxies: Local Group, galaxies: stellar content, galaxies: evolution

1. A Working Definition of "Origin"

It is important for this talk that I clarify my use of the word origin. Frequently, when one is referring to the origin of a galaxy, one means the time when a dark matter halo established its gravitational identity. However, in this talk, the origin which I am referring to is the point when a dwarf galaxy ceases to form stars.

At some point early in its history, a galaxy which we observe to be a dE today was an actively star forming galaxy; thus, it had cold gas. Structurally, dE galaxies are quite similar to present day star forming dwarf galaxies (dIs, see next section). Thus, the defining moment of the creation of a dE is when it loses its cold gas. Since many dE galaxies show the presence of an intermediate age population, the origins of dE galaxies, as defined here, are not constrained to a single epoch. Some dEs are consistent with no intermediate age stars, and thus, were created quite early in the history of the Universe (e.g., Ursa Minor; Olszewski & Aaronson 1985), while others show star formation up to very recent times (e.g., Leo I; Gallart *et al.* 1999). Thus, the process or processes which convert actively star forming dwarf galaxies into dE galaxies have been taking place over the entire history of our Universe.

2. Similarities Between dEs and dIs

2.1. *Background*

The primary distinction between dE and dI galaxies is the presence or absence of cold gas. This is usually measured through HI emission at 21 cm. The typical values of M(HI)/L for dIs range from 0.1 to 10, while most dEs are non-detections in HI or show ratios of 10^{-3} or less (Skillman 1996). Otherwise, dE and dI galaxies have many similar properties. Most importantly, it has been known for a long time now that dEs and dIs have similar structures (Faber & Lin 1983; Kormendy 1985; Caldwell & Bothun 1987; Binggeli & Cameron 1991).

When trying to understand the possible relationships between dE and dI galaxies, there are several observations to consider. Perhaps most important is the strong morphology – density relationships observed in both the group (e.g., Einasto *et al.* 1974) and cluster

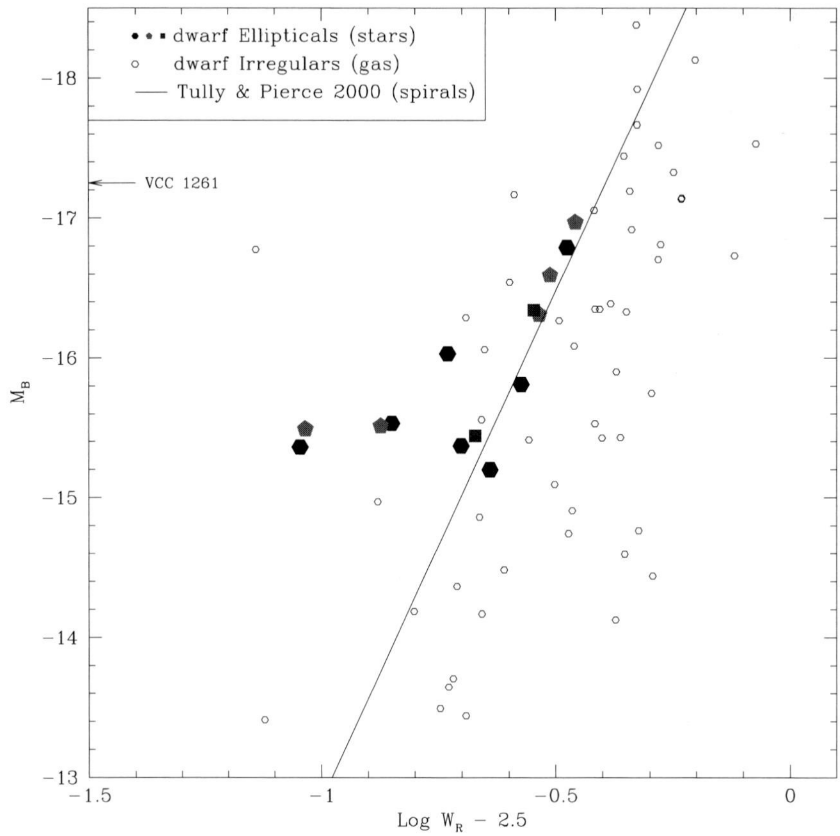

Figure 1. The luminosity-line width relation for dwarf elliptical galaxies [note that the values shown are the observed maximum linewidths, which are possible underestimates of the full rotation velocity; see van Zee *et al.* (2004a) for a full discussion of possible correction factors to the observed widths)]. Dwarf ellipticals with blue cores (squares) and red cores (pentagons) follow the same relation as spiral galaxies (line; Tully & Pierce 2000) and dwarf irregular galaxies (open dots; van Zee 2001). (From van Zee *et al.* 2004b)

(e.g., Binggeli *et al.* 1990) environments. Because of their low masses, both are recognized as fragile systems (e.g., Dekel & Silk 1986). For a long time, dEs were considered to be non-rotating systems, and lack of rotational support distinguished them from the dIs (e.g., Bender *et al.* 1991). Many dEs in the cluster environment have nuclei (e.g., Caldwell 1983). In contrast, the typical dI does not show the presence of a nucleus, but centrally concentrated star formation is common (and defines the class of blue compact dwarf galaxies). Finally, almost by definition, dEs and dIs must have different star formation histories (as dEs have no current star formation and almost all dIs do). Although it has been known for quite a while that the Milky Way dSph companions show a great variety of star formation histories (e.g., Mateo 1998), it is now emerging that dIs may too.

2.2. *Recent Results*

2.2.1. *Rotation in dEs*

One of the properties which was long thought to separate the dEs from the dIs was rotation. Bender and collaborators found very little evidence of rotation in the few

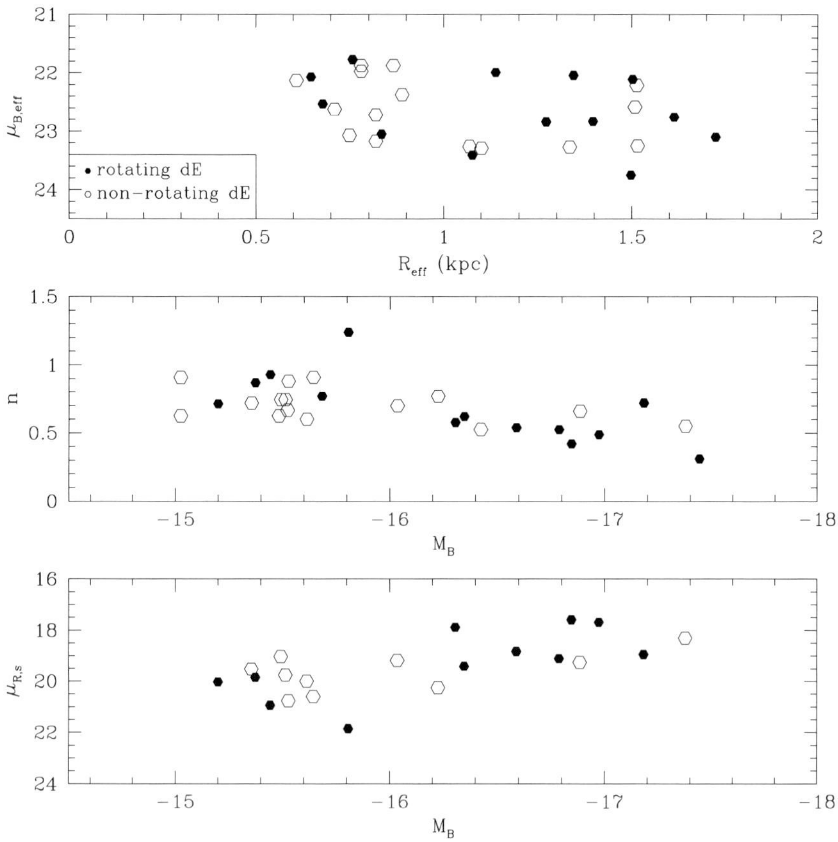

Figure 2. Structural parameters for Sérsic fits to dwarf elliptical galaxies in Virgo with measured kinematic properties (Pedraz *et al.* 2002; Geha *et al.* 2003; van Zee *et al.* 2004a); rotating (filled hexagons) and non-rotating (open hexagons) dE galaxies have similar structural properties. (a) Model independent parameters of half-light radius and effective surface brightness. (b) Absolute magnitude and Sérsic shape parameter. (c) Absolute magnitude and central surface brightness from a Sérsic fit to the R-band surface brightness profile. The non-rotating and rotating dwarf elliptical galaxies cannot be distinguished based on structural parameters. (From van Zee *et al.* 2004b)

systems that they observed. However, recent programs to observe dEs have discovered that a significant fraction of dEs show significant rotational support (de Rijke *et al.* 2001; 2003; Pedraz *et al.* 2002; Geha *et al.* 2002; 2003; van Zee *et al.* 2004a). Interestingly, van Zee *et al.* (2004b) find a Tully-Fisher relationship between luminosity and velocity width for the Virgo dEs which do show rotation (Figure 1). It is intriguing that the rotating and non-rotating dEs are not distinguished structurally (Figure 2).

2.2.2. *dI Star Formation Histories*

Recently, my collaborators and I have published a star formation history (SFH) for a halo field in IC 1613, a Local Group dI, based on relatively deep HST WFPC2 observations (Figure 3; Skillman *et al.* 2003). This is the deepest color-magnitude diagram (CMD) of an isolated dI. Although detailed SFH studies exist for other dIs, no dIs at distances beyond the Magellanic Clouds have been observed to the depth of the present

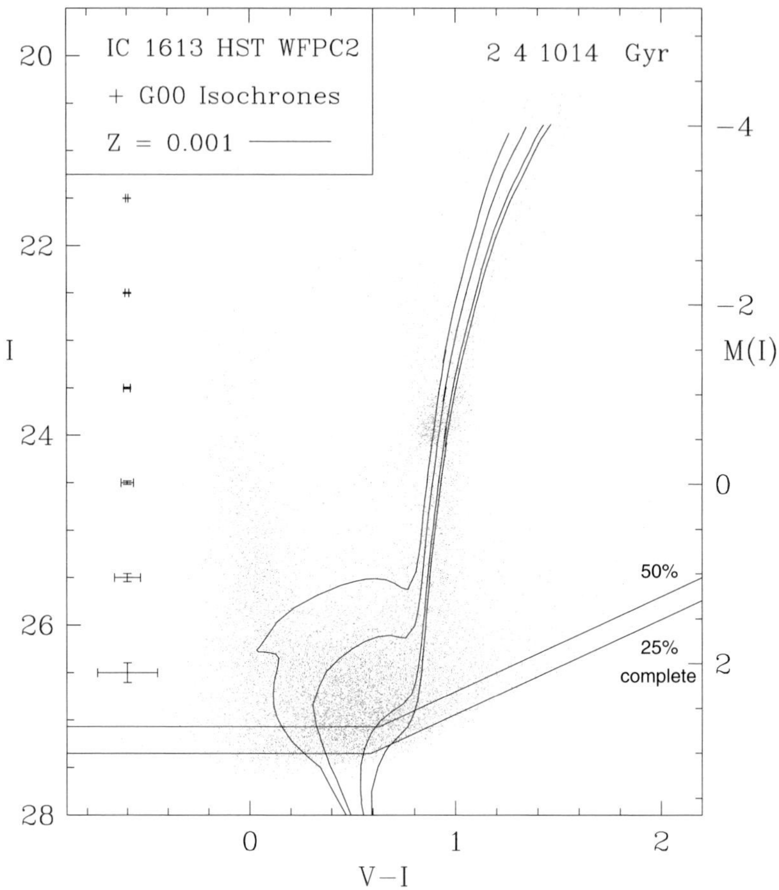

Figure 3. CMD of IC 1613 derived from HST WFPC2 observations. Isochrones for a metallicity of $Z = 0.001$ and ages of 2, 4, 10, and 14 Gyr from Girardi *et al.* (2000) have been added in order to show the limitations of the observations in terms of MS ages. MS turnoffs back to intermediate ages (~5 Gyr) are well represented in the observations, but the oldest MS turnoffs (~10 Gyr) fall below the 50% completeness limit and are not represented. Thus, constraints on the oldest populations will need to come from the evolved stars.

study of IC 1613 ($M_V \simeq +3.4$). Given the density-morphology relationship in the Local Group, the dI galaxies are at much greater distances than the dSphs, and, thus, have correspondingly shallower CMDs. As a result, it has been difficult to make direct comparisons between the CMDs of dI and dSph galaxies.

The complete SFH for the halo field in IC 1613 is shown in Figure 4. The main feature seen in the SFH is an extended event from ~2 Gyr ago until ~6 Gyr ago. While there has been star formation since that event (a significant amount coming 0.5 Gyr ago), the bulk of the stars in this region of IC 1613 come from the earlier age. Although Dolphin *et al.* 2001 found RR Lyraes in this field, the ancient ($\geqslant$10 Gyr) SFR was well below the lifetime average. Thus, to first order, star formation in IC 1613 appears to have occurred predominantly at intermediate ages. This is also characteristic of several of the outer MW dSph satellites (specifically Carina, Fornax, Leo II and Leo I). A particularly interesting (and to me surprising) comparison is between IC 1613 and Leo I; both appear

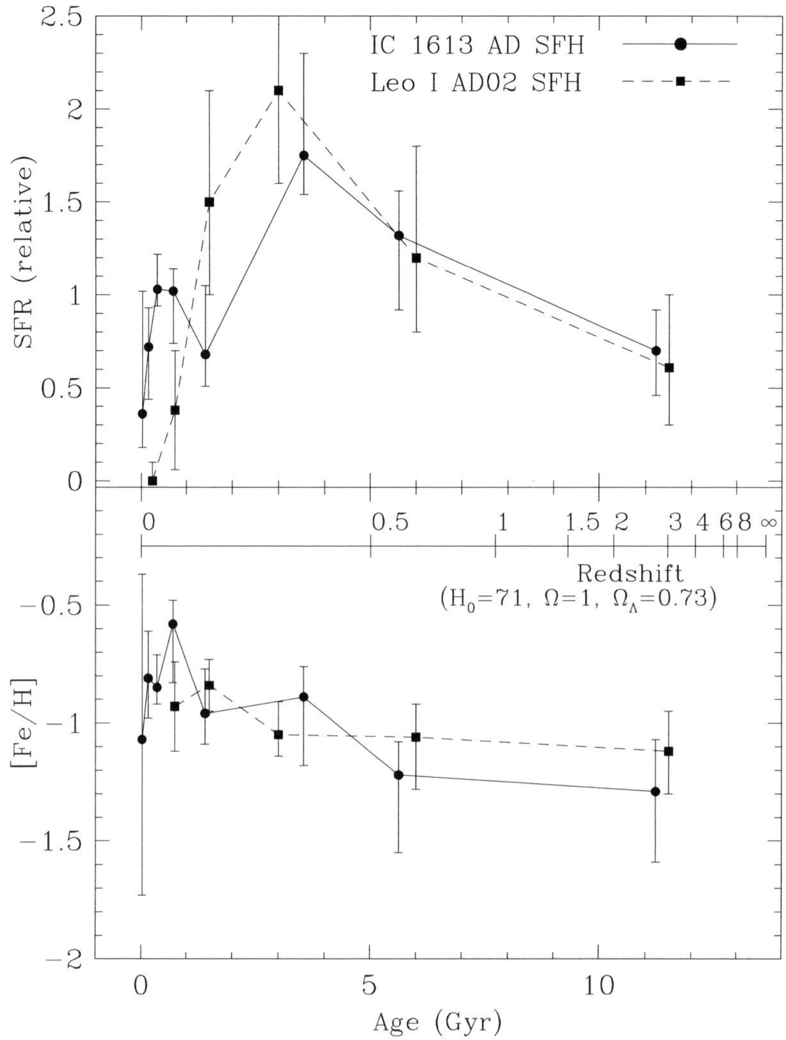

Figure 4. Comparison of SFHs and metal enrichment histories for IC 1613 and the dSph Leo I (both derived via the Dolphin method, Dolphin 2002; Skillman *et al.* 2003). The star formation and metal enrichment histories appear nearly identical. A timeline comparing redshift to real time has been added for the noted cosmology. Note that the bulk of the star formation and chemical enrichment has occurred at $z < 1.0$.

to be dominated by star formation at the same intermediate ages (Gallart 1999; Dolphin 2002). In Figure 4 we compare the SFH and age metallicity relationship for IC 1613 and Leo I (both using the method of Dolphin 2002). Figure 4 shows that the SFHs and AMRs for IC 1613 and Leo I, when derived via identical methodology, are nearly identical.

One possible interpretation of Figure 4 is that, absent the youngest stars, is it possible that there are no differences between the stellar populations of isolated dI and dSphs which are more distant from their parent galaxies. The implication is that some dI and dSph galaxies have similar progenitors; the differences which we see today are due to environmental influences during the lifetimes of the galaxies which allow one type of

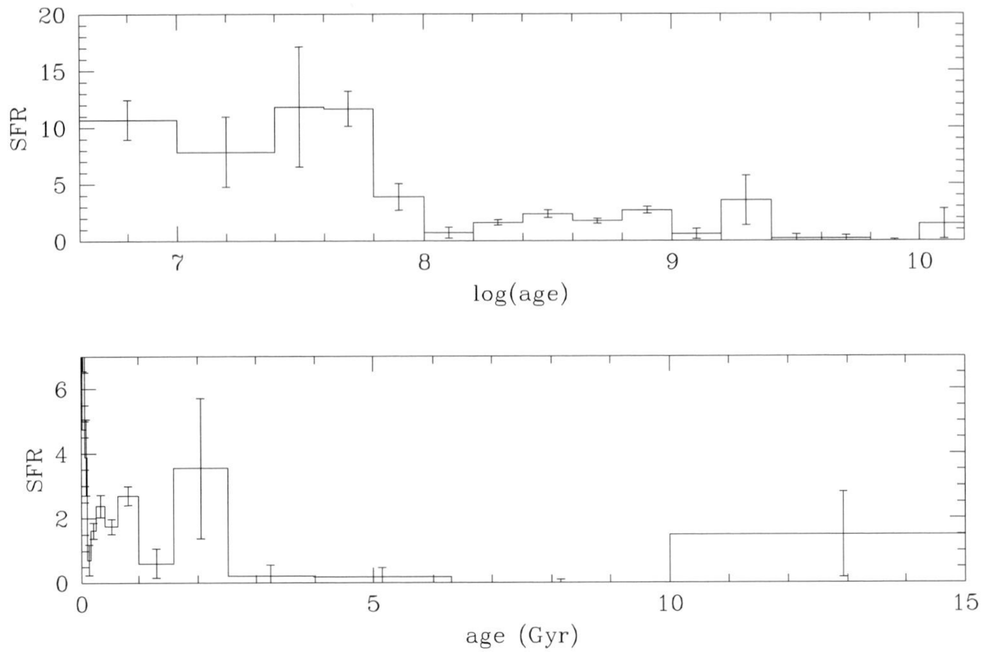

Figure 5. The reconstructed star formation history of Sextans A (from Dolphin *et al.* 2003). The top panel is on a logarithmic time scale; the bottom panel has a linear scale. Both panels show the star formation history normalized to a lifetime average of 1.0. Note the relative paucity of star formation at intermediate ages.

galaxy to retain its gas and form stars up until the present and another not. Certainly the morphological census of the Local Group has evolved with time. One possible environmental factor is the radiation background. The ionizing background can heat the ISM in low mass halos and reduce the rate of radiative cooling by reducing the number of neutral atoms. The photoionizing background will reduce the fraction of gas which collapses, suppressing star formation until later times ($z \leqslant 1$) when then background decreases and the gas can cool (see discussion in Skillman *et al.* 2003).

There are other dI galaxies for which it has been suggested that most of the stars have been formed relatively recently. Our earlier study of Leo A (Tolstoy *et al.* 1998) suggested that the majority of the star formation had occurred in the last 2 Gyr. Dolphin *et al.* (2002) discovered RR Lyraes in Leo A, and thus the presence of very early star formation, but converting the number of RR Lyrae stars to an early SFR is very uncertain. A similar SFH is found for Sextans A. From a relatively deep ($M_V \simeq +2.0$) CMD, Dolphin *et al.* (2002) find that while there is evidence for very old stars in Sextans A, the SFR at intermediate ages (3–10 Gyr) was quite low, and the SFR has been the highest in the last 2 Gyr (see Figure 5).

3. Three Channels for dE Formation

From the above discussion, we can conclude that there are at least three channels available for the formation of a dE galaxy from a dI galaxy, and all three are quite likely:

• Stripping of gas from dIs to form dEs in the cluster environment seems to be inevitable and would naturally explain the density-morphology relationship. (Perhaps the same is true in the group environment?) Given the result of the ACS Virgo Cluster Survey presented by Cote at this conference that a very high fraction of the Virgo dEs are nucleated, it would appear that the formation of a nucleus is a natural by-product of this process.

• Blow-away is an energetically favorable process, but it does not produce the density-morphology relationship. Nonetheless, it is likely to happen some times (preferentially in the lowest mass systems, Mac Low & Ferrara 1999; Ferrara & Tolstoy 2000).

• Although some dEs are now known to show significant rotational support, tidal stirring (Mayer *et al.* 2001a,b) represents another promising channel for the conversion of dIs to dEs. This process may be the most likely for the non-rotating dEs.

4. Why it Doesn't Matter

Why isn't the signature of the process which removes the gas obvious in the structural parameters of a dE? Perhaps the uniformity of the dark matter halos dominates all other factors, so, in the end, all low mass systems without cold gas end up looking the same (Dekel & Silk 1986)? Over the last two decades there have been a number of studies aimed at understanding the relationship between the dE and dI galaxies. Our inability to unambiguously identify the gas removal process on a galaxy-by-galaxy basis is the fundamental impediment to solving this long lived problem.

5. Conclusions

All dwarfs which have been suitably observed show evidence of a population of stars with ages ~10 Gyr. The star formation histories for these galaxies show a large variety. This has been known for some time now for the Milky Way dSph companions, but it now appears to be true also for dIs. The star formation histories may be revealing the effects of the x-ray background radiation responsible for re-ionization (i.e, delayed galaxy formation, squelching, suppression). Since there are probably several channels for dI to dE conversion, the comparison of properties of dEs and dIs hold great promise for telling us about the effects of environment on galaxy evolution.

Acknowledgements

I would like to thank the conference organizers for inviting me to give this talk and for organizing a very lively conference. This review covers much of the work of very fruitful collaborations, for which I am grateful to have been part of. I gratefully acknowledge partial support of my research from a NASA LTSARP grant No. NAG5-9221 and the University of Minnesota.

References

Bender, R., Paquet, A. & Nieto, J.-L. 1991, *A&A* 246, 349
Binggeli, B. & Cameron, L.M. 1991, *A&A* 252, 27
Binggeli, B., Tarenghi, M. & Sandage, A. 1990, *A&A* 228, 42
Caldwell, N. 1983, *AJ* 88, 804
Caldwell, N. & Bothun, G.D. 1987, *AJ* 94, 1126
Dekel, A. & Silk, J. 1986, *ApJ* 303, 39
De Rijcke, S., Dejonghe, H., Zeilinger, W.W. & Hau, G.K.T. 2001, *ApJ* 559, L21
De Rijcke, S., Dejonghe, H., Zeilinger, W.W. & Hau, G.K.T. 2003, *A&A* 400, 119
Dolphin, A.E. 2002, *MNRAS* 332, 91
Dolphin, A.E., *et al.* 2001, *ApJ* 550, 554

Dolphin, A.E., *et al.* 2002, *AJ* 123, 3154
Dolphin, A.E., *et al.* 2003, *AJ* 126, 187
Einasto, J., Kaasik, A. & Saar, E. 1974, *Nature* 250, 309
Faber, S.M. & Lin, D.N.C. 1983, *ApJL* 266, L17
Ferrara, A. & Tolstoy, E. 2000, *MNRAS* 313, 291
Gallart, C., Freedman,W., Aparicio, A., Bertelli, G. & Chiosi, C. 1999, *AJ* 118, 2245
Geha, M., Guhathakurta, P. & van der Marel, R. P. 2002, *AJ* 124, 3073
Geha, M., Guhathakurta, P. & van der Marel, R. P. 2003, *AJ* 126, 1794
Girardi, L., Bressan, A., Bertelli, G. & Chiosi, C. 2000, *A&AS* 141, 371
Kormendy, J. 1985, *ApJ* 295, 73
Mac Low, M. & Ferrara, A. 1999, *ApJ* 513, 142
Mateo, M.L. 1998, *ARA&A* 36, 435
Mayer, L., Governato, F., Colpi, M., Moore, B., Quinn, T., Wadsley, J., Stadel, J. & Lake, G. 2001a, *ApJ* 547, L123
Mayer, L., Governato, F., Colpi, M., Moore, B., Quinn, T., Wadsley, J., Stadel, J. & Lake, G. 2001b, *ApJ* 559, 754
Olszewski, E.W. & Aaronson, M. 1985, *AJ*, 90, 2221
Pedraz, S., Gorgas, J., Cardiel, N., Sánchez-Blázquez, P. & Guzmán, R. 2002, *MNRAS* 332, L59
Skillman, E.D. 1996, *ASP Conf.Ser.* 106: The Minnesota Lectures on Extragalactic Neutral Hydrogen, 106, 208
Skillman, E.D., Tolstoy, E., Cole, A.A., Dolphin, A.E., Saha, A., Gallagher, J.S., Dohm-Palmer, R.C. & Mateo, M. 2003, *ApJ* 596, 253
Tolstoy, E., *et al.* 1998, *AJ* 116, 1244
Tully, R.B. & Pierce, M.J. 2000, *ApJ* 533, 744
van Zee, L. 2001, *AJ* 121, 2003
van Zee, L., Barton, E.J. & Skillman, E.D. 2004b, *AJ* 128, 2797
van Zee, L., Skillman, E.D. & Haynes, M.P. 2004a, *AJ* 128, 121

Discussion

READ: To what extent is it fair to say that the Local Group dEs are the same objects as the cluster dEs, in particular with regard to the issue of rotation?

SKILLMAN: I think it is likely that they are very different populations.

DRINKWATER: You described the similarity in size and surface brightness between dI and dE galaxies. If the dIs become dEs by gas removal and hence reduction in star formation, shouldn't the dEs have lower surface brightness than the dIs?

SKILLMAN: If the H I mass is a small fraction of total mass the surface mass density will not be strongly affected. Given the dispersion in the relationships, I think that reasonable fading vectors can be proposed.

HENSLER: How do tidal-tail dwarfs fit into your picture of evolution, if they are void of dark matter?

SKILLMAN: If they have no dark matter, I would expect them to be short-lived, and therefore not observable outside of the tidal-tail environment.

Near-Field Cosmology with Dwarf Elliptical Galaxies
Proceedings IAU Colloquium No. 198, 2005
H. Jerjen and B. Binggeli, eds.

doi:10.1017/S1743921305003650

The effects of early reionization on the formation of dwarf galaxies

H. Susa[1] & M. Umemura[2]

[1]Department of Physics, College of Science, Rikkyo University, Nishi-Ikebukuro, Toshimaku, Tokyo, Japan
email: susa@rikkyo.ac.jp

[2]Center for Computational Science, University of Tsukuba, Tsukuba, Japan
email: umemura@rccp.tsukuba.ac.jp

Abstract. We consider the feedback effects of the early reionization on the formation of small galaxies. For this purpose, we perform 3D radiation hydrodynamic simulations with incorporating the radiative transfer for ionizing photons. As a result, it is found that the early reionization is so devastating for low mass systems with $M_{\rm vir} \lesssim 10^8 M_\odot$ or $v_{\rm circ} \lesssim 20$km/s, and almost all gas is photo-evaporated in more than 95% of low mass systems. These results indicate that the low mass dwarf galaxies are not formed directly from isolated CDM density perturbations.

Keywords. radiative transfer,methods: numerical,galaxies: dwarf,formation

1. Introduction

According to the latest cosmology, first galactic systems have formed at $z \sim 10$–30, and the masses of the systems are as large as those of low mass dwarf galaxies, i.e. $10^6 M_\odot \lesssim M \lesssim 10^8 M_\odot$. Because of such early formation epoch, the remnants of these first galactic objects should have very old stellar populations, which are in fact found in all of the dwarf galaxies in Local Group. Therefore, first generation galaxies directly formed from CDM density fluctuations are expected to be the most plausible candidate of the progenitor of present dwarf galaxies observed in our neighborhood.

On the other hand, the results from WMAP satellite tell us that the universe was reionized at very early epoch. Because of the negative feedback effects of the ionizing radiation field in very early universe, formation of low mass galactic systems might have suppressed severely.

In order to assess the effects of reionization on the formation of such systems, we have to take into account the effects of radiative transfer of ionizing photons, otherwise, we overestimate the negative feedback effects.

In this contribution, we present the 3D radiation SPH simulation on the formation of such low mass systems, especially taking into account the effects of radiative transfer and early reionization.

2. Numerical Methods

The detail of numerical method is provided in Susa & Umemura(2004a). In this section, we briefly describe the simulations done in this work.

Hydrodynamics is calculated by Smoothed Particle Hydrodynamics (SPH) method. We use the version of SPH by Umemura(1993) with the modification according to Steinmetz & Müller(1993), and also we adopt the particle resizing formalism by Thacker *et al.* (2000). The gravitational force is calculated by a special purpose processor for gravity calculation, GRAPE-6 (Makino 2002). In order to access GRAPE boards, we utilize the Heterogeneous Multi-Computer System (HMCS) (Boku *et al.* 2002) which enables us to use GRAPE in parallel processors such as PC clusters.

The softening length for gravity is set to be 20pc for all SPH and CDM particles. This value is determined by the convergence test. The non-equilibrium chemistry and radiative cooling for primordial gas are calculated by the code developed by Susa & Kitayama (2000), where H_2 cooling and reaction rates are taken from Galli & Palla (1998). Radiation transfer equation for ionizing photons is solved based upon the method proposed by Kessel-Deynet & Burkert (2000), which utilizes the neighbor lists of SPH particles to assess the optical depth from a certain source to an SPH particle.

The "star formation" conditions employed in this work is basically the same as Susa & Umemura(2004a), except that $c_* = 1$ is adopted in this paper. The assumption of $c_* = 1$ allows the star formation at physically acceptable maximal rate, because no stars should not form faster than local free-fall time scale.

As for ionizing flux, we employ the UV intensity $I_{21} = 0.01$ for $z < 17$, that allows the early reionization of the universe inferred by *WMAP* (Susa & Umemura(2004b)).

3. Results & Discussion

The summary of the numerical runs is shown in Figure 1, where M is the mass of virialized halo and z_c is the collapse epoch. In order to distinguish the fate of baryonic component, we evaluate the final mass-to-light ratio (M/L) of the formed small galaxies by assuming $M_{star}/L = 3$ for formed stars in solar units. Open circles denote $M/L > 100$ and filled circles $M/L < 100$. The dotted lines represent the collapse epoch of halos formed from $1\sigma, 2\sigma$ and 3σ CDM density fluctuations. Roughly 95% of fluctuations collapse after the epoch predicted by the 2σ line. Dashed lines denote the loci of constant circular velocities (5 km/s, 10km/s, 20km/s) of the collapsed halos.

Open symbols ($M/L > 100$) should be regarded as "failed" halos, because the maximal M/L in Local Group dSphs is roughly 100 (van den Bergh(1999), Mateo (1998), Hirashita, Takeuchi & Tamura(1998)). It is noted, however, the specific choice of the critical mass-to-light ratio of 100 is not so important, since the gradient of M/L on $z_c - M$ plane is quite steep.

From these results, it is clear that most of the baryonic component in the halos formed from relatively low density peaks ($\delta\rho/\rho \lesssim 2\sigma$) is blown away by the thermal pressure of photo-heated gas. On the other hand, in the halos at high density peaks ($\delta\rho/\rho \gtrsim 2\sigma$), gas components are efficiently converted to stars, avoiding the complete photo-evaporation caused by the reionization of the universe. Since the assumed star formation time scale (local free-fall time) is the shortest among the physically acceptable estimates, the mass of luminous matter in the halos should be regarded as maximal one. In other words, the obtained mass-to-light ratios of galaxies designate the lower limits. Therefore, the area in which the galaxies are "failed" in Figure 1 can be underestimated. If we adopt more realistic star formation scheme, the "failed" region would extend to higher redshift. However, it should be also noted that the gas in the halos with circular velocity $\gtrsim 20\text{kms}^{-1}$ does not evaporate even when it is photo-heated to $\sim 10^4$K, since the gravitational potential is deep enough to retain the ionized gas. Thus, it is concluded that, in more than 95% of the halos with $v_{circ} \lesssim 20\text{kms}^{-1}$, the galaxy formation is prohibited due to the early reionization feedback.

These results indicate that if we accept early reionization at $z > 17$, it is quite difficult for low mass dwarf galaxies to form ($M \lesssim 10^8 M_\odot$, especially dSphs) directly from isolated

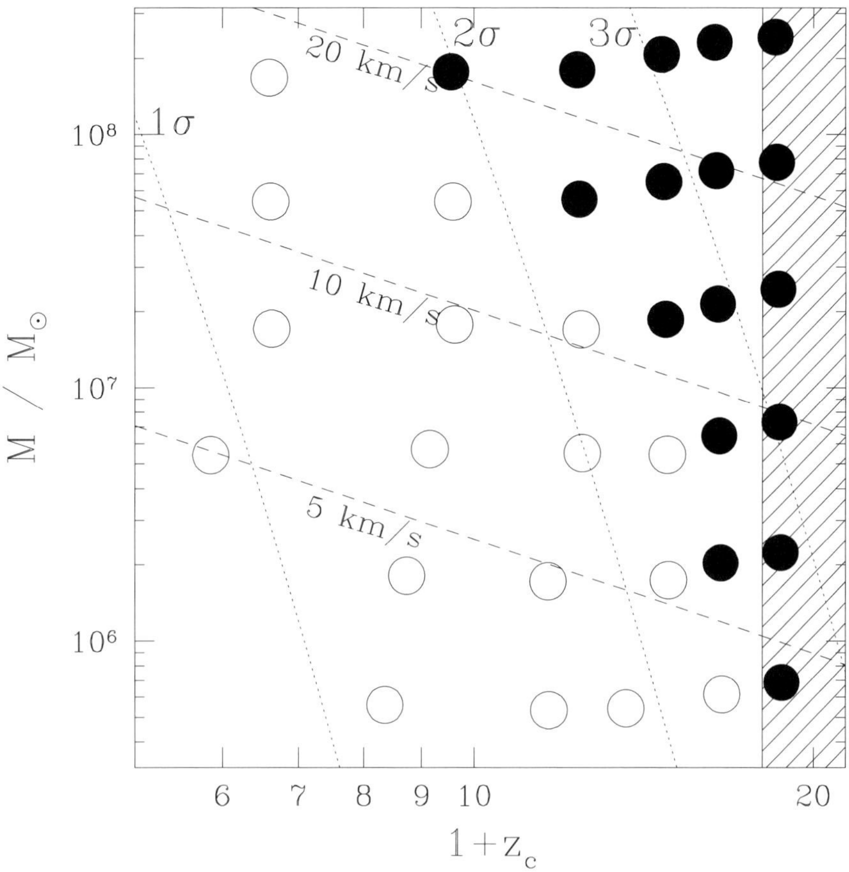

Figure 1. Summary of numerical runs are shown on collapse epoch v.s. virialized mass plane. Filled circles denote the runs in which the stars are formed effectively, and significant amount of baryonic matter remain in the halo potential($M/L < 100$). Open circles denote the runs in which almost all of the baryonic matters are lost($M/L > 100$). Dashed lines denote the loci of constant circular velocities (5 km/s, 10km/s, 20km/s) of the collapsed clouds, and the dotted lines denote the collapse epoch of CDM density perturbations whose amplitudes are 1σ, 2σ and 3σ. Hatched region in the right of the panel shows the pre-reionization era.

CDM density fluctuations, because 1) they are too small to retain the reionized gas, and 2) the evidence of star formation activity at $z \lesssim 5$ are found for these galaxies.

There are a few logical scenarios for the formation of such galaxies:

1. Reionization epoch is much later ($z \lesssim 10$). Thus, the radiative feedback is not important for the halos formed at $z \gtrsim 10$. However, even with such late reionization assumption, we need some other mechanism which can allow the star formation activity well after the reionization.

2. The circular velocities of dSphs are much larger as claimed by N-body simulations (Stoehr *et al.* 2002, Hayashi *et al.* 2003). If this is the case, photo-evaporation mechanism give the low mass limit of small galaxies ($M > 10^8 M_\odot$).

3. dSphs are embedded in tidally stripped halos, that is originally more massive (Kravtsov, Gnedin & Klypin 2004). Kravtsov *et al.* found that 10% of subgalactic halos

were formed through the tidal stripping of rather massive halos ($M \sim 10^9 M_\odot$). They suggest that dSphs are formed in such halos, not from the small scale ($M \lesssim 10^8 M_\odot$) density perturbations imprinted in the very early universe. This suggestion could be consistent with the strong negative feedback effects of early reionization and the observed star formation history of dSphs.

Acknowledgements

The HMCS has been developed in a project which Center for Computational Physics, University of Tsukuba propelled in the course of JSPS Research-for-the-Future program of Computational Science and Engineering. The analysis has been made with computational facilities at Center for Computational Science in University of Tsukuba and Rikkyo University. We acknowledge Research Grant from Japan Society for the Promotion of Science (15740122:HS, 16002003:MU).

References

Boku, T., Makino, J., Susa, H., Umemura, M., Fukushige, T. & Ukawa, A. 2002, *IPSJ Transactions on High Performance Computing Systems* 41, 5
Galli D. & Palla F. 1998, *A&A* 335, 403
Hayashi, E., Navarro, J., Taylor, J.,Stadel, J. & Quinn, T. 2003, *ApJ* 584, 541
Hirashita, H., Takeuchi, T.T. & Tamura, N. 1998, *ApJL*, 504, L83
Kessel-Deynet, O. & Burkert, A. 2000, *MNRAS* 315, 713
Kravtsov,A., Gnedin,O. & Klypin, A., 2004, *astro-ph(0401088)*
Makino, J. 2002, *ASP Conf. Ser. 263: Stellar Collisions, Mergers and their Consequences*, 389
Mateo, M. 1998, *ARA&A* 36, 435
Steinmetz, M. & Müller, E. 1993, *A&A* 268, 391
Susa, H. & Kitayama, T. 2000, *MNRAS* 317, 175
Susa, H. & Umemura, M. 2004, *ApJ* 600, 1
Susa, H. & Umemura, M. 2004, *ApJ* 610, L5
Stoehr, F., White, S.D.M., Tormen, G. & Springel, V., 2002, *MNRAS* 335, 84
Thacker, J., Tittley, R., Pearce, R., Couchman, P. & Thomas, A. 2000, *MNRAS* 319, 619
Umemura, M. 1993, *ApJ* 406, 361
van den Bergh, S. 1999, *A&ARv* 9, 273

Discussion

FERGUSON: How does the ionizing photon density from the dwarf galaxies in your model compare to the ionizing background you assumed at the start (from WMAP)?

SUSA: In my model the intensity of ionizing photons is set high enough to reionize the universe at $z \approx 17$. Because the radiation field is treated as external radiation field, the ionizing photons from the dwarfs are not included. At present we haven't compared the intensity from dwarfs with the assumed background intensity. But it might be possible.

LIN: If the background UV radiation is supplied by massive stars, its flux would be self-regulated. The evaporated gas may be episodically cooled to fall back to the minihalos of dwarf galaxies. Have you considered the case where individual dwarf ellipticals are not isolated but embedded in a much larger intergalactic medium complex?

SUSA: If the source of ionizing radiation is massive stars, self-regulation might be possible. However, if the radiation is supplied by black holes or other mechanisms, self-regulation doesn't work. Concerning the last part of your question, the answer is no, I haven't.

Near-Field Cosmology with Dwarf Elliptical Galaxies
Proceedings IAU Colloquium No. 198, 2005
H. Jerjen and B. Binggeli, eds.

doi:10.1017/S1743921305003662

Making compact elliptical satellite galaxies: a conceptual model for M32

John S. Gallagher, III[1], Eva K. Grebel[2] and Linda J. Smith[3]

[1]Dept. Astronomy, U. Wisconsin, 475 N. Charter St. Madison, WI 53760, USA
email: jsg@astro.wisc.edu
[2]Astronomisches Institut, Venusstrasse 7, CH-4102, Binningen, Switzerland
[3]Dept. Physics & Astronomy, Gower St., U. College London, London WC1E 6BT, UK

Abstract. M32 and other compact dwarf galaxies present interesting evolutionary puzzles. Why are most dwarf galaxies low density systems while a very few, like M32, have the highest stellar densities of any galaxies? We suggest that gas capture should be added to the tidal stirring mix as a way to make high density M32-type dwarfs starting with relatively ordinary satellite galaxies. This paper presents some of the basic features and issue associated with gas capture models.

Keywords. galaxies: dwarf, galaxies: evolution, galaxies: individual (M32, NGC 1510), galaxies: structure

---

1. Introduction

The formation of galaxies requires the production of appropriately massive gravitationally bound structures and gas dissipation to make the stars which inhabit all galaxies. In this regard lower mass galaxies present an interesting diversity, as they commonly include the least dense stellar systems, which are inferred to have experienced minimal dissipation. Thus the dSphs may sit on the boundary between objects which become galaxies, and those which fail. Even though the dEs, such as the NGC 147-185-205 satellites of M31 have higher surface densities than the dSphs, they are still relatively puny systems. The rare compact dwarf ellipticals, such as M32, however, sit at the other extreme and are in effect high density galactic cannon balls. This paper briefly presents some ideas about how such high density galaxies might evolve and discusses why they are important.

M32 and the dE NGC205 have similar optical luminosities ($M_V \approx -16$) and thus contain roughly the same stellar masses ($\sim 10^9$ $M_\odot$). M32 is *much* more compact with an effective radius that is 0.18 that of NGC 205. The central density of M32 exceeds that of the NGC 205 star cluster nucleus by a factor of ~100 (Choi *et al.* 2002) and it even lies above the central densities of M31 and M33 (Lauer *et al.* 1998). M32 likely falls near the maximum stellar density limit for galaxies, and probably has a $\sim 10^6$ $M_\odot$ nuclear black hole (Joseph *et al.* 2001).

Despite its unusual structure, M32 seems to have had a comparatively normal star formation history. Star formation in its central region probably continued until 3–5 Gyr ago, while the outer stellar populations tend to be older (Rose *et al.* 2005, Worthey 2004, Worthey *et al.* 2004). A range in mean stellar metallicity exists, falling from near solar at the nucleus to modestly subsolar at larger radii. The abundance ratio of the α-element Mg to Fe is solar or below. The extended period of star formation in M32 is typical of dEs, with the caveat that M32 has higher stellar metal abundances relative to its current stellar mass.

How M32, a galactic "swan," ended up among the ducks and geese of the M31 companion dSphs and dEs is a long standing puzzle. One constraint is set by the rarity of M32-type galaxies; their production should be uncommon and/or they should be short-lived. The existing models can be roughly divided into three general classes:

- Formation oddities: M32 owes its existence to highly unusual conditions at the time of its formation. For example, Burkert (1994) suggested that it could be the result of the collapse of a galaxy within the pre-existing tidal field of a giant galaxy.
- Transformation of a normal galaxy: This has historically been the most popular model for M32 (Faber 1973; see also Nieto & Prugniel 1987, Choi *et al.* 2002 and references therein). Recent extensions of this model include the harassed spiral picture developed by Bekki *et al.* (2001).
- Mergers of star clusters: If dense star clusters formed in a starburst clump merge, then they can make some classes of ultra-compact dwarfs (e.g., Fellhauer & Kroupa 2005). However, M32 seems too massive for this path to apply; one would need to merge thousands of super star clusters to make M32.

Here we start from the fundamental concepts of the Nieto & Prugniel (1987) model and also consider the role of gas capture in the evolution of dwarf companion galaxies. At the outset we should emphasize that more than one path may exist to the production of compact E dwarfs; one model may not fit all cases (see §4 and Bekki *et al.* 2001 for discussion of issues).

2. Gas Accretion and Starbursts in Dwarf Satellites

While gas loss from dwarfs orbiting in the halos of giant galaxies are frequently discussed, the possibility that some dwarf companions can capture gas from their giant hosts receives less attention. Hawarden *et al.* (1979) invoked gas capture by NGC 1510, a satellite interacting with the spiral NGC 1512, to explain the disturbed nature of NGC 1510. Their model assumed Bondi-Hoyle accretion, $dM_g/dt \propto M_{sat}^2 \overline{\rho_{ISM}}/V_{rel}^3$ where $\overline{\rho_{ISM}}$ is the mean galactic ISM gas density and V_{rel} is the velocity of the satellite through the ISM. Thus high gas densities and, especially, low V_{rel} favor accretion. However, galaxies moving with V_{rel} that are characteristic of galactic orbital velocities probably will be stripped rather than accrete gas (e.g., the Magellanic Stream).

The optical structure of the NGC 1510/12 system shows that the gas-accretion situation may be more favorable than suggested by the Bondi-Hoyle models. A tidal arm appears to extend from NGC 1512 to NGC 1510 (Figure 1), a circumstance that is seen in other interacting starburst satellites, including NGC 3077, the starburst companion to M81. Presumably these interacting satellites are on eccentric prograde orbits with low inclinations where coupling to the disk of the giant system is enhanced (e.g., Walker *et al.* 1996). The observations further suggest that matter from the arm is ending up in the dwarf. This process resembles the assembly of a new dwarf galaxy in a tidal tail as described by Duc *et al.* (2004), with the added advantage that a gravitationally bound object is already present to collect any gas that flows into the area.

A process that fosters intense modes of star formation also is needed if we are to achieve high stellar densities. Fortunately the observations suggest that star formation in accreting dwarfs is usually intense and centrally concentrated (e.g., Figure 2). This effect may be aided by the tendency for interacting disk galaxies to form transient bars (Miwa & Noguchi 1997), which funnel gas inwards and feed nuclear starbursts. Furthermore, in orbiting galaxies we expect multiple starbursts associated with peri-galacticon passages (Kravtsov *et al.* 2004). The combination of the possible loss of ejecta from SNe II (e.g.,

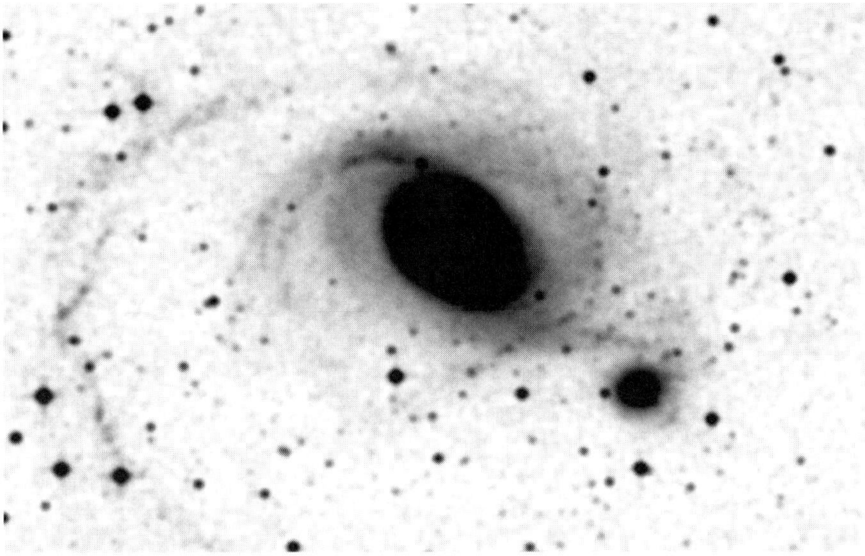

Figure 1. Digitized Sky Survey (DSS) image of the NGC 1512 spiral and NGC 1510 companion galaxy (below right) showing the luminous bridge connecting the two systems. Similar bridges are seen in HI and stars in other starburst satellites.

Ott *et al.* 2003) and averaging over multiple bursts (Gilmore & Wyse 1991) can lead to solar or even subsolar α-element to iron ratios, as observed in M32.

3. Evolution of the System

The long term evolution of the accreting dwarfs should be quite interesting. They initially will grow in mass, and their dynamical friction time scales ($t_{df} \propto M_{sat}^{-1}$) at first get shorter (e.g., Taylor & Babul 2001). But eventually competing effects must come into play, which may include exhaustion of the outer disk gas supply for accretion and the beginnings of tidal mass loss as the orbital radius shrinks. At this point we expect that mass growth will cease, and the companion galaxy would enter a phase of its evolution primarily driven by dynamical processes: tidal friction and tidally induced mass loss (see Kravtsov *et al.* 2004).

Satellite galaxies orbiting with low inclinations to a disk in prograde directions are not only well positioned to capture gas, but also lose angular momentum relatively rapidly due to resonant coupling to the disk (Walker *et al.* 1996). As a result, their dynamical friction time scales are reduced. If the satellite moves inwards, then the kinds of tidal effects invoked in galaxy stripping models (Nieto & Prugniel 1987) become increasingly important. However, in cases where earlier star formation produced high density central regions, the galaxy will be quite durable and a well-defined remnant can persist into the final phases of the capture process.

This extension of the transformed galaxy model differs from the purely tidal models in suggesting that under some circumstances companion galaxies can grow to high central densities that are fed by gas capture and subsequent starburst. Thus the system that enters into the final capture by its host is already unusually dense, a requirement for the M32 progenitor (Choi *et al.* 2002). This factor then joins the time scale arguments put forward by Nieto & Prugniel to explain the rarity of M32-type compact dwarf ellipticals. Moderate mass galaxies will have orbital lifetimes of at most a few Gyr after orbiting

Figure 2. WFPC2 I-band image of the center of NGC 1510 showing the central concentration of star formation. The vertical space between the lines is 14 arcsec corresponding to 680 pc for D=10 Mpc to the system. This centrally condensed star forming pattern is also seen in other interacting companion galaxies.

close enough to their massive hosts to be heavily tidally stripped in their outer regions, thereby assuming compact morphologies.

4. Discussion and Open Issues

This M32 model resembles that of Bekki *et al.* (2001) in invoking special circumstances to drive gas inwards as fuel for a central starburst. The main difference is our assertion that gas capture is likely to play a role, which we believe is observationally supported. Indeed it seems reasonable that any model that produces concentrated starbursts in satellite galaxies sets the stage for the eventual formation of M32-like tidal remnants.

A second factor in favor of the capture model is the amount of gas that would be required to make the high density M32 galaxy. If we assume that the gas lost via winds about equals the amount of gas put into stars, then we need to have fed $\geqslant 3\times10^9$ $M_\odot$ of gas into the center of the progenitor system. This is a challenge for any model, though the outer gas reservoirs of major spirals, especially in their youth, may have been up to the task.

Naturally several issues remain to be resolved, including: (1) The central mass densities and stellar velocity dispersions in starburst satellite galaxies should be measured and compared with those in compact dwarf Es. We also need star formation rate determinations to assess time scales for stellar mass growth in their inner regions. (2) These are the galaxies with the highest baryonic densities. What sets this limit? And does the central black hole in M32 play a role in this? (3) The impact of extensive mass exchanges between satellite galaxies and the outer regions of their spiral galaxy hosts has yet to be considered. (4) What mass hosts are necessary to produce M32-type remnants under this model? For example, the Bekki *et al.* (2001) models required relatively massive spiral precursors, while M82 could be an example of a moderate mass precursor that has grown through gas capture. (5) Is the orbit of M32 suitable for gas capture, or nearly polar as adopted by Bekki *et al.* (2001)? (6) Additional numerical modeling of the circumstances under which dwarf galaxies can efficiently capture gas would be very helpful. In particular, these might allow us to understand the conditions, if any, under which late gas accretion could occur into the predicted large numbers of CDM mini-halos.

Acknowledgements

Thanks go to Bruno Binggeli and Helmut Jerjen for arranging this excellent workshop. We thank Tabetha Hole for helpful comments on this paper. JSG's research on dwarf galaxies is supported in part by NASA through grants administered by the Space Telescope Science Institute in conjunction with U.S. general observer programs, and by the National Science Foundation through grant AST-9803018 to the University of Wisconsin.

References

Bekki, K., Couch, W.J., Drinkwater, M.J. & Gregg, M.D. 2001, *ApJ* (Letters) 557, L39
Burkert, A. 1994, *MNRAS* 266, 877
Choi, P.I., Guhathakurta, P. & Johnston, K.V. 2002, *AJ* 124, 310
Duc, P.-A., Bournaud, F. & Masset, F. 2004, *A&A* 427, 803
Faber, S.M. 1973, *ApJ* 179, 423
Fellhauer, M. & Kroupa, P. 2005, *MNRAS* 359, 223
Gilmore, G. & Wyse, R.F.G. 1991, *ApJ* (Letters) 367, L55
Hawarden, T.G., van Woerden, H., Goss, W.M., Mebold, U. & Peterson, B.A. 1979, *A&A* 76, 230
Joseph, C.L. *et al.* 2001, *ApJ* 550, 668
Lauer, T.R., Faber, S.M., Ajhar, E.A., Grillmair, C.J. & Scowen, P.A. 1998, *AJ* 116, 2263
Kravtsov, A.V., Gnedin, O.Y. & Klypin, A.A. 2004, *ApJ* 609, 482
Miwa, T. & Noguchi, M. 1998, *ApJ* 499, 149
Nieto, J.-L. & Prugniel, P. 1987, *A&A* 186, 30
Ott, J., Martin, C. & Walter, F. 2003, *ApJ* 594, 776
Rose, J.A., Arimoto, N., Caldwell, N., Schiavon, R.P., Vazdekis, A. & Yamada, Y. 2005,*AJ* 129, 712
Taylor, J.E. & Babul, A. 2001, *ApJ*, 559, 716
Walker, I.R., Mihos, C. & Hernquist, L. 1996 *ApJ* 460, 121
Worthey, G. 2004, *AJ* 128, 2826
Worthey, G., Mateo, M., Alonso-García, J. & España, A.L. 2004, *PASP* 116, 295
Young, L. 2000, *AJ* 119, 188

Discussion

LIN: With the shallow potential of satellite galaxies it is difficult to accrete gas unless the relative speed between them and the gas is less than the internal velocity dispersion.

GALLAGHER: Agreed. Only satellites on prograde orbits, presumable near the disk plane, should be able to accrete much gas. One caveat is how dwarfs interact with tidal spiral arms – this may increase accretion efficiencies.

MIESKE: Would you expect formation of compact dEs via gas accretion in very massive clusters, where you have high relative velocities?

GALLAGHER: No, because you need low relative velocities. Ram-pressure stripping seems to be the key factor for evolution of cdEs in massive clusters.

Near-Field Cosmology with Dwarf Elliptical Galaxies
Proceedings IAU Colloquium No. 198, 2005
H. Jerjen and B. Binggeli, eds.

doi:10.1017/S1743921305003674

Faint blue galaxies revisited

Henry C. Ferguson

Space Telescope Science Institute, 3700 San Martin Drive, Baltimore, MD 21218, USA
email: ferguson@stsci.edu

Abstract. If dwarf-elliptical galaxies formed their stars very rapdily (on timescales of less than 1 Gyr), they may in principle be detectable out to high redshift. Prior to the discovery of cosmic acceleration, it appeared that rapid and late formation dwarf elliptical galaxies might be *required* to explain the number counts of faint galaxies. A plausible hypothesis emerged: that photoionization by the UV background prevents gas cooling in low-mass halos until $z \lesssim 1.5$. The discovery of cosmic acceleration eased the tension between predicted galaxy number counts and galaxy-evolution models. Nevertheless, there is some evidence for relatively late star formation in nearby dE's, and the photoionization delay mechanism still appears to have some merit. It is thus of interest to look back in time to see if we can find starbursting dwarf galaxies at moderate redshift. We review the connection between faint-blue galaxies and bursting-dwarf galaxies and discuss some attempts to identify progenitors to dE galaxies in the Hubble Ultra Deep Field (HUDF) observations. We find roughly 85 galaxies in the HUDF with redshifts $0.6 < z < 1.1$ that appear to have formed most of their stars at $z < 1.5$. Of these, 70% have half-light radii less than 1.5 kpc. These are thus "smoking gun" candidates for dwarf galaxies that are either collapsing for the first time at moderate redshifts or have otherwise been unable to form stars for more than 1/3 of a Hubble time.

Keywords. galaxies: dwarf, galaxies: evolution

1. Introduction

A variety of interesting cosmological issues converge when we consider the evolution of dwarf elliptical galaxies. There are some strong tensions between observations and theory, and some serious unknowns in the physical processes that might be shaping these galaxies. This of course makes the topic interesting and provides fodder for excellent conferences in beautiful places!

Among the various tensions, a few are relevant to studies of distant galaxies because they determine how many we faint galaxies should expect to see and how bright they should be at high redshift.

(*a*) The dE galaxies that we obseve in the local universe show a luminosity function $\phi(L)dL$ continues to rise to the limits of our ability to detect galaxies (Trentham, Sampson & Banerji 2005). However, the slope of this function is significantly shallower than the predicted slope of the mass-function of dark-matter halos (e.g. Moore *et al.* 1999). Either something has supressed star formation in low-mass halos, or the halo mass function itself deviates from the predictions. This missing-dwarf problem is one aspect of the "substructure crisis" that currently challenges Cold-Dark Matter (CDM) theory.

(*b*) Certain observations favor rapid star-formation in dE galaxies. In particular, the reasonably well-defined metallicity–luminosity relation (summarized by Dekel & Woo 2003) is easiest to explain if star formation in dE galaxies was truncated by supernova-driven winds. Such winds are efficient if the star-formation episode is rapid enough (a few $\times 10^7$ yr) to allow overlap of expanding supernova remnants (Dekel & Silk 1986). However, recent observations suggest that local-group dSph galaxies have α/Fe element ratios indicative of significant enrichment by type Ia supernovae (e.g. Venn *et al.* 2004). The characteristic timescale for enrichment by type Ia SNe is generally thought to be $\sim$1 Gyr. Furthermore, evidence for rotational support and/or cold gas in some dE's suggests that star formation was not rapid and may not have run to completion (e.g. Geha *et al.* 2003; Conselice *et al.* 2003; van Zee *et al.* 2004; Michielsen *et al.* 2004).

(*c*) One possible solution to the missing dwarf problem is that star-formation in low-mass halos has been "squelched" due to photoionization by the metagalactic UV background (Efstathiou 1992; Babul & Rees 1992; Babul & Ferguson 1996; Somerville 2002). Photoionization prevents cooling of HI and the subsequent H_2 formation needed to form stars. The UV background becomes important at redshifts $6 < z < 20$, and drops significantly from $z = 1$ to the present (Scott *et al.* 2002). A possible result is that at least some low-mass halos retain their gas, but do not form many stars until 4–7 Gyr after reionization. Unfortunately, while the theory seems well founded, there is very little observational evidence to support it. The star-formation histories of local-group dSph galaxies do not suggest a preferred epoch to their formation either prior to reionization or at low redshifts. Stellar population indicators for cluster dE's show more diversity than expected if such galaxies formed all their stars *before* reionization, but do not show a clear indication of a long, coherent hiatus in star formation. Dwarf irregular galaxies of course do show evidence for recent star formation, but it is difficult to determine star-forming histories over the timespan 6–12 Gyr in the past.

In summary, while dE (and dSph) galaxies appear to be the most common type of galaxy in the local universe, there do not appear to be enough of them. Those that we see have some well-defined systematic trends, suggestive of some common mechanism that regulates star formation and feedback. But there is considerable diversity in ages and evidence increasingly disfavors models where the star formation occured in short bursts.

Deep surveys may help sort things out by allow us to detect and study the progenitors of today's dE galaxies at significant lookback times.

1.1. *A brief history of faint-blue galaxies*

The idea of testing world models by counting galaxies goes back to the early part of the 20th century. Counting galaxies became an industry in the late 1970s to late 1990s as bigger telescopes, better detectors, and increasingly sophisticated measurement techniques provided large gains in sensitivity (Kron 1978; Tyson 1988; Koo & Kron 1992; Ellis 1997). These surveys revealed that the number counts of galaxies were in excess of expectations from prevailing cosmological models (with $\Omega_M = \Omega_{\rm tot} = 1$), and that this excess was most pronounced at blue wavelengths (Yoshii & Takahara 1988; Guiderdoni & Rocca-Volmerange 1991). The redshift distribution was relatively broad and peaked at redshifts $z < 1$ (Cowie *et al.* 1991; Crampton *et al.* 1995). The faint-blue galaxies did not appear to be strongly clustered (Efstathiou 1995).

The steep CDM halo mass function presented a possible solution. If star-formation in low-mass halos were delayed until $z < 1$ by the UV background, and if stars subsequently formed on timescales $\sim 10^7$ yr, then these small dwarfs could in principle dominate the faint-blue galaxy number counts (Efstathiou 1992; Babul & Rees 1992). However in the $\Omega_m = 1$ SCDM model, Ferguson & Babul (1998) showed that these dwarfs would fade too slowly to be consistent with the counts and colors of galaxies in the Hubble Deep Field. Thus, while late formation of dE galaxies had the potential to solve the faint-blue galaxy problem, the extreme model required to reconcile SCDM with the number counts did not appear to work without some additional tweaking (e.g. adopting a steep IMF).

The current ΛCDM concordance cosmological model with $\Omega_{\rm tot}, \Omega_m, \Omega_\Lambda = 1.0, 0.3, 0.7$ has both more volume and more time at high redshifts than the previously-favored model.

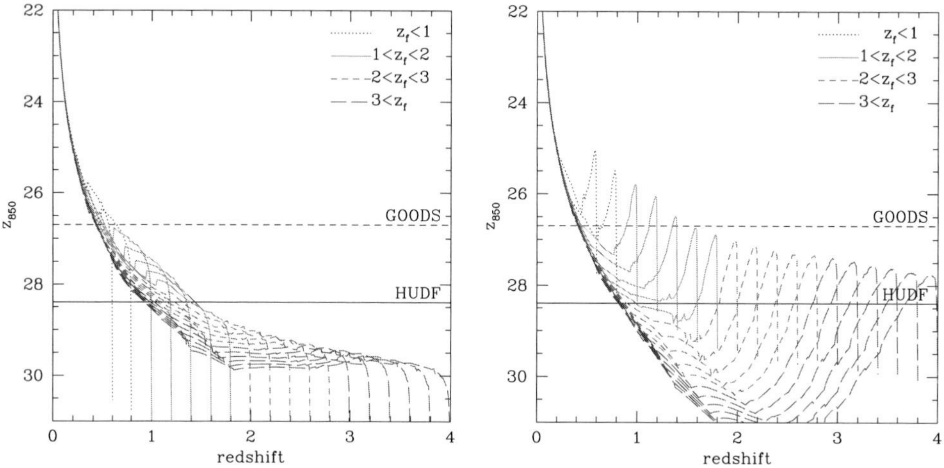

Figure 1. (Left) Apparent HST z_{850} magnitude as a function of redshift for dE galaxies forming at a variety of redshifts with an exponential star-formation timescale $\tau_{\rm SF} = 10^9$ yr. The dashed line shows the magnitude limit for GOODS for a 0.2 square arcsecond aperture. The solid line shows the detection limit for the HUDF. (Right). Same for galaxies with $\tau_{\rm SF} = 10^8$ yr.

Galaxy counts are therefore not nearly as interesting as they used to be. Few papers have been published in the past five years comparing modern ΛCDM hierarchical models to galaxy counts. Nagashima *et al.* (2002) show a comparison to one set of models indicating good, but not perfect, agreement.

1.2. *The visibility of dE galaxies*

Consider a dE galaxy with a present-day absolute magnitude $M_B = -14$. The roughly $10^9 M_\odot$ of stars could have formed within the $\sim 10^7$ yr timescale of supernova driven winds, in which case it might have had a peak star-formation rate of $100 M_\odot$ yr^{-1}, comparable to typical Lyman-break galaxies seen at redshift $z \sim 3$. Alternatively, the galaxy might form stars on the 10^8 to 10^9 yr timescales predicted by models with a gentler form of self regulation (Lin & Murray 1992; Hensler *et al.* 2004). In Fig. 1 we show apparent magnitude vs. redshift for dwarf galaxies forming at a variety of redshifts with star-formation timescales of 10^9 yr (left) and 10^8 yr (right), all of which end up at $M_B = -14$ at the present epoch. The HST Advanced Camera for Surveys (ACS) z-band magnitude limits are shown for Great Observatories Origins Deep Survey (GOODS) and the Hubble Ultra-Deep Field (HUDF). Independent of the star-forming history, a dE galaxy would be visible in GOODS to $z \sim 0.4$ and in the HUDF to $z \sim 0.8$. If the star-formation timescale is longer than 10^9 yr, even the HUDF would not be deep enough to see the formation phase of this hypothetical dE if it takes place at $z > 1.5$. On the other hand, if the star-formation timescale is 10^8 yr, the HUDF is deep enough to see the formation phase out to $z > 4$.

1.3. *Emission-line searches bursting dwarfs*

The magnitude limit of unbiased redshift surveys with Keck and the VLT is $R \sim 24$, which is too bright for our proto-typical dE galaxy at $z > 0.5$ even with star-formation timescales as short as 10^7 yr. We must rely on emission lines and/or photometric redshifts to find these galaxies and constrain their redshifts. For redshifts $z > 0.4$, Hα line becomes increasingly difficult to detect from the ground through the forest of night-sky lines. The HST NICMOS grism survey found about 0.5 Hα emitters per square arc minute with typical star-formation rates $\sim$20 $M_\odot \mathrm{yr}^{-1}$ in the redshift range $0.7 < z < 1.9$ (McCarthy *et al.* 1999). In visual appearance, at least some of these galaxies look compact enough to be dwarfs. What we do not see in this survey is a pileup of emission line galaxies at $z < 1$, which might have signaled a renewed epoch of dwarf-galaxy formation.

The [OII] 3727Å line, while a less directly correlated with star-formation, is accessible in the optical out to $z \sim 1.4$. Lotz, Martin & Ferguson (2003) carried out an [OII] emission-line search for star-forming galaxies in two clusters at $z \sim 0.4$. Interestingly, one cluster, A851, showed a factor of 3–4 excess of emission-line objects relative to the field, while the other cluster MS1512.4+3647, showed no excess. The detection limits were sufficient to detect galaxies with star-formation rates as low as $0.13 M_\odot \mathrm{yr}^{-1}$. Many of fainter emission-line galaxies detected in these clusters are either in the field, or infalling into the clusters. While they could be progenitors of present-day dE galaxies, it is unlikely that such accretion of dwarfs since $z = 0.4$ could produce the numbers or spatial distribution of dwarfs seen locally in places like the Virgo and Fornax Clusters. Thus, we are probably seeing at $z = 0.4$ a snapshot of an ongoing process.

Lilly, Carollo & Stockton (2003) used the R_{23} method to measure metallicities of galaxies in the redshift range $0.5 < z < 0.9$ with magnitudes $I_{AB} < 22.5$. They found that 75% of the galaxies had metallicities comparable to nearby galaxies of similar luminosities. Of the remaining 25%, only few appeared have the sizes consistent with being dwarf galaxies. The implication is that dE galaxies probably formed stars on moderately long timescales ($>10^8$yr) or before $z = 1$.

1.4. *Dwarfs in the Hubble Ultra Deep Field*

For most galaxies fainter than $R \sim 24$, we must rely on indirect techniques to estimate redshifts. Photometric redshift estimates typically involve fitting spectral templates to galaxy broad-band photometry. The most important feature that is shifting through the optical and near-IR bandpasses out to redshifts $z \sim 2$ is the Balmer or 4000 Å break. Instead of fitting templates, we can define photometric quantities (differences of colors) that identify the spectral inflection around this break. This "spectral inflection selection" technique is akin to the Lyman-break technique in that careful selection of colors can isolate star-forming galaxies in well-defined redshift intervals. The detailed selection criteria involve multiple bands and would take too much space to describe here. As an example, if we define three HST colors as $c_1 = B_{435} - V_{606}$ and $c_2 = V_{606} - i_{775}$, and $c_3 = i_{775} - z_{850}$, then a good selection criterion for galaxies at $z \sim 0.8$ is to require $-0.7 < c_1 - c_2 < 0.1$ and $0 < c_2 - c_3 < 0.7$. Figure 2 (left panel) shows the redshift selection functions for three different redshift intervals derived from simulations. The righthand panel shows that there is reasonable agreement with the spectroscopic redshift distribution for GOODS galaxies that meet the selection criteria.

Sersic profiles — $I(r) \propto \exp[(-r/r_e)^{1/n}]$ — have been fit to the HUDF images (Ravindranath, private communication). A large fraction of galaxies in our photometrically identified redshift slices are well fit by profiles that are nearly exponential ($n < 1.5$). While this is reminiscent of local dE galaxies, disk galaxies also have $n \sim 1$, and both inspection of the images and axial-ratio measurements indicate that at least a small fraction of our sample has disk-like morphologies.

Table 1 shows the absolute magnitude limits and approximate number densities. The volume estimates are very rough, and are formally model dependent because the redshift selection-function depends on the spectral-energy distributions of the evolving dE galaxies. Nevertheless, these densities are low enough that it appears unlikely that a large fraction of the dE population formed most of their stars in any of these redshift intervals. Consider two extreme hypotheses: (a) most dE galaxies form most of their stars above

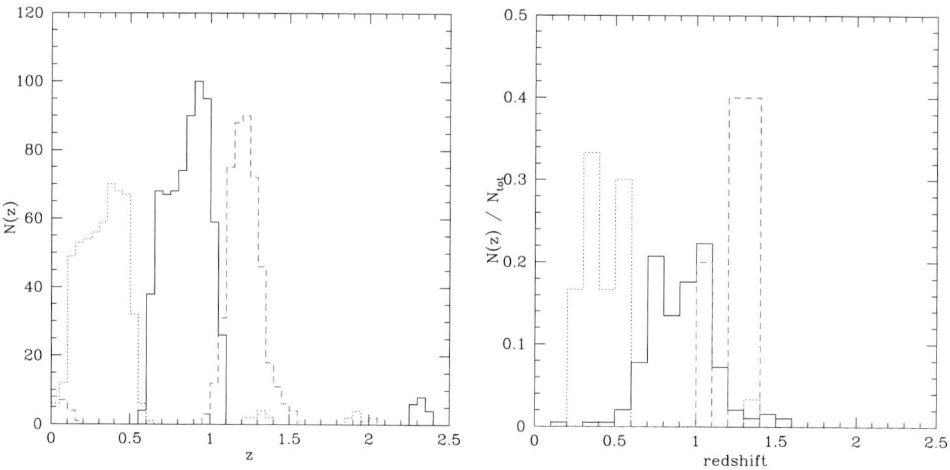

Figure 2. (Left) Redshift selection functions for searching for moderate-redshift dE galaxies. The spectral-inflection selection is efficient at identifying star-forming galaxies or galaxies with strong 4000 Å breaks. It selects against galaxies that are either outside the redshift intervals of interest, or highly reddened. The selection functions shown here were derived by estimating colors for a wide variety of SED models. (Right) Measured redshift distribution for galaxies in the GOODS CDF-S field that meet the same selection criteria.

Table 1. HUDF redshift slices

Redshift range	mag limit (i_{775})	M_B (rest) limit	N	Density ($\times 10^{-3}$ Mpc$^-3$)
$0 - 0.5$	27	-15.2	4	4.4
$0.6 - 1.0$	27.5	-16.2	144	21
$1.0 - 1.3$	27	-18.3	20	2.7

$z = 3$ on short timescales and fade to the present day, and (b) most dE galaxies form their stars at $z < 1.2$ with star-formation timescales $\tau_{SF} \sim 1$ Gyr. To estimate how many dE we should expect to see in these two scenarios, we need to adopt a local luminosity function (and force both scenarios to match it). The local dE LF in the field is not well constrained. We adopt the Blanton *et al.* (2005) LF here in order to have some plausible numbers to work with. (Strictly speaking, their LF is for all dwarfs, not just dE.) Using this LF, in case (a) we should be able to detect to $0.01L^*$ in the redshift interval $0.6 < z < 1$. Integrating from $0.01L^*$ to $0.18L^*$, we expect roughly 200 dwarfs in this redshift range in the HUDF. However, the SED's in case (a) are sufficiently smooth that these galaxies would generally not fulfill the selection criteria that we have adopted. On the other had, if the galaxies formed at $z \sim 1.2$, they would meet the selection criteria, and the galaxies would be about 2.2 mag brighter at $z = 1$ than in case (a). Thus number densities would increase dramatically because we probe further down the intrinsic mass function. Integrating from $0.0016L^*$ to $0.025L^*$ we would predict roughly 600 galaxies in the HUDF, meeting our selection critiera for $0.6 < z < 1.1$. This is significantly higher than the number of galaxies that we find in the HUDF (144) or the number (69) that have half-light radii less than 1.5 kpc, and hence are likely to be dwarfs (see Fig. 3).

It is interesting to examine the colors of these faint, moderate-redshift galaxies. An example color-color diagram is shown in Fig. 4. This shows $B_{435} - V_{606}$ vs. $z_{850} - H_{160}$ for the sample of galaxies selected by spectral inflection to lie in the redshift range

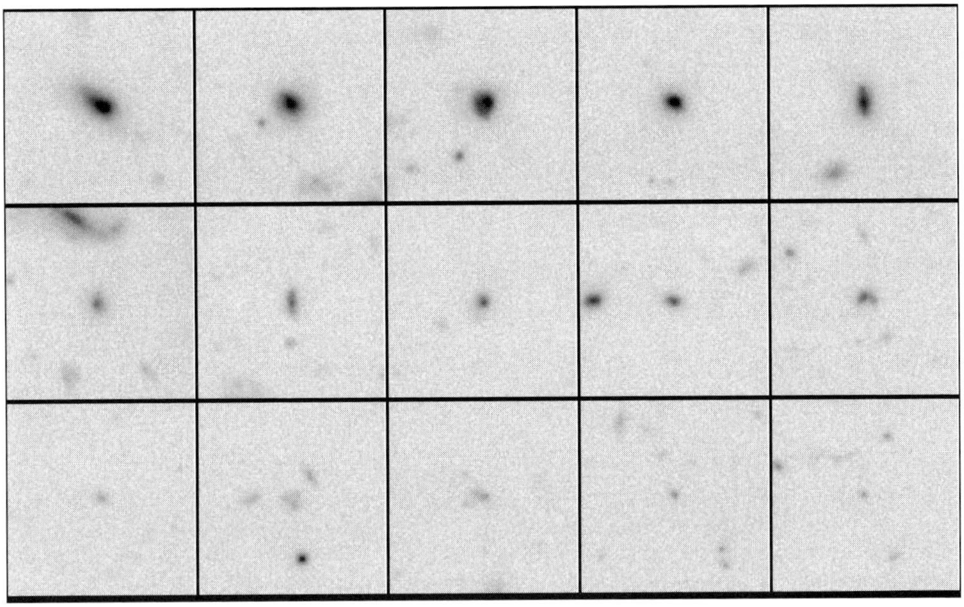

Figure 3. Montage of galaxies selected via spectral inflection to lie in the redshift range $0.6 < z < 1.1$. The candidates shown all have half-light radii $r_h < 1.5$ kpc and we show in the top panel the brightest five candidates, in the middle panel some candidates near the mid-point in luminosity, and the bottom panel the five faintest candidates. While some of the galaxies have disk-like morphologies, many are reminiscent of dE galaxies. Most have exponential surface-brightness profiles. The galaxy at the top left has a star-formation rate of approximately $1.4 M_\odot\,\mathrm{yr}^{-1}$; the galaxy at the lower right has $SFR \sim 0.03 M_\odot\,\mathrm{yr}^{-1}$.

$0.6 < z < 1.1$. We have computed colors for a large suite of models, spanning formation redshifts 0.1–5 in steps $\Delta z = 0.1$ metallicity [Fe/H]$= -1.3$, burst durations $\log(\tau_{SF}) = 6, 7, 8, 9, 10.5$ and Calzetti *et al.* (2000) extinctions $E(B-V) = 0.2, 0.4$. In the left panel we show all models that began forming stars at redshift $z < 1.5$. In the right panel we show all models that began forming stars at $z > 1.5$. There is a clear separation of the models, and it is evident that over half of the galaxies appear to be incompatible with redshifts of formation $z_f > 1.5$. To be precise, 84 out of 148 galaxies in this redshift interval lie bluewards of $B_{435} - V_{606} = 0.3$, a color which separates all models with $z_f > 1.5$ from most models with $z_f < 1.5$. Of these 84 galaxies, most (70%) have half-light radii less than 1.5 kpc. A similar behavior is seen in the next higher redshift slice $1 < z < 1.3$, where we find an additional 22 young galaxies ($z_f < 1.5$) of which 60% have $r_h < 1.5$ kpc. While these are clearly interesting galaxies, the total population of bursting-dwarf candidates that we have identifed between $z = 0.6$ and $z = 1.3$ represents only ~11% of the galaxy counts in the magnitude range $H_{606} = 22$–26. We have no constraints on bursting dwarfs in our low-redshift slice ($z < 0.5$), because the high-z_f and lower-z_f occupy the same locus in this color-color diagram.

In summary, the number counts of faint-blue galaxies are not in serious conflict with ΛCDM cosmological models. Nevertheless, there are interesting faint-blue galaxies in deep images, at least some of which may be progenitors of present-day dE galaxies. Spectral-inflection selection in the HUDF reveals in intriguing population of faint, compact galaxies that appear to have waited to $z < 1.5$ to form stars. While these may be progenitors of some dE galaxies, if *most* dE galaxies formed their stars at $z < 1.5$, we would have expected to find many more such candidates.

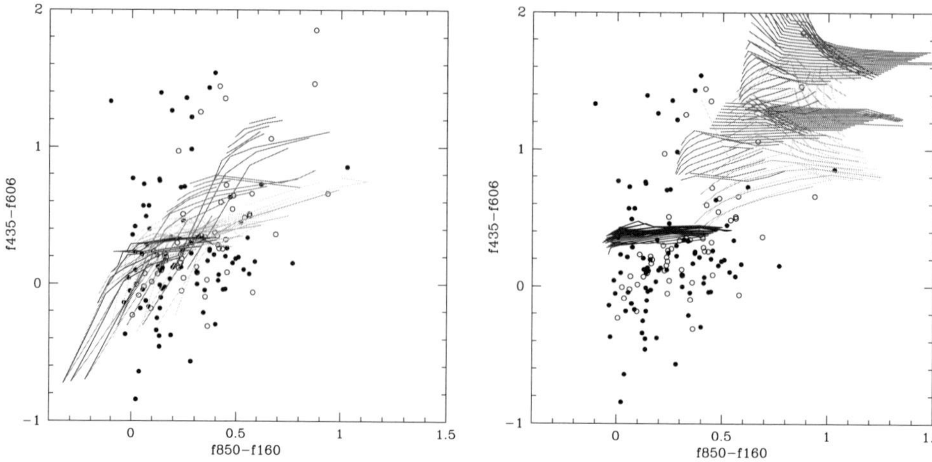

Figure 4. Observed colors of galaxies at $0.6 < z < 1.1$ compared to models. In both panels, the solid points are galaxies with $r_h < 1.5\,\mathrm{kpc}$, and the open circles are galaxies with larger half-light radii. All galaxies have been identified via spectral inflection, as described in the text. The **left** panel shows a suite of models (described in the text), all of which began forming stars at $z_f < 1.5$. The models are viewed at a finely sampled set of redshifts $0 < z < z_f$, but the colors are only plotted when the model spectral inflections meet the selection criteria used for the galaxies. The models with short τ_{SF} tend to congregate toward the lower right in this diagram. (For readers viewing these diagrams in color: yellow curves have $\log \tau_{SF} = 6, 7$, green has $\log \tau_{SF} = 8$, cyan has $\log \tau_{SF} = 9$, blue has $\log \tau_{SF} = 10.5$, all with no dust; magenta has $\log \tau_{SF} = 9$ and $E(B-V) = 0.2$ and red has $\log \tau_{SF} = 9$ and $E(B-V) = 0.4$.) To explain the data points that lie below and to the right of most of the models may require lower metallicites, emission lines, or an unusual IMF. In the **right** panel, we show the suite of models with $z_f > 1.5$. Even the bluest models in this set (those with nearly constant star formation over a Hubble time) are redder than more than half of the observed galaxies. This suggests that many of the galaxies in this sample are bursting dwarf galaxies that could evolve into present-day dE galaxies.

Acknowledgements

We would like to acknowledge the support and collaboration of the GOODS and HUDF teams in this endeavor. Thanks in particular to Swara Ravindranath for providing GALFIT results for the HUDF. Support for this work was provided by NASA through grant GO09583.01-96A from the Space Telescope Science Institute, which is operated by the AURA, Inc., under NASA contract NAS5-26555.

References

Babul, A. & Rees, M.J. 1992, *MNRAS* 255, 346

Babul, A. & Ferguson, H.C. 1996, *ApJ* 458, 100

Blanton, M., Lupton, R.H., Schlegel, D.J., Strauss, M.A. Brinchman, J., Fukugita, M. & Loveday, J. 2005, *ApJ* submitted, astro-ph/0410164

Calzetti, D., Armus, L., Bohlin, R.C., Kinney, Al, Koorneef, J. & Storchi-Bergmann, T. 2000 *ApJ* 533, 682

Conselice, C.J., O'Niel, K., Gallagher, J.S. & Wyse, R.F.G. 2003 *ApJ* 591, 167

Cowie, L.L., Songaila, A. & Hu, E.M. 1991 *Nature* 354, 460

Crampton, D., Le Fevŕe, O., Lilly, S.J. & Hammer, F. 1995 *ApJ* 455, 96

Dekel, A. & Silk, J. 1986 *ApJ* 303, 39

Dekel, A. & Woo, J. 2003 *MNRAS* 344, 1131

Efstathiou, G. 1992, *MNRAS* 256, 43

Efstathiou, G. 1995, *MNRAS* 272, L25
Ellis, R.S. 1997, *ARAA* 35, 389
Ferguson, H.C. & Babul, A. 1998, *MNRAS* 296, 585
Geha, M., Guhathakurta, P. & van der Marel, R.P. 2003, *AJ* 126, 1794
Guiderdoni, B. & Rocca-Volmerange, B. 1991 *A&A* 252, 435
Hensler, G., Theis, Ch. & Gallager, J.S. III. 2004 *A&A* 426, 25
Koo, D. & Kron, R. 1992 *ARAA* 30, 613
Kron, R. 1978, *Thesis, University of California at Berkeley*
Lin, D.N. & Murray, S.D. 1992 *ApJ* 394, 523
Lilly, S.J., Carollo, C.M. & Stockton, A. 2003 *ApJ* 597, 730
Lotz, J.M., Martin, C.L. & Ferguson, H.C. 2003 *ApJ* 596, 143
McCarthy, P.J., *et al.* 1999 *ApJ* 520, 548
Michielsen, D., de Rijcke, S., Zeilinger, W.W., Prugniel, P., Dejonghe, H. & Roberts, S. 2004 *MNRAS* 353, 1293
Moore, B., *et al.* 1999 *ApJ* 524, L19
Nagashima, M., Yoshii, Y., Totani, T. & Gouda, N. 2002, *ApJ* 578, 675
Scott, J., Bechtold, J., Morita, M., Dobrzycki, A. & Kulkarni, V. P. 2002, *ApJ* 571, 665
Somerville, R.S. 2002, *ApJ.* 572, L23
Trentham, N., Sampson, L. & Banerji, M. 2005, *MNRAS* 357, 783
Tyson, J. A. 1988, *AJ* 96, 1
van Zee, L., Skillman, E.D. & Hanes, M.P. 2004, *AJ* 128, 121
Venn K.A., Irwin, M., Shetrone, M.D., Tout, C.A., Hill, V. & Tolstoy, E. 2004, *AJ* 128, 1177
Yoshii Y. & Takahara, F. 1988, *ApJ* 326, 1

Discussion

CONSELICE: What fraction of your dwarf sample is dE and dIrr?

FERGUSON: Good question. The visual impression is that most of the small galaxies look like late-type dwarfs or dEs, but we have not yet looked at the statistical distribution of concentration indices, Sersic parameters, or ellipticities.

ZAGGIA: Have you searched for any spatial clustering around possible galaxy parents?

FERGUSON: not yet.

Near-Field Cosmology with Dwarf Elliptical Galaxies
Proceedings IAU Colloquium No. 198, 2005
H. Jerjen and B. Binggeli, eds.

doi:10.1017/S1743921305003686

Galaxies at high redshifts: where are the progenitors of today's dwarf ellipticals?

Max Pettini

Institute of Astronomy, University of Cambridge, Madingley Road,
Cambridge CB3 0HA, England
email: pettini@ast.cam.ac.uk

Abstract. The epochs when many dwarf ellipticals were forming their stellar populations have recently become accessible to direct observation, as surveys at $z = 1$–5 have now uncovered thousands of galaxies. I briefly review the properties of various categories of high redshift galaxies in search of clues for possible links to today's galaxy populations. While much of the emphasis up to now has been on massive elliptical galaxies, with several groups claiming to have identified their progenitors with a variety of different techniques, dwarf galaxies have generally been overlooked—no doubt because the observational difficulties at high redshift are still considerable. Based on available data, it is possible that at least some of the 'Lyα emitters' and the damped Lyα systems may be on evolutionary paths which will transform them into the galaxies which are the subject of this meeting.

1. Introduction

I should like to echo Eva Grebel's words in her excellent review which opened this Colloquium and emphasized the importance of local studies as the key to interpreting the increasingly detailed picture of the high redshift universe which has been emerging in recent years. The wealth of data being gathered in galaxy surveys which stretch from the present-day universe (e.g. Croton *et al.* 2005), through intermediate (e.g. Madgwick *et al.* 2003; Abraham *et al.* 2004) and high (e.g. Steidel *et al.* 2004, 2005) redshifts, to the epoch when the universe became reionised (e.g. Bunker *et al.* 2004) is truly staggering. The thread which is common to all of these efforts is the desire to follow the evolution of the Hubble sequence of galaxies backwards in time so as to identify the progenitors of today's elliptical and spiral galaxies, and thereby throw light on the physical processes driving their formation and which must be at the root of their morphological differences.

Of relevance to the theme of this meeting is therefore the question "Have we found any clues to the origin of dwarf elliptical galaxies in the high-redshift galaxy population?" By their very nature, dwarf galaxies are obviously the most difficult to see in any magnitude-limited survey of the distant universe. Nevertheless, it is worthwhile confronting the properties of the galaxies being targeted by high redshift surveys with those of the dwarf ellipticals whose stellar populations, as we have learnt from Eva's work, in many cases formed at redshifts $z = 2$–6, that is at the same early cosmic epochs now accessible to direct observation.

I shall address this question for the two classes of high redshift galaxies (without active galactic nuclei) which have been most extensively studied up to now: the Lyman break galaxies and the damped Lyman α systems.

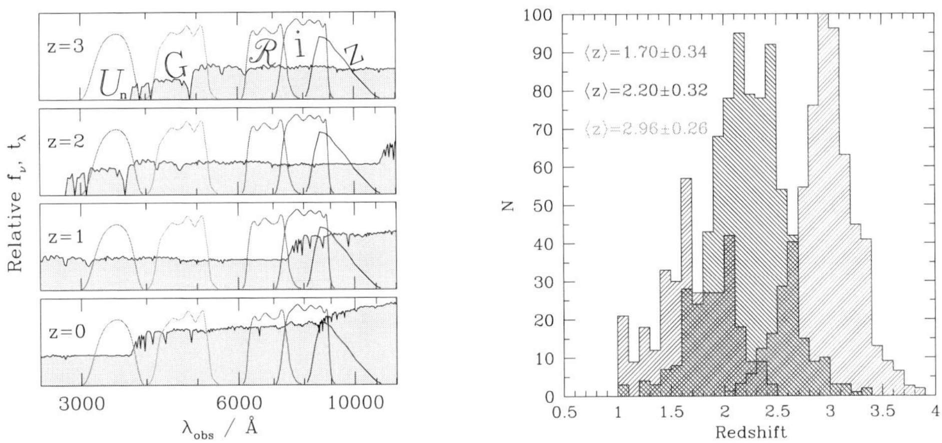

Figure 1. *Left Panel* (reproduced from Adelberger *et al.* 2004): The $U_n G\mathcal{R}iz$ filter passbands straddle different spectral features at different redshifts in the spectral energy distribution of a typical star-forming galaxy. Consequently, galaxies can be photometrically preselected within reasonably well-defined redshift intervals which are well separated on a colour-colour plane, as shown by the redshift distributions of spectroscopically confirmed galaxies from the surveys by Steidel *et al.* 2004 (shown in the *Right Panel*).

2. Star-forming galaxies at $z = 1.5$–3.5

Lyman break galaxies are just one example of UV-bright—and therefore star-forming—galaxies selected within a predetermined redshift interval by the application of appropriate broad-band photometric criteria. The original $U_n G\mathcal{R}$-based colours of Steidel, Pettini & Hamilton (1995) have more recently been 'tuned' (see Adelberger *et al.* 2004) to select galaxies at mean redshifts $\langle z \rangle = 3.0$ (LBGs), 2.2 (BX galaxies) and 1.7 (BM galaxies), in each case with a redshift dispersion of about 0.3 (see Figure 1).

The BzK criteria of Daddi *et al.* (2004) can be used to select either galaxies which overlap significantly with the BX/BM galaxies of Steidel *et al.* (2004) or, by focusing on a different portion of the BzK colour-colour plane, galaxies which by virtue of their red optical-to-infrared colour must be either passively evolving or heavily reddened. Similarly red galaxies are targeted by the FIRES survey of Franx *et al.* (2003). The way these different searches overlap and complement each other has recently been assessed in a comprehensive study by Reddy *et al.* (2005). All these populations of high redshift galaxies span a wide range in luminosity (see Adelberger & Steidel 2000; Shapley *et al.* 2001), and it is the ones at the brighter end of the distribution, with $L \geqslant L^*$, which have naturally been the subject of most spectroscopic studies so far.

There are now several lines of evidence that lead us to associate these bright galaxies with the progenitors of today's massive ellipticals and bulges (an interpretation favoured by nearly all of the works cited in the preceding paragraph), rather than with low mass galaxies briefly lit up by a short-lived starburst episode. These lines of evidence include:

Stellar masses. Analysis of the spectral energy distributions of BX galaxies from the rest-frame UV to the rest-frame near-IR (now possible by combining ground-based deep imaging in the optical and near-IR with IR multi-band data from the *Spitzer* satellite) indicates assembled stellar masses in the range $M_* = 10^{9.5}$–$10^{11.5}\,M_\odot$, with a median value $\langle \log(M_*/M_\odot) \rangle = 10.3 \pm 0.5$ (see Figure 2).

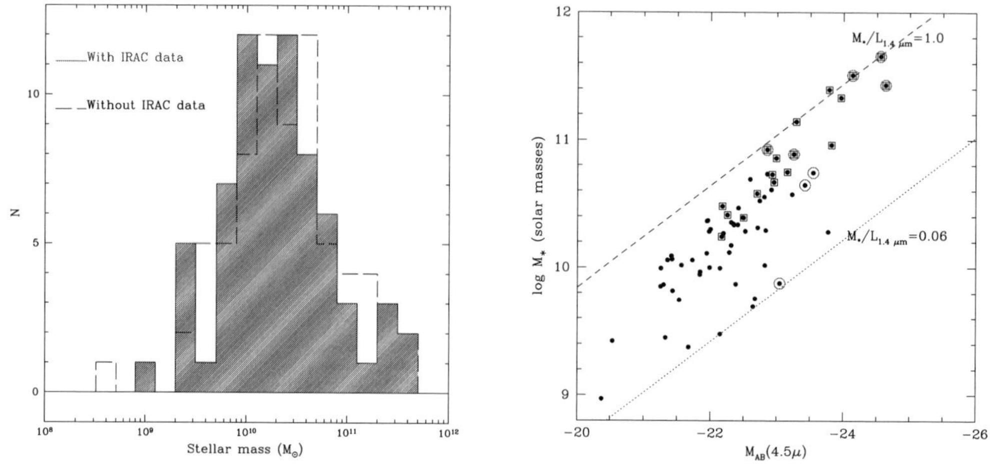

Figure 2. Stellar masses of BX galaxies in the field of the QSO HS1700+643 (for which *Spitzer* IRAC $3.6-8.0\,\mu$m photometry is available). In the right-hand panel, the stellar masses are plotted versus the absolute magnitude at $4.5\,\mu$m, equivalent to rest-frame $1.4\,\mu$m luminosity at the mean $\langle z \rangle = 2.2$ of these galaxies. It can be seen that even in the near-IR the mass-to-light ratio of these galaxies spans a factor of $\sim$15. Evidently, the most massive galaxies had already attained values comparable to those of present-day galaxies, while more typically $(M/L)_{1.4\,\mu m} \simeq 0.4$, a factor of $\sim$2.5 smaller. Circles indicate galaxies brighter than $K_s = 20.0$, while open squares denote red galaxies with $\mathcal{R} - K_s > 3.5$. (Figures reproduced from Shapley *et al.* 2005)

Clustering properties. These galaxies already exhibit a high degree of clustering by $z = 2$–3, with correlation lengths $r_0 \simeq (4 \pm 0.5)\,h^{-1}$ comoving Mpc (Adelberger *et al.* 2004). The brightest galaxies, with $K_s < 20$, are more strongly clustered with $r_0 = (10 \pm 3)\,h^{-1}$ Mpc (Adelberger *et al.* 2005; Daddi *et al.* 2004). In ΛCDM N-body simulations (e.g. Frenk *et al.* 2000), this is the degree of clustering exhibited by massive halos, with $\log(M_{\rm DM}/M_\odot) \sim$ 11–12, which by the present day evolve to have number densities and correlation lengths typical of elliptical galaxies (see Figure 3).

Metallicities. A variety of techniques—among which the ratio of the [N II]λ6583 and Hα emission lines has had the widest application so far (see Figure 4)—give a concordant picture showing that these galaxies had already reached solar, or near-solar, abundances at $z = 2$–3. While they are undoubtedly bright for their metallicity compared to the present-day metallicity-luminosity relation (another indication of their low M/L ratios), it is hard to conceive of a plausible scenario by which these galaxies would evolve into today's metal-poor dwarf galaxies.

Presumably, and not surprisingly, the progenitors of today's dwarf galaxies are to be found among objects fainter than L^* in the luminosity function of high redshift galaxies. Up to now these candidates have been little studied spectroscopically, as they are beyond the reach of current instrumentation, except for the fraction which shows a strong Lyman α emission line (e.g. Dawson *et al.* 2004; Hu *et al.* 2004). Lyα emitters have been identified in large numbers at redshifts up to $z = 6.6$ (Taniguchi *et al.* 2005), and among them there may well be galaxies on an evolutionary path which will lead them to become today's dwarf ellipticals.

3. Damped Lyman α absorption systems

With column densities of neutral hydrogen in excess of $N(\mbox{H I}) = 2 \times 10^{20}\,\mathrm{cm}^{-2}$, DLAs account for most of the neutral gas seen in absorption along random sight-lines to

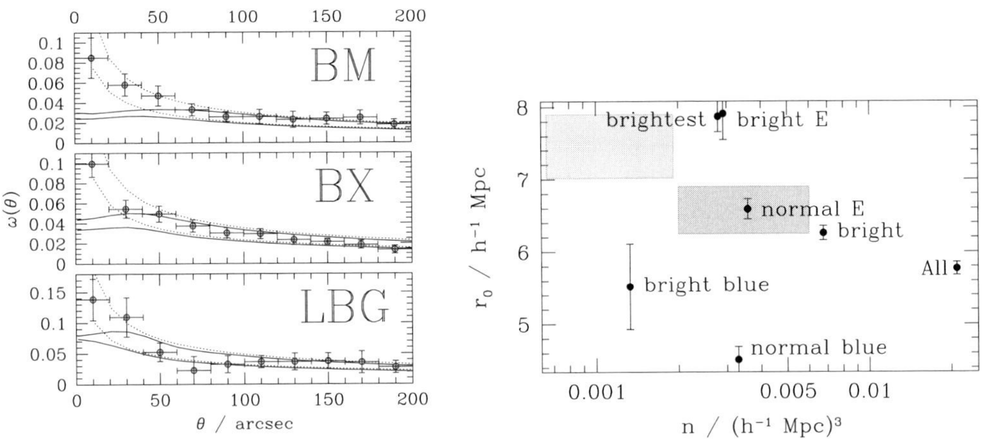

Figure 3. *Left Panel:* The measured angular correlation functions of LBG, BX, and BM galaxies (points with error bars) match those of dark matter halos with masses in the range $\log(M_{\rm DM}/M_{\odot}) \sim 11$–12 in N-body simulations such as that released by the Virgo consortium (Frenk *et al.* 2000) and indicated by the continuous lines. *Right Panel:* Following the evolution of these massive halos forward in time shows that, by $z = 0.2$, LBG, BX, and BM galaxies will have attained correlation lengths $r_0 \simeq 7h^{-1}$ Mpc, similar to those of elliptical galaxies. The shaded rectangles show the ranges in the values of r_0 and n (number density per unit volume) expected for the descendents of 'typical' (near the centre of the plot), and the brighter ($L > L^*$; top left-hand corner) $z = 3$ Lyman break galaxies. The points with error bars are values of r_0 and n measured by Budavari *et al.* (2003) for various samples of galaxies from the *Sloan Digital Sky Survey*, as labelled. (Figures reproduced from Adelberger *et al.* 2004).

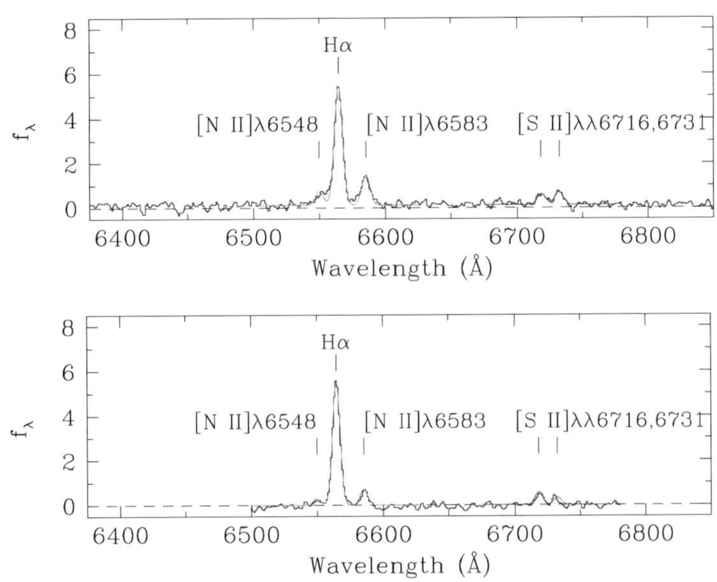

Figure 4. *Upper Panel:* Composite spectrum of 21 BX galaxies brighter than $K_s = 20$ obtained with NIRSPEC on the Keck II telescope (Erb *et al.* 2005). *Lower Panel:* Composite spectrum of 66 BX galaxies fainter than $K_s = 20$. Using the calibration by Pettini & Pagel (2004), the measured [N II]λ6583/Hα ratios imply approximately solar metallicities for the $K_s < 20$ galaxies, and about half-solar for the fainter ones.

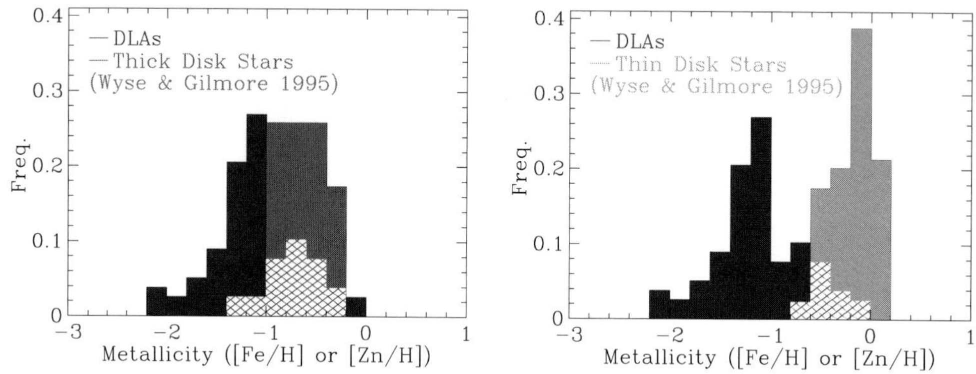

Figure 5. Metallicity distributions of DLAs and Milky Way thick (left panel) and thin (right panel) disk stars from the work by Akerman *et al.* (2005). The metallicity is measured from Zn in DLAs and Fe in Galactic stars. Approximately one half of the DLA measurements of [Zn/H] are upper limits; they have been included in the histograms as if they were detections.

distant QSOs. After two decades of extensive observational work, we know many of their properties in exquisite detail—such is the power of QSO absorption line spectroscopy, but somehow this makes it all the more frustrating that their connection to present-day galaxies is still a matter of debate. H I surveys of the local universe now seem to indicate that a wide range of galaxy luminosities (from $M_B \simeq -21$ to -15) and morphological types can contribute to the cross-section for DLA absorption (Zwaan *et al.* 2005). At higher redshifts, $z = 1.5$–3.5, it is unclear how DLAs relate to other galaxy populations. The working hypothesis since the early days of this subject has been that DLAs are the progenitors of spiral galaxies, observed at a time when most of their baryons were in the interstellar medium rather than in stars, but this picture may be at odds with theoretical ideas of a slow build-up of disks over most of the Hubble time (e.g. Naab & Ostriker 2005). The possibility that at least some high-z DLAs may be associated with dwarf galaxies is, on the other hand, supported by several observational results:

Luminosities. The galaxies which host $z \sim 3$ DLAs are generally fainter than L^* LBGs (Cooke *et al.* 2005).

Metallicities. DLAs are metal-poor, with typical abundances between 1/10 and 1/100 of solar (Akerman *et al.* 2005). As can be seen from Figure 5, the metallicity distribution of DLAs peaks at lower values than those of Milky Way thick and thin disk stars, although there is some overlap, particularly with the thick disk. The difficulty here for the spiral galaxy interpretation of DLAs is that the period of time during which the Milky Way disk was as metal-poor as a typical DLA was probably short-lived, as evidenced by the near-flat age-metallicity relation of its stars. A possible way out may be a scenario where the (relatively) metal-rich, star-forming, inner disk is surrounded by a larger (thick) disk of mostly unprocessed gas, as proposed by Wolfe (2005). More generally, steep radial abundance gradients in young disks (Naab & Ostriker 2005) may be the explanation for the low metallicity of most DLAs.

Element ratios. There has been much discussion of recent findings that stars in dwarf spheroidals of the Local Group may have lower α/Fe ratios than Galactic halo stars of similar Fe/H, and of what the implications of such differences might be for hierarchical models in which the halo was assembled by the mergers of many such sub-units (see,

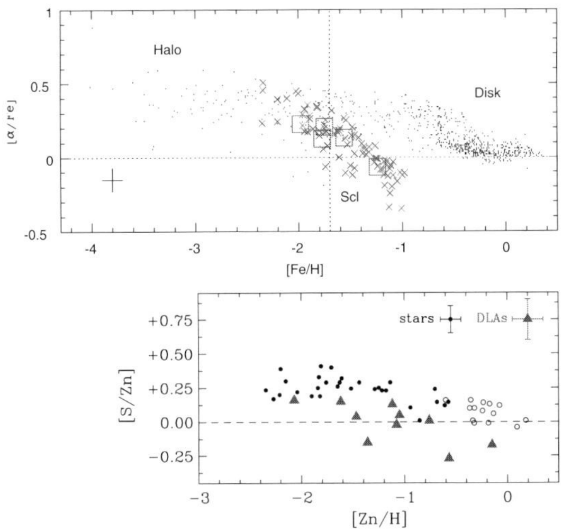

Figure 6. *Upper Panel:* α-element/Fe ratios of individual stars in the Sculptor dwarf spheroidal galaxy (crosses and large open squares) compared with the values measured in Milky Way stars (diagram reproduced from Tolstoy 2005). *Lower Panel:* Ratio of S (an α-capture element) to Zn (an Fe-peak element) in Galactic stars and DLAs (reproduced from Nissen *et al.* 2004). While some Sculptor stars and some DLAs conform to the well-known α-element enhancement typical of the metal-poor Galactic halo, in general there appears to be a wider spread of $[\alpha/\text{Fe}]$ values in the former environments compared to the latter. Such differences bring into question the working hypotheses that: (a) the Galactic halo was assembled by many smaller sub-units similar to the Local Group dwarf galaxies which survive today, and (b) DLAs arise in the progenitors of disk galaxies like the Milky Way.

for example, Eline Tolstoy's contribution to this volume). It is thus intriguing that even from this point of view DLAs resemble more closely dwarf galaxies than the Milky Way (see Figure 6), although the number of DLAs in which the ratio α/Fe-peak has been measured so far is still small.

ISM temperature. By combining values of $N(\text{H I})$ measured from the damping wings of the Lyα absorption line with the optical depth in 21 cm absorption, spin temperatures (or upper limits) have now been determined for 28 DLAs lying in front of radio-loud QSOs (see Figure 7). It is intriguing that at $z > 2$ only high spin temperatures have been measured so far, whereas at lower redshifts both high and low values of T_s are encountered. Furthermore, all DLAs with low T_s have been identified with spiral galaxies, while most of the low-z, high T_s, DLAs are associated with $L \lesssim 0.1L^*$ galaxies. If we are not being misled by the different spatial scales probed by absorption and emission line measurements, this shift in the distribution of ISM temperatures with redshift could be taken as evidence that the population of DLA absorbers is dominated by dwarf galaxies at high z, and that it evolves at lower redshifts to include more massive, metal-rich spirals. Presumably, the high values of T_s of most DLAs are just another manifestation of their low metallicities, since cooling is via heavy element atomic transitions. Thus, if we could clarify why most DLAs are metal-poor at an epoch when the Milky Way disk had already been enriched to $Z \sim 1/3 Z_\odot$, and what this means for the DLA-spiral galaxy connection, we would also solve the riddle posed by their high ISM temperatures.

Of course, there are two fundamental differences between DLAs and dwarf ellipticals: DLAs are gas-rich and are large. Radii of order 10 kpc are indicated by: (a) the frequency

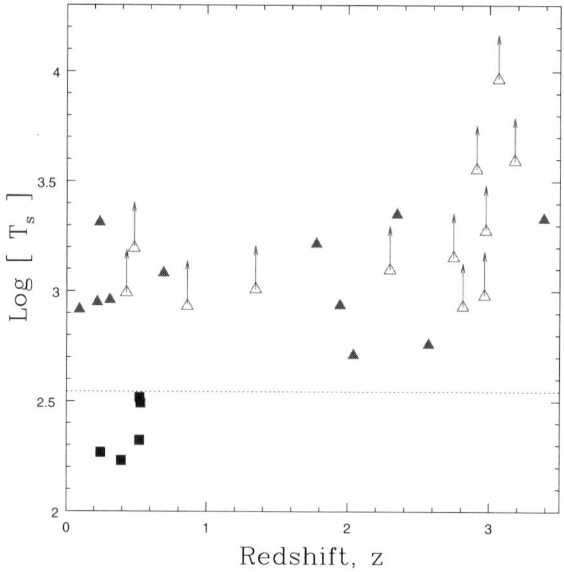

Figure 7. Spin temperatures of DLAs as a function of redshift (Kanekar & Chengalur 2003; Kanekar *et al.* 2005). Squares represent DLAs identified with spiral galaxies. Filled triangles show other detections of 21 cm absorption, open triangles are 3σ lower limits to the spin temperature. The horizontal dashed line indicates $T_s = 350\,\mathrm{K}$.

with which a random sight-line to a background QSO intersects a DLA at $z = 2$–3 (albeit under the unrealistic assumption of a non-evolving galaxy luminosity function, e.g. Steidel 1993), and (b) the typical impact parameters encountered at $z < 1$ (which however only refer to the DLA host galaxies which have been successfully imaged; Zwaan *et al.* 2005). What is of interest in the context of the present discussion is whether some DLAs could turn into dwarf ellipticals if they lost their gaseous reservoirs early on in their evolutionary history. Plausible local examples of this scenario do exist (see Liese van Zee's contribution to this volume), but it remains to be established whether these rare cases were in fact more common at high redshift.

4. Conclusions

While large-scale surveys at a variety of wavelengths are providing an increasingly detailed picture of the high redshift universe, the connection between its denizens and today's morphological classes of galaxies has yet to be established with any degree of confidence. Many of the properties of luminous star-forming galaxies at $z = 1.5$ to 3.5—particularly their high stellar masses, metallicities and clustering lengths—point to today's massive ellipticals and bulges as their likely descendants. Presumably the high redshift counterparts of today's dwarf ellipticals are to be found among galaxies fainter than L^*. In this faint regime, we know too little as yet about the 'Lyα emitters' to draw meaningful comparisons, but we have identified several properties of the damped Lyα systems which are suggestive of a possible link, at least for some of the DLAs. These are of course only the first steps. Tracing back the evolutionary history of dwarf galaxies, whether ellipticals, irregulars, or Local Group spheroidals, still relies on the clues provided by their stellar populations—as the title of this colloquium implies, this is the realm of 'near-field cosmology'.

Acknowledgements

I should like the thank the organisers of this stimulating and well planned Colloquium, and the International Astronomical Union for the award of a travel grant. I am grateful to Chris Akerman and Nissim Kanekar for permission to show data in advance of publication.

References

Abraham, R.G. *et al.* 2004, *AJ* 127, 2455

Adelberger, K.L., Erb, D.K., Steidel, C.C., Reddy, N.A., Pettini, M. & Shapley, A.E. 2005, *ApJ* 620, L75

Adelberger, K.L. & Steidel, C.C. 2000, *ApJ* 544, 218

Adelberger, K.L., Steidel, C.C., Shapley, A.E., Hunt, M.P., Erb, D.K., Reddy, N.A. & Pettini, M. 2004, *ApJ* 607, 226

Akerman, C.J., Ellison, S.L., Pettini, M. & Steidel, C.C. 2005, *A&A* in press

Budavari, T. *et al.* 2003, *ApJ* 595, 59

Bunker, A.J., Stanway, E.R., Ellis, R.S. & McMahon, R.G. 2004, *MNRAS* 355, 374

Cooke, J., Wolfe, A.M., Prochaska, J.X. & Gawiser, E. 2005, *ApJ* 621, 596

Croton, D.J. *et al.* 2005, *MNRAS* 356, 1155

Daddi, E., Cimatti, A., Renzini, A., Fontana, A., Mignoli, M., Pozzetti, L., Tozzi, P. & Zamorani, G. 2004, *ApJ* 617, 746

Dawson, S. *et al.* 2004, *ApJ* 617, 707

Erb, D.K. *et al.* 2005, in preparation

Franx, M. *et al.* 2003, *ApJ* 587, L79

Frenk, C.S. *et al.* 2000, astro-ph/0007362

Hu, E. M., Cowie, L.L., Capak, P., McMahon, R.G., Hayashino, T. & Komiyama, Y. 2004, *AJ* 127, 563

Ibata, R., Chapman, S., Ferguson, A.M.N., Lewis, G., Irwin, M. & Tanvir, N. 2005, *ApJ*, submitted (astro-ph/0504164)

Kanekar, N. & Chengalur, J.N. 2003, *A&A* 399, 857

Kanekar, N. *et al.* 2005, in preparation

Madgwick, D.S. *et al.* 2003, *ApJ* 599, 997

Naab, T. & Ostriker, J.P. 2005, *MNRAS*, submitted

Nissen, P.E., Chen, Y.Q., Asplund, M. & Pettini, M. 2004, *A&A* 415, 993

Pettini, M. & Pagel, B.E.J. 2004, *MNRAS* 348, L59

Reddy, N.A., Erb, D.K., Steidel, C.C., Shapley, A.E., Adelberger, K.L. & Pettini, M. 2005, *ApJ* in press

Shapley, A.E., Steidel, C.C., Adelberger, K.L., Dickinson, M., Giavalisco, M. & Pettini, M. 2001, *ApJ* 562, 95

Shapley, A.E., Steidel, C.C., Erb, D.K., Reddy, N.A., Adelberger, K.L., Pettini, M., Barmby, P. & Huang, J. 2005, *ApJ* in press (astro-ph/0503485)

Steidel, C.C. 1993, "The Environment and Evolution of Galaxies", J.M. Shull & H.A. Thronson eds., (Dordrecht: Kluwer Academic Publishers), 263

Steidel, C.C., Adelberger, K.L., Shapley, A.E., Pettini, M., Dickinson, M. & Giavalisco, M. 2003, *ApJ* 592, 728

Steidel, C.C., Pettini, M. & Hamilton, D. 1995, *AJ* 110, 2519

Steidel, C.C., Shapley, A.E., Pettini, M., Adelberger, K.L., Erb, D.K., Reddy, N.A. & Hunt, M.P. 2004, *ApJ* 604, 534

Taniguchi, Y. *et al.* 2005, *PASJ* 57, 165

Tolstoy, E. 2005, ESO workshop on "Chemical Abundances and Mixing in Stars", L. Pasquini & S. Randich eds., in press (astro-ph/0412067)

Wolfe, A.M. 2005, *Proc. IAU Coll. 199*, P.R. Williams, C. Shu, & B. Méenard eds., in press
Zwaan, M.A., van der Hulst, J.M., Briggs, F.H., Verheijen, M.A.W. & Ryan-Weber, E.V. 2005, *MNRAS*, submitted

Discussion

READ: Are the kinematics of DLAs consistent with the low velocity dispersions of most dwarf ellipticals?

PETTINI: The short answer is "Yes, in some cases". But interpreting the velocity structure of interstellar absorption lines in terms of the underlying gravitational potential is far from straightforward, because small parcels of gas accelerated by energetic events such as supernova explosions, can add significantly to the overall velocity spread of saturated absorption lines. Opinions differ on how much weight to place on such kinematical evidence.

SKILLMAN: Given that dEs are normally found with companion large galaxies, wouldn't the damped Lyα systems be more likely to be associated with a nearby L^* galaxy?

PETTINI: You would think so, wouldn't you? But it all boils down to how the cross-section for H I absorption scales with the galaxy luminosity. If each L^* galaxy had many dwarf companions, and if the scaling of H I radius with luminosity was flatter than the Holmberg relation, then dwarf galaxies may well contribute significantly to the DLA population. This at least seems to be the case in the nearby universe, where H I surveys show that 87% of the DLA cross-section is in galaxies fainter than L^*, and 45% in galaxies with $L < 1/10L^*$ (Zwaan *et al.* 2005). The metallicity evidence I have presented is also more in line with an origin in galaxies less massive than the Milky Way. However, at this meeting we also heard from Rodrigo Ibata of the recent discovery of a very extensive disk of metal-poor stars in M31 (Ibata *et al.* 2005). Such a structure would match very closely the known properties of DLAs at $z = 2 - 3$.

CONSELICE: Since the mass function of Lyman break galaxies is so steep, could these be the progenitors of dwarf galaxies?

PETTINI: That was the gist of my presentation. Most detailed studies of Lyman break galaxies up to now have focussed on the bright end of the luminosity function. At present we know too little about sub-L^* LBGs to make a definite connection with today's galaxy populations, although I drew attention to Lyα emitters and DLAs as possible progenitors of dwarf ellipticals, at least in some cases. Both LAEs and DLAs are to be found at the steep faint end of the luminosity function of high redshift galaxies.

Near-Field Cosmology with Dwarf Elliptical Galaxies
Proceedings IAU Colloquium No. 198, 2005
H. Jerjen and B. Binggeli, eds.

doi:10.1017/S1743921305003698

Does the counter-streaming instability produce lopsided dE,Ns?

Victor P. Debattista[1] **and Sven De Rijcke**[2]

[1]Department of Astronomy, University of Washington, Seattle, WA 98195, USA
email: debattis@astro.washington.edu

[2]Sterrenkundig Observatorium, Universiteit Gent, Krijgslaan 281, S9, B-9000, Gent, Belgium

Abstract. We present evidence that FCC 046, a dE,N galaxy with an offset nucleus, suffered a counter-streaming instability. This instability may explain the presence of lopsided nuclei in other dE,Ns if these galaxies are weakly rotating. The counter-streaming needed for the instability may result from the destruction of box orbits during the assembly of the nucleus.

Keywords. instabilities, stellar dynamics, galaxies: dwarf, galaxies: evolution, galaxies: kinematics and dynamics, galaxies: structure

---

1. Introduction

About half of all dwarf elliptical (dE) galaxies harbor a bright central nucleus. Tidal stripping of dE,N galaxies in a dense environment may liberate these nuclei to form both the ultracompact dwarfs (Phillipps *et al.* 2001; Drinkwater these proceedings) and massive globular clusters, such as ω Cen in the Milky Way (Gnedin *et al.* 2002) and G1 in Andromeda (Meylan *et al.* 2001). In a fraction of dE,Ns, the nucleus is displaced with respect to the centers of the outer isophotes of their host galaxies. Based on photographic plate material, Binggeli *et al.* (2000) estimate a typical offset of $\sim 1''$ ($\sim$100 pc) in the Virgo cluster for $\sim$20% of dE,Ns. Various models have been proposed for explaining the offset/lopsided nuclei. One possibility is that the lopsided systems are produced by the counterstreaming instability (Zang & Hohl 1978; Sawamura 1988; Palmer & Papaloizou 1990; Merritt & Stiavelli 1990; Levison *et al.* 1990; Sellwood & Merritt 1994; Sellwood & Valluri 1997). This instability can occur whenever two coincident stellar populations counter-rotate relative to each other. The result is a lopsided distribution, i.e. a significant $m = 1$ Fourier component in the density distribution (for comparison, the more familiar bar instability corresponds to an $m = 2$ Fourier component). N-body simulations have found that the lopsidedness produced by the counter-streaming instability are robust and long-lived. When the system has zero net angular momentum then the lopsidedness is stationary while it rotates slowly if the system has some angular momentum.

2. FCC 046

Observations of FCC 046 have produced suggestive evidence for counter-streaming in FCC 046 (De Rijcke & Debattista 2004). FCC 046 is a dE,N on the outskirts of the Fornax cluster; its nearest neighbor is a dwarf over 100 kpc away from it (in projection). Photometry of FCC 046 (De Rijcke *et al.* 2003) and VLT FORS2 long-slit spectra with an instrumental resolution $\sigma_{instr} = 30$ km s^{-1} were available to us, as well as *Hubble Space Telescope* (*HST*) archival images in the F814W and F555W filters.

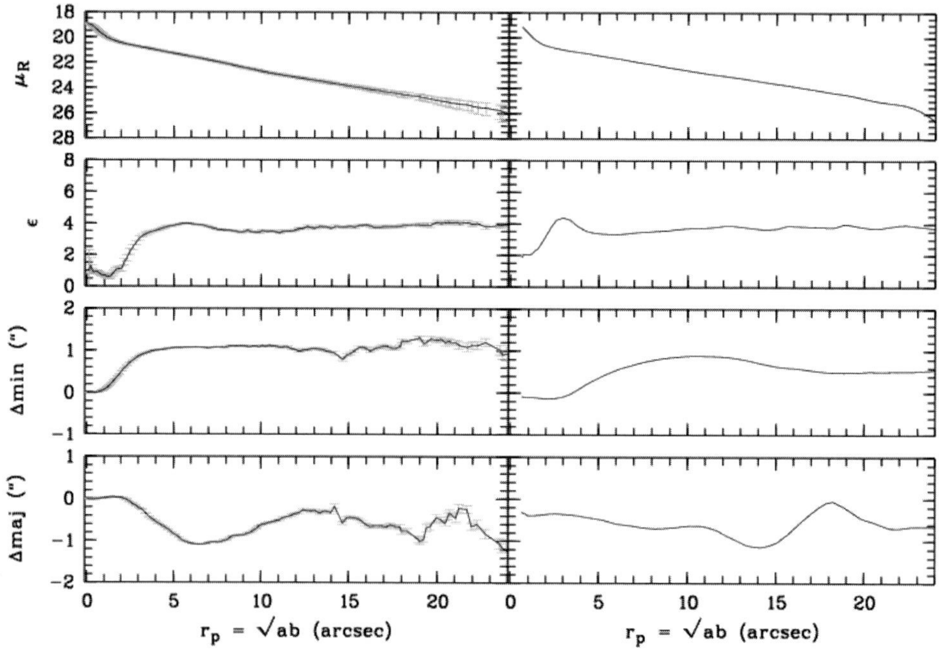

Figure 1. The parameters of ellipse fits for FCC 046 (left panels) and of the N-body simulation (right panels). The centers of the ellipses were allowed to vary. The central surface brightness of the N-body simulation was set to match that of FCC 046. The simulation can reproduce the gross structural properties of FCC 046.

The nucleus of FCC 046, which comprises some 10% of the B-band light, is offset from the center of the outer isophotes by some $1.2''$. In the *HST* images, the nucleus is clearly resolved, with FWHM $= 0.27''$, ruling out that it is a foreground star. Moreover, the spectra show that its velocity is identical to the main galaxy, which excludes a background galaxy.

Photometry (Figure 1) reveals an important property of the lopsidedness in FCC 046, namely that it extends beyond the nucleus out to $\sim 4''$ or a full exponential scale-length of the outer light profile. *Thus the lopsidedness is not due merely to a massive globular cluster displaced from the galaxy center.* It is a coherent, large-scale feature of the galaxy. The relative isolation of FCC 046 makes it unlikely that the lopsidedness was tidally induced, unless a long-lived mode was excited (Weinberg 1994). On the other hand, the absence of rotation in FCC 046 (De Rijcke & Debattista 2004) makes it possible that the counter-streaming instability generated the lopsidedness.

To explore this possibility, we compared N-body simulations with slit observations of FCC 046. Here we discuss one such simulation which produced a reasonable facsimile of the global structure of FCC 046. Briefly, the simulation consisted of nucleus inside a flattened (E8) unrotating disk with mass ratio 0.08 : 0.92. The model went through the counter-streaming instability. The resulting system reproduces the gross features of FCC 046 (Figure 1). We then used this N-body simulation to produce a model spectrum consisting of a flat continuum and a δ-function absorption line broadened by the instrumental profile. On the major-axis, the model predicts no evidence of counter-rotation within $8''$ of the nucleus, and increasing line splitting at larger radii. On the minor-axis, the model predicts no line splitting. All these predictions are borne out by the spectra, albeit at low signal-to-noise (see Figure 2).

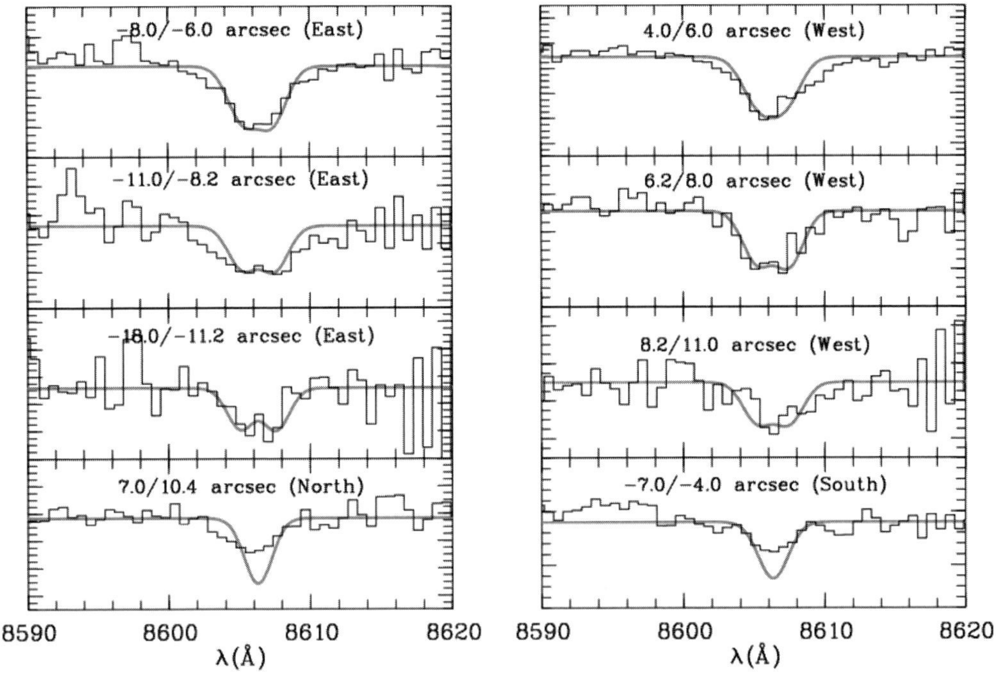

Figure 2. The strongest Ca II absorption line in FORS2 spectra of FCC 046 (histograms) compared with the prediction from the N-body simulation (gray line). The top three panels are along the major-axis centered on the nucleus, while the bottom panel is on the minor axis. Some evidence of counter-streaming is seen on the major-axis where it is predicted by the simulation, and absent where it should be.

3. What causes counter-streaming?

If the counter-streaming instability is producing these lopsided dE,Ns, what causes the counter-streaming in the first place? One possible mechanism is the destruction of box orbits in a slowly, or non-, rotating triaxial galaxy. Box orbits can be destroyed by a growing central mass, which can be the nucleus itself in the case of dE,Ns. If the nucleus grows to more than $\sim$2% in mass, then destruction of box orbits takes a few crossing times ($\sim 10^7$–10^8 yr for a dE,N). Because the colors of nuclei are very similar to those of their host galaxies (Lisker, these proceedings), gas infall is likely not the way that nuclei grow. A more promising possibility is that sinking globular clusters merge at the center to form the nucleus.

4. Conclusions

Although FCC 046 appears to have counter-streaming, the evidence at this point is merely suggestive, not definitive. Testing the counter-streaming hypothesis only requires high spectral resolution and high S/N observations.

If the nuclei grow via the merger of globular clusters, then the nuclei of dE,Ns may provide a channel by which super-massive black holes (SMBHs) grow. We note that a SMBH has been detected in G1, a giant globular cluster believed to be the tidally stripped nuclear remnant of a dE,N.

Acknowledgements

Victor Debattista is supported by a Brooks Fellowship at the University of Washington.

References

Binggeli, B., Barazza, F., Jerjen, H. 2000, *A&A* 359, 447
De Rijcke, S. & Debattista, V.P. 2004, *ApJ* 603, L25
De Rijcke, S., Zeilinger, W.W., Dejonghe, H. & Hau, G.K.T. 2003, *MNRAS* 339, 225
Evans, N.W. & Collett, J.L. 1994, *ApJ* 420, L67
Gnedin, O.Y., Zhao, H.-S., Pringle, J.E., Fall, S.M., Livio, M. & Meylan, G. 2002, *ApJ* 568, L23
Goodman, J. & Binney, J. 1984, *MNRAS* 207, 511
Levison, H.F., Duncan, M.J., & Smith, B.F. 1990, *ApJ* 363, 66
Merritt, D. & Quinlan, G.D. 1998, *ApJ* 498, 625
Merritt, D. & Stiavelli, M. 1990, *ApJ* 358, 399
Meylan, G., Sarajedini, A., Jablonka, P., Djorgovski, S.G., Bridges, T. & Rich, R.M. 2001, *AJ* 122, 830
Palmer, P.L., & Papaloizou, J. 1990, *MNRAS*, 243, 263
Phillipps, S., Drinkwater, M., Gregg, M., & Jones, J. 2001, *ApJ* 560, 201
Sawamura, M. 1988, *PASJ* 40, 279
Sellwood, J.A., & Merritt, D. 1994, *ApJ* 425, 530
Sellwood, J.A., & Valluri, M. 1997, *MNRAS* 287, 124
Weinberg, M.D. 1994, *ApJ* 421, 481
Zang, T.A., & Hohl, F. 1978, *ApJ*, 226, 521

Discussion

MOORE: How does destruction of the box orbits leave you with a counter-streaming population? Would a cuspy halo stabilise the system?

DEBATTISTA: If the system starts out with no angular momentum, by conservation of angular momentum, the scattering of the box orbits must leave a system with no angular momentum. Therefore box orbits will scatter to loop orbits in equal numbers for the two rotation senses. The effect of central mass growth and box orbit scattering is to generate a tangentially biased distribution (Goodman & Binney 1984; Merritt & Quinlan 1998).

In our simulations we find the scale of the lopsidedness decreases as the inner halo becomes increasingly massive. The instability can still occur in fairly massive halos but presumably massive enough cusps would inhibit the instability. However, as with bars, tidal interactions may provoke it in otherwise stable systems. Once a lopsidedness forms, because its motion is generally slow at best, the instability is not efficiently damped.

READ: To what extent does your model rely on the assumption of box orbits initially — would these orbits be expected in the initial conditions?

DEBATTISTA: Not very much. Counter-streaming can be generated in at least one other way: secondary infall. We did not consider this possibility very likely, simply because dwarf ellipticals may not be as efficient at accreting equal-sized objects. But this is admittedly a bias, and may be wrong. Any source of counter-streaming in dE,Ns is a candidate for generating the initial conditions which lead to the lopsidedness. The possibility that counter-streaming is generating these lopsided dE,Ns can be observationally tested independent of the hypothesis that box orbit scattering is giving rise to counter-streaming.

Near-Field Cosmology with Dwarf Elliptical Galaxies
Proceedings IAU Colloquium No. 198, 2005
H. Jerjen and B. Binggeli, eds.

doi:10.1017/S1743921305003704

Evolution of field dwarf galaxies with GEMS

Fabio D. Barazza[1]† and Shardha Jogee[2]

[1]Space Telescope Science Institute, 3700 San Martin Drive, Baltimore, MD 21218, USA
email: barazza@astro.as.utexas.edu

[2]Department of Astronomy, University of Texas at Austin, RLM 15.308, Austin, TX 78712, USA

Abstract. We present a study of the evolution of late-type field dwarfs over the last 1.5 Gyr, based on HST ACS observations carried out as part of the GEMS survey. This study is amongst the first to probe the evolution of dwarfs over such a large timescale. The comparison of structural properties, particularly size and scale length, indicates that the dwarfs in the redshift range $z \sim 0.01$ to 0.15 (look-back times up to 1.8 Gyr) are more extended than local dwarfs. We argue that this difference is due to the star formation activity becoming more centrally concentrated in late-type dwarfs over the last ~1.5 Gyr. We discuss several possible causes for this evolution. We also find a lack of blue compact dwarfs in the GEMS sample and interpret this as indication of the fact that strong, centrally concentrated star formation is a feature of evolved dwarfs that are entering their final stages of evolution.

Keywords. galaxies: evolution, galaxies: dwarf, galaxies: structure

1. Introduction

The study of dwarf galaxies outside of galaxy clusters has, so far, mainly been restricted to the Local Group and the local volume (<10 Mpc). In these regimes the dwarf population is dominated by late-type dwarfs (henceforth referred to as dwarf irregulars, dIs), many of which are isolated, i.e. they are not associated with a giant galaxy. On the other hand, early-type dwarfs (henceforth referred to as dwarf spheroidals, dSphs) are almost absent outside of groups, but they are frequently satellites of giants and therefore quite numerous in groups. Except this morphology-density relation no structural differences have been found between group and field dwarfs. It is therefore still unclear, for instance, whether dIs are the progenitors of dSphs and what processes might lead to their transformation. Detailed studies of field dIs indicate that they generally have an older stellar population and that they would be able to produce stars for another Hubble time, if they would maintain their current low star formation rates (van Zee 2001, Grebel 2004). Hence, if no environmental interaction occurs, these objects will not end up as dSphs by passive evolution.

However, the samples of dwarf galaxies studied so far are, in general, still too small in order to draw firm conclusions. It is therefore desirable to enlarge the database not only at $z = 0$, which might be realized with the Sloan Digital Sky Survey, but also for earlier epochs, where signatures of a possible evolution might be found. In this paper, we present first results based on a study of dwarfs at intermediate redshifts ($z \sim 0.01$–0.15, corresponding to a look-back time of ~1.8 Gyr)‡, observed as part of the Galaxy Evolution from Morphology and SEDs (GEMS, Rix *et al.* (2004)) survey. GEMS consists

† Present address: Department of Astronomy, University of Texas at Austin, RLM 15.308, Austin, TX 78712, USA
‡ We assume in this paper a flat cosmology with $\Omega_M = 1 - \Omega_\Lambda = 0.3$ and $H_0 = 65$ km s^{-1} Mpc^{-1}.

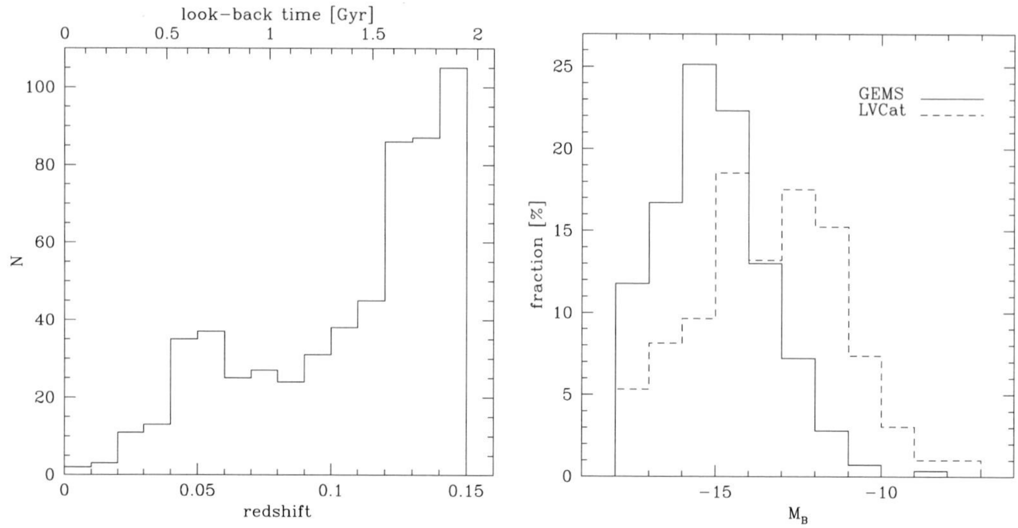

Figure 1. *left:* Redshift distribution of the final GEMS dwarf sample. *right:* Luminosity distributions of the GEMS sample and the catalog of neighboring galaxies (LVCat).

of the largest-area multi-filter survey carried out to date with the Advances Camera for Surveys (ACS) on the *Hubble Space Telescope* (*HST*), and might, therefore, offer the best premise for the study of dwarfs at higher redshifts. The GEMS survey produced high-resolution ($0\farcs05$ correspond to 168 pc) images for $\sim$ 8300 galaxies out to $z \sim 1.2$, for which accurate redshifts ($\delta_z/(1+z) \sim 0.02$ down to $R_{Vega} = 24$) are available from the COMBO-17 project (Wolf *et al.* 2004).

2. The GEMS sample and the local comparison samples

The largest local dwarf sample is the catalog of neighboring galaxies (Karachentsev *et al.* 2004), which exhibits a peak around $M_B \sim -14$ mag in the luminosity distribution of galaxies. We use this catalog for the comparison of the luminosity distribution of the GEMS sample. However, since this catalog does not provide additional structural parameters, we compiled a second sample of local galaxies from the literature, using the following three sources: Parodi, Barazza, & Binggeli (2002, and references therein), a sample of 80 field and group late-type dwarfs (LV); van Zee (2000), a sample of 43 isolated dIs (ISO); Cairós *et al.* (2001) a sample of 15 blue compact dwarfs (BCD). These samples have been selected more or less randomly, i.e. they are not magnitude limited. But together they comprise all classes of local dwarfs commonly studied and they cover all environments outside of clusters. and are, therefore, appropriate to compare to the GEMS sample.

In order to be complete for galaxies with $M_B < -14$ mag, we had to restrict the GEMS sample to $z < 0.15$. This limit was determined by considering the detection limit of the COMBO-17 survey, i.e. at a redshift of $z \sim 0.15$ the faintest galaxies observed have $M_B \sim -14$ mag. (Note that we did not apply a K-correction.) For all galaxies we determined the surface brightness profile and fitted an exponential model to it. Based on a rough visual inspection we excluded all objects with signs of interaction or overlapping with other objects. The left panel of Figure 1 shows the redshift distribution of the dwarf sample. More than 70% of the objects are at look-back times of >1 Gyr. The right panel of Figure 1 shows the luminosity distributions of the GEMS sample and the catalog of

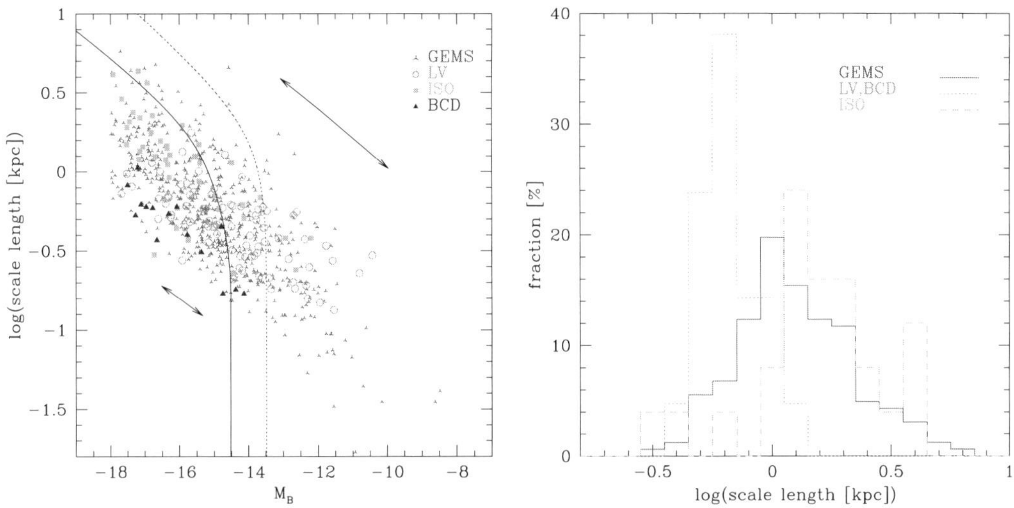

Figure 2. *left:* Exponential scale length versus the absolute B-band magnitude. The solid and dashed lines are the 100% and 50% completeness limits, respectively, of the GEMS sample. The bold line with arrows represent the typical error bars caused by redshift errors and show the direction along which such errors would move the data points. Note that the measurements for the GEMS dwarfs have been performed on the V-band image, which corresponds to a rest-frame band somewhere between B and V depending on the redshift. *right:* Histograms of the scale lengths for objects with $M_B < -16$ mag.

neighboring galaxies. The agreement is quite good (keeping in mind that the GEMS sample is roughly complete for objects with $M_B < -14$ mag), which means that out study targets systems of similar luminosity across the GEMS and local samples.

3. Results

We compared structural parameters for the local and GEMS dwarfs. Figure 2 shows the most striking result. The GEMS dwarfs with $M_B < -16$ mag in the redshift range 0.01 to 0.15 seem to be more extended than the local dwarfs, particularly than LV and BCD dwarfs, which tend to be located in galaxy groups or are even satellites of giant galaxies. In contrast, no significant difference is seen between GEMS dwarfs and the ISO sample of more isolated local dwarfs. Since our study uses filter bands mainly tracing the younger stellar populations, we argue that the difference in scale lengths is the result of differences in the spatial extent of star formation sites. In GEMS dwarfs the regions of active star formation are less centrally concentrated than in the local dwarfs.

It remains to be determined whether the different extent of star formation activity in the two samples is caused by differences in environment or is simply a temporal sequence. In the latter scenario, dwarfs at earlier epochs (as typified by the GEMS dwarfs) might evolve with time into local-type dwarfs, where the extended stellar population ages passively, while star formation is sustained in the central regions. The replenishment of the central gas supply could be caused by spontaneously or tidally induced non-axisymmetric features, such as bars or through accretion of satellites. We note that the irregular appearance of dIs does not make such an alternative unlikely and that a majority of late-type disk galaxies have bars. Moreover, this interpretation receives support from the fact that a majority of local dwarfs exhibit red color gradients, i.e. they become redder with increasing radius. In addition, in all Local Group dwarfs, for which this information is known, the older stellar population is more extended than the younger one (van den Bergh 1999). It is rather unlikely that these old stars have been formed near the centers of the galaxies and have subsequently been dispersed to larger radii, since the required mass losses (Ferrara & Tolstoy 2000) would have to be caused by strong star bursts, which are not observed in the star formation histories of typical dwarfs. Thus, if they have been formed in situ, the star formation activity must have been more extended in these dwarfs in the past.

However, the better agreement between the GEMS sample and the ISO sample provides support for the scenario, in which the finding is caused by differences in the environment. This might, moreover, suggest that small galaxy groups, which are common in the local volume, only formed in the last 1–2 Gyr.

Another interesting result is the apparent lack of BCDs in the GEMS sample. This finding implicates that it is very unlikely that BCDs represent the initial stages of the evolution of late-type dwarfs or that late-type dwarfs frequently undergo a phase of strong, centrally concentrated star formation. We rather suggest that BCD-like structures are a feature of evolved dwarfs, entering their final stages of evolution.

However, it has to be kept in mind that the local samples are incomplete and that, therefore, systematics in the GEMS or local samples can affect the result. In a future paper (Barazza *et al.* 2005, in prep.) we will therefore compare our GEMS sample also to a sample of local dwarfs drawn from the SDSS.

Acknowledgements

F.D.B. and S.J. acknowledge support from the National Aeronautics and Space Administration (NASA) under LTSA Grant NAG5-13063 issued through the Office of Space Science. Support for GEMS was provided by NASA through GO-9500 from the Space Telescope Science Institute, STScI, which is operated by the Association of Universities for Research in Astronomy, Inc. AURA, Inc., for NASA, under NAS5-26555.

References

Barazza, F.D. *et al.* 2005, in prep.

Cairós, L.M., Caon, N., Vílchez, J.M., González-Pérez, J.N. & Muñoz-Tuñón, C. 2001, *ApJS* 136, 393

Ferrara, A. & Tolstoy, E. 2000, *MNRAS* 313, 291

Grebel, E.K. 2004, in Origin and Evolution of the Elements, *Carnegie Observatories Astrophysics Series*, Cambridge University Press, 237

Karachentsev, I.D., Karachentseva, V.E., Huchtmeier, W.K. & Makarov, D.I. 2004, *AJ* 127, 2031

Parodi, B.R., Barazza, F.D. & Binggeli, B. 2002, *A&A* 388, 29

Rix, H. *et al.* 2004, *ApJS* 152, 163

van den Bergh, S. 1999, *A&AR*, 9, 273

van Zee, L 2000, *AJ* 119, 2757

van Zee, L. 2001, *AJ* 121, 2003

Wolf, C., *et al.* 2004, *A&A* 421, 913

Discussion

Tolstoy: The very faint galaxies in your GEMS sample, being of comparable luminosity to local dSphs, do they look like them?

Barazza: No. They appear very compact, are forming stars, have a high surface brightness, and are also very blue.

Near-Field Cosmology with Dwarf Elliptical Galaxies
Proceedings IAU Colloquium No. 198, 2005
H. Jerjen and B. Binggeli, eds.

doi:10.1017/S1743921305003716

Giant halos in dwarf irregular galaxies vs. dwarf elliptical galaxies

Myung Gyoon Lee and Narae Hwang

Astronomy Program, SEES, Seoul National University, Seoul, 151-742, Korea
email: mglee@astrog.snu.ac.kr, nhwang@astro.snu.ac.kr

Abstract. We present a discovery of a giant stellar halo in NGC 6822, a dwarf irregular galaxy in the Local Group. This halo is mostly made of old red giants, showing striking features: 1) it is several times larger than the main body of the galaxy seen in the optical images, and 2) it is elongated in the direction almost perpendicular to the HI disk of NGC 6822. The structure of this stellar halo looks similar to the shape of dwarf elliptical galaxies, indicating that the halos of dwarf irregular galaxies share the same origin with those of the dwarf elliptical galaxies.

Keywords. galaxies:halos, galaxies: individual (NGC 6822), galaxies: dwarf

1. Introduction

It has been known long that old stellar populations exist in the dwarf irregular galaxies and dwarf elliptical galaxies and the I-band magnitude of the tip of these old populations (red giant branch) has been often used to estimate their distances (Lee *et al.* 1993). However, little is known about the size and morphological structure of the halos of these old stellar populations in the dwarf galaxies. A study of the morphological structure of the halos in dwarf galaxies may provide some clues to understanding the origin of the dwarf irregular galaxies and the dwarf elliptical galaxies. We have carried out a project to investigate the halos in dwarf irregular galaxies using the MegaPrime camera at the CFHT. Here we present a case study of NGC 6822 which is a famous dwarf irregular galaxy in the Local Group. NGC 6822 is known to have a huge rotation HI disk (de Blok & Walter 2000) and a wide distribution of carbon stars (Letarte *et al.* 2002). It is an well isolated galaxy belonging neither to our Galaxy nor to M31. NGC 6822 is located at the distance of 500 kpc, and one arcmin at this distance corresponds to 145.4 pc.

2. Color-Magnitude Diagram of the Dwarf Irregular Galaxy NGC 6822

We display in Figure 1 the $i - (g - i)$ color-magnitude diagram of three regions at $R < 5'$, $5' < R < 10'$, and $10' < R < 15'$ where R is the angular distance from the center of NGC 6822. This is part of the data covering a 1 deg × 1 deg field including NGC 6822 (corresponding to a field of 8.7 kpc × 8.7 kpc). Figure 1 shows several distinguishable stellar populations in NGC 6822 as marked by the boxes: blue main sequence stars and blue supergiants (MS), red supergiants (RSG), old red giant branch stars (RGB), asymptotic giant stars (AGB), and young red giants (RG). A broad vertical structure seen at $(g - i) \approx 0.9$ represent foreground stars.

Young stars are mostly seen at $R < 10'$, while the old RGB stars are abundant even at $R > 10'$. This old population may belong to the halo of NGC 6822. We have used the stars inside each box for the following analysis to investigate the morphological structure of each population.

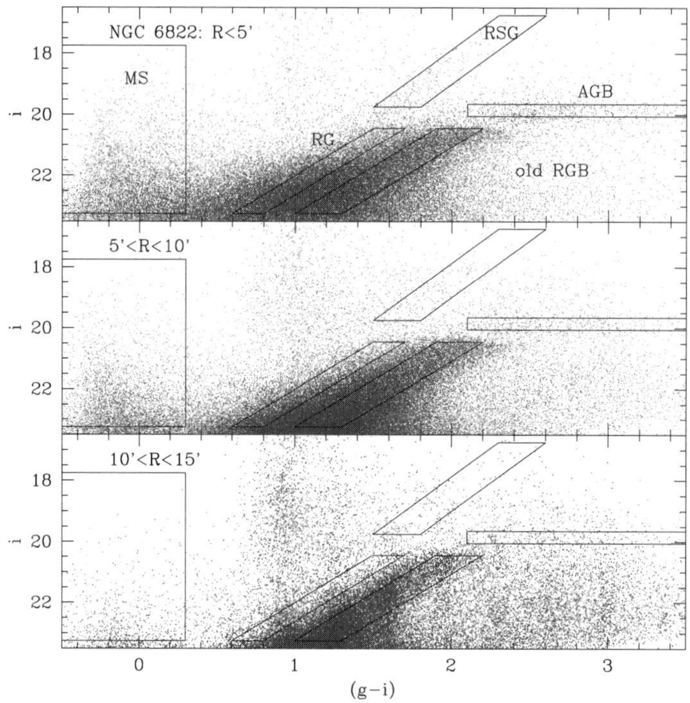

Figure 1. Color-magnitude diagrams of the stars at $R < 5'$ (top panel), $5' < R < 10'$ (middle panel), and $10' < R < 15'$ (bottom panel) where R is the angular distance from the center of NGC 6822. Thick solid line boxes represent the regions selected for the analysis of the different stellar populations.

3. Spatial Distribution of Stellar Populations in NGC 6822

We have derived the surface number density maps for the MS, young RG and old RGB in the field of 1 deg × 1 deg to investigate their spatial distribution, displaying them in Figure 2 in comparison with the greyscale map of the digitized POSS map of NGC 6822. Figure 2 shows several striking features.

First, the structure of the old RGB (halo) is dramatically different from the main body ($10' \times 20'$) of NGC 6822 seen in the POSS map. It is surprisingly much larger than the POSS image, exceeding one squaredegree field. Its inner region is elongated horizontally, while its outer region is elongated along the NE-SW direction. Its inner region looks elliptical, very similar to the structure of dwarf elliptical galaxies. However, its outer part shows some distorted structure in the NE-SW end indicating some tidal interaction. The existence of this tidal feature is intriguing, because NGC 6822 has no known companion galaxies in the nearby distance.

Second, the spatial distribution of the young MS in the central region is similar to the shape of the main body of NGC 6822 seen in the POSS map, but it is much more extended and elongated along the NW-SE direction. This structure is consistent with the HI structure (de Blok & Walter 2000).

Third, the spatial distribution of the young RG looks round and similar to that of the central part of the old RGB distribution, but its center is offset from the center of the

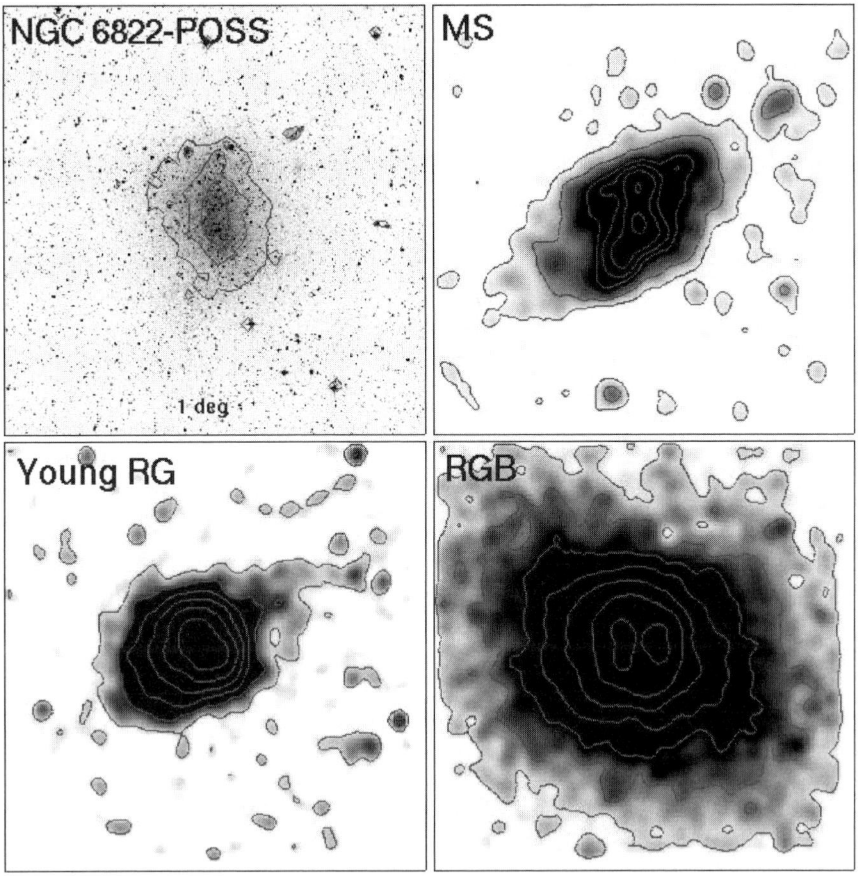

Figure 2. A grayscale map of the digitized POSS of NGC 6822 (upper left) and the surface number density maps of the different populations in NGC 6822 (upper right: young main sequence stars; lower left: young red giants, and lower right: red giant branch stars). Each panel covers a field of 1 deg × 1 deg. North is up, and east to the left.

RGB distribution. In addition, it shows stronger central concentration than that of the old RGB.

We have also investigated the radial variation of the surface number density for the RGB, RG, MS, AGB and carbon stars (Letarte *et al.* 2002) as plotted in Figure 3. Figure 3 shows that the young populations (MS and RG) are more centrally concentrated than the older populations (RGB, AGB and carbon stars).

4. Conclusion

Through the deep wide-field imaging we have found that NGC 6822 has a giant halo made of old red giants, extending much farther than the optical body and HI disk of this galaxy. The structure of this stellar halo looks similar to those of dwarf elliptical galaxies so that we conclude that the halos of dwarf irregular galaxies may share the same origin with those of the dwarf elliptical galaxies. NGC 6822 also is known to have an old globular cluster at as far as 13 kpc from the center, supporting that its halo is really huge (Hwang *et al.* 2005). The existence of the giant halos may be common among dwarf irregular galaxies, waiting to be discovered.

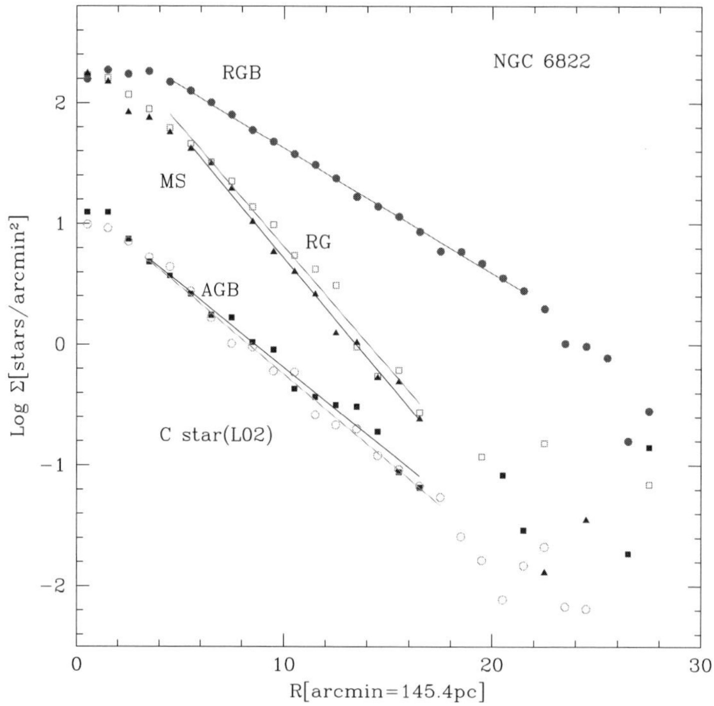

Figure 3. Radial variation of the stellar surface number density of the different populations in NGC 6822 (as marked in Figure 1). Filled circles: old red giants (RGB), open squares: young red giants (RG), filled triangles: young main sequence stars and blue supergiants (MS), filled squares: asymptotic giant branch (AGB), and open circles: carbon stars (Letarte *et al.* 2002). Solid lines represent power-law fits to the data.

Acknowledgements

This work is supported in part by the ABRL grant (R14-2002-058-010000-0) and the BK21 program.

References

de Blok, W.J.G. & Walter, F. 2000, *ApJ* 537, L95
Hwang, N. *et al.* 2005, in preparation
Lee, M.G., Freedman, W.L. & Madore, B.F. 1993, *ApJ* 417, 553
Letarte, B., Demers, S., Battinelli, P. & Kunkel, W.E. 2002, *AJ* 123, 832

Discussion

GALLAGHER: It is known that there is significant differential reddening over the field of NGC 6822. Is there any possibility that the offset between the centers of the old RGB and young RG distribution may be due to differential reddening effect?

LEE: There is some differential reddening over the field of NGC 6822, but it is not large enough to affect our results.

Near-Field Cosmology with Dwarf Elliptical Galaxies
Proceedings IAU Colloquium No. 198, 2005
H. Jerjen and B. Binggeli, eds.

doi:10.1017/S1743921305003728

Clues to dwarf galaxy formation from clustering and kinematics

James E. Taylor[1], Joe Silk[2] and Arif Babul[3]

[1]Astronomy Department, California Institute of Technology, mc 105-24, 1200 E. California Blvd., Pasadena, CA 91125, USA
email: jet@astro.caltech.edu

[2]Astrophysics, University of Oxford, Denys Wilkinson Building, Keble Road, Oxford, OX1 3RH, UK
email: silk@astro.ox.ac.uk

[3]Department of Physics and Astronomy, University of Victoria, Elliott Building, 3800 Finnerty Road, Victoria, B.C., V8P 1A1, Canada
email: babul@uvic.ca

Abstract. If the CDM paradigm is correct, then dwarf galaxies are far rarer than dark matter haloes of comparable circular velocity at the present day. Some process or processes must suppress galaxy formation almost completely on these small scales. Several plausible mechanisms may explain the scarcity of dwarf galaxies in the local universe, but it has been hard to distinguish between the different models observationally. The spatial distribution and kinematics of dwarf galaxies contain additional information about the age and formation history of these systems. Interpreted in the CDM framework, they provide powerful new constraints on the physics of dwarf suppression, ruling out at high significance the simplest models, in which the halo occupation probability is constant above some fixed mass or velocity threshold. We describe recent tests of small-scale galaxy formation models using the satellites of the Local Group and other nearby groups, and discuss the implications of these results for studies of the stellar halo, sub-galactic dark matter and the reionization of the universe at high redshift.

Keywords. dark matter; galaxies: formation, dwarf, Local Group; Galaxy: halo

---

1. Introduction

In the current cosmological model, cold dark matter (CDM) haloes are the sites of all galaxy formation. Haloes build up progressively through hierarchical merging, such that the most massive haloes at any epoch are also the 'youngest', having assembled their mass into a single structure most recently. A naïve application of this trend to galaxy formation suggests that massive galaxies like the Milky Way should have formed recently, while small galaxies like the dwarf spheroidals of the Local Group may be 'building blocks' left over from an earlier era of structure formation. This picture seems consistent with some observations – the chemical evidence for continuous infall of new gas into the Milky Way, for instance – but fails to explain many others, notably the lack of active star formation in cluster ellipticals, the most massive galaxies in the present-day universe, and its continued presence in field dwarf galaxies.

At least part of this apparent disagreement stems from neglecting the other strong correlation observed in CDM simulations, between age and environment. Dense environments produce structure on any given scale faster than low-density environments, and consequently we expect the oldest galaxy-mass systems to be in rich clusters and the youngest to be in the field and in voids. Applying the same logic to smaller scales, dwarf galaxies *within* virialized systems, such as the dozen small satellites orbiting in the halo of the Milk Way, should be older than similar objects seen in the field, with a continuous gradient in age extending from one environment to the other.

This trend in age is particularly interesting when combined with the other puzzling prediction from CDM models. Simulations of CDM halo formation suggest that the Milky Way should be surrounded by hundreds of dark matter cores dense enough to host dwarf galaxies, whereas in practice only a dozen luminous satellites are observed. While there are several plausible ways of suppressing galaxy formation on these scales – the heating effects of reionization or internal feedback from supernovae, for instance – it has been hard to test and distinguish between these different scenarios observationally.

The age information inherent in the positions and kinematics of the local dwarfs provides a new way of addressing this problem. Comparing the observed distributions to semi-analytic models of halo substructure, we will show evidence that dwarf galaxies populate the oldest subcomponents of the dark matter halo around the Milk Way, rather than the most massive ones. This implies a change in the efficiency of dwarf galaxy formation with redshift, favouring external feedback models over internal ones.

2. Modelling the Local Satellite Distribution

We model the evolution of substructure in the dark matter halo of a system like the Milky Way using the semi-analytic model described in Taylor & Babul (2004). This model combines Press-Schechter merger histories, which predict when subhaloes merge into the halo of the main system, with an analytic description of orbital evolution and mass loss, which determines how they evolve subsequently. While the model is approximate in many ways, it has shown good agreement with self-consistent numerical simulations at the highest resolution (Taylor & Babul 2005a,b). The model produces catalogues of subhaloes orbiting within the final system, with full information about their current and prior mass, structure and orbit.

Using this model we generated a set of 144 merger trees for a concordance LCDM cosmology ($\Lambda = 0.7$, $\Omega_{\rm m} = 0.3$, $h = 0.7$, $\sigma_8 = 0.9$). Each model halo had a final mass of 1.6×10^{12}, consistent with mass estimates for the Milky Way, and in each system we also modelled the dynamical effects of a growing galactic component, letting a disk and bulge build up following the prescriptions described in Islam, Taylor & Silk 2003. In each halo, we then identified the 'first-ranked' subhaloes at the present day, ranking subhaloes by their *original* peak circular velocity $V_{\rm peak,0}$, that is the *maximum* value this quantity attained at any point in the system's past.

Figure 1 compares the positions and velocities of first-ranked systems ($V_{\rm peak,0} > 45\,{\rm km\,s^{-1}}$) in a few of our model haloes with the same quantities for the known satellites. The spatial spatial distribution of the first-ranked subhaloes is clearly very different from that of the observed satellites, which lie at smaller galactocentric radii and at negative velocities. Overall, fewer than 1% of our model haloes have their dozen first-ranked satellites as tightly clustered as the satellites of the Milk Way. Given the age gradient in substructure with galactocentric distance, the simplest way to account for the different distributions is to suppose that dwarf galaxies occupy the oldest subhaloes preferentially.

3. Implications

Dark Satellites: Figure 1 shows that the visible satellites are more clustered around the Milky Way than its most massive subhaloes should be. One immediate implication is that some of the largest dark matter structures nearby contain no visible stars. Such

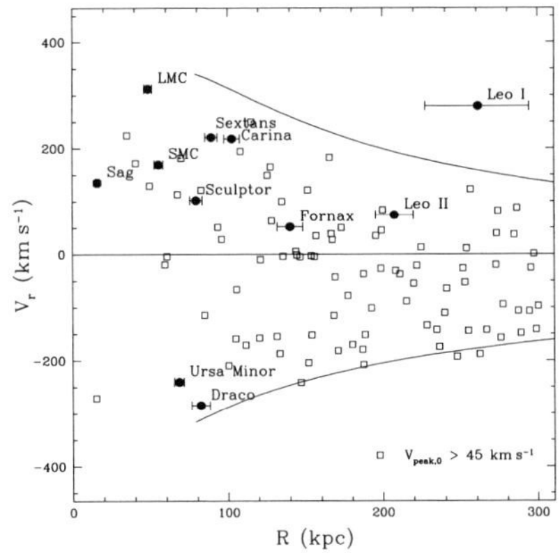

Figure 1. Radial velocity vs. position for the known galactic satellites (solid points with error bars) and massive subhaloes in a set of model haloes (filled squares).

massive ($\sim 10^9$–$10^{10} M_\odot$) dark satellites might be detectable around distant galaxies that are strong gravitational lenses (Chen, Dalal & Taylor in preparation).

Reionization: If local dwarf galaxies preferentially occupy the oldest substructures, this implies that dwarf formation was more efficient at high redshift, possibly because of the absence of a strong ionizing background. In fact, dwarf galaxies like those in the Local Group may be the source of much of this ionizing background (Yan & Windhorst 2004), and enhanced dwarf galaxy formation may explain why the universe was reionized at a relatively high redshift ($z \gtrsim 10$–15? Spergel *et al.* 2003) and how it remained ionized during the pre-quasar era ($z \sim 4$–6).

Tidal Streams: If we assume an efficiency for galaxy formation that decreases with time in such a way as to reproduce the number and spatial distribution of the observed dwarfs, then we can also comment on how many visible satellites may have merged with the Milky Way in the past. We set a threshold in the (original) peak circular velocity $V_{\rm peak,min} = 135\ {\rm km\,s^{-1}}(1+z)^{2/3}$ above which all haloes are occupied by galaxies, where the normalization is fixed to get the right number of surviving systems and the slope was motivated by the results of Kravtsov Gnedin & Klypin (2004). Figure 2a shows the properties of the luminous satellites selected using this prescription. We find that if a galaxy like the Milky Way has 10 surviving luminous satellites at the present day, there should be an equal number of heavily stripped or disrupting systems in its halo, together with the debris from ten times as many completely disrupted systems. This suggests that future surveys of the stellar halo could detect abundant substructure. One the other hand, we also find that most of these satellites first merged with the main system more than 10 Gyr ago, so their debris may heavily mixed and hard to isolate in practice.

Satellites in Other Nearby Groups: Our conclusions are based on the properties of the satellites of the Milky Way, which may not be a typical case. It is essential to build up observations of other similar systems to confirm our general conclusions about dwarf galaxy formation. While there are a number of well-studied groups in the local universe (Karachentsev *et al.* 2005), distances to their satellites are uncertain by hundreds of kpc, leaving only a weaker, projected signature of clustering. Intriguingly, projected distributions of satellites around the Milky Way do appear slightly more clustered than those of other nearby systems (figure 2b). Dwarf galaxies are faint, low surface-brightness objects, however, so catalogues of these objects may be incomplete, especially for objects so close to their parent system they are projected near or on top of it. Further surveys

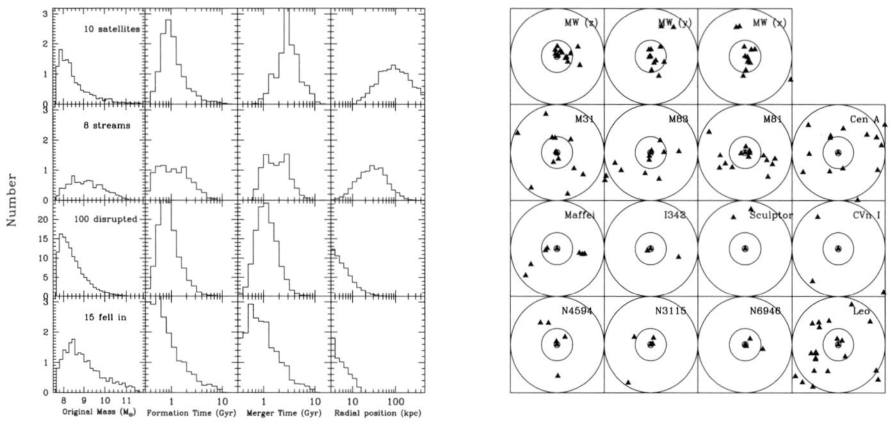

Figure 2. a) Original mass, formation time, merger time, and galactocentric distance (four columns, left to right) for luminous satellites that are intact, heavily stripped, disrupted and centrally merged (four rows, top to bottom) in the model haloes at the present day. **b)** The dwarf satellites of the Milky Way, projected along three axes (top three panels), compared with the projected distribution of satellites in a number of other local systems (Karachentsev 2005). The small and large circles in each panel indicate 0.3 and 1.0 virial radii respectively, assuming halo mass scales with galaxy mass.

of nearby groups should help fix the strength of dwarf galaxy clustering, giving stronger constraints on dwarf formation models.

Acknowledgements

The authors gratefully acknowledge helpful discussions with A. Ferguson, A. Kravtsov and I. Karachentsev, as well as financial support from the U.K. Particle Physics and Astronomy Research Council (PPARC), the U.S. National Science Foundation (grant AST-0307859) and the U.S. Department of Energy (contract DE-FG02-04ER41316).

References

Islam, R.R., Taylor, J.E. & Silk, J. 2003, *MNRAS* 340, 647
Karachentsev, I.D. 2005, *AJ* 129, 178
Kravtsov, A.V., Gnedin, O.Y. & Klypin, A.A. 2004, *ApJ* 609, 482
Spergel, D.N. *et al.* 2003, *ApJS* 148, 175
Taylor, J.E. & Babul, A. 2004, *MNRAS* 348, 811
Taylor, J.E. & Babul, A. 2005a, *MNRAS* submitted (astro-ph/0410048)
Taylor, J.E. & Babul, A. 2005b, *MNRAS* submitted (astro-ph/0410048)
Yan, H. & Windhorst, R.A. 2004, *ApJ* 600, L1

Discussion

DEKEL: How model-dependent is your argument concerning the clustering of dark halos versus MW dwarfs? How does it compare to N-body simulations?

TAYLOR: The argument is generally model-independent. The only major uncertainty in the modelling concerns highly stripped halos; it is not clear whether these would correspond to dwarfs, tidal streams, or both (e.g., Sagittarius). The N-body simulations find the same result (cf. Diemand's talk).

Near-Field Cosmology with Dwarf Elliptical Galaxies
Proceedings IAU Colloquium No. 198, 2005
H. Jerjen and B. Binggeli, eds.

doi:10.1017/S174392130500373X

Modified newtonian dynamics: success and problems on globular cluster scale

HongSheng Zhao†

University of St Andrews, School of Physics and Astronomy, KY16 9SS, UK
National Astronomical Observatory of China, Chinese Academy of Science, Datun Road A20, Chaoyang District, Beijing, 100012, China

Abstract. Many past attempts to kill MOND have only strengthened the theory. Better data on galaxy velocity curves clearly favor MOND (without fine-tuning) over cold dark matter. The usual critism on the incompleteness of classical MOND has spurred a Modified Relativity (MR) by Bekenstein. After outlining cosmology and lensing in MOND, we review MOND on small scales. We point out some potential problems of MOND in two-body relaxation and tidal truncation. We argue that the tidal field in any MOND-like gravity theory predicts that the Roche lobe sizes of a binary system are simply proportional to the binary baryonic mass ratio to the power 1/3. An immediate application of this result is that the tidal field and tidal truncation radii of million-star globular clusters and million-star dwarf galaxies (e.g., the Milky Way satellites NGC2419 and Carina) would be very similar because of the one-to-one relation between gravity and baryon distribution. This prediction appears, however, inconsistent with the fact that *all* globulars are truncated to much smaller sizes than *all* dwarf galaxies. Whether tide is uniquely determined by baryons can also be used to falsify any MOND-like gravity theory, whether classical or relativistic.

Keywords. dark matter

1. Introduction

A large amount of astronomical data suggests that the GM/r^2 Newtonian gravity from the baryonic material in galaxies fails to explain the large accelerations implied by motions in spiral and elliptical galaxies. However, Zwicky's proposal to introduce unseen gravitating (dark) mass with a very flexible density distribution wherever needed for the motions of galaxies and stars has been troubled by a lack of experimental confirmation of its elementary particle counterparts for more than 60 years. Evidence is also summarized in recent articles about the difficulties facing dark matter (DM) theories on galaxy scales (Spergel & Steinhardt 2000, Ostriker & Steinhardt 2003). Surprisingly, an alternative gravity theory with Modified Newtonian Dynamics (MOND, Milgrom 1983) has been doing very well exactly where the dark matter theory is doing poorly. The predictive power of this 20-year-old classical theory with virtually no free parameters (Bekenstein & Milgrom 1984) is recently highlighted by the astonishingly good fits to contemporary kinematic data of a wide variety of high and low surface brightness spiral and elliptical galaxies; even the fine details of the ups and downs of velocity curves are elegently reproduced without fine tuning of the baryonic model (Sanders & McGaugh 2002, Milgrom & Sanders 2003). Originally it was proposed that galaxy rotation curves could be fit by motions in a modified gravity $\mathbf{g}$, which is significantly stronger than Newtonian gravity

† Overseas Outstanding Young Scholar in Beijing Observatory and PPARC Advanced Fellow

in the weak regime defined by a gravitational field energy density

$$\frac{|\mathbf{g}|^2}{8\pi G} \leqslant \frac{a_0^2}{8\pi G} \approx \rho_b(0)c^2, \tag{1.1}$$

where G is the gravitational constant, and $\rho_b(0)c^2$ is the present universal mean baryon energy density (commonly the weak regime is defined as where gravity $g \leqslant a_0 \sim 1.2 \times 10^{-8}\mathrm{cm\,sec}^{-2}$). Far away from the baryons, the modified gravity $\mathbf{g}$ satisfies

$$\frac{\mathbf{g}}{\tilde{G}(g)} \approx \frac{\mathbf{g}_N}{G} \equiv -\sum_i \frac{M_i \hat{\mathbf{R}}_i}{|R_i|^2} \tag{1.2}$$

where $\mathbf{g}_N$ is the vector sum of Newtonian gravity (distance-inverse-squared attractive force) of all baryonic particles with masses M_i at distances R_i, where the effective gravitational constant $\tilde{G}$ can be chosen as, e.g., $\frac{G}{\tilde{G}(g)} = 1 - \exp[-(\frac{g}{a_0})^\alpha]$, where $\alpha = 1$ gives the MOND gravity $\mathbf{g}$. We get a MOND-like theory for any positive α. In these theories, the only matter that matters is the luminous (baryonic) matter, and the gravitational field is a *unique* function of the luminous matter distribution.

These alternative theories based on modifying the law of gravity in a purely baryonic universe have been gaining ground rapidly. Milgrom's empirical formula is a simplified treatment of a classical theory with a curl-free gravitational field by Bekenstein & Milgrom (1984), which is now extendable to a Modified Relativistic (MR) theory (Bekenstein 2004) which passes standard and cosmological tests used to check General Relativity (GR). This recent breakthrough by Bekenstein transforms the empirical MOND formula to a falsifiable theory with respectable theoretical underpinnings comparable almost to the rival dark matter theory. Here we examine various aspects of the MOND theory after outlining cosmology and lensing in the relativistic MOND.

1.1. *Cosmology and Lensing*

Bekenstein (2004) has shown that MONDian cosmology and lensing follow the standard GR formulation. The physical metric of the FRW universe in MOND is written as,

$$g_{\mu\nu}dx^\mu dx^\nu = c^2dt^2 - R(t)^2\left[d\chi^2 + f(\chi)^2\left(d\theta^2 + \sin^2\theta d\phi^2\right)\right] \tag{1.3}$$

where $f(\chi) = \sin\chi, \chi, \sinh\chi$ for closed, flat and open universe as in FRW GR, and the physical scale factor is $R(t) = R_0/(1+z)$. Note that we have removed the tilde sign that Bekenstein uses to denote physical coordinates. if we introduce the vaccum energy and neglect the scale field ϕ as in Bekenstein, then the MONDian universe expands with a Hubble parameter $H(z(t)) = \frac{d\ln R(t)}{dt}$ given approximately by

$$\frac{H^2(z)}{H_0^2} \approx \Omega_m(1+z)^3 + \Omega_r(1+z)^4 + \Omega_K(1+z)^2 + \Omega_\Lambda, \Omega_K \equiv 1 - \Omega_m - \Omega_r - \Omega_\Lambda, \tag{1.4}$$

where $\Omega_r \sim 10^{-4}h^{-2}$ is normalised by the 2.7K CMB, and $\Omega_m \sim 0.05h^{-2}$ from all observable baryons, and Ω_K is the curvature term. The matter-radiation equality happens at redshift of about $\Omega_m/\Omega_r \sim 1000$; this is much lower than in CDM models.

To do lensing, Bekenstein shows (in eq. 116) that for small perturbations of the metric by a potential Φ, light travels on geodesic given by

$$ds^2 = \tilde{g}_{\mu\nu}dx^\mu dx^\nu \approx -\left(1 + \frac{2\Phi}{c^2}\right)c^2dt^2 + \left(1 - \frac{2\Phi}{c^2}\right)dl^2 = 0, \tag{1.5}$$

where dl is the proper length. Hence the light travel time

$$dt \approx \left(1 - \frac{2\Phi}{c^2}\right)\frac{dl}{c}, \tag{1.6}$$

same as in Einstein universe. The gravititional potential satisfies approximately the MOND eq.

$$-\nabla \cdot \left[\mu\left(\frac{g}{a_0}\right)\mathbf{g}\right] = 4\pi G\rho, \qquad \mathbf{g} = -\nabla\Phi, \tag{1.7}$$

where ρ is the gravitational mass density of baryonic matter, and the μ is the usual dimensionless function of the rescaled gravity $\frac{g}{a_0}$ introduced by Milgrom.

The deflection for a light array of the closest approach (roughly impact parmameter, physical length) b is calculated by

$$\alpha = \int \frac{2g_\perp dt}{c} \tag{1.8}$$

where $g_\perp = g(r) \cdot \frac{b}{r}$ is the perpendicular component of the gravity $g(r)$ at radius $r = \sqrt{(ct)^2 + b^2}$, and ct is the path length along the line of sight counted from the point of closest approach. Hence the deflection angle

$$\alpha = \int_{-\infty}^{\infty} \frac{2g(r)dt}{c} \frac{b}{\sqrt{(ct)^2 + b^2}}. \tag{1.9}$$

This is essentially the same as eq. (109) of Bekenstein.

2. Two-body Relaxation Time in MOND

As pointed by Ciotti & Binney (2004), relaxation in MOND is a non-trivial calculation because it is a long-range gravity with a divergent potential at large distance. However, by and large, MOND gravity is effectively a stronger-than-Newtonian gravity. This suggests that we can rescale Newtonian gravity by a factor $\frac{\tilde{G}}{G}$. To balance this stronger gravity, a bound object of certain size r acquires a higher internal velocity (than Newtonian case with pure baryons), hence shorter dynamical time, and faster relaxation since the relaxation time is proportional to dynamical time with

$$t_x \propto N t_{\rm dyn} \sim \frac{Nr}{CV}, \tag{2.1}$$

where $C \sim 7 \ln N \sim 50$ is a large constant related to the Coulomb logarithm, and the dynamical time

$$t_{\rm dyn} = \left(\frac{r}{V}\right) \sim 1\,{\rm Myr}\frac{r}{\rm pc}\frac{1\,{\rm km\,s}^{-1}}{V} \tag{2.2}$$

where r is the half-mass radius, and V is characteristic circular velocity. So for $N \sim 10^{4-5}$ globular cluster, it relaxs on time scales of 100-1000 dynamical times. Typical dynamical time is of order Myr, hence a globular is generally relaxed in a fraction of the Hubble time. In MOND the circular velocity is generally given by

$$V = (GMa_0)^{1/4} = 3.5\,{\rm km\,s}^{-1}\left(\frac{M}{10^4 M_\odot}\right)^{1/4} \tag{2.3}$$

if in deep-MOND where $r \geqslant \sqrt{\frac{GM}{a_0}} = 3.5\,{\rm pc}\sqrt{\frac{M}{10^4 M_\odot}}$.

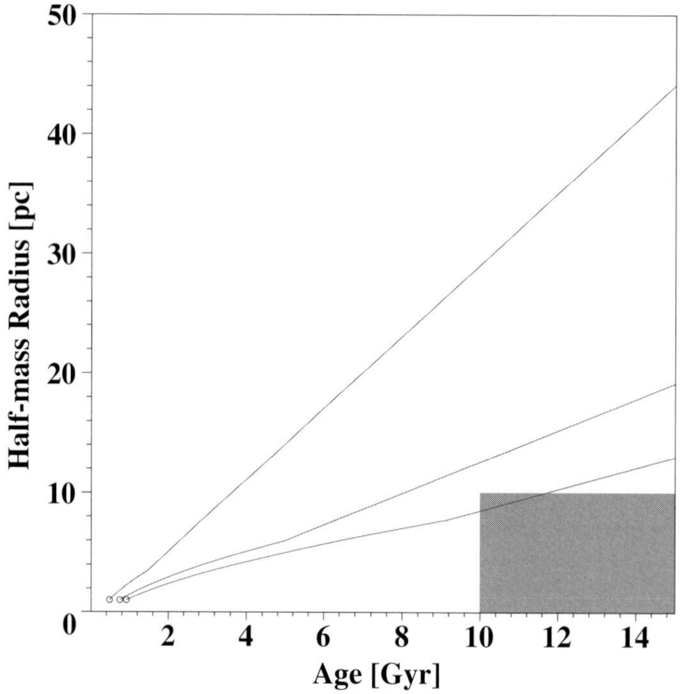

Figure 1. shows the relaxation-driven expansion of the half-mass radius of an isolated globular cluster of $N = 2 \times 10^4, 6 \times 10^4, 1 \times 10^5$ (curves from top to bottom) half-solar-mass stars with a pre-core-collapse size of 1pc (marked by the blue circles). The change from strong to weak gravity is indicated by the change from thin lines to thick lines. Observed globulars typically occupy the green shaded zone in the age vs. size plane.

Relaxation drives a globular to expand. Star clusters might have been born very compact, and in the strong-gravity regime. However, fast relaxation means that globular clusters cannot stay small, will eventually expand and enter the weak regime. The relaxation time can be expressed in terms of some kind of rate of expansion per relaxation time

$$\frac{r}{t_x} = \frac{50\,\mathrm{pc}}{10\,\mathrm{Gyr}}\frac{10^4}{N}\left(\frac{V}{1\,\mathrm{km\,s^{-1}}}\right), \qquad V = 3.5\,\mathrm{km\,s^{-1}}\left(\frac{M}{10^4 M_\odot}\right)^{1/4}, \tag{2.4}$$

where we have assumed $C \sim 7\ln N \sim 50$, and have used the circular velocity in the MOND regime. The above implies that a cluster should expand to a size of ∼175pc over a Hubble time in MOND.

To illustrate this, we give a few examples in Fig. 1. Consider, for example, a hypothetical cluster born fully isolated with $N = 2 \times 10^4$ half-solar-mass stars inside a half-mass radius of 1pc. The cluster is initially dense and in the strong gravity regime. Rapid relaxation leads to core collapse after about $t_0 = 0.4$ Gyrs where we take t_0 as 10 times the initial relaxation time t_x. Afterwards the cluster expands as $t^{2/3}$ in strong gravity till r_h reaches about 3pc at about $t = 1.4$ Gyr, after which the cluster enters the weak gravity regime, r_h expands linearly with time according to MOND, and reaches a size $r_h = 45$pc after a Hubble time. Systems with larger N grows slower (Fig. 1). Nevertheless, clusters with $N = 2 \times 10^4$ to 1×10^5 half-solar-mass stars all grow to a half-mass size 12–45 pc, too big to be consistent with observed globular clusters.

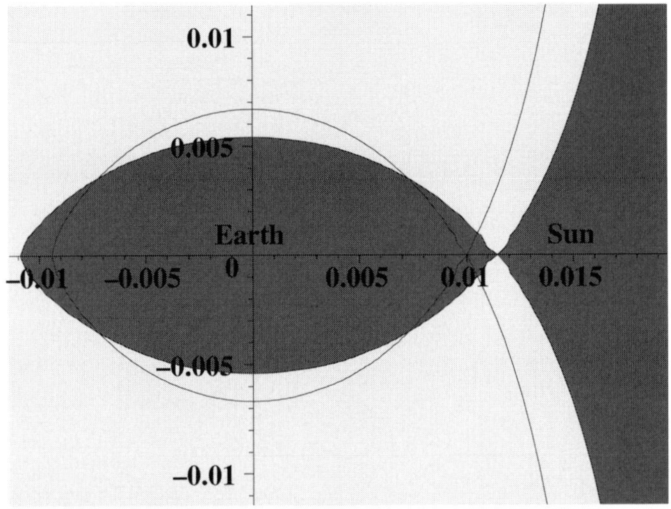

Figure 2. shows the rescaled Roche lobes (contours of the effective potential) in the equatorial plane of a hypothetical isolated Earth-Sun binary mass ratio 3×10^{-6} in the strong gravity regime (say with the separation 1AU, thin blue lines), and in the weak gravity regime (say with separation 1pc, and $GM_\odot/1\,\mathrm{pc}^2 \sim 0.001a_0 \ll a_0$, shaded areas). The Earth is at origin and the Sun is at rescaled unit length to the right. The inner Lagrangian point is a saddle point between the Earth and the Sun, which is slightly further away from the Earth in the deep-MOND regime than in strong gravity regime.

3. Tidal truncations or Roche lobes of binary systems

First we note that at large distances from the baryonic particles Gauss's theorem predicts that a MOND-like gravity $\mathbf{g}$ with an effective gravitational constant $\tilde{G}(g)$ is approximately parallel to the direction of the Newtonian gravity $\mathbf{g}_N$ with the amplitude satisfying eq. (1), so the gravity $\mathbf{g}$ is determined by the total mass of the baryons $\sum_i M_i$ and the mean distance r, independent of the spatial distribution of the baryons. This is true even in the rigorous classical counterpart (Bekenstein & Milgrom 1984) of the Bekenstein theory. So, e.g., a dense globular cluster and a fluffy dwarf galaxy of $10^5 L_\odot$ should have comparable baryonic mass, hence the similar gravity field on scales of kpc, nearly independent of the internal density profile. Indeed MOND nicely accounts for the very small scatter of dynamical properties of structures of similar baryonic distribution on galaxy scale (Sanders McGaugh 2002, McGaugh 2000, 2004). However, this "nice feature" of baryonic gravity also means that the dynamics of wide binary stars would be a scaled-down version of the Antenna-like merging galaxies.

Theoretically in a binary system, the Roche radius (or tidal radius or virial radius when referring to galaxies sometimes) is where the average density of the mass-losing satellite is comparable to the average density of the mass-receiving host at some point R along the orbit of satellite, i.e.,

$$\frac{Gm}{\frac{4\pi}{3}r_t^3} \approx k\frac{GM}{\frac{4\pi}{3}R^3}, \tag{3.1}$$

with the prefactor k ranging from 1 to 3 in different definitions in the literature (Binney & Tremaine 1987); e.g., $k = 2$ according to von Hoerner (1957). This well-known tidal criterion in Newtonian gravity is in fact a result of basic dynamics, independent of the force law parameter $\tilde{G}(g)$ and applies to any gravity. It is shown elsewhere (Zhao *et al.* 2005) that the prefactor $k = 2$ rigourously in the deep-MOND regime. The shape of the

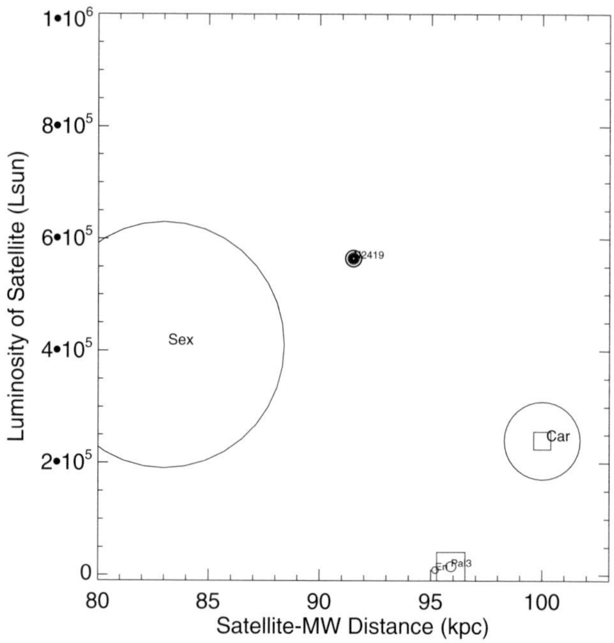

Figure 3. shows the satellite distance vs. luminosity scatter diagram for the dwarf galaxies (labeled with larger font) and globular clusters (labeled with smaller font) orbiting in Milky Way's very outer halo (with orbital periods about one quarter of a Hubble time). The sizes of the symbols are scaled to the physical sizes of the satellites (as a reference, the faint Eridanius cluster at the bottom is 100pc, and Sextans dwarf is 3000pc). The sizes of all globulars of the Milky Way are all drawn as smaller rings inside NGC2419, which is the largest, and most distant bright globular cluster. Predicted sizes of the faint Pal3 cluster and Carina dwarf are also shown as squares based on their proper motion inferred pericentres in a $10^{11} L_{\odot}$ baryonic Milky Way.

Roche lobes is somewhat more squashed than the Newtonian counterpart. An example of Roche lobes is given in Fig. 2.

3.1. *Application to satellite globular clusters and dwarf galaxies*

The Roche criteria (eq. 3.1) predicts that objects should have same sizes if they have similar baryonic content in similar enviornment. This, however, is inconsistent with basic observation data, and we see a large scatter for the truncation radii of outer globular and a dozen dwarf galaxy satellites accelerating in the tide and very weak gravity of a luminous host galaxy, like the Milky Way (see Fig. 3). So the observed limiting radii of these satellites (Harris 1996, Mateo 1998) are no longer simply and uniquely scaled with the binary baryonic mass ratio as predicted by a MOND-like theory (eq. 3.1).

For example, NGC 2419, the largest cluster of the Milky Way, is as distant as and 1–2 times as luminous as the $(2.5 - 4) \times 10^5 L_{\odot}$ dwarfs Carina/Sextans, which has an equilibrium of stars extending to 1000-3000pc radii at a distance of about 90kpc from the Milky Way. A baryonic universe predicts that NGC 2419 should resist the tide of the Milky Way as well as the Sextans dwarf does on 3000pc scale; Sextans shows no sign of being disrupted (Grebel *et al.* 2000). Nevertheless, even this largest globular is truncated on 200pc scale. Such large discrepancy does not go away if we take into account of different orbits or mass of the satellites, e.g., a baryonic universe would predict that the cluster Pal3 (faint $10^4 L_{\odot}$ and presently near pericentre at 90 kpc, Dinescu *et al.* 1999) is bigger than the Carina dwarf (presently near apocentre, Piatek *et al.* 2003), but in reality as shown in Fig. 3, Pal3 (Sohn *et al.* 2003) is much smaller than Carina (Walcher *et al.* 2003). Note that orbits in MOND is not much different from orbits in isothermal halo potentials (Read & Moore 2005).

This example points out a generic problem general to galaxies with resolved satellite globulars vs. dwarfs in any baryonic universe including Bekenstein's cosmology: why there is a clear size gap between 250pc–400pc and nature forms neither very big globulars nor very small dwarfs despite a good mix of the baryonic mass and enviornment variables (orbital phase and pericentric tide) of these satellites. In a baryonic universe "globulars" and "dwarf ellipticals" are practically similar assembly of stars described by the King models: they are pressure-supported low-dispersion gas-poor low-luminosity elliptical equilibria immersed in and sculptured by the tidal field of a common host galaxy.

3.2. *Application: the sizes of the globular clusters in Fornax*

Could it be that the globulars were born small and stayed small? It is plausible that globulars might have formed from denser and cooler gas clouds than stars in some dwarfs. However, a very dense core of a globular is theoretically unstable to stellar encounters and must evolve rapidly, and on a Hubble time typically 10–100 percent of the initial mass of a star cluster are unbound due to relaxation-driven stellar evaporation (Johnstone 1993). Relaxation also drives the half-light radius of a globular to expand as a power law of the time $t^{2/3}$ if we assume globulars are born infinitely compact and dense so we can neglect the delay for the core to collapse. Using eq. (1) and Fig. 1 of the numerical simulations of (Giersz & Heggie 1996), we estimate the radius containing 90 percent of the mass of a typical globular cluster of 10^5 half-solar mass stars expands as $100\mathrm{pc}(\frac{t}{10\mathrm{Gyr}})^{2/3}$.

Relaxation is even faster in MOND (Ciotti & Binney 2004) because of the stronger gravity, higher velocity dispersion and faster dynamical time (Baumgardt *et al.* 2005). A detailed calculation of the expansion rate in MOND is given elsewhere (Zhao *et al.* 2005). Essentially a globular cluster born very compact in MOND cannot stay compact.

There are five bright old globular clusters orbiting around the Fornax dwarf elliptical, 140kpc away from the Milky Way centre. These globulars have a typical baryonic mass of $10^{4-5}M_\odot$ enclosed in a limiting radius of 30–50 pc with a half-mass radius of 4–12 pc (using data compiled in Mackey & Gilmore 2003 and Rodgers & Roberts 1994). One of clusters, No. 1, has a distorted profile and appears to be tidally disturbed. The clusters No. 3,4,5 appear to have excess stars near their limiting radii. The clusters No. 1,2,5 are on the outskirts of Fornax, between 1–1.5kpc from the Fornax centre. Our model of relaxation-driven expansion predicts (cf. Fig. 1) that Fornax No. 1 ($1 \times 10^4 L_\odot$) should have grown to a half-mass size of about 40pc, and more massive clusters No. 2 and No. 5 ($5 \times 10^4 L_\odot$) should grow to 12pc. The predicted sizes are significantly greater than the observed half-light radius of 12pc (No.1) and 6pc (No. 2 and No. 5).

Our model also predicts (cf. eq. 3.1) that the Roche lobe of these Fornax clusters should have a radius of (650–1000) pc if they are on a 140kpc orbit around the Milky Way ($5 \times 10^{10} L_\odot$), or a Roche lobe radius of (100–130) pc if they are on a 1.5 kpc orbit around Fornax ($2 \times 10^7 L_\odot$). Both of these predictions are much greater than the observed limiting radius of (40–50)pc for all Fornax clusters.

3.3. *Discussion*

The limiting size to which a globular can maintain internal equilibrium are most likely limitted by the tidal force of the Milky Way in general. This is supported by the observation of a characteristic sudden change of outer density profile of most globulars and transient non-spherical outer features, as illustrated by the beautiful stream of Pal 5 found

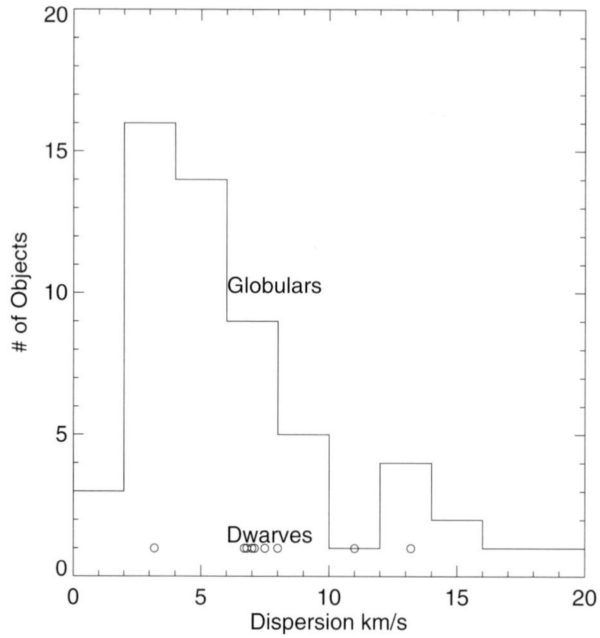

Figure 4. shows the central line of sight velocity dispersion of Galactic globular clusters (in histogram), and the velocity dispersion of Galactic dwarf galaxies (circles). Data taken from Gnedin *et al.* (2002) and Mateo (1998).

by Sloan Digital Sky Survey, and many distant clusters, e.g., NGC5694 and NGC1904, when deep observations have been taken along lines of sight with least background confusion (Leon *et al.* 2000, Odenkirchen *et al.* 2003, Grillmair *et al.* 1995,1996, Lehmann & Scholz 1997). These features are most likely due to stars leaking from a boundary set by the Roche lobe much like the donor star in a binary evolving off the main sequence and transfering mass to its companion. The MOND gravity has a $\ln(r)$ infinite potential well, and so if in isolation a globular will grow indefinitely due to two-body relaxation. So both two-body relaxation and violent relaxation tend to smooth out any sharp initial boundaries. It is hard to imagine a non-tidal mechanism to sustain sudden changes of the density profiles. A globular must have a finite potential well or an outer truncation also because large amount of recycled gas from aging and evolved stars must flow out of the tidal radii of a globular as old globulars keep less than $1M_{\odot}$ gas inside (Spergel 1991).

The cores of many Galactic globulars are as hot as dwarf galaxies with central velocity dispersion $\geqslant$8 km/s (cf. Fig. 4), and some dwarf galaxies, e.g., Sextans (Kleyna *et al.* 2004) are colder than most globulars. An isolated object enters weak regime if bigger than $\sqrt{GM/a_0} \sim 3.5\,\mathrm{pc}\sqrt{M/10^4 M_{\odot}}$, hence dwarf galaxies, and outer globulars like NGC2419 and Palomar clusters are all partly or entirely in the weak gravity regime. Nature sometimes put star clusters inside a small galaxy (as in Fornax, WLM and Sagittarius dwarf galaxies), but never vice versa. In fact there are evidences that some globulars are stripped out from the envelopes or carved from the cores of dwarf galaxies by galactic tide, which explains the similar orbits and smooth transition of baryonic mass of these two populations of satellites. It seems that while MOND-like gravity theories can account rotation curves of galaxy-scale structures without fine tuning, they do not genericly come with an important ingredient or sub-galaxy scale parameter that distinguishes dwarfs from globulars. Having dark matter in dwarves but not in globulars is perhaps one way out of the dilemma of a baryonic universe.

Acknowledgements

This proceeding was partly written during an extended visit to Beijing Observatory, funded through the Overseas Outstanding Young Scholarship program of the Chinese National Science Fundation. I would like to thank the Beijing Observatory for hospitality. I also acknowledge discussions with David Bacon, Xuelei Chen, Douglas Heggie, Keith Horne and Andy Taylor.

References

Baumgardt H., Grebel E.K. & Kroupa P. 2004, *MNRAS* (astro-ph/0501272)
Bekenstein J. 2004, *PhRvD*, 70, 3509 (astro-ph/0403694)
Bekenstein J. & Milgrom M. 1984, *ApJ* 286, 7
Binney J. & Tremaine S. 1987, *Galaxy dynamics*, Princeton University Press.
Ciotti L. & Binney, J. 2004, *MNRAS* 351, 285
Dinescu, D.I., Girard, T.M. & van Altena, W.F. 1999, *AJ* 117, 1792
Grebel E.K., Odenkirchen M. & Harbeck, D. 2000, *AAS* 200, 4607
Grillmair C.J., Ajhar E.A., Faber S.M., *et al.* 1996, *AJ* 111, 2293,
Harris, W.E. 1996, *AJ* 112, 1487
Giersz M. & Heggie D.C. 1996, *MNRAS* 279, 1037
Gnedin O., Zhao H., *et al.* 2002, *ApJ* 568, L23
Johnstone, D. 1993, *AJ* 105, 155,
Kleyna J., Wilkinson M., Evans W. & Gilmore G. 2004, *MNRAS* 354, L66
Lehmann I. & Scholz R.D. 1997, *A&A* 320, 776
Leon S., Meylan G. & Combes F. 2000, *A&A* 359, 907
Mateo M. 1998, *ARA&A* 36, 435
McGaugh, S., *et al.* 2000, *ApJ* 553, 99
McGaugh S. 2004, *ApJ* 609, 652
Milgrom M. 1983, *ApJ* 270, 365
Milgrom M. & Sanders R. H. 2003, *ApJ* 599, L25
Odenkirchen M. *et al.* 2003, *AJ* 126, 2385
Ostriker J. & Steinhardt P. 2003, *Science* 300, 1909
Piatek S. *et al.* 2003, *AJ* 126, 2346
Read J. & Moore B., 2005, *astro-ph/0501273*
Spergel D.N. 1991, *Nature* 352, 221
Spergel D.N. & Steinhardt P. 2000, *PRL* 84, 3760
Sanders, R. & McGaugh S. 2002, *ARA&A* 40, 263
Sohn Y. *et al.* 2003, *AJ* 126, 803
von Hoerner, S. 1957, *ApJ* 125, 451
Walcher C.J., Fried J.W., Burkert A. & Klessen R.S. 2003, *A&A* 406, 847
Zhao H.S., *et al.* 2005, in preparation.

Near-Field Cosmology with Dwarf Elliptical Galaxies
Proceedings IAU Colloquium No. 198, 2005
H. Jerjen and B. Binggeli, eds.

doi:10.1017/S1743921305003741

3D Simulations of the ISM pollution by SNII and SNI in dwarf spheroidal galaxies

Andrea Marcolini[1], Fabrizio Brighenti[1] and Annibale D'Ercole[2]

[1] Dipartimento di Astronomia, Università di Bologna, via Ranzani 1, 40127 Bologna, Italy
[2] Osservatorio Astronomico di Bologna, via Ranzani 1, 40127 Bologna, Italy
email: andrea.marcolini@bo.astro.it

Abstract. We present preliminary results of 3-D hydro simulations of the interstellar medium evolution in dwarf spheroidal galaxies undergoing star formation for the first time. The star formation is assumed to occur in a sequence of instantaneus bursts separated by quiescent periods. Different models are made changing the number and the intensity of the bursts in such a way that the final mass of the formed stars remains the same. We followed the enrichment of the ISM taking into account the contribution of both type Ia and II supernovae. The aim of our models is to find a star formation history compatible with the observed spread of stellar age and metallicity in such galaxies and to reproduce the observed mass-metallicity relation.

Keywords. methods: numerical – galaxies: dwarf – galaxies: ISM – ISM: evolution

1. Introduction

The nearby dwarf spheroidal (dSph) galaxies offer an unique opportunity to study the dynamical and chemical feedback of star formation on the ISM and the resulting stellar population. The color-magnitude diagrams of these galaxies have revealed an unexpectedly complex star formation history (SFH). Many of these galaxies have formed stars at an approximately constant rate for a long time, of the order of a few (2–4) Gyr (e.g. Dolphin 2002). The large metallicity spread within individual galaxies (Shetrone, Cote' & Sargent 2001) and evidence of Type Ia supernova (SN) enrichment also indicate a prolonged SFH. These observational facts are puzzling and difficult to explain. The energy released by the SNe during the SF episodes largely exceeds the ISM binding energy and yet the delicate ISM is not disrupted or removed from the galaxy. Surprisingly, the SNe (and SNIa) heated the ISM with very low efficiency and this put strong constraints on the SFH and on the nature of the SN feedback. We are currently calculating a series of 3D hydrodynamical simulations of starforming dSph galaxies in order to find a plausible SFH and SN feedback which consistently explains the general properties of the stellar population.

2. The model

We have built a one parameter galaxy family (the parameter being the stellar mass, $M_\star$) which agrees with many observed relation (Peterson & Caldwell 1993, Mateo 1998). Each galaxy has initially two components: a spherical quasi isothermal dark halo and an isothermal ($T = T_{\rm virial}$) ISM in hydrostatic equilibrium in the gravitational potential of the dark matter. The initial gas mass is $M_{\rm ISM} \sim 0.15 M_{\rm dark}$, a value in reasonable agreement with the cosmological one (Spergel *et al.* 2003).

Table 1. Galaxy parameters

Model	$M_{\rm dark}$ ($10^7\ M_\odot$)	$\rho_{0,\rm d}$ (g cm^{-3})	$R_{\rm c,d}$ (pc)	$R_{\rm t,d}$ (kpc)	$M_{\rm ISM}$ ($10^6\ M_\odot$)	$\rho_{0,\rm ISM}$ (g cm^{-3})	$T_{\rm ISM}$ (K)	$E_{\rm bind}$ (10^{51} erg)	$M/L_{\rm V}$
Dra I	2.2	6.5×10^{-24}	160	1.0	3.0	2.3×10^{-24}	2×10^3	11	80
Dra II	7.7	4.3×10^{-24}	300	1.2	11.0	0.4×10^{-24}	1×10^4	66	280

The meaning of the simbols is as follows: $M_{\rm dark}$, $\rho_{0,\rm d}$, $R_{\rm c,d}$ and $R_{\rm t,d}$ are the mass, the central density, the core radius and the truncation radius of the dark matter halo; $M_{\rm ISM}$, $\rho_{0,\rm ISM}$, $T_{\rm ISM}$ and $E_{\rm bind}$ are the mass, the central density, the initial temperature and the binding energy of the ISM, respectively; $M/L_{\rm V}$ is the total mass to light ratio of the galaxy in the V band.

Table 2. Stellar and star formation parameters

Model	$M_\star$ ($10^6\ M_\odot$)	$\rho_{0,\rm s}$ (g cm^{-3})	$R_{\rm eff,s}$ (pc)	$R_{\rm c,s}$ (pc)	$R_{\rm t,s}$ (pc)	$N_{\rm SNII}$	$N_{\rm SNII}/N_{\rm bur}$	$\Delta t_{\rm bur}$ (Myr)	$N_{\rm SNI}$
Dra-50	0.56	1.0×10^{-24}	210	130	630	5.6×10^3	112	60	254
Dra-30	0.56	1.0×10^{-24}	210	130	630	5.6×10^3	224	120	257
Dra-10	0.56	1.0×10^{-24}	210	130	630	5.6×10^3	560	300	266

Here $M_\star$, $\rho_{0,\rm s}$, $R_{\rm eff,s}$, $R_{\rm c,s}$ and $R_{\rm t,s}$ are the final mass, central density, effective radius, core radius and truncation radius of the stellar component, respectively; $N_{\rm SNII}$ is the total number of SNII; $N_{\rm SNII}/N_{\rm burst}$ is the number of SNII in each burst; $\Delta t_{\rm burst}$ is the time interval between two succesive bursts and $N_{\rm SNI}$ is the total number of SNI at the end of the simulation.

We assume that stars form in a sequence of $N_{\rm burst}$ instantaneous bursts, separated by quiescent periods, in such a way that after 3 Gyr (the end of our simulation) the stellar mass agrees with the initially chosen one ($M_\star$). Several $N_{\rm burst}$ have been considered (see table 2). For simplicity, the stellar mass (a small fraction of the dark halo mass) does not contribute to the gravitational potential. With a typical stellar mass to light ratio of $M_\star/L_V = 2$ the total mass to light ratio fall in the observed range (see below). The SNII explode at a constant rate for a period of 30 Myr (the lifetime of a 8 $M_\odot$ star, the less massive SNII progenitor) after the occurrence of each stellar burst. Type Ia supernovae in each burst start to explode after 30 Myr and the rate decreases in time after a quick initial rise (Matteucci & Recchi 2001). Both type of SNe are stochastically placed according a spatial distribution given by a modified King profile (see Table 2 for details).

Here we present preliminary results relative to two galactic models taylored on a galaxy similar to Draco. The two models differ in the dark matter (and gas) content as illustrated in Table 1. For each of these models we considered three different star formation histories as illustrated in Table 2.

We solve the usual hydrodynamical equations through a second order espicit hydro code developed by the Bologna Hydrodynamics group. The numerical grid consistes of 160^3 mesh point. The central 100^3 mesh points (covering the stellar volume) have an uniform spatial separation of 13 pc, while the outer ones are distributed logarithmically. Outflow conditions are imposed at every boundary. Each supernova event is simulated by adding 10^{51} erg of thermal energy to the gas within a region with radius of 2 zones. In order to prevent the SN gas to cool too quickly the same region is devoided of its ISM content.

3. Results

Model Dra-I, with $M_{\rm dark} = 2.2 \times 10^7\ M_\odot$ is rapidly ($t < 200$ Myr) devoided of gas after few bursts for any of the three star formation histories adopted. Therefore this

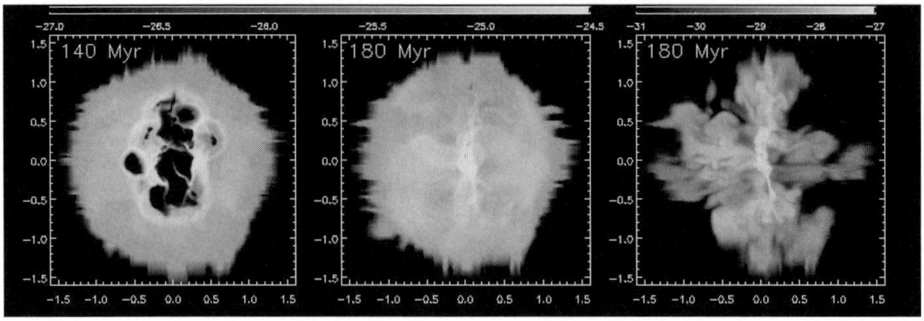

Figure 1. The two panel on the left show the density distribution of the ISM after 20 Myr and 60 Myr after the beginning of the third burst. The last panel illustrate the distribution of the metals ejected by the SNII at the same time of the central panel, immediately before the occurence of the new burst. Distances are given in kpc and density in g cm^{-3}.

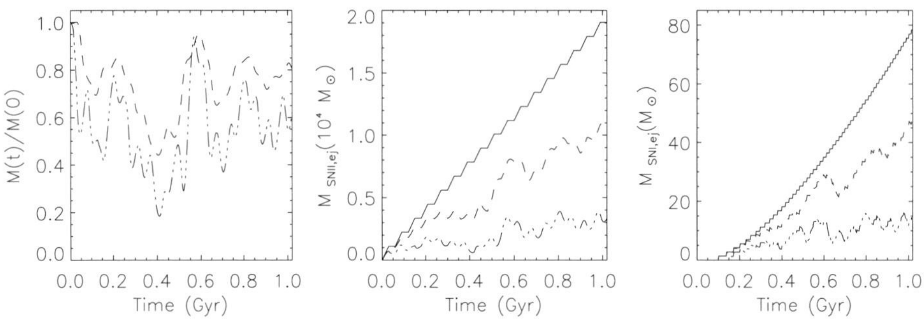

Figure 2. The first panel illustrates the evolution of the ISM mass within the dark matter and stellar volume normalized to the initial ISM mass. The second and third panel show the evolution of the SNII and SNI ejecta, respectively. Solid lines refer to the total ejecta mass produced by SNs, while dashed and point-dashed lines refer to the amount of mass in the dark halo and in the stellar region, respectively.

model is inconsistent with a prolonged SFH. Likely, larger dark matter halos are needed to retain the ISM for longer times. We thus describe below in some detail model Dra-II with $M_{\rm dark} = 7.7 \times 10^7\ M_\odot$ (corrisponding to a total $M/L_V = 280$). Note that this model has a $M/L_V = 80$ within the stellar region. Preliminary results show that in this case supernovae are not able to remove the gas from the galaxy for the model DRA-50 and DRA-25.

We focus here on model DRA-50. When the SNIIs start to explode, they quickly collide and form a central very hot (up to $T = 10^8$ K) and tenuous medium. The cold and dense SNR shells interact each other forming filaments, and the mixture of the hot and cool gas is in a sort of turbulent equilibrium as long as SNII keep exploding (see Fig. 1). Then, during the quiescent time before the next starburst, radiative losses are no more balanced by SN energy input and the ISM cools down collapsing toward the centre. Figure 1 shows the ISM density distribution for this model at time $\Delta t = 20$ Myr (first panel) and $\Delta t = 60$ Myr (second panel) after the beginning of the third burst. As the secular evolution of the ISM is rather slow compared to the time interval between bursts, these shots are representative of the ISM evolution after any other starburst episode. The third panel in the figure shows the density of the SNII ejecta, at the same time of the central panel, immediately before the occurence of the next starburst. While most of the ejecta (~60% at 1 Gyr) is retained by the galactic potential well, only $1 - 4 \times 10^3\ M_\odot$ (~15% at 1 Gyr) are present into the stellar volume at any time. This is apparent in

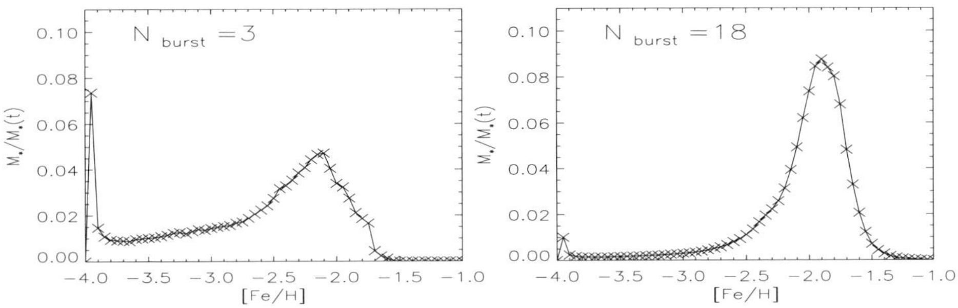

Figure 3. Metallicity distribution of the forming stars at two different times for the model DRA-50

Fig. 2 which shows the gas content evolution of the galaxy. The first panel illustrates the behavior of the ISM mass, while the central panel and the third panel illustrate the evolution of the content of the SNII and SNI ejecta, respectively.

In the first panel we see that almost all (∼80%) the gas remains bounded around the galaxy (at least for the first Gyr of star formation). The same is true for the SN ejecta. Note, however, the amount of metals into the stellar region grows very slowly, because each new burst "pushes" the present metals at larger radii.

In Fig. 3 we plot the mass of the forming stars versus their iron metallicity. This is obtained calculating the metallicity of the ISM immediately before each burst, and assuming that the density of the forming stars is proportional to the stellar King profile. As it is apparent, after few bursts, the stars tend to distribute in metallicity around [Fe/H] = −1.9 with a spread of 1 dex. Both the mean value and the spread are consistent with the observations (e.g. Aparicio, Carrera & Martinez-Delgado 2001).

References

Aparicio A., Carrera R. & Martnez-Delgado D. 2001, *AJ* 122, 2524
Dolphin A.E. 2002, *MNRAS* 332, 91
Fragile P.C., Murray S.D., Anninos P. & Lin D.N. 2003, *ApJ* 590, 778
Mateo M. 1998, *ARA&A* 36, 435
Matteucci F. & Recchi S. 2001, *ApJ* 558, 351
Peterson, R.C. & Caldwell, N. 1993, *AJ* 105, 1411
Shetrone, M.D., Côté, P. & Sargent WS.L.W. 2001, *ApJ* 548, 592
Spergel D.N. *et al.* 2003, *ApJS* 148, 175

Discussion

HENSLER: What is the process in your numerical scheme that accounts for transfer of metals from the hot SN gas to the cool star-forming gas phase? Is it only cooling?

MARCOLINI: Yes. However, beside the metal dependent cooling implemented in the code, the numerical diffusion contributes to mix the SN ejecta with the cold gas. Altough such a diffusion mimics phisical mechanisms such as thermal conduction and mixing layers, we can not reproduce them quantitatively.

LIN: Is the SNe energy lost via radiative cooling, or is it dispersed into the IGM?

MARCOLINI: Even if a small fraction can be dispersed into the IGM, the majority of the energy is radiated away in this model.

Near-Field Cosmology with Dwarf Elliptical Galaxies
Proceedings IAU Colloquium No. 198, 2005
H. Jerjen and B. Binggeli, eds.

doi:10.1017/S1743921305003753

The near-IR luminosity-metallicity relation of dwarf irregular galaxies

Ivo Saviane[1], Enrico V. Held[2], Valentin Ivanov[1], Danielle Alloin[3], Fabio Bresolin[4], R. Michael Rich[5], Luca Rizzi[4] and Yazan Momany[6]

[1]ESO, [2]OAPD, [3]CEA, [4]IfA Hawaii, [5]UCLA, [6]UniPD

Abstract. We report on the recent developments of our long-term investigation of the near-IR luminosity-metallicity relation for dwarf irregular galaxies in nearby groups. A very well-defined relation is emerging from our observational database, and a preliminary discussion of its implications is given.

Keywords. ISM: abundances, (ISM:) HII regions, galaxies: abundances, galaxies: dwarf, galaxies: evolution, galaxies: fundamental parameters, galaxies: ISM, galaxies: irregular, infrared: galaxies

---

1. Introduction

Since a few years, we are carrying out a program to collect near-IR imaging of dwarf irregular (dIrr) galaxies in galaxy groups, and optical spectra of their HII regions, to test the existence of a luminosity-metallicity (L-Z) relation. The main purpose of the study is to provide a sound starting point to discuss how the relation is created, which gives insight into the interplay between processes like mass loss on galactic scales, and galactic chemical evolution. In turn, this has implications on the identification of the sources of metals in the intergalactic medium (IGM), and on the cosmic chemical evolution. The existence of a L-Z relation is in fact usually explained, at least in the case of spheroids, as the result of mass loss through galactic winds triggered by SN explosions (Larson 1974). Notwithstanding its importance, the existence of a L-Z relation for dIrr is controversial, with some studies finding very well defined correlations (Skillman *et al.* 1989; Richer & McCall 1995; Pilyugin 2001), some finding mild relations with substantial scatter (Skillman *et al.* 2003b), and some finding no correlation at all (Hidalgo-Gámez & Olofsson 1998; Hunter & Hoffman 1999). So we decided to improve the situation by devising an observational campaign to overcome some of the existing limitations. The idea is (a) to put together homogeneous samples of oxygen abundances; to consider only galaxies in (b) well-defined environments, and (c) with reduced distance ranges; and finally (d) to image galaxies in the near-IR. The campaign started as a pilot program at 4-m class telescopes, so it had been restricted to the three nearest groups of galaxies, namely M81, Centaurus A, and Sculptor, whose barycenters lie at similar distances (~3Mpc) from the Sun. They also let us probe diverse environments, since members of the Cen A and M81 groups show more interactions than those of the Scl group. A progress report based on data for the Sculptor group has been presented in Saviane *et al.* (2004). More recently we have been able to obtain and reduce data for a few galaxies in the northern M81 group, so here we present an update of the project including data for three more galaxies.

Figure 1. From left to right, we show the morphology of one of our targets (ESO 245-005) in the optical (DSS), infrared H (SOFI), and Hα+R (EFOSC). The reddish, diffused nebulae in the rightmost image are HII regions, and the two arrows mark those for which spectra have been obtained. The images cover $\sim 5 \times 5$ arcmin^2.

2. The data sets

Nine galaxies of the Sculptor group have been observed so far, mostly in the year 2002; ESO 245-005 was re-observed in October 2003, to get a spectrum of a region closer to the optical center of the galaxy (see Fig. 1). All data have been obtained at the La Silla observatory using SOFI at the ESO/NTT and EFOSC at the ESO/3.6m. Typical exposure times were of 1h for spectroscopy and 1 hr on target + 1 hr on sky for the near-IR imaging in the H band. For the M81 group, ten galaxies have been observed in the course of the years 2001 to 2003: interestingly, while only one galaxy of the Scl group did not show Hα emission, in this case 50% of them did not show any emission. For this northern group, the KAST spectrograph at the Lick/3m telescope, and the INGRID camera at the ING/WHT telescope in La Palma were used, with exposure times comparable to those listed above. If the suppressed SF activity (compared to the Scl group) is confirmed, it will be telling us something about the effect of an environment that is more 'risky' for dwarf galaxies. The reductions and calibrations of the data are thoroughly described in Saviane *et al.* (2005, in preparation), and it is worth mentioning that, for the galaxies in common, our abundances are consistent with those of Skillman et al. (2003b; see also Saviane *et al.* 2004).

3. The near-IR luminosity-metallicity relation

To construct our near-IR luminosity-metallicity relation, ideally one would like to include only those galaxies for which abundances based on the direct method could be obtained. This would mean only two galaxies for the M81 group, thus, to improve the situation, we measured oxygen abundances also using the indirect method proposed by Pilyugin (2000, P00), which allows to include all three galaxies for which reliable spectra could be obtained.

Oxygen abundances vs. apparent H magnitudes are presented in Fig. 2: the left panel shows the whole data set, while the right panel shows a fiducial subsample. Since the galaxies are at a similar average distance, we do not need to correct for distance modulus: in any case the horizontal error bars show the variation in luminosity due to a 1 Mpc distance uncertainty at 3 Mpc. The external region of ESO 245-005 (at $H = 11$) is under-abundant by $\sim$0.5 dex compared to the central one. The region we measured in DDO 42 (near $H = 10$) is also external, but we cannot make a comparison with a central one. Finally, metallicities estimated with the P00 method differ by at most ± 0.2 dex from

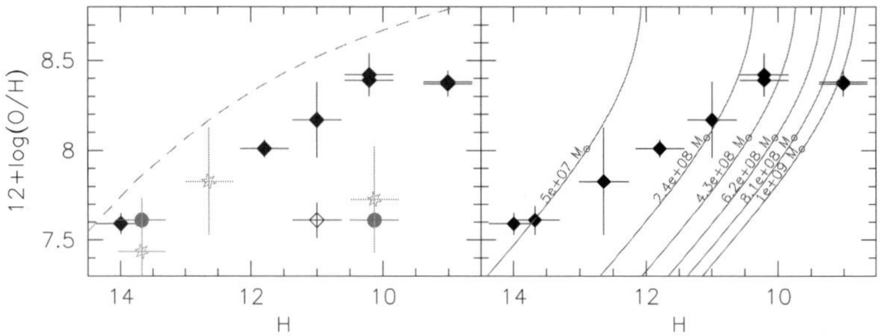

Figure 2. Metallicity of HII regions vs. apparent H luminosity of galaxies in the Scl and M81 groups. In the left panel, the diamonds represent HII regions of the Scl group galaxies: filled diamonds are regions near the center of each galaxy, while the open diamond is the external region of ESO 245-005 (see also Fig. 1). M81 galaxies are represented by filled circles (abundances obtained through the direct method) and stars (abundances measured with the P00 method). The dashed line marks the metallicity reached if the chemical evolution is truncated when the remaining gas mass goes below $2 \times 10^7\ m_\odot$. In the right panel, our best data are compared with evolutionary tracks for closed-box models with reduced yields (see text).

those measured using the direct method, in agreement with the findings of Skillman *et al.* (2003b). In the right panel of the figure we plot only the central HII regions, including the one having the abundance estimated with the P00 method, to which we assign a 50% error bar. With this selection, a clear relation emerges, in the usual sense of having higher oxygen abundances for more massive galaxies. Our strategy has allowed to construct a relation that is much better defined than the existing ones. It also seems like the dependence on luminosity 'saturates' when the brightest galaxy in our sample is reached (NGC 625): excluding this galaxy, a linear regression has a correlation coefficient $r = -0.996$, and a slope of -0.22 ± 0.01 dex/mag.

The interpretation of the relation is not straightforward, though. As it was recalled in the Introduction, in the case of dE/dSph galaxies the L-Z relation is explained in the scenario of mass-loss through galactic winds. However, dIrrs are still evolving systems, so if we want to maintain the mass-loss scenario, then we have to assume that it leaves behind a L-Z relation very early in the history of galaxies. Recently Skillman *et al.* (2003b) and Pilyugin *et al.* (2004) have found that the chemical evolution of dIrrs can be approximated by a closed-box, provided that a low effective yield (≈1/3 of the standard) is adopted: this is a typical signature of gas exchange with the environment. Pushing this hypothesis further, we plot such closed-box models† in the right panel of Fig. 2, for total masses $m_{\rm tot} = m_{\rm gas} + m_{\rm stars}$ varying between $5 \times 10^7\ m_\odot$ and $10^9\ m_\odot$. We adopted the proposed low yield $p_{\rm (O/H)} = 1.6 \times 10^{-4}$, and we assumed $M/L = 1.2$, $M_{H,\odot} = 3.1$ and $(m - M) = 28$ for both groups. The gas mass fraction $\mu = m_{\rm gas}/m_{\rm tot}$ decreases monotonically along each track as abundance increases, so if one assumes that all galaxies were born at the same time, then the figure is telling us that more massive galaxies evolve faster along their tracks. This is confirmed by plotting $d\mu/dt$ vs. total mass from data published in Skillman *et al.* (2003a): one can see a general trend of $d\mu/dt$ increasing with total mass, i.e. larger galaxies being more effective in converting their gas into stars. A similar conclusion was reached by Pilyugin & Ferrini (2000), who found that the L-Z relation is a combined effect of smaller gas loss and higher astration level as the mass increases. More detailed modeling is needed to clarify the role of other processes such like infall (e.g. Hidalgo-Gámez *et al.* 2003).

† More appropriately, one should perhaps call these models "open-boxes simulating low-yield closed-boxes"

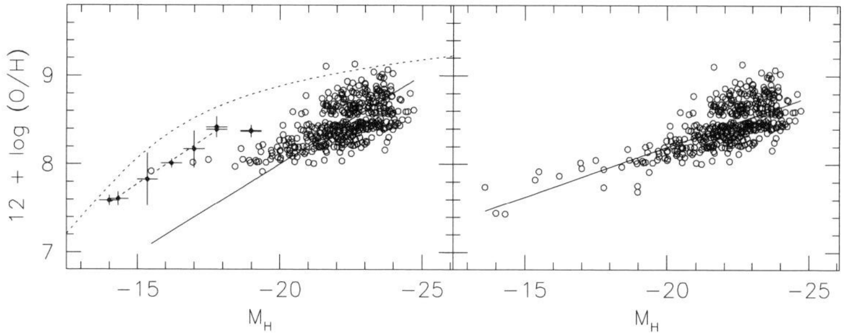

Figure 3. Oxygen abundances vs. *H* luminosity for both the S05 emission-line galaxies (KBG calibration) and our dwarf irregulars. In the left panel we plot the direct abundances, while abundances from the P00 method are used in the right panel. The left panel shows also the L-Z relation that S05 fit to their galaxies, and the fit we obtain for our dwarfs. The dotted line is the same curve shown in Fig. 2, and in the right panel the solid line is an unweighted fit through the whole sample.

It is also interesting to note that, although the closed-box predicts arbitrarily large values of Z as μ goes to zero ($Z = p\ln(1/\mu)$; Searle & Sargent 1972), if we assume that the mass of each new generation of stars, which we can call $m_{\rm SF}$, is roughly constant, then the maximum metallicity we can measure will be $Z_{\rm max} = p\ \ln(m_{\rm tot}/m_{\rm SF}) = p\ \ln m_{\rm tot} - Z_0$. This dependence on the mass of the galaxy is shown by the dashed curve in Fig. 2, where we assumed $m_{\rm SF} = 2 \times 10^7 m_\odot$, a mass of stars that can be created in $\sim 10^8$yr for a typical SFR $\sim 0.1\ m_\odot {\rm yr}^{-1}$.

4. A dwarf vs. giant galaxy dichotomy?

The only study of the L-Z relation of giant galaxies that includes near-IR magnitudes is that of Salzer *et al.* (2005; hereafter S05). They took spectra of emission-line galaxies in fixed apertures of 1.5″ or 2″, and computed oxygen abundances with a reduced number of emission lines, due to the limited spectral coverage of their data: using additional spectra, they first obtained metallicities for a subsample of galaxies using both the direct and the R_{23} methods, and then calibrated them vs. [N II]λ6583/Hα and [O III]λ5007/Hβ. For the R_{23} method they used the P00 calibration for the lower branch of the 12 + log(O/H) vs. R_{23} relation, and three calibrations for the upper branch: Edmunds & Pagel (1984; EP), Kennicutt, Bresolin, & Garnett (2003; KBG), and Tremonti et al. (2004). Assuming again an average distance modulus $(m - M) = 28$, in Fig. 3 we plot our data together with Salzer's *et al.* data. The figure shows that the scatter in the L-Z relation for giants is very large, compared to that of dwarfs, perhaps due to fixed-aperture effects and uncertainties intrinsic to the empirical methods (see also the discussion in S05). The straight lines in the left panel are the fits obtained by S05 and by us, and taken at face value, the panel would suggest a well-defined offset between the L-Z relation of giants and dwarfs. The slope of the S05 L-Z relation is in fact close to what we find for dwarfs alone, namely -0.215 ± 0.003 and -0.201 ± 0.004 for abundances obtained with the EP and KBG calibrations, respectively. Now to be consistent with S05, in the right panel we plot our abundances computed with the P00 method (which allows to add AM 106-382). The offset between galaxies in the two mass ranges seems to disappear: it is in fact possible to do a linear fit of the whole sample (solid line), although a change of slope at $M_H \approx -20$ would seem a better choice. The question of a dwarf-giant dichotomy is then open: moving from the empirical to the direct abundances, the L-Z relation for dwarfs becomes much better defined, parallel to the one for larger galaxies, and with a substantial offset. It remains to be seen what would happen to the L-Z relation of giant galaxies: we plan to measure direct abundances for some of the S05 galaxies, to see if the offset is confirmed or not. If the scenario of the top panel were confirmed, then we should conclude that dwarf galaxies are able to make more metals than giants, which at first sight is puzzling. A possible explanation is the following: if we assumed that the fundamental relation is that of dIrr galaxies, then we could say that the amount of metals produced by a giant galaxy is the same as that produced by a dwarf galaxy ca. 10 times less luminous ($\sim$2.7 mag). In other words, only 10% of the disk of a giant galaxy would participate in making its metals, or perhaps only 1% if we considered that the yield in giant galaxies could be some 3 times higher than in dwarfs. Metaphorically, one could think that metals in the central regions of star-forming galaxies are made in universal cells, and that large cells make metals faster than small cells. And while small cells are found in isolation (dwarf galaxies), larger cells are found in groups of 10–100 in the centers of giant galaxies. This argument is certainly appealing, but it needs confirmation with a follow-up of the S05 study. In fact at the moment we can somewhat reconcile the two L-Z relation only at the price of a degradation of our data. It is finally worth mentioning that, since the S05 sample includes star-forming galaxies, one could suspect that giant galaxies are offset in luminosity due to the presence of a star-burst: however, Lee *et al.* (2004) find that this offset is a few tenths of magnitude in B, compared to quiescent galaxies. We expect that the effect in the IR must be even lower, and certainly not comparable to the one we observe.

Acknowledgements

We thank John Salzer for useful discussions and for providing his data in electronic format, and Jay Gallagher for important remarks in the course of the meeting. I.S. wishes to thank Ana Maria Hidalgo-Gámez for her kind invitation to UNAM, where some issues of this work were discussed.

References

Hidalgo-Gámez, A.M. & Olofsson, K. 1998, *A&A* 334, 45
Hidalgo-Gámez, A.M., Sánchez-Salcedo, F.J. & Olofsson, K. 2003, *A&A* 399, 63
Hunter, D.A. & Hoffman, L. 1999, *AJ* 117, 2789
Edmunds, M.G. & Pagel, B.E.J. 1984, *MNRAS* 211, 507 (EP)
Kennicutt, R.C.Jr., Bresolin, F. & Garnett, D.R. 2003, *ApJ* 591, 801 (KBG)
Larson, R.B. 1974, *MNRAS* 169, 229
Lee, J.C., Salzer, J.J. & Melbourne, J. 2004, *ApJ* 616, 752
Pilyugin, L.S. 2000, *A&A* 362, 325 (P00)
Pilyugin, L.S. 2001, *A&A* 374, 412
Pilyugin, L.S. & Ferrini, F. 2000, *A&A* 358, 72
Pilyugin, L.S., Vílchez, J.M. & Contini, T. 2004, *A&A* 425, 849
Richer, M.G. & McCall, M.L. 1995, *ApJ* 445, 642
Salzer, J.J. *et al.* 2005 *ApJ* 624, 661 (S05)
Saviane, I. *et al.* 2004, *IAUS* 217, 200
Searle, L. & Sargent, W.L.W. 1972, *ApJ* 173, 25
Skillman, E.D., Kennicutt, R.C. & Hodge,P.W. 1989, *ApJ* 347, 875
Skillman, E.D., Côté, S. & Miller, B.W. 2003a, *AJ* 125, 593
Skillman, E.D., Côté, S. & Miller, B.W. 2003b, *AJ* 125, 610
Tremonti, C.A. *et al.* 2004, *ApJ* 613, 898

Near-Field Cosmology with Dwarf Elliptical Galaxies
Proceedings IAU Colloquium No. 198, 2005
H. Jerjen and B. Binggeli, eds.

doi:10.1017/S1743921305003765

Can satellite dwarfs excite galactic warps?

Jeremy Bailin[1] and Matthias Steinmetz[2]

[1]Centre for Astrophysics and Supercomputing, Swinburne University of Technology, Mail H39, PO Box 218, Hawthorn, Victoria, 3122, Australia; email: jbailin@astro.swin.edu.au

[2]Astrophysikalisches Institut Potsdam, An der Sternwarte 16, D-14482, Potsdam, Germany; email: msteinmetz@aip.de

Abstract. We present a series of N-body simulations examining the tidal impact of satellite dwarfs on galactic disks, in particular on the abilities of satellites to excite warps like the one seen in the disk of the Milky Way. We have examined satellites with masses spanning 1.5 orders of magnitude, ranging from the Sgr dSph to the LMC. Orbits similar to the eccentric polar orbit of Sgr were examined, along with orbits of intermediate inclination and of lower eccentricity. The fundamental physical processes which govern the transfer of angular momentum between the satellite and disk, and within the disk, are identified to be precession of the satellite orbit, shrinking of the satellite orbit due to dynamical friction, warping of the disk within a slewing potential, and differential precession. We find that although very large satellites can tidally excite warps shortly after each perigalactic passage, the effects of such satellites are quite destructive and the resulting disks have smaller warps and appear more disturbed than the disks of the Milky Way and other strongly-warped galaxies. Smaller satellites, however, may be capable of producing the minor warps typical of many disk galaxies.

Keywords. stellar dynamics, methods: n-body simulations, Galaxy: disk, galaxies: kinematics and dynamics, galaxies: dwarf, galaxies: interactions, galaxies: individual (Sgr dSph)

1. Introduction

The disk of the Milky Way is warped like an integral sign, rising above the plane on one side and falling below the plane on the other. This warp is seen both in maps of neutral hydrogen (e.g., Diplas & Savage(1991)) and in the stellar distribution (e.g., Drimmel *et al.*(2000)). The Milky Way is far from unique in this regard; the planes of most external galaxies are warped to some degree (Reshetnikov & Combes(1998), Schwarzkopf & Dettmar(2001), García-Ruiz *et al.*(2002b)). While many of these warps are of the dramatic variety normally envisioned by the phrase "warped galaxy" and exemplified by the Milky Way, 1/3 of warps are of a much more minor variety, with warp angles less than 5° (García-Ruiz *et al.*(2002b)).

The ubiquity of warped galaxies poses a puzzle. Isolated warps have a strong tendency to damp and disperse due to the differential precession of the disk (Kahn & Woltjer(1959), Hunter & Toomre(1969)). There are two possible solutions to this puzzle: either some phenomenon stabilizes warps against differential precession, or warps are dynamically transitory but are excited so often that the probability of observing any particular galaxy in a warped state is high. Many authors have investigated the first possibility; most of these studies rely on either the galactic dark matter halo (Sparke & Casertano(1988), Kuijken(1991), Debattista & Sellwood(1999), Ideta *et al.*(2000)) or infalling intergalactic gas (Kahn & Woltjer(1959), López-Corredoira *et al.*(2002)) to stabilize the warp. In this contribution, we investigate the second possibility, that there exists a method of easily exciting warps.

The particular perturbations whose effects we examine are the dwarf satellites surrounding disk galaxies. Ideally, a description of warped galaxies should at least be able to explain the Milky Way warp, about which we have more information than about any other galaxy. While the Magellanic Clouds have often been proposed as the perturbation responsible for the Milky Way's warp (Weinberg(1998), Tsuchiya(2002)), García-Ruiz *et al.*(2002a) demonstrated that a warp caused by the Magellanic Clouds would have its line of nodes oriented orthogonal to the observed line of nodes. Therefore, the Magellanic Clouds are not suitable candidates for producing the Milky Way warp.

The orbital plane of the Sagittarius dwarf spheroid galaxy, on the other hand, is oriented correctly with respect to the Galactic warp. It has a small perigalactic distance, resulting in a very strong tidal torque. The radius of its paragalacticon matches where the warp is strongest. Moreover, there is a remarkable coincidence between the orbital angular momentum of Sagittarius and the misaligned angular momentum in the disk responsible for the warp, which suggests that the systems are coupled, i.e. that Sagittarius is responsible for the warp (Bailin(2003)).

Simulations of satellite-disk interactions that have been performed to date cannot properly test this hypothesis. Tsuchiya(2002) tested the warping effects of the Magellanic Clouds, but did not investigate a satellite with the properties of Sgr, while Ibata & Razoumov(1998) studied the effects of Sgr assuming it had a mass of $5 \times 10^9\ M_\odot$, much larger than current estimates of $(2\text{–}5) \times 10^8\ M_\odot$ (Law *et al.*(2005)), and did not include a live Galactic halo, which may amplify the torque of the satellite (Weinberg(1998)).

We have performed a set of N-body simulations designed to study the interactions between satellites and warps, and in particular to test the hypothesis that the Sgr dSph galaxy is responsible for the warp of the Milky Way.

2. The Simulations

The simulations were performed using the parallel version of the GADGET N-body code (Springel *et al.*(2001)), and only take into account gravitational forces. The disk galaxy model is based on the Milky Way models of Dehnen & Binney(1998), particularly model 3. It consists of 1048576 halo particles, 969438 disk particles and 79138 bulge particles. The satellite consists of a single particle containing the entire satellite mass. We investigated the effects of a satellite with the estimated mass of Sgr, $4\times10^8\ M_\odot$, along with more massive satellites with masses of $2 \times 10^9\ M_\odot$ and $1 \times 10^{10}\ M_\odot$ respectively in order to better understand how the satellite mass affects the strength of the coupling between satellites and the warp.

Each simulation was evolved from $t = 0$ to $t = 4$ Gyr. We have adopted a very conservative timestep with parameter $\eta = 0.01$ to ensure that vertical oscillations were not artificially amplified by integration errors. This resulted in an average timestep of 8×10^{-5} Gyr.

The initial orbital parameters were chosen to mimic those of Sgr: a polar orbit with an apogalacticon of 70 kpc and a perigalacticon of 13 kpc (Helmi & White(2001)). However, a wide range of additional orbits were simulated in order to understand the effect of the orbital eccentricity and inclination.

Because the satellite is represented by a single particle, its internal degrees of freedom are not taken into account. The internal rotation of the satellite cannot be an important reservoir of angular momentum in the system unless the satellite is disrupted; therefore, our results are expected to be valid during the time before which a fully self-consistent satellite would disrupt or lose most of its mass, i.e. for a couple of orbits.

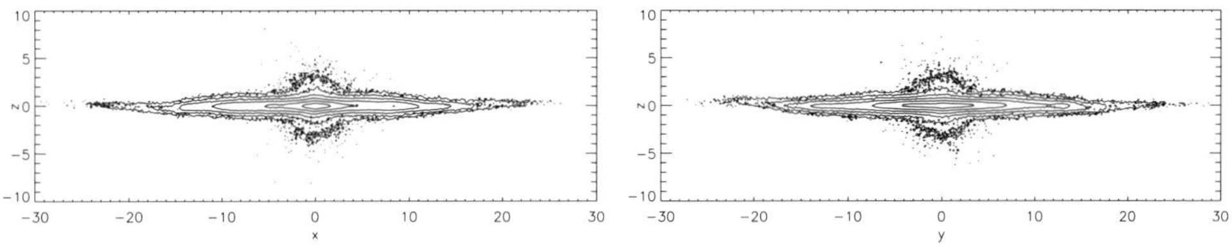

Figure 1. Log-spaced contours of the projected density of the bulge and disk at the time of maximum warp (250 Myr after the perigalactic passage) for a satellite of mass $2 \times 10^9\ M_\odot$ on a polar orbit. The left panel is the x-z projection, while the right panel is the y-z projection. The global tilt of the disk has been removed.

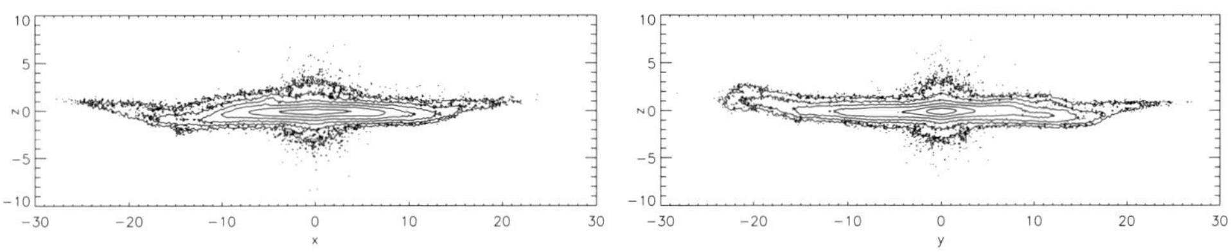

Figure 2. As in Figure 1, but for a satellite of mass $1 \times 10^{10}\ M_\odot$. The maximum warp occurs 200 Myr after the perigalactic passage.

3. Results

There are four fundamental processes that govern the internal angular momentum of the disk in the presence of the satellite. Firstly, the satellite precesses in the flattened potential of the disk. The change in the satellite's orbital angular momentum is balanced by a corresponding change in the angular momentum of the disk, causing it to tilt. Secondly, dynamical friction causes the satellite orbit to sink. This loss of angular momentum is transferred to the halo and to the disk, especially the outer disk, causing it to tilt and warp (Huang & Carlberg(1997)). Thirdly, Ostriker & Binney(1989) determined that if the potential of a disk slews, the angular momenta of rings within the disk tilt differentially toward the slewing vector, causing the disk to warp. Finally, differential precession in a warped disk quickly damps the warp in the absence of the above processes.

Our least massive satellite excites no visible warp in the galactic disk. Therefore, it appears that Sgr with its current mass cannot be responsible for the warp of the Milky Way. More massive satellites excites small warps in the disk. Figure 1 demonstrates the warp in the disk 250 Myr after the perigalactic passage of our intermediate-mass satellite. The warp is barely visible at the left edge of the y-z projection. Our most massive satellite excites a much stronger warp, as seen in Figure 2. This simulation corresponds to a satellite with the mass of the LMC on the orbit of Sgr. Even this warp is not as strong as the warp of the Milky Way, and moreover the disk appears much more disturbed than simply warped. Since it is unlikely that there is an unknown satellite of the Milky Way at least as massive as the LMC on an orbit at least as close as that of Sgr, the warp of the Milky Way is unlikely to be caused by an interaction with a satellite dwarf.

We find that the warp is strongest one disk rotation period after the perigalactic passage of the satellite. The warps are often asymmetric. We have performed simulations with satellites on orbits inclined 45° to the disk, both on prograde and retrograde orbits, and found that the orbital orientation affects the orientation of the line of nodes of the warp, but does not have a strong effect on the warp amplitude. The angular momentum of the warp is typically $\approx$1% of the orbital angular momentum of the satellite.

4. Conclusions

Very massive satellites are able to tidally excite small warps in disk galaxies. The warp reaches its maximum amplitude one disk rotation period after perigalacticon. The simulated warps are too small to explain strongly warped galaxies; in particular, the tidal torque of Sgr appears insufficient to cause the warp of the Milky Way. However, minor warps with amplitudes less than 5° are typical of many disk galaxies, and large satellites are capable of exciting warps of this magnitude.

References

Bailin, J. 2003, *ApJ* 583, L79
Debattista, V.P. & Sellwood, J.A. 1999, *ApJ* 513, L107
Dehnen, W. & Binney, J. 1998, *MNRAS* 294, 429
Diplas, A. & Savage, B.D. 1991, *ApJ* 377, 126
Drimmel, R., Smart, R.L. & Lattanzi, M.G. 2000, *A&A* 354, 67
García-Ruiz, I., Kuijken, K. & Dubinski, J. 2002a, *MNRAS* 337, 459
García-Ruiz, I., Sancisi, R. & Kuijken, K. 2002b, *A&A* 394, 769
Helmi, A. & White, S.D.M. 2001, *MNRAS* 323, 529
Huang, S. & Carlberg, R.G. 1997, *ApJ* 480, 503
Hunter, C. & Toomre, A. 1969, *ApJ* 155, 747
Ibata, R.A. & Razoumov, A.O. 1998, *A&A* 336, 130
Ideta, M., Hozumi, S., Tsuchiya, T. & Takizawa, M. 2000, *MNRAS* 311, 733
Kahn, F.D. & Woltjer, L. 1959, *ApJ* 130, 705
Kuijken, K. 1991, *ApJ* 376, 467
López-Corredoira, M., Betancort-Rijo, J. & Beckman, J.E. 2002, *A&A* 386, 169
Law, D.R., Johnston, K.V. & Majewski, S.R. 2005, *ApJ* 619, 807
Ostriker, E.C. & Binney, J.J. 1989, *MNRAS* 237, 785
Reshetnikov, V. & Combes, F. 1998, *A&A* 337, 9
Schwarzkopf, U. & Dettmar, R.-J. 2001, *A&A* 373, 402
Sparke, L.S. & Casertano, S. 1988, *MNRAS* 234, 873
Springel, V., Yoshida, N. & White, S.D.M. 2001, *New Astronomy* 6, 79
Tsuchiya, T. 2002, *New Astronomy* 7, 293
Weinberg, M.D. 1998, *MNRAS* 299, 499

Discussion

MARTINEZ-DELGADO: You have shown the prominent warp of NGC 5907. It is interesting to note that this galaxy has one of the brightest tidal streams ever discovered in an external galaxy. Do you think this is a coincidence?

BAILIN: No, probably not. The mass in the NGC 5907 stream is very large, probably more than Sgr. Also, the effects of a completely disrupted satellite, like in NGC 5907, may be different than with a single satellite.

DEBATTISTA: What is the velocity dispersion profile of your disks? In real galaxies warps are observed in gas which has a low Q. As Q gets larger warps get smaller because gravitational stresses are spread over larger regions and lead to more coherent tilting than warping. I suspect that with a lower Q in the outer parts you can get warps with smaller satellite mass.

BAILIN: We use $Q = 1.5$. We find that if we reduce Q, the disk becomes quite unstable to bar formation, which induces $m = 2$ vertical modes that dominate over the $m = 1$ mode of the warp.

Near-Field Cosmology with Dwarf Elliptical Galaxies
Proceedings IAU Colloquium No. 198, 2005
H. Jerjen and B. Binggeli, eds.

doi:10.1017/S1743921305003777

Evolution of CDM halos and the Milky Way satellites

Jürg Diemand

Department of Astronomy and Astrophysics, University of California,
1156 High Street, Santa Cruz, CA 95064, USA
email: diemand@ucolick.org

Abstract. A cold dark matter halo big enough to host the Milky Way contains hundreds of subhalos massive enough to host dwarf galaxies. The difference to the much smaller observed number of satellite galaxies seems to be a problem for CDM. The galaxy number density profile and disk like configuration are also different form the total subhalo populations in CDM simulations. A number of different models of dwarf galaxy formation which are able to reproduce the right number of luminous subhalos have been proposed. Some of them also give the right radial distributions and make disk like configurations more probable. Additional information about the typical formation times and sites of dwarf galaxies can be found in the stellar halo of the Milky Way, i.e. from the stellar debris of tidally disrupted dwarfs. A stellar halo with a realistic concentration is obtained when most dwarfs form early (before redshift 10) in small halos (virial mass above 10^8 solar masses). This mass scale found from simulations of dark matter structure formation coincides with the virial temperature of $10^4 K$ which is needed for efficient atomic cooling.

Keywords. methods: n-body simulations,methods: numerical, Galaxy: halo galaxies: dwarf, galaxies: evolution, galaxies: formation, dark matter

---

1. Introduction

In a ΛCDM universe (0.3, 0.7, $\sigma_8 = 0.9$) a dark matter halo big enough to host the Milky Way contains 300 ±100 subhalos with circular velocities above 10 km/s (Diemand, Moore & Stadel 2004b). These subhalos would all be massive enough to contain a dwarf galaxy but there are only about 10 observed satellites around our Galaxy. This discrepancy, the so called 'missing satellites problem' (Moore *et al.* 1999; Klypin *et al.* 1999), is one of the challenges faced by ΛCDM. Other open issues also concern the internal halo structure predicted by CDM: the steep density profiles (Moore 1994; Flores & Primack 1994) and that the distribution of angular momentum is different form the one observed in disk galaxies (e.g. Primack 2003). There are two different ways of dealing with the missing satellites problem, the first one is to change structure formation in a way which dramatically reduces the number of subhalos (e.g. warm dark matter). Here we consider a second class of solutions: The CDM subhalo predictions are not altered but galaxy formation somehow does not take place in all subhalos and leaves most of them dark.

2. Do only the most massive subhalos host dwarfs?

Stoehr *et al.* (2002) suggested that the stellar velocity dispersions of dwarfs are much smaller than the maximum of the circular velocity of the subhalo they live in and that the most massive subhalos correspond to the observed satellites. Small stellar dispersions in the inner part of massive subhalos are possible if the subhalo density profile is shallow or even flat in the inner part. Stoehr *et al.* (2002) suggested a subhalo profile which is much shallower than the NFW-like profiles typically found in field halos. However Kazantzidis *et al.* (2004) found that subhalo density profiles in high resolution cosmological simulations resemble those of field halos over the entire resolved range. While mass is removed at all radii, tides are more effective further out and the density of dark matter bound to a subhalos is lowered much more in the outer part. The inner profile slopes are unchanged after the tidal stripping (Kazantzidis *et al.* 2004). Therefore we expect similar inner profiles for subhalos and field halos. In field halos highest resolution cosmological simulations now start to resolve an inner cusp, i.e. a region with constant logarithmic slope, for an average cluster halo an asymptotic slope of -1.2 is found (Diemand *et al.* 2005a).

One has to assume an anisotropy profile $\beta(r)$ for the stars to convert the subhalo density profile into a stellar line of sight velocity dispersion profile with the Jeans equation. For the $\beta(r)$ which give realistic velocity dispersion profiles one finds v_c values which are not much higher than the observed dispersions. For example for the subhalo around Fornax one finds a v_c of 20–30 km/s (Kazantzidis *et al.* 2004) which is lower than the v_c values of the ten most massive subhalos.

The radial distribution of the most massive subhalos at $z=0$ is more extended than the dark matter halo (Ghigna *et al.* 1998; Ghigna *et al.* 2000; De Lucia *et al.* 2004; Diemand, Moore & Stadel 2004b). The number density of the observed satellites follows roughly the dark halo density, i.e. the most massive subhalos are too far out to host our dwarf galaxies (Moore 2001;Taylor, Silk & Babul 2003).

3. Only early forming subhalos host dwarfs

Due to tidal stripping the present bound mass of subhalos can be much smaller than the mass before accretion. Dwarf galaxies probably have formed before accretion therefore a selection of luminous subhalos should rather depend on their mass (or v_c or virial temperature) at some time before accretion. Unlike selections by bound mass at z=0 selections before accretion give the right radial extent of luminous subhalos (Moore 2001; Nagai & Kravtsov 2005; Moore *et al.* 2005).

There are several plausible physical processes which may suppress gas accretion and star formation in small dark matter halos. One possibility is that small halos accrete gas efficiently only before reionisation. In this case only the earliest forming small halos from stars. It is possible to get the right number of surviving luminous subhalos for a number of plausible reionisation redshifts ($z_{RE} = 5$–12) and star formation efficiencies (Bullock, Kravtsov & Weinberg 2000). After reionisation the increasing UV background heats up the intergalactic medium and the minimal halo mass needed for efficient gas accretion grows with time, i.e. there is a filtering mass scale M(z) which decides if a halo is massive enough to become luminous (e.g. Gnedin 2000). Such a filtering mass function was implemented to populate the subhalos of high resolution cosmological dark matter simulations by Kravtsov, Gnedin & Klypin (2004; hereafter KGK04). For each surviving subhalo the simulations were used to extract the detailed histories of mass accretion, mass loss due to tidal stripping, tidal forces and orbits. In this model most luminous subhalos were much more massive before accretion, about 90 percent had a virial mass larger than $10^{10} M_\odot$ and $v_c > 40$km/s. Tidal stripping reduces both the bound mass and the maximal circular velocity since the density is lowered at all radii (Kazantzidis *et al.* 2004). This point was neglected in semi-analytic models like the one by Bullock, Kravtsov & Weinberg (2000) and it is essential to get the correct velocity function in the model by KGK04.

The smallest luminous halos ($v_c < 30$ km/s) could loose their gas due to photo-evaporation after reionisation or also due to supernova feedback (e.g. Dekel, Lin, these

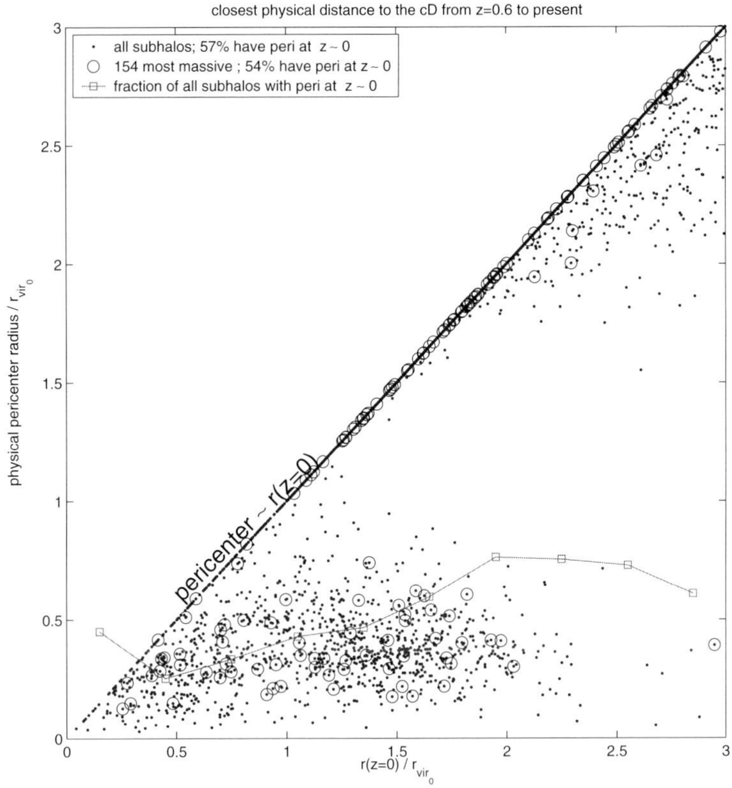

Figure 1. Closest physical distance to the halo center form z=0.6 to present vs. distance form the halo center today. The red line gives the fraction of (sub)halos that have their peri-center today (within the time resolution of 0.6 Gyrs). The points in the lower right corner show that many halos well outside the primary halo today passed deep through it earlier.

proceedings). The smaller dwarfs have only old stellar populations but the larger satellite galaxies manage to hold on to their gas and they from stars also at low redshift, maybe in bursts triggered by tidal forces (Mayer *et al.* 2001). Tidal forces can also change the morphologies of galaxies (Moore *et al.* 1996a; Mastropietro *et al.* 2005), in the Local Group they transform dwarf irregulars to dwarf spheroidals (Mayer *et al.* 2001) explaining the morphological segregation. Exceptions (dwarf spheroidals far away form their host like Cetus and Tucana) are allowed since the entire history of all tidal forces endured determines todays appearance. Some halos beyond 2 virial radii at z=0 have been closer than half a virial radius at some earlier time (see Figure 1 and Moore, Diemand & Stadel 2004), such halos could host remote dwarf spheroidals like Cetus and Tucana.

4. Destroyed dwarfs build up the stellar halo

Most dwarf galaxies which ever existed in the Local Group were destroyed by the tidal forces of the MW and M31 (or their main progenitors). Today their stars and globular cluster make up the halos of the MW and M31. The surviving number, velocity function and radial distribution of dwarfs can be explained with a time dependent filtering mass M(z) which rises after the reionisation time z_{RE}. Such models work for a wide range of plausible z_{RE} values (Bullock, Kravtsov & Weinberg 2000; KGK04). In this section we examine the additional constraints form the spatial distribution of the disrupted dwarfs, i.e. the stellar halo. For simplicity we assume a step function for M(z): halos have to reach some minimal mass $M_{\min}$ at the reionisation redshift z_{RE} in order to accrete gas and form stars. For a large number of $(M_{\min}, z_{RE})$ pairs we identify these early star forming proto-galaxies in a cosmological simulation and trace them to z=0. Then we analyze both the component bound inside surviving subhalos and the disrupted stellar halo component (see Figure 2).

Marking halos above some $M_{\min}$ at some z_{RE} in four galaxy halos (each resolved with a few million particles in a cosmological simulation) as luminous we get realistic satellite populations for the $(M_{\min}, z_{RE})$ pairs connected with the dashed line in Figure 3†. Also plotted are the numbers of sigmas a peak of some mass collapsing at this redshift corresponds to‡. The concentration of the stellar halo depends only on how rare (above how many sigmas) the small high redshift building blocks are, the rare ones form more biased towards the center and give colder, more concentrated and more radially anisotropic z = 0 distributions (Diemand, Madau & Moore 2005b)¶. To match the number density profile of the stellar halo and the metal poor globular clusters of the Galaxy its building blocks must form in collapsed fluctuations above 2.5 σ (Moore *et al.* 2005). This constraint intersects with the dashed line for producing a realistic satellite galaxy population at $M = 10^8 M_\odot$ and $z_{RE} = 12$. This mass-redshift pair inferred from structure formation arguments alone corresponds an important physical scale for neutral gas: Such halos have viral temperatures just above $10^4 K$ (dotted line), i.e. the infalling gas gets hot enough to allow efficient atomic line cooling.

In our scenario the filtering mass function follows the $10^4 K$ line up to $z_{RE} = 12$ and then rises steeply along the 2.5 σ line (or steeper) up to a few times $10^{10} M_\odot$ where the UV background cannot suppress gas accretion anymore. There are a few hundred luminous building blocks above $M = 10^8 M_\odot$ at $z_{RE} = 12$ per galaxy halo, these are relatively rare peaks ($\geqslant 2.5\sigma$) and their formation sites are strongly biased towards the center of the galaxy forming region. Most of them merge together quite rapidly and about 95% of them are completely disrupted in this process. The outermost building blocks (see upper panel of Figure 2) fall in later and with larger orbital energy, they have good chances to survive the tidal forces and to become one of the satellite galaxies at z = 0.

5. Is the disk like configuration a serious problem for CDM?

The satellite galaxies around the Milky Way lie close to a plane and the probability to draw such a configuration from a spherical distribution is practically zero (Kroupa, Theis & Boily 2005). But CDM halos are triaxial and the luminous subhalos in the KGK04 scenario trace the underlying mass distribution (Zentner *et al.* 2005). Also the small number of observed satellites and the fact that most of them are nearby makes it easier to find a plane like configuration like the observed one (Kang *et al.* 2005; Zentner *et al.* 2005). The Milky Way resembles their modeled CDM galaxy-satellite systems in about 10% of the cases, which is still a remarkably low value but not low enough to indicate a serious problem for CDM.

† After z = 12 this line is similar to the filtering mass function for z_{RE}=10 model by KGK04.

‡ Using the dashed line as filtering function includes some rare, early peaks, but the lowest σ allowed will dominate. That is why in the KGK04 model most luminous subhalos are above $10^{10} M_\odot$ before accretion.

¶ At lower redshifts ($z < 3$) the relevant building blocks are more massive $M \simeq 10^{10} M_\odot$ and the z=0 distributions start do depend also on which parts of them are marked and traced to z=0.

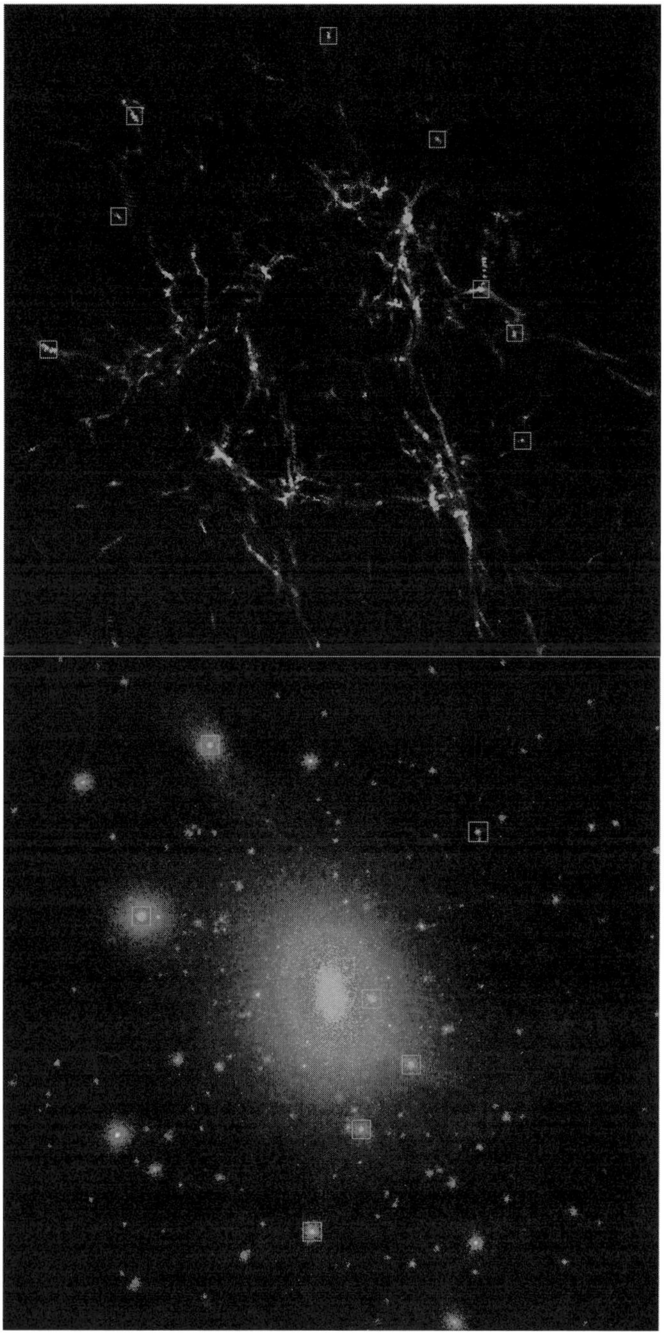

Figure 2. The upper panel shows the dark matter density distribution at a redshift $z = 12$ from a region that will form a single galaxy at $z = 0$ (lower panel) in a cosmological simulation. The green regions show the halos with mass above $10^8 M_\odot$ at $z = 12$. The same marked particles are plotted green also at $z = 0$. Most of the rare peaks are located near the center of the galaxy today. The squares in both panels indicate those first objects that survive as satellite galaxies today orbiting within the final galactic mass halo. Most proto-galaxies were destroyed and have built up a concentrated stellar halo by $z = 0$.

Near-Field Cosmology with Dwarf Elliptical Galaxies
Proceedings IAU Colloquium No. 198, 2005
H. Jerjen and B. Binggeli, eds.

doi:10.1017/S1743921305003789

The environment of dwarf spheroidal satellites; ram pressure, tides and external radiation fields

Lucio Mayer

Institute of Theoretical Physics, University of Zürich, Winterthurerstrasse 190, 8057 Zurich, Switzerland
email: lucio@physik.unizh.ch

Abstract. We discuss the role of environmental mechanisms in the evolution of dwarf galaxy satellites using high-resolution N-Body+SPH simulations that include simultaneously tidal forces, ram pressure and heating from ionizing radiation fields. Tidally induced bar-buckling instabilities can transform a rotating disky dwarf into pressure supported spheroidals. Efficient gas removal requires instead a combination of tidal mass loss and ram pressure stripping in a diffuse gaseous corona around the primary system. The efficiency of ram pressure depends strongly on how extended the gas remains during the evolution. Bar driven inflows that tend to drive the gas to the bottom of the potential well can be opposed by the heating from external radiation fields. We show that even fairly massive dwarfs ($V_{peak} > 30$ km/s) would be stripped of their gas over a few Gyr if they enter the Milky Way halo at $z > 2$ thanks to the effect of the cosmic UV background. Gas mass loss can be much faster, occurring in less than 1 Gyr, if dwarf satellites have their first close approach to the primary at the epoch of bulge formation. Indeed at that time the primary galaxy should have a FUV luminosity comparable to that of major present-day starbursts, resulting in a local UV field even more intense than the cosmic background.

Keywords. galaxies:dwarf;galaxies: evolution; cosmology: dark matter

1. Introduction

Dwarf spheroidals (dSphs) are the faintest galaxies known. They are gas poor and have pressure supported stellar components (Mateo 1998). Among them some stopped forming stars about 10 Gyr ago and other have extended star formation histories (Hernandez *et al.* 2000). They are typically clustered around the largest galaxy in a group. Both mass loss from supernovae winds (Dekel & Silk 1986) and environmental mechanisms like tidal and ram pressure stipping (Einasto *et al.* 1974) have been invoked to explain their properties. Suppression of gas accretion and/or photoevaporation during the reionization epoch likely played a role as well (Bullock *et al.* 2000). In this paper we describe the results of N-Body/SPH simulations of disky dwarf galaxies subject to the combined action of the main environmental mechanisms, tides, ram pressure, and both a cosmic and local ultraviolet radiation field. The model galaxies and the choice of the orbits are consistent with the predictions of LCDM models (Mayer *et al.* 2002, 2005, hereafter MA05). These simulations allow for the first time to explore the overall effect of the environment.

2. The masses of dwarf spheroidals

Knowing the present and past mass of dSphs is crucial in order to compute the effects of both environmental and internal mechanisms that might affect their evolution. Moreover,

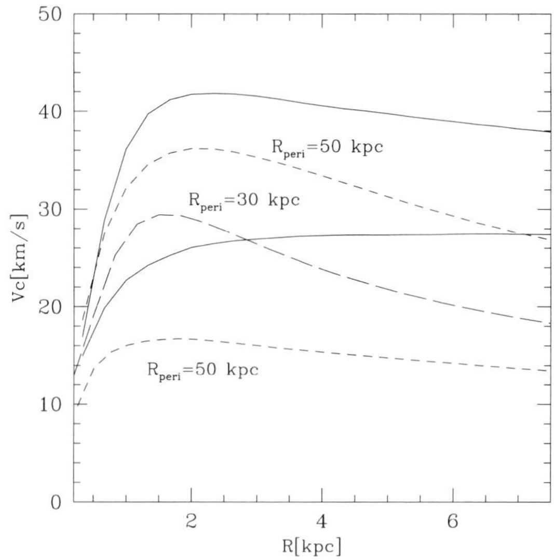

Figure 1. Evolution of the rotation curves of dwarf models with V_{peak} = 40 km/s (thick lines)and V_{peak} = 28 km/s (thin lines). The solid lines represent the initial conditions while the short-dashed show the curves after 3 orbits. The thick long-dashed line shows the curve for the same dwarf model with V_{peak} = 40 km/s on a different orbit. Pericenter distances for the various orbits are indicated in the plot (see MA05)

the mass of dSphs has important implications on the missing satellites problem (Moore *et al.* 1999), namely the fact that CDM predicts an order of magnitude more satellites than observed around the Milky Way or M31. White (2001) has suggested that this problem could be alleviated if the true circular velocities of the halos of dSphs, and thus their masses, are higher than those obtained from their central velocity dispersion under the assumption of a flat (isothermal) rotation curve. This is expected if dSphs sit at the center of an extended halo with an non-isothermal (e.g. NFW) profile, thus having circular velocity that keep rising well beyond the apparent tidal radius of the galaxy. Stoehr *et al.* (2002) fitted the kinematics of dSphs inside subhaloes of a cosmological ΛCDM simulation using Jeans modeling and King profiles for the light distribution of dSphs and found that these galaxies likely live in halos having a peak circular velocity V_{peak} =40–60 km/s. The halo density profiles employed in their analysis are much flatter than and NFW profile at the scale of the core radius of dSphs.

Kazantzidis *et al.* (2004) have studied the tidal disruption of individual CDM satellites with as many as 10^7 particles and of subhaloes in a hi-res LCDM simulation. They repeated the analysis of Stoehr *et al.* (2002) and found that the observed velocity dispersions of Draco and Fornax can only be reproduced in subhaloes with V_{peak} = 20–30 km/s. In their simulations halos maintain cuspy profiles down to the force resolution, which is smaller than the core radius of dSphs (∼100 pc). As a result the rotation curves rise more steeply than those of Stoehr *et al.* (2002), explaining the discrepancy. The masses of the tidally truncated subhaloes of Kazantzidis *et al.* (2004) are in the range 10^8–$10^9 M_{\odot}$, in agreement with more recent analysis based on newer data on stellar velocities (e.g. Wilkinson *et al.* (2001). These masses are still quite high, yielding $M/L > 100$ for Draco.

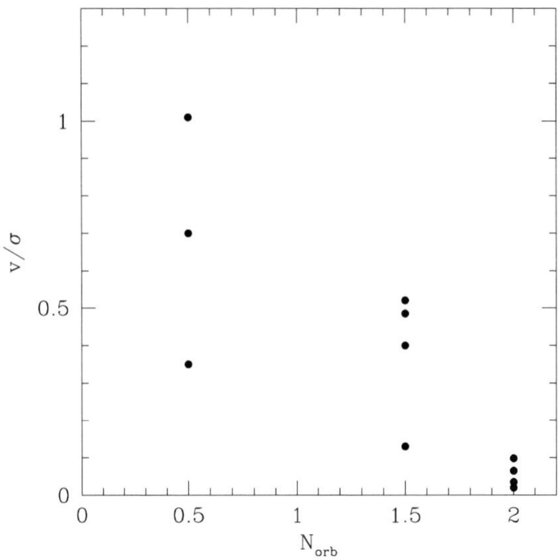

Figure 2. v/σ as a function of the number of orbits for the satellites in the cosmological hydrodynamical simulation of Governato, Mayer *et al.* (2004). v/σ is measured within the half mass radius of the dwarfs. The mean number of orbits for each of the three groups of dwarfs appears on the horizontal axis.

However peak circular velocities are at most a factor of 2 higher than those used in Moore *et al.* (1999), with the consequence that little changes for the missing satellite problem.

The initial V_{peak} of dSphs was larger than 30 km/s since mass loss produced by tidal shocks can lower V_{peak} (Ghigna *et al.* 1998). Kravtsov *et al.* (2004)), using hi-res cosmological simulations without baryons, find that V_{peak} can drop by a factor of 2 in 10 Gyr or so. Mayer *et al.* (2002) and MA05 always find a smaller decrease in V_{peak} in their large set of SPH simulations that follow the evolution of satellites with baryonic disks embedded in CDM haloes (Figure 1). The different results might be due to the fact that some of the satellites in Kravtsov *et al.* (2004) have orbits with pericenters smaller than those considered by MA05. However, the inclusion of baryons also makes the satellites more robust to the tidal disruption of the inner region, which is what matters for the evolution of V_{peak}. Whether or not the baryonic disk becomes bar unstable also is important, since bar-driven inflows can deepen the central potential of the dwarf (Kazantzidis, Mayer *et al.*, in preparation). High-resolution cosmological simulations with hydrodynamics will eventually be able to quantify the overall significance of these effects. In any case, the results of MA05 suggest that the initial V_{peak} of dSphs was in the range 35–50 km/s. This has important implications. First, it means that photoevaporation did not play a major role during the early evolution of satellites at high z since it is only effective for $V_{peak} < 20$ km/s (Susa & Umemura 2004; Shaviv & Dekel 2003). This is consistent with the lack of a clear signature of the reionization epoch in the star formation histories of dSphs (Grebel & Gallagher 2004). Second, blow-out of most of the gas mass due to supernovae winds cannot have occurred since it requires $V_{peak} < 30$ km/s (e.g. MacLow & Ferrara 1999).

3. Morphological evolution of disky dwarfs into dSphs; tidal stirring

Aside from the different gas content, three other facts have to be considered when comparing dSphs and gas-rich, similarly faint dwarf galaxies known as dwarf irregulars (dIrrs). First, except for the faintest among them all dirrs in the Local Group and nearby groups exhibit substantial rotation while dSphs do not. Second, stellar profiles in the two classes of galaxies are similarly close to exponential. Third, a morphology density relation exists such that dSphs are clustered around the primary galaxies while dIrrs are found at much larger distances from the them. Interestingly, Karachentsev (these proceedings) finds that the fraction of dSphs over the total number of dwarf in several nearby groups (including the Local Group) decreases drastically at distances larger than about 250 kpc from the primary galaxies, these being comparable to the virial radius of haloes hosting bright spiral galaxies in LCDM models. This clearly suggests that only dSphs are bound satellites of the primaries and thus the environment must be playing a crucial role in differentiating dIrrs from dSphs. Of course there are outliers like the Local Group dSphs Cetus and Tucana, located at more than 500 kpc from, respectively, M31 and the MW, but this is seen also in cosmological simulations, where a few satellites on very plunging orbits can have apocenters exceeding the virial radius of the primary (Ghigna *et al.* 1998).

Mayer *et al.* (2001a,b) and MA05 have shown that repeated tidal shocks at pericenters of their orbits within the halo of a massive spirals can transform disky dwarfs into objects resembling dSphs. The timescale of the transformation is a few orbital times (several Gyr). The mechanism behind the transformation has to do with non-axisymmetric instabilities of stellar disks. First, tidal shocks induce strong bar instabilities in otherwise stable, light disks resembling those of present-day dIrrs. Second, the bar buckles due to the amplification of vertical bending modes and turns into a spheroidal component. Finally, tidal heating/tidal mass loss thicken/remove the disk outside the buckled bar. We stress that stellar mass loss can be minimal when the disk sits deep in the potential well of a massive halo (see the GR8 run in Mayer *et al.* 2001b) and yet the transformation occurs. The remnants have nearly exponential profiles; the brightest among them, those coming from progenitors having relatively massive disks ($>10^8 M_\odot$), develop a central steepening of the profile due to the particularly strong bar, and overall their profile resembles that of the bright dwarf elliptical satellites of M31, like NGC205. Most importantly, the bar sheds angular momentum outwards and as a result the systems end up with low rotation, with a typical final $v/\sigma < 0.5$. The transformation of course needs the dwarf to be on a bound orbit; for the mean apocenter/pericenter ratio found in cosmological simulations, $\sim$5, Mayer *et al.* (2001b) determine that an apocenter distance comparable to the distance of Leo I (250 kpc), the farthest dSph satellite of the MW, is a limiting case for producing a dSph from a disky dwarf in less than 10 Gyr (Tucana and Cetus should thus be on radial orbits).

In brief this "tidal stirring" accounts for most of the similarities and differences between dIrrs and dSphs, including the existence of the morphology density relation, by postulating progenitors of dSphs with light, low surface brightness disks embedded in massive halos like dIrrs. This does not mean that such progenitors were identical to present-day dIrrs. Since they formed at high z they likely had assembly histories quite different from present-day dIrrs, and thus their stellar populations and metallicities were probably different. The missing piece now is how to place this model in the context of hierarchical structure formation. Unfortunately, to date cosmological simulations with hydrodynamics have not allowed a robust analysis of the structural evolution of satellites due to their limited resolution (typically the force resolution is 500-1 kpc, so the tiniest dSphs like Draco are not resolved at all). Searching for evidence of tidal stirring

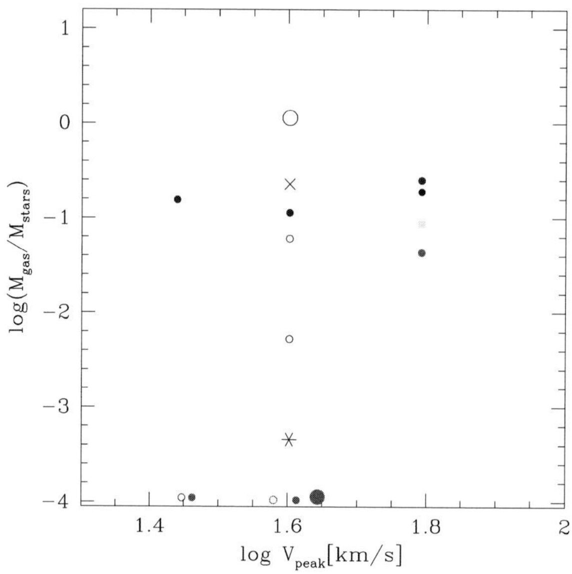

Figure 3. Gas mass fraction (relative to the stellar mass) of simulated dwarf galaxy satellites plotted againts the initial peak velocity of the model galaxies. Data are taken from the N-Body/SPH simulations described in MA05 plus new runs employing a local, time-dependent UV radiation field (see section 4). Measurements are done after 2 orbits within the Milky Way halo. The lower limit on the vertical axis corresponds to the resolution limit in the simulations (therefore points at the very bottom correspond to zero gas mass as measured in the simulation). Some symbols have been displaced along the horizontal axis to avoid overlaps. Filled symbols refer to runs with pericenters of 50 kpc, open symbols refer to runs with pericenters of 30 kpc. Large symbols are for galaxies with a disk gas mass fraction of 0.9, small symbols are for those with a disk gas mass fraction of 0.3 (see section 4). Red is used for adiabatic runs, blue for runs with radiative cooling, black for runs with radiative cooling+cosmic UV background and yellow for runs with cooling, UV and star formation. The cross and star symbols are used for two runs with radiative cooling and a local UV radiation field yielding a maximum heating rate of, respectively, 5 times and 10 times the cosmic one.

in these simulations is thus quite hard, but it can be done for a few well resolved dwarfs. We selected the brightest satellites ($M_B > -14$) of the large spiral galaxy in the cosmological hydro simulation of Governato, Mayer *et al.* (2004, hereafter GM04) as well as similarly bright dwarfs outside the virial radius of the same galaxy. Figure 2 shows their v/σ within the effective radius versus the number of orbits performed within the main system. Clearly v/σ correlates well with the number of orbits as expected within the tidal stirring scenario. We also found that the shape of the stellar components of the dwarfs goes from more disky to more spheroidal with increasing number of orbits. The absolute values of the v/σ have to be taken with caution because numerical two-body heating (e.g. Mayer 2004) is certainly an issue for these objects (halos have only a few thousand particles at this scale). However the trend is evident.

4. Gas stripping; ram pressure, tides and ultraviolet radiation

Mayer *et al.* (2001b) showed that tidal stripping alone could not produce the low gas fractions found in dSphs starting from a gas-rich disky dwarf. Gas consumption by star formation does not change the results significantly for mean star formation rates consistent with those inferred for dSphs (Hernandez *et al.* 2000). MA05 study the combined effect of ram pressure and tidal stripping. They construct two-component models for the Milky Way halo in which a dark matter halo consistent with the results of LCDM simulations has an embedded diffuse gaseous component with a temperature of $\sim 10^6$ K and a density of about $\sim 10^{-4}$ atoms/cm^3 at 30 kpc from the center, consistent with the values inferred from OVI absorption measurements and the existence of the Magellanic Stream (Sembach *et al.* 2003). Dwarf galaxies are placed on eccentric orbits with pericenters of 30 or 50 kpc. In these simulations gas-rich disky dwarfs ($M_{gas}/M_{stars} \geqslant 0.4$) embedded in massive dark haloes typically lose 90% of their gas content. In fact ram pressure increases by a factor up to 10 the amount of stripped gas mass compared to the case in which only tides are included. Ram pressure strongly depends on the depth of the potential well of the dwarfs. While for dwarfs with $V_{peak} \leqslant 30$ km/s most of the gas content is easily removed, for more massive dwarfs the end result depends a lot on the orbit and on the temperature evolution of their gas (see Figure 3). The pericenter distance sets the strength of the ram pressure force mostly through the dependence on the orbital speed of the galaxy at pericenter. The temperature evolution determines whether the gas component stays extended or becomes concentrated in the central, deeper part of the potential well as a result of tidally induced bar-driven inflows. Compression from the outer medium heats the gas. The gas is rapidly heated to 10^5 K, where the cooling function peaks (the initial temperature is $\sim$8000 K) at which point it cools radiatively to 10^4 K in a fraction of the dynamical time; this cold gas easily sinks towards the center due to the torque exerted by the bar and cannot be removed by ram pressure. Instead, if compressional heating is not radiated away, as when the gas evolves adiabatically, the increased pressure opposes the bar-driven inflow, keeping more gas at larger radii where the it is more easily stripped.

For stripping to be very effective for dwarfs with initial $V_{peak} \sim 40$–50 km/s it is sufficient that the temperature of the gas is kept above 10^4 K (the virial temperature of these halos is $3-5\times 10^4$ K). This requires some heating source to counteract radiative cooling. MA05 find that the (uniform) cosmic UV background at $z > 2$ (Haardt & Madau 1996) can achieve this. MA05 find that if the progenitors of Draco or Ursa Minor fell into the Milky Way at $z > 2$ then ram pressure combined with tides was able to remove their entire gas content in a couple of orbits. This translates into a timescale of about 2-3 Gyr for a reasonable orbital time (namely one consistent with the current orbital distance). In this scenario the observed truncation of the star formation in Draco and Ursa Minor more than 10 Gyr ago occurred as a consequence of the infall of these galaxies into the Milky Way halo. Draco would have formed most of its stars before infall since most of the gas becomes ionized while approaching pericenter for the first time.

However, MA05 also find that the exact gas fraction of gas that becomes ionized and then stripped is quite sensitive on the initial gas density (Figure 3). Starting from initial conditions having an exponential disk comprising about 4% of the virial mass(comparable to disk mass fractions of dIrrs), a disk with 30% gas becomes entirely ionized while one with 90% remains neutral in the center and is not stripped. The mass of neutral gas could produce a stellar mass comparable to the present-day luminosity of Draco assuming star formation goes on for another few Gyr. This would yield a star formation history more extended than that of Draco, possibly closer to that of Carina and Fornax. On the other end a progenitor with a mostly gaseous disk that undergoes complete stripping is an attractive case since it would naturally produce the large M/L of Draco without the need of invoking an extremely low disk mass fraction in the initial conditions (MA05). One obvious solution is to make the case for a higher heating and photoionization rate; no a dramatic increase of the rates would be needed since the neutral gas has a temperature only 20% lower than 10^4 K in the gas rich models. The local UV flux coming from the primary galaxy during a phase of intense star formation could have been stronger than the average cosmic background. Mashchenko *et al.* (2004) calculate that the FUV flux of M31 derived from its H_α luminosity is higher than the present-day value of the cosmic UV background out to 10–20 kpc from its center. Current starbursts have FUV luminosities 10^3–10^4 higher than the MW and M31, in the range 10^{44}–10^{45} erg/s (Leitherer *et al.* 2002). According to cosmological simulations at $z = 2$–3 the Milky Way was hosting a major central starburst as the bulge was being assembled from a big merging event (GM04). A starburst is also predicted by models of the star formation history in the galactic bulge, which yield peak star formation rates of up to $100 M_\odot$/yr (Elmegreen 1999), comparable to the most spectacular of present-day starbursts. Lower quantities of dust expected given the lower metallicity of the gas at high redshift imply an escape fraction of ultraviolet photons higher than that in present-day starbursts. But let us be conservative and just assume that the FUV luminosity of the bulge was comparable to that of a major present-day starburst ($L_{FUV} = 10^{45}$ erg/s). Under this hypothesis one obtains that at 30 kpc from the bulge the heating rate was 10 times higher than that associated with the metagalactic UV background at $z = 2$–3. We have run new simulations including such a local, time-dependent UV background (the intensity of the flux is modulated by the orbital distance of the dwarf) and found that ram pressure stripping at the first pericenter passage can be greatly enhanced in a gas-rich dwarf. Even the most dark matter dominated among our models, with $V_{peak} = 40$ km/s, loses more than 90% of its gas along the first orbit and the remaining gas is completely ionized. The ionized gas is then stripped along the second orbit (Figure 3). This simulation reproduces naturally both the early truncation of the star formation in Draco and its present-day high M/L ratio (> 100).

5. Towards a coherent picture; from gas stripping and gas inflows to the star formation histories

A coherent picture is emerging. Dwarf spheroidals live in tidally truncated halos with moderate masses, 10^8–$10^9 M_\odot$, yet this implies $M/L > 100$ in some of them (e.g. Draco and Ursa Minor). However, since simulations show that most of the dark matter halo was stripped as the dwarfs entered the primary halos, their original dark matter content was much higher. Tidal stripping can remove mass even quite close to the center, reducing their V_{peak} by 30–50% (see Figure 1). The tidal interaction with the massive halos of the primaries is also responsible for the morphological transformation of small disks into spheroidal systems. Gas stripping instead is the result of the combination of tidal and ram pressure stripping. However a heating source is required in order to keep the gas extended and allow efficient stripping for galaxies with $V_{peak} > 30$ km/s. The more attractive scenario is one in which the heating was provided by the cosmic UV background at high redshift aided by an even stronger local FUV field produced by the primary galaxies during an epoch of intense star formation. According to galaxy formation models this epoch should coincide with the time of bulge formation, at $z \geqslant 2$. This would explain the sudden truncation of star formation that occurred in Draco and Ursa Minor about 10 Gyr ago. Dwarfs that entered the Milky Way halo later would have kept some gas for at least one or two orbits and thus undergo extended star formation. Our simulations show that the remaining gas falls towards the center as a result of bar driven inflows at pericenter passages. The central gas density there increases and a burst of star formation will occur, as shown in Mayer *et al.* (2001b). This explains the bursty star formation histories of

Carina or Fornax. Ram pressure and tidal stripping are reduced significantly at more recent epochs for two reasons; the first is that in ΛCDM models orbits of satellites have on average larger pericenters and longer orbital times at more recent epochs (see the cosmological simulations of Ghigna *et al.* (1998) for example), the second is that both the local and cosmic UV background fade by orders of magnitude at $z < 2$. In summary we predict that dSphs with extended or truncated star formation histories come from similar progenitors that fell into the primary halos at different epochs. One variable whose evolution with time is not well known is the density of the hot gaseous corona of the Milky Way which sets in part the strength of ram pressure. However new simulations of galaxy formation currently in progress suggest that the density of such halo was higher at higher redshift, which would reinforce our scenario.

Acknowledgements

I thank Stelios Kazantzidis, Chiara Mastropietro, Ben Moore and Fabio Governato for the many stimulating discussions on dwarf galaxies and Cristiano Porciani for his help on issues regarding the effects of the ionizing radiation.

References

Bullock, J.S., Kravtsov, A.V. & Weinberg, D.H. 2000, *ApJ* 539, 517
Dekel, A. & Silk, J. 1986, *ApJ* 303, 39
Einasto, J., Saar, E., Kaasik, A. & Chemin, A.D. 1974, *Nature* 252, 111
Elmegreen, B. 1999, *ApJ* 517, 103
Ghigna, S., Moore, B., Governato, F., Lake, G., Quinn, T. & Stadel, J. 1998, *MNRAS* 300, 146
Governato, F., Mayer, L., Wadsley, J., Gardner, J.P., Willman, B., Hayashi, E., Quinn, T., Stadel, J. & Lake, G. 2004, *ApJ* 607, 688
Grebel, E.K. & Gallagher, J.S. 2004, *ApJ* 610, L89
Haardt, F. & Madau, P., 1996, *ApJ* 461, 20
Hernandez, X., Gilmore, G. & Valls-Gabaud, D. 2000, *MNRAS* 317, 831
Kazantzidis, S., Mayer, L., Mastropietro, C., Diemand, J., Stadel, J. & Moore, B. 2004, *ApJ* 608, 663
Kravtsov, A.V., Gnedin, O.Y. & Klypin, A.A. 2004, *ApJ* 609, 482
Leitherer, C., I-Hui, L., Calzetti, D. & Heckman, T.M. 2002, *ApJ* 303
Mac Low, M. & Ferrara, A. 1999, *ApJ* 513, 142
Mashchenko, S., Carignan, C. & Bouchard A. 2004, *MNRAS* 352, 168
Mateo, M. 1998, *ARA&A* 36, 435
Mayer, L., Governato, F., Colpi, M., Moore, B., Quinn, T., Wadsley, J., Stadel, J. & Lake G. 2001a, *ApJ* 547, L123
Mayer, L., Governato, F., Colpi, M., Moore, B., Quinn, T., Wadsley, J., Stadel, J. & Lake G. 2001b, *ApJ* 559, 754
Mayer, L., Moore, B., Quinn, T., Governato, F. & Stadel, J. 2002, *MNRAS* 336, 119
Mayer, L., Mastropietro, C., Wadsley, J., Moore, B., & Stadel, J. 2005, *MNRAS* submitted, astro-ph/0504277
Moore, B., Ghigna, S., Governato, F., Lake, G., Quinn, T., Stadel, J. & Tozzi, P. 1999, *ApJ* 524, L19
Sembach, K.R. *et al.* 2003, *ApJS*146, 165
Stoehr, F., White, S.D.M., Tormen, G. & Springel, V. 2002, *MNRAS* 335, L84
Susa, H. & Umemura, M. 2004 *ApJ* 600, 1
Wilkinson, M., Kleyna, J.T., Evans, W.N., Gilmore, G.F., Irwin, M.J. & Grebel, E.K. 2004, *ApJ* 611, L21

Discussion

DEKEL: How do you explain Tucana and Cetus? 2) How do you explain the continuous SFH in some dwarfs?

MAYER: We don't know the orbits of Tucana and Cetus. They could be nearly radial, and one perigalactic passage is enough to strip most of the gas with tides and ram pressure. You see a few subhalos on such orbits in CDM simulations. Also, Kravtsov *et al.* (2004) have shown that you can get tidal stirring also *before* dwarfs enter the MW halo in a few cases. 2) Tidal stirring *does* predict periodic, extended SFHs. You need to add ram pressure stripping for those satellites that do not have extended SF, like Draco and UMi. So, if you have tides and ram pressure you can have a truncated or an extended SFH depending on (i) the initial mass of the satellite and (ii) its orbit. therefore I think you can get the different SF histories of dSphs quite naturally.

SKILLMAN: Eline Tolstoy has measured a large radial velocity for Tucana implying it would have been close to the Local Group barycenter about 5 Gyr ago (about when star formation ceased), in which case stripping is a possibility.

MAYER: This would be consistent with CDM simulations as well. You always see a few satellites on very plunging orbits as Tucana. If Tucana has a DM halo of $< 10^9 M_{sun}$ then one passage would be enough for tides and ram pressure to strip most of the gas.

KROUPA: Can your model also account for the lack of MW-to-LMC-sized satellites in fossil groups, as found by D'Onghia & Lake (2004)? 2) Can your model remove the DM cusps in MW satellites? Kazantzidis *et al.* (2004) find this is not possible.

MAYER: I think the luminosity function of the fossil groups is not known well enough. Only a few objects have been studied and I do not think we know the substructure mass function down to the LMC scale. 2) No, cusps always stay.

Near-Field Cosmology with Dwarf Elliptical Galaxies
Proceedings IAU Colloquium No. 198, 2005
H. Jerjen and B. Binggeli, eds.

doi:10.1017/S1743921305003790

Dwarf galaxies are among the best places to probe the nature of dark matter

S. Colafrancesco

INAF - Osservatorio Astronomico di Roma, Via Frascati 33, Monteporzio, I00040, Italy.
email: cola@mporzio.astro.it

Abstract. I discuss here the main physical constraints on Dark Matter particles that can be obtained by the study of the astrophysical signals produced by their annihilation within Dwarf Galaxies. The present results are obtained in the framework of neutralino Dark Matter scenarios and under the assumption that Dwarf Galaxies are dominated by Dark Matter.

Keywords. Dwarf galaxies, Dark Matter, Cosmology

1. Signals from the Dark Universe

Dark Matter (DM) is everywhere in the universe. This is a lessons we learned from cosmology and from models of structure formation. The first teacher who pointed out the existence of DM in large-scale structures was F.Zwicky (1933) who realized that the Coma cluster of galaxies is dominated by DM from the study of the galactic dynamics. The evidence for the presence of DM in systems of galaxies accumulated since Zwicky's age, and we know today that dark halos are found around all galaxy systems, from galaxy clusters to giant ellipticals and to dwarf galaxies.

The lesson we received from observational cosmology (WMAP vs. distant SN Ia) indicated that the matter content of the universe is $\Omega_m h^2 = 0.135^{+0.008}_{-0.009}$ with a baryon density of $\Omega_b h^2 = 0.0224 \pm 0.0009$ (Spergel *et al.* (2003)). These values translate to matter density parameters $\Omega_m \approx 0.27 \pm 0.04$ and $\Omega_b \approx 0.044 \pm 0.004$ for the value $h = 0.71^{+0.04}_{-0.03}$ (Page (2003)). This means that the amount of material which is in the form of a Dark non-baryonic Matter should be $\Omega_{DM} \approx 0.23$ (range 0.21–0.25) assuming $h = 0.7$.

Although we know that DM is the most abundant and common form of matter in the universe, we do not know yet what its nature. Dark Matter has to be in the form of cold particles in order to reproduce the observed features of the large-scale distribution of matter in the universe. Among the viable competitors for having a cosmologically relevant DM species, the leading candidate is the lightest supersymmetric (SUSY) particle, plausibly the neutralino χ, with a mass M_χ in the range between a few GeV to a few hundreds of GeV (Jungman *et al.* (1996)).

The dark side of the universe sends us signals of the presence and of the nature of its constituents that can be recorded using different astrophysical probes. These probes are of *inference* and *physical* character.

Inference probes of DM (e.g., the CMB anisotropy spectrum, the dynamics of galaxies, the hydrodynamics of the hot intra-cluster gas and the gravitational lensing distortion of background galaxies by the intervening potential wells of galaxy clusters) can be used to assess the presence, the amount and the spatial distribution of DM in large scale structures but are not able to provide information on its nature.

Physical probes tell us about the nature and the physical properties of the DM particles and come thru the analysis of the astrophysical effects of DM interaction, which can be directly or indirectly observed by astrophysical techniques. These signals involve, in the case of a neutralino (χ) DM, emission of gamma-rays, neutrinos, synchrotron and bremsstrahlung radiation together with the Inverse Compton Scattering (ICS) of the CMB (and other background) photons and the SZ effect produced by the secondary electrons produced in the DM annihilation process (see Fig.1 and Colafrancesco (2003), Colafrancesco (2004a), Colafrancesco (2005b) for details).

2. Dark Matter signals from Dwarf Galaxies

The best places to probe DM nature in the universe are galaxy clusters and dwarf galaxies: the former are the largest bound containers of DM in the universe while the latter are the smaller and closer approximation to a *pure* DM halo. We will discuss here the case of dwarf galaxies, while the case of galaxy clusters has been discussed elsewhere (e.g., Colafrancesco & Mele (2001), Colafrancesco (2003), 2004b).

Dwarf galaxies, in fact, seem to be dominated by DM both in their central regions and in their outskirts (see, e.g., W.Ewans, M.Mateo, at this Meeting) and many of them have undergone an efficient star formation which consumed most of their gas content. Moreover, these systems contain little or no hot gas, as shown by absence of X-ray emission, and a lack of relativistic particles (cosmic rays), as shown by the absence of radio emission. Furthermore, we have evidence of very close dwarf galaxies (a dozen of which are in the MW environment or around nearby galaxies like M31) which hence allow us to study in details the astrophysical features of the DM annihilation signals. Here we consider specifically the case of the Draco dwarf galaxy, whose structural and physical properties have been discussed extensively at this Meeting.

2.1. *DM annihilation in dwarf galaxies*

The χ annihilation rate in a DM halo is $R = n_\chi(r)\langle\sigma V\rangle_A$, where $n_\chi(r) = n_{\chi,0}g(r)$ is the neutralino number density and $\langle\sigma V\rangle_A$ is the $\chi\chi$ annihilation cross section averaged over a thermal velocity distribution at freeze-out temperature (Jungman *et al.* (1996)). Here we assume a cosmologically relevant DM with a cuspy DM density profile $g(r) = (r/r_s)^{-1}(1+r/r_s)^{-2}$ (e.g., Navarro, Frenk & White (1997)) consistent with both numerical simulations and with velocity dispersion profiles. Although the $\chi\chi$ annihilation cross section is a non-trivial function of the mass and physical composition of the neutralino, to our purpose it suffices to recall that the χ relic density is approximately given by

$$\Omega_\chi h^2 \simeq \frac{3\times 10^{-27}\ \mathrm{cm}^3\ \mathrm{s}^{-1}}{\langle\sigma V\rangle_A}. \qquad (2.1)$$

Enhancing (suppressing) the χ annihilation rate will have the effect of rescaling the final secondary products ($e^\pm, \pi^{0,\pm}$) spectra by the same enhancement (suppression) factor. Neutralinos which annihilate inside a DM halo produce quarks, leptons, vector bosons and Higgs bosons, depending on their mass and physical composition. Electrons are then produced from the decay of the final heavy fermions and bosons (monochromatic electrons, with energy about M_χ, coming from the direct channel $\chi\chi \to ee$, are in general much suppressed). The different composition of the $\chi\chi$ annihilation final state will in general affect the form of the final electron spectrum. We use here DM source spectra which are obtained from Monte Carlo simulations (e.g., DarkSUSY, Gondolo *et al.* (2004)). Analytical, approximated expressions for the $e^\pm$ spectrum which resume the relevant aspects of detailed numerical studies has been given by Colafrancesco & Mele (2001).

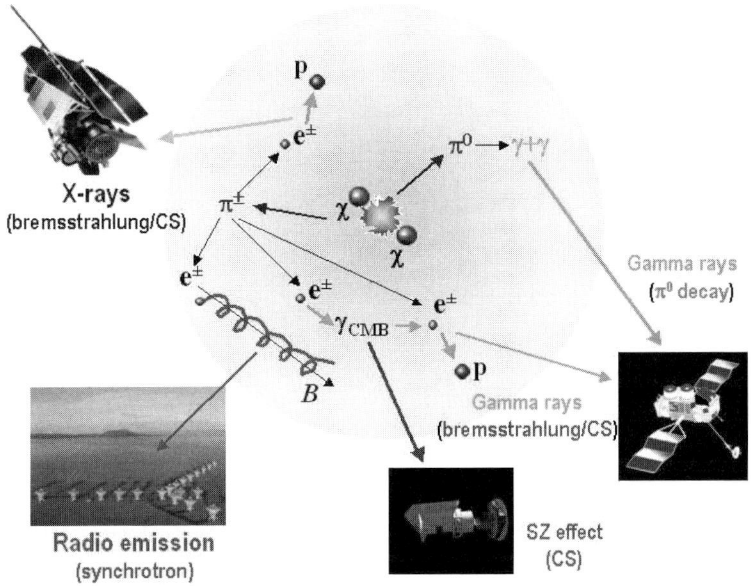

Figure 1. A simple model which shows the basic astrophysical mechanisms underlying the search for the nature of DM particles (χ) through the emission features occurring in large-scale structures (e.g., galaxy clusters and galaxies). These astrophysical mechanisms are, among others: gamma-ray emission from $\pi^0 \to \gamma+\gamma$, relativistic bremsstrahlung of secondary electrons and CS of CMB photons by secondary electrons; X-ray/UV emission due to non-thermal bremsstrahlung and CS of background photons by secondary electrons; synchrotron emission by secondary electrons diffusing in the intra-cluster magnetic field; SZ_{DM} (CS of CMB photons by secondary electrons) effect. Figure from Colafrancesco (2003).

The time evolution of the secondary electron spectrum in a DM halo is described by the transport equation:

$$\frac{\partial n_e(E,r)}{\partial t} - \mathrm{div}[D\nabla n_e(E,r)] - \frac{\partial}{\partial E}\left[n_e(E,r)b(E)\right] = Q_e(E,r) \tag{2.2}$$

where $n_e(E,r)$ is the equilibrium spectrum at distance r from the DM halo center for the electrons with energy E. The source spectrum, $Q_e(E,r)$, reaches its equilibrium configuration mainly due to the diffusion process and to energy losses (mainly synchrotron and Compton Scattering losses at energies $E \gtrsim 150$ MeV and to Coulomb losses at smaller energies). Since DM annihilation produces a source spectrum $Q_e(E,r)$ which is constant in time, the population of high-energy electrons can be described by a quasi-stationary ($\partial n_e/\partial t \approx 0$) transport equation, from which the equilibrium spectrum $n_e(E,r)$ can be derived (see Colafrancesco, Profumo & Ullio (2005) for details). The following qualitative solution (Colafrancesco (2005b)) for the average electron density

$$n_e(E,r) \approx [Q_e(E,r)\tau_{loss}] \times \frac{V_s}{V_s+V_o} \times \frac{\tau_D}{\tau_D+\tau_{loss}} \tag{2.3}$$

resumes the relevant aspects of the transport eq.(2.2). Here, $V_s \propto R_h^3$ and $V_o \propto \lambda^3(E)$ are the volumes occupied by the DM source (we assume a spherical dwarf galaxy with radius R_h) and the one occupied by the diffusing electrons which travel a distance $\lambda(E) \approx [D(E)\cdot\tau_{loss}(E)]^{1/2}$ before loosing much of their initial energy. The relevant time scales in eq.(2.2) are the diffusion time-scale, $\tau_D \approx R_h^2/D(E)$, and the energy loss time-scale $\tau_{loss} = E/b_e(E)$. The diffusion coefficient $D(E) = D_0(E/E_0)^\gamma B^{-\gamma}$ is assumed to be

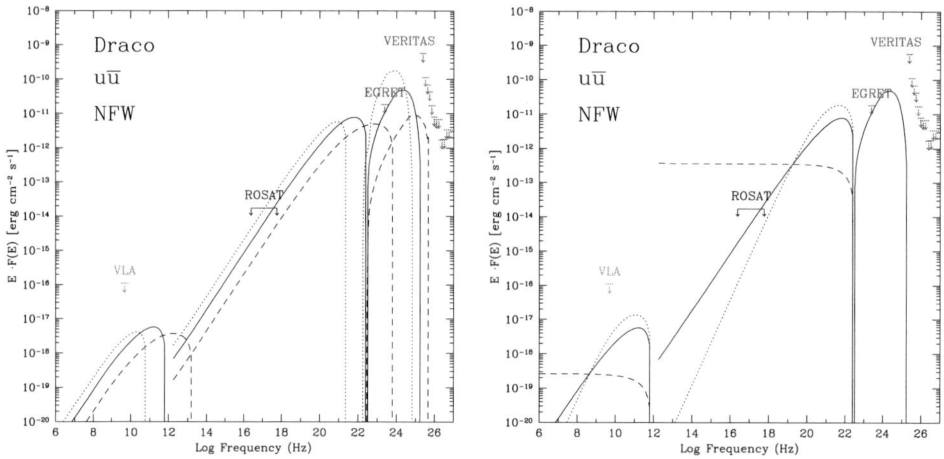

Figure 2. Left. The SED of Draco as produced by $\chi\chi$ annihilation for masses $M_\chi = 50$ GeV (dotted), 100 GeV (solid) and 500 GeV (dashed). The three sets of curves refer (from low to high frequencies) to synchrotron (we assume a uniform magnetic field $B_\mu = 3$), ICS and γ-rays from $\pi^0 \to \gamma\gamma$ decay). A diffusion coefficient $D \sim E^\gamma$ with $\gamma = 1/3$ is assumed here. **Right**. We show here the effect of a different energy dependence of the diffusion coefficient on the Draco SED produced from $\chi\chi$ annihilation: $\gamma = 0$ (dotted), $\gamma = 1/3$ (solid), $\gamma = 1$ (dashed).

only dependent on the electron energy and on the galaxy magnetic field B_μ here given in μG. The function $b_e(E)$ gives the electron energy loss per unit time at energy E: $b_e(E) = b_0(B_\mu)(E/GeV)^2 + b_{Coul}$, where $b_0(B_\mu) \approx (2.5 \times 10^{-17} + 2.54 \times 10^{-18} B_\mu^2)$ and $b_{Coul} \approx 7 \times 10^{-16} [n(r)/1cm^{-3}]$, if b_e is given in units of GeV/s.

For $E > E_* = (D_0 E_0 / R_h^2 b_0 B_\mu^\gamma)^{1/(1-\gamma)}$, the condition $\tau_D > \tau_{loss}$ (and consistently $\lambda(E) < R_h$) holds, the diffusion is not relevant and the solution of eq.(2.2) is $n_e \sim Q_e(E,r)\tau_{loss}$ and shows an energy spectrum $\sim Q(E) \cdot E^{-1}$. For $E < E_*$, the condition $\tau_D < \tau_{loss}$ (and consistently $\lambda(E) > R_h$) holds, the diffusion is relevant and the solution of eq.(2.2) is $n_e \sim [Q_e(E,r)\tau_D] \times (V_s/V_o)$ and shows an energy spectrum $\sim Q(E) \cdot E^{(2-5\gamma)/2}$ which is flatter or equal to the previous case for reasonable values $\gamma = 1/3 - 1$. This last situation ($\lambda(E) > R_h$, $\tau_D < \tau_{loss}$) applies to the regime of dwarf galaxies and, specifically, to the case of Draco.

2.2. *Multi-ν emission from DM annihilation*

The astrophysical signals of DM annihilation can be evaluated from the equilibrium spectrum $n_e(E,r)$ and populate the entire e.m. spectrum, from radio-waves to gamma-rays (see Colafrancesco (2005b), Colafrancesco, Profumo & Ullio (2005) for details). We show in Fig.2 the overall spectrum of Draco as produced by the decay and emission of secondary products of $\chi\chi$ annihilation. We briefly discuss here the main features of the overall spectral energy distribution (SED) of Draco derived from DM annihilation and we refer the interested reader to Colafrancesco (2005b) for the discussion of the technical aspects of the illumination of dwarf galaxies from DM annihilation. Gamma-ray emission is predominantly due to the hadronization of the decay products of $\chi\chi$ with the production of a continuum spectrum due to the decay $\pi^0 \to \gamma + \gamma$ (see Colafrancesco & Mele (2001), Colafrancesco (2005a)), even though the direct neutralino annihilation results in a line emission feature at an energy $\sim M_\chi$. Gamma-ray emission is the most direct signal of DM annihilation and it could be revealed provided that i) sufficient spectral and spatial resolution will be achieved by the next coming experiments and, ii) a clear understanding of other competing gamma-ray emission mechanisms (like cosmic-ray acceleration) expected to be at work in DM halos (see Colafrancesco (2003), Colafrancesco (2005b)) will be obtained. Gamma-ray emission is also expected from secondary electrons produced by $\chi\chi$ annihilation via $\chi\chi \to \pi^{\pm} + X$, $\pi^{\pm} \to \mu^{\pm}\nu_{\mu}(\bar{\nu}_{\mu})$, $\mu^{\pm} \to e^{\pm} + \bar{\nu}_{\mu}(\nu_{\mu}) + \nu_e(\bar{\nu}_e)$. These secondary electrons may produce gamma-rays through bremsstrahlung and ICS of CMB photons up to high energies (see Colafrancesco (2003), Colafrancesco (2005a)). We show in Fig.2 the γ-ray flux from ICS and π^0 decay predicted for Draco. The present γ-ray limits from EGRET and from VERITAS are also shown.

The ICS of CMB photons by means of secondary electrons produces a diffuse emission in the DM halo of Draco which extends in energy even down to the IR, optical, UV and X-ray frequency bands with the same spectrum (see Fig.2). An X-ray upper limit for Draco has been obtained from ROSAT PSPC observations (Zang & Meurs (2001)) and is already able to put severe constraints on the DM particle nature, as well as on the diffusion coefficient of relativistic particles in dwarf galaxies.

The same secondary electrons produced by $\chi\chi$ annihilation unavoidably produce synchrotron emission (if a magnetized atmosphere is present) which can be observed at radio frequencies as a diffuse and extended radio-halo centered on the center of the DM halo. Observations of Draco at 4.9 GHz (Fomalont & Geldzhaler(1979)) only provides an upper limit.

The production of secondary electrons from DM annihilation provides other interesting and relevant effects like i) an heating of the residual gas (left in the galaxy atmosphere after the star formation phase) through Coulomb scattering (Colafrancesco (2005b) and ii) the ICS distortion of the CMB spectrum (Colafrancesco (2004a)). In this context, it has been recently proposed (Colafrancesco (2004a)) that the unavoidable ICS of CMB photons by the secondary electrons produced in $\chi\chi$ annihilation provides a source of SZ effect, that we call here SZ_{DM}, along the line of sight through the dwarf galaxy. Such SZ_{DM} effect has very specific spectral and spatial features that allow its detection in future high-sensitivity CMB experiments.

2.3. *Constraints on DM properties*

The illumination of dwarf galaxies produced by DM annihilation is able to provide strong constraints to the χ physical properties. Fig.3 shows, in fact, the constraints set by the spectral energy distribution of Draco on the $\langle\sigma V\rangle_A$-M_χ plane combined with the WMAP and BBN constraints. The present analysis shows that even the available upper limits for Draco are able to strongly constrain χ DM models. The main astrophysical constraint so far comes from the ICS emission of secondary electrons compared to the X-ray and SZ experimental limits.

The results of this combined analysis are only preliminary since we concentrated on $u\bar{u}$ and W^-W^+ neutralino models (see Colafrancesco & Profumo (2005) for details), and a more refined analysis of the exclusion plots shown in Fig.3 is being performed for several specific models of neutralinos predicted in supersymmetric theories for a wide choice of masses and physical compositions of neutralinos that can be relevant as CDM candidates (Colafrancesco, Profumo & Ullio (2005)). Nonetheless, the results presented here are already indicative of the constraining power of astrophysical observations of DM dominated structures (like dwarf galaxies). Further restrictions of the previous exclusion plot may be obtained by comparing astrophysical constrains to those coming from both accelerator physics and from other cosmological probes which are sensitive to the amount and nature of the DM.

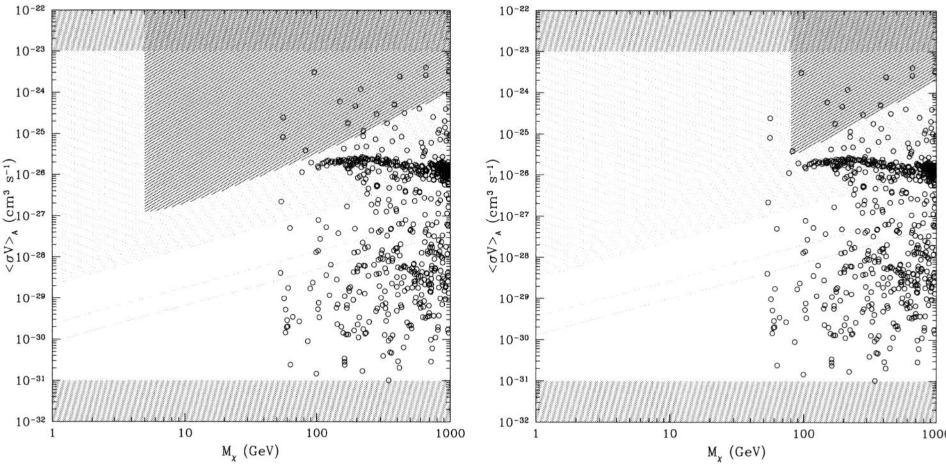

Figure 3. The exclusion regions obtained from Draco (yellow areas, a diffusion coefficient $D \propto E^\gamma$ with $\gamma = 1$ is used here), from Big Bang Nucleosynthesis (BBN) (red areas, see Jedamzik (2004)) and from WMAP are shown for the two neutralino models here considered: the $u\bar{u}$ (left panel) and the W^-W^+ models (right panel). The dotted circles in each panel represent specific supersymmetric models in these two classes which have relic abundance consistent with the WMAP limits (see Colafrancesco & Profumo (2005) dor details). The dashed and dot-dashed curves represents the limits from Draco with an enhancement factor due to DM clumpiness of $f = 3$ and 5, respectively.

3. Conclusion

The nature of DM, despite its overwhelming evidence, is still intangible. It is nonetheless appealing, in these respects, that some astrophysical features of dwarf galaxies might give information on the fundamental properties of the DM particles.

References

Colafrancesco, S. 2003, *ChJAA* 3, 126
Colafrancesco, S. 2004, *A&A* 422, L23
Colafrancesco, S. 2005, in: F.Giovannelli & G. Mannocchi (eds.), *Frontier Objects in Astrophysics and Particle Physics* in press
Colafrancesco, S. 2005, *A&A* submitted
Colafrancesco, S. & Mele, B. 2001, *ApJ* 562, 24
Colafrancesco, S. & Profumo, S. 2005, *A&A* submitted
Colafrancesco, S., Profumo, S. & Ullio, P. 2005, *A&A* submitted
Fomalont, E. & Geldzhaler, B. 1979, *AJ* 84, 12
Gondolo, P. *et al.* 2004, *JCAP* 0407, 008
Jedamzik, K. 2004, *Ph.Rev.D* 70, 3510
Jungman, G. *et al.* 1996, *Phys.Rep.* 267, 195
Navarro, J., Frenk, C. & White, S.D.M. 1997, *ApJ* 490, 493
Page, L. 2003, preprint astro-ph/0306381
Spergel, D.N. *et al.* 2003, *ApJS* 148, 175
Zang, Z. & Meurs, E.J.A. 2001, *ApJ* 556, 24
Zwicky, F. 1933, *Helv.Phys.Acta* 6, 110

Near-Field Cosmology with Dwarf Elliptical Galaxies
Proceedings IAU Colloquium No. 198, 2005
H. Jerjen and B. Binggeli, eds.

doi:10.1017/S1743921305003807

The mass of dwarf spheroidal galaxies and the missing satellite problem

J. I. Read[1]† M. I. Wilkinson[1] N. Wyn Evans[1] G. Gilmore[1] & Jan T. Kleyna[2]
[1]Institute of Astronomy, Cambridge University
[2]Institute for Astronomy, University of Hawaii

Abstract. We present the results from a suite of N-body simulations of the tidal stripping of two-component dwarf galaxies comprising some stars and dark matter. We show that recent kinematic data from the local group dwarf spheroidal (dSph) galaxies suggests that dSph galaxies must be sufficiently massive (10^9–10^{10}M$_\odot$) that tidal stripping is of little importance for the stars. We discuss the implications of these massive dSph galaxies for cosmology and galaxy formation.

Keywords. dwarf, dark matter, Local Group, kinematics and dynamics, interactions

1. Introduction

In the current cosmological paradigm all structure forms from the successive mergers of smaller substructures (see e.g. White & Rees 1978). Cosmological N-body simulations demonstrate that much of this substructure will survive up to the present epoch in the form of dwarf galaxies (Navarro *et al.* 1996). While it is well known that cold dark matter (CDM) theories over-produce this substructure, this 'sub-structure problem' can be solved by either suggesting that only the most massive substructure halos form stars (Stoehr *et al.* 2002); or by suppressing small scale power (Avila-Reese *et al.* 2001 and Zentner & Bullock 2002).

Either way, these solutions suggest that dSph galaxies in the local group must be massive – some $\sim 10^9$–10^{10}M$_\odot$ in total mass. At this mass, for all but the most extreme plunging orbits, tidal stripping would not affect the central stars and gas (Hayashi *et al.* 2003).

In contrast to this picture, there is compelling evidence to suggest that tidal stripping could be extremely important for the formation and evolution of the local group dSph galaxies. Walcher *et al.* (2003) and Wilkinson *et al.* (2004) point out that all of the local group dSph galaxies observed with deep enough photometry have a break in the light profile which is similar to that observed in tidal stripping simulations (see e.g. Aguilar & White 1986, Johnston *et al.* 2002 and Gómez-Flechoso & Martínez-Delgado 2003). Furthermore, Mayer *et al.* (2001b) and Mayer *et al.* (2001a) have suggested that the very low internal angular momenta and spheroidal morphologies observed in local group dSph galaxies could be explained by tidal effects. However, to experience strong tidal effects, the local group dSph galaxies must reside in low mass dark matter halos ($\lesssim 10^9$M$_\odot$) which appears to be at odds with naive cosmological predictions.

Kazantzidis *et al.* (2004) and Kravtsov *et al.* (2004) have recently argued that dSph galaxies might not inhabit the most massive substructure halos. Kravtsov *et al.* (2004) suggest that reionisation sets an *epoch* at which the smallest collapsed structures cannot form stars, rather than having a mass scale below which star formation ceases. This can

† email: jir22@ast.cam.ac.uk

go some way towards solving the mystery, but not all of the way: even in the scheme proposed by Kravtsov *et al.* (2004) dSphs must inhabit the more massive halos on average.

In order to resolve these issues we really require more observational constraints from the local group dSph galaxies. Recently Wilkinson *et al.* (2004) have found a sharp fall-off at ~1 kpc in the projected velocity dispersion of both the Draco and Ursa Minor (UMi) dSph galaxies. At first glance, these data would seem to rule out the tidal stripping model. It is well known that tidal stripping produces heating of stars within the tidal radius (see e.g. Oh *et al.* 1995 and Gómez-Flechoso & Martínez-Delgado 2003). Thus the existence of a cold population at ~1 kpc suggests that the tidal radius must lie beyond this point and that dSph are then 'high mass' galaxies.

In this paper we examine this argument more carefully. We present the results from a suite of N-body numerical simulations of the tidal stripping of two component dwarf galaxies comprising some stars and dark matter, in orbit around a Milky Way-like galaxy.

2. Results

The initial conditions for the dwarf galaxies were set up as in Kazantzidis *et al.* (2004), but for two-component spherical galaxies comprising a dark matter halo and some stars. The satellite galaxies were well resolved with 10^5 star and 10^6 dark matter particles.

The N-body integration for the orbit of the dwarf galaxy about the Milky Way was then performed using a modified version of the GADGET N-body code Springel *et al.* (2001). The code allowed for a fixed potential to be introduced to model the Milky Way. We used a Miyamoto-Nagai potential for the Milky Way disc and bulge and a logarithmic potential for the Milky Way dark matter halo. The simulations, initial conditions and model parameters are described in more detail in Read *et al.* (2005).

Figure 1 shows the results for three very different dwarf galaxies on orbits around the Milky Way after ~10 Gyrs. In each case the orbiting galaxy is currently at apocentre.

The first model (model A) is one in which the satellite galaxy contains *no dark matter.* It was a Plummer sphere Binney & Tremaine (1987) with total mass and scale length of $10^7 M_\odot$ and 0.23 kpc. The galaxy was placed on a plunging orbit with apocentre and pericentre of 80 and 20 kpc respectively. Notice, as published by previous authors, that the isodensity contours (left panel) show the tidal 'S-shape' caused by the leading and trailing tidal debris. Several authors have argued that there is evidence for such shapes in the isodensity contours of the local group dSph galaxies (see e.g. Gómez-Flechoso & Martínez-Delgado 2003 and Munoz *et al.* 2005); and that this, therefore, implies that the local group dSph galaxies must be strongly tidally distorted. Consider, however, the middle panel for the line of sight velocities. There is a strong velocity gradient across the the galaxy which is not observed in Draco or Ursa Minor (UMi) (Wilkinson *et al.* 2004). Furthermore, the projected velocity dispersion is rising beyond the tidal radius, whereas in both Draco and Ursa Minor it is observed to drop sharply (Wilkinson *et al.* 2004). Projections which give distorted isodensity contours, small velocity gradients and near-flat projected velocity dispersions are very rare Read *et al.* (2005). In *no* projection was the projected velocity dispersion of the stars found to fall.

Even if we are viewing the local group dSph galaxies from a projection which minimises any kinematic trace of tidal stripping, notice that the central velocity dispersion in model A is much smaller than that observed in the Draco dSph galaxy (see data points with large crosses). As shown by Kleyna *et al.* (2001), the only way to reproduce such a large central velocity dispersion with the stars we see in Draco is to introduce a significant dark matter component (and see also Klessen *et al.* 2003).

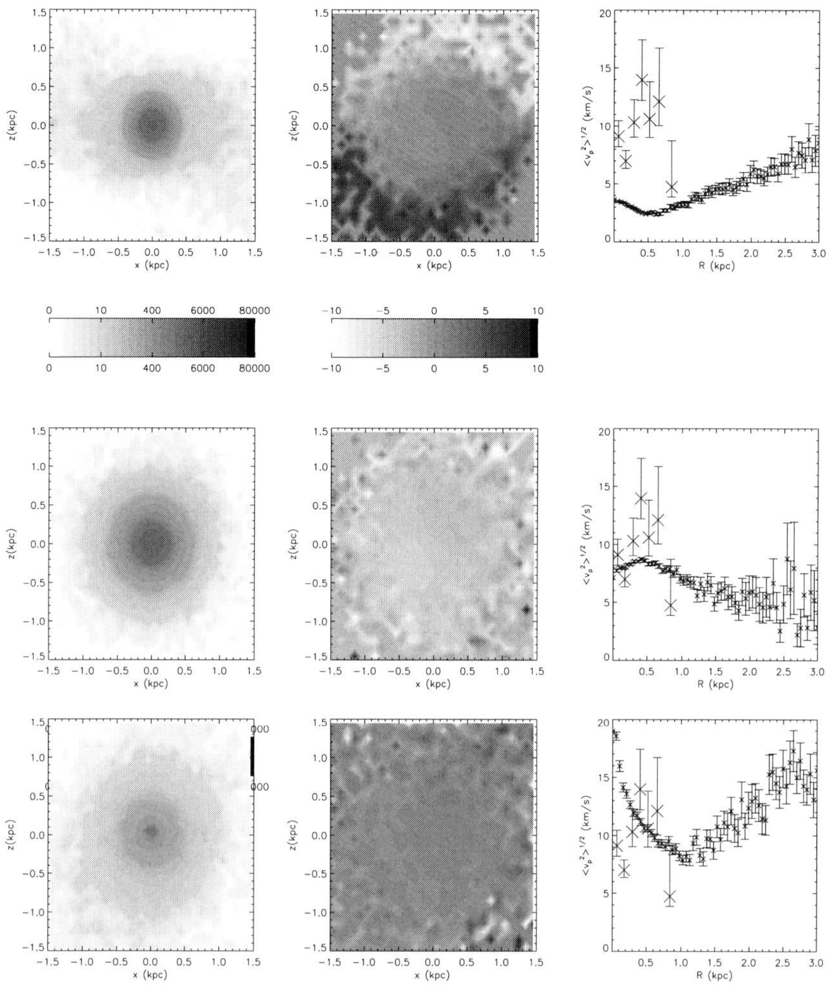

Figure 1. Results from three models: A (top); B (middle) and C (bottom). Left panels show the projected surface density of the stars, the middle panels show the line of sight velocities of the stars and the right panels show the projected velocity dispersion of the stars as a function of radius. The contour bars beneath the top left and middle panels show the number of star particles and the projected velocity in km/s respectively. Over-plotted on the right panels (large crosses) are the data points from the Draco dSph galaxy (Wilkinson *et al.* 2004).

Model B contained a significant Plummer dark matter component with mass $14 \times 10^7 M_\odot$ and scale length 0.5 kpc; its orbit was identical to model A. Even after 10 Gyrs the galaxy has regular isodensity contours, no velocity gradient and a falling projected velocity dispersion. This is because the initial mass and concentration of dark matter was chosen to be sufficiently high that tidal stripping produced almost no effect on the stars over 10 Gyrs. Notice that for this model, the central velocity dispersion is much closer to that of Draco.

Finally, model C was chosen to have a dark matter halo consistent with those observed in cosmological simulations. We used a Hernquist profile with mass $10^{10} M_\odot$ and scale length 1.956 kpc. In order for it to have a final velocity dispersion low enough given this starting mass we had to choose an extreme orbit with a pericentre of just 6 kpc†. At

† The central velocity dispersion is much lower than in the initial conditions for this model

this radius the galaxy would likely affect the Milky Way disc (and dynamical friction would become important) and so already the orbit is quite unrealistic. However, notice that even for such an extreme orbit, the central velocity dispersion is still too large after 10 Gyrs to be consistent with Draco.

In conclusion, tidal stripping cannot be very strong for many, if not all, of the local group dSphs. Strong tidal stripping, which would produce distorted isodensity contours, also leads to velocity gradients and flat or rising projected velocity dispersions - neither of which are observed in the local group dSphs for which we have good kinematic data (but see also Munoz *et al.* 2005). This suggests that dSph galaxies must be sufficiently massive such that tidal stripping is of little importance for the stars. Either they are on orbits with large pericentres, in which case they can have masses as low as $\sim 10^8 M_\odot$ Kleyna *et al.* (2001); or they are on more extreme orbits in which case they must be $\sim 10^9$–$10^{10} M_\odot$ depending on the extremity of the orbit. Our current cosmological paradigm would favour the latter hypothesis, but this leaves us with a puzzle: if the dSph are really as massive as $\sim 10^{10} M_\odot$ and have dark matter densities which are cosmologically consistent then they would have central velocity dispersions which are too large to be consistent with Draco or UMi - even after significant tidal stripping and shocking.

References

Aguilar, L.A. & White, S.D.M. 1986, *ApJ* 307, 97

Avila-Reese, V., Colín, P., Valenzuela, O., D'Onghia, E. & Firmani, C. 2001, *ApJ* 559, 516

Binney, J. & Tremaine, S., 1987, *Galactic dynamics* Princeton, NJ, Princeton University Press, 1987, 747

Gómez-Flechoso, M.Á. & Martínez-Delgado, D. 2003, *ApJ* 586, L123

Hayashi, E., Navarro, J.F., Taylor, J.E., Stadel, J. & Quinn, T. 2003, *ApJ* 584, 541

Johnston, K.V., Choi, P.I. & Guhathakurta, P. 2002, *AJ* 124, 127

Kazantzidis, S., Magorrian, J. & Moore, B., 2004, *ApJ* 601, 37

Kazantzidis, S., Mayer, L., Mastropietro, C., Diemand, J., Stadel, J. & Moore, B. 2004, *ApJ* 608, 663

Klessen, R., Grebel, E. & Harbeck, D. 2003, Draco – a failure of the tidal model . astro-ph/0302287

Kleyna, J.T., Wilkinson, M.I., Evans, N.W. & Gilmore, G. 2001, *ApJ* , 563, L115

Kravtsov A.V., Gnedin O.Y., Klypin A.A., 2004, *ApJ* , 609, 482

Mayer L., Governato F., Colpi M., Moore B., Quinn T., Wadsley J., Stadel J., Lake G., 2001a, *ApJ* , 559, 754

Mayer, L., Governato, F., Colpi, M., Moore, B., Quinn, T., Wadsley, J., Stadel, J. & Lake, G. 2001b, *ApJ* 547, L123

Munoz, R.R. *et al.*, 2005

Navarro, J.F., Frenk, C.S. & White, S.D.M. 1996, *ApJ* 462, 563

Oh, K.S., Lin, D.N.C. & Aarseth, S.J. 1995, *ApJ* 442, 142

Read, J.I., Wilkinson, M.I., Evans, N.W., Gilmore, G. & Kleyna, J.T. 2005, in preparation.

Springel, V., Yoshida, N. & White, S.D.M. 2001, *New Astronomy* Volume 6, 79

Stoehr, F., White, S.D.M., Tormen, G., Springel, V. 2002, *MNRAS* 335, L84

Walcher, C.J., Fried, J.W., Burkert, A. & Klessen, R.S. 2003, *A&A* 406, 847

White, S.D.M. & Rees, M.J. 1978, *MNRAS* 183, 341

Wilkinson, M.I., Kleyna, J.T., Evans, N.W., Gilmore, G.F., Irwin, M.J. & Grebel, E.K. 2004, *ApJ* 611, L21

Zentner, A.R. & Bullock, J.S. 2002, *Phys. Rev. D* 66, 43003

due to disc shocking and tidal stripping (see Read *et al.* (2005) for more details). It does tend to zero at the centre as it should, but this is disguised somewhat by the choice of bin size.

Discussion

DIEMAND: In the NFW model the projected stellar velocity dispersion droped to about 0.5 of the initial value. Does $v_{c,max}$ of the dark matter profile go down by a similar factor?

READ: Good question. Certainly the change in the central density is comparable to results from Kazantzidis *et al.* I couldn't say how much v_c drops, but I will certainly look at that.

Near-Field Cosmology with Dwarf Elliptical Galaxies
Proceedings IAU Colloquium No. 198, 2005
H. Jerjen and B. Binggeli, eds.

doi:10.1017/S1743921305003819

Substructure in dwarf spheroidals – a star cluster connection?

Mark I. Wilkinson[1], Jan T. Kleyna[2], N. Wyn Evans[1], Gerry F. Gilmore[1], Eva K. Grebel[3], Andreas Koch[3], Justin Read[1], Roland Young[1]

[1] Institute of Astronomy, Madingley Road, Cambridge CB3 0HA, UK
email: markw@ast.cam.ac.uk

[2]Institute for Astronomy, 2680 Woodlawn Drive, Honolulu, Hawaii 96822-1897, USA

[3] Astronomisches Institut, Universität Basel, Venusstrasse 7, CH 4102 Binningen, Switzerland

Abstract. The observational evidence for kinematic substructure in Local Group dSphs is reviewed. The properties of these substructures are consistent with their being disrupted star clusters. The persistence of cold substructure argues strongly against the presence of dark matter cusps in the haloes of dSphs. A formation scenario for dSphs is described involving the merger of star clusters in the potential well of a low-mass dark matter halo.

Keywords. stars: kinematics,galaxies: dwarf,galaxies: formation, galaxies: halos, galaxies: kinematics and dynamics, Local Group,galaxies: star clusters

---

1. Observations of kinematic substructure in Local Group dSphs

Dwarf spheroidal galaxies (dSphs) provide a valuable opportunity to test theories of dark matter and of galaxy formation and evolution. An increasing body of data on the kinematics of individual stars in Local Group dSphs appears to confirm early suggestions that these objects contain significant quantities of dark matter (e.g. Aaronson 1983; Evans – this proceedings). However, recent observations have shown that the internal kinematics of these apparently simple stellar systems are more complicated than was first thought (e.g. Wilkinson *et al.* 2004; Tolstoy *et al.* 2004) with distinct stellar populations exhibiting different kinematic signatures.

Observations of Ursa Minor and Sextans in particular have revealed the presence of kinematically cold substructures near the centres of both these dSphs. The left panel of Figure 1 shows the stellar isophotes of Ursa Minor which display a secondary peak offset from the centre of the galaxy (corresponding to the origin of this plot). The velocity distribution of the stars in the vicinity of this secondary peak is significantly narrower than that of the main body of Ursa Minor. If we assume that the velocity distribution in this region can be represented as the combination of two Gaussian components we find that the cold population can be represented by a Gaussian of width $0.5\mathrm{km\,s^{-1}}$ and velocity offset of $1.0\mathrm{km\,s^{-1}}$. In the aperture in which the likelihood ratio of a two-component model compared to a single component model is maximal, approximately 70 per cent of the stars are found to be associated with the cold component (see Kleyna *et al.* 2003 for details). The total luminosity of the component is then roughly $1.5 \times 10^4 \mathrm{L}_\odot$, which corresponds to the luminosity of a low-mass star cluster.

A similar kinematic feature has been identified in the centre of the Sextans dSph, where the velocity distribution inside 10 arcmin (250 pc) is dominated by a kinematically cold component (see Figure 1 and Kleyna *et al.* 2004). The total luminosity of this feature is about $1.3 \times 10^5 \mathrm{L}_\odot$, comparable to the luminosity of Milky Way globular clusters. Kleyna

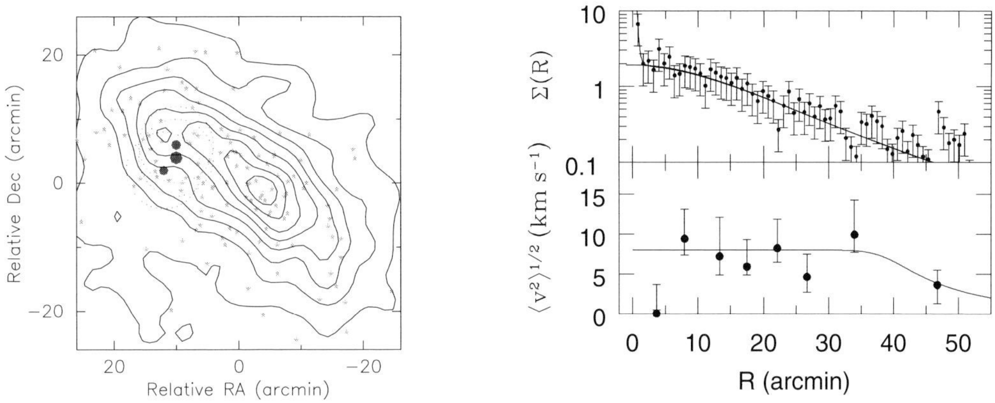

Figure 1. Left: Isophotes of Ursa Minor. The largest dot indicates the aperture in which a velocity distribution which includes a kinematically cold component is 4.7×10^4 times more likely than a single component model (statistical significance 99.94 per cent). (From Kleyna *et al.* 2003). Right: Surface brightness profile (top) and velocity dispersion profile (bottom) of the Sextans dSph. The low dispersion at small radii indicates the presence of cold substructure. (From Kleyna *et al.* 2004).

et al. (2004) suggested that this feature is the remnant of a disrupted star cluster. In addition to explaining the cold kinematics of the centre of Sextans, this picture also explains the surprising observation by Lee *et al.* (2003) that the blue straggler population of Sextans appears to be mass segregated. While this is difficult to understand in the context of a dSph with surface brightness as low as that of Sextans, the disruption of a star cluster within which mass segregation has already taken place would naturally produce the observed difference in the spatial distributions of low and high mass blue stragglers as the high mass objects would typically be found close to the centre of the star cluster and would therefore be the last objects to be removed by tidal disruption.

In Sextans, the spatial scale of the velocity feature matches closely the scale on which the nature of the stellar populations changes, as indicated by the radial variation of the horizontal branch ratio (see e.g. Lee *et al.* 2003; Harbeck *et al.* 2001). Recently, Tolstoy *et al.* (2004) have found that in the Sculptor dSph, differences in the spatial distribution of the stellar populations are mirrored by variations in the velocity dispersions of the populations. A preliminary analysis of our VLT-FLAMES data for the Carina dSph (see Koch – this proceedings – for a discussion of these data) provides hints of velocity substructure in this galaxy also. In the core of Carina, the velocity distribution of the more metal-rich population is noticeably narrower than that of the metal-poor population. This may indicate that at small radii the stellar distribution (and hence the observed kinematics) is dominated by a more concentrated population with smaller velocity dispersion, possibly corresponding to the red clump population identified by Harbeck *et al.* (2001). Thus, kinematic substructure appears to be a common feature in dSphs.

2. Implications of substructure I: cored haloes

The presence of long-lived, cold substructure in Ursa Minor has important implications for the shape of its dark matter halo. If the halo has a cusped density distribution, as expected on the basis of cosmological simulations, then any cold substructure would disperse on a short time-scale. Kleyna *et al.* (2003) presented the evolution of such a substructure on a radial orbit and showed that after 1 Gyr no discernible structure remained. By contrast, substructure in a cored halo can persist for up to 12 Gyrs, due

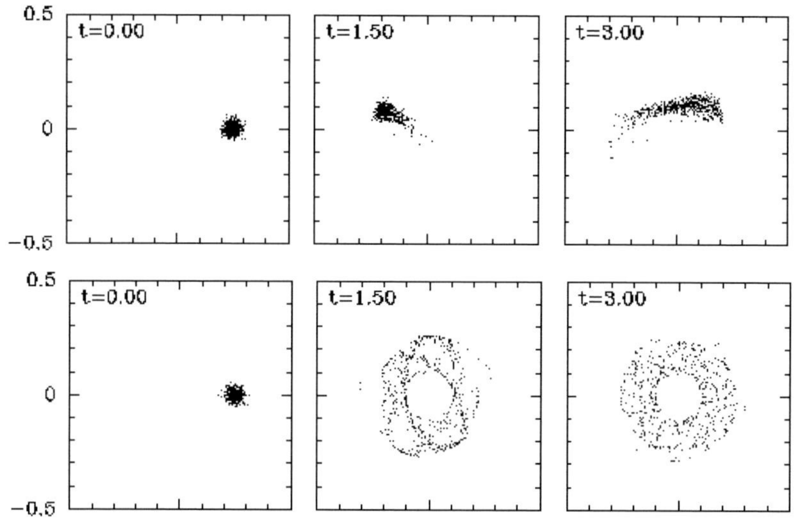

Figure 2. Evolution of cold substructure on orbits with initial tangential velocity equal to half the local circular speed. Results for a cored halo (core radius 0.6 units: top) and a cusped halo ($\rho \sim 1/r$: bottom) up to 3 Gyr are shown. The initial velocity dispersion for the substructure is 1 $\mathrm{km\,s^{-1}}$ in the cored halo and 0.1 $\mathrm{km\,s^{-1}}$ in the cusped halo.

to the fact that orbits in a harmonic gravitational potential are harmonically bounded and therefore orbits remain coherent for a long time. Figure 2 presents the evolution for substructures on more circular orbits, for which the tangential velocity at apocentre is half the local circular velocity. As in the case of a radial orbit, the localised substructure disperses in under 3 Gyr in the cusped potential while in a cored potential localised structure remains (up to 12 Gyr).

The orbits shown in Figure 2, and those presented in Kleyna *et al.* (2003), are all in the plane of the sky. This is driven by the observation that the substructure in Ursa Minor has no significant offset in mean velocity from the bulk velocity of the dSph. Independent of the shape of the potential, any orbit has to be close to face-on in order to avoid giving rise to a measurable line of sight velocity component. The radii of the near-circular orbits were chosen to correspond to the approximate current location of the observed Ursa Minor structure. Given that the timescale for a cluster at large radii to move close to the centre due to dynamical friction is less than a Hubble time, we would typically expect any substructure to have spent several Gyr close to its current location. If the halo had a central cusp there has thus been sufficient time for disruption to occur.

3. Implications of substructure II: origin of dSphs

We have seen that the properties of the kinematic substructure identified in Ursa Minor and Sextans are consistent with those expected for disrupted star clusters. This leads us to consider the possibility that dSphs were built up from the remnants of disrupted clusters. In the Local Universe, it is now well established that all stars form in cluster environments, which range in size from OB associations up to masses comparable to the old Milky Way globular clusters and most of which dissociate within a short time of their formation due to a combination of gas loss and external tidal fields. Given the broad range of star forming environments in which clusters are observed, it is reasonable to suppose that this was also the dominant mode of star formation at the time when the dSphs formed most of their stars. For example, Fellhauer & Kroupa (2002) investigated the merger of star clusters as a possible origin for ultra-compact dwarf galaxies. The merger time for star clusters formed in a small volume is very short – if this merger takes place within the gravitational potential of a dSph galaxy, then dynamical friction will further reduce this time scale. A violent merger at the centre of a halo is likely to produce a significant population of stars on highly radial orbits. In the pure-stellar models of Fellhauer & Kroupa (2002), such stars would be removed by the tidal field of a parent galaxy when they moved to the outer parts of the satellite. Within a dSph halo, however, a large fraction of these stars may remain bound to the galaxy. This population might then explain the sudden fall-off seen in the velocity dispersion profiles of Ursa Minor, Draco and Sextans which would then be properly viewed as arising from the presence of two distinct populations in each dSph each with their own spatial distribution and velocity anisotropy. Surviving cold substructures correspond to later accretions.

4. Conclusions

Recent kinematic observations in Local Group dSphs have shown us that kinematic substructure is relatively common and that the different stellar populations within dSphs often have differing velocity distributions. The survival of cold substructure can be used to infer the shape of the dark matter density distribution and, in particular, can rule out the presence of a cusp in the halo of Ursa Minor. The properties of the substructure are also consistent with a formation scenario in which dSphs are formed through the merger of star clusters in a dark matter halo. Further numerical simulations are underway to confirm that this picture is fully consistent with the observed properties of the dSphs.

References

Aaronson, M. 1983, *ApJ* 266, L11
Ashman, K.A., Bird, C.M. & Zepf, S.E. 1994, *AJ* 108, 2348
Fellhauer, M. & Kroupa, P. 2002, *MNRAS* 330, 642
Harbeck, D. *et al.* 2001, *AJ* 122, 3092
Kleyna, J.T., Wilkinson, M.I., Gilmore, G.F. & Evans, N.W. 2003, *ApJ* 588, L21
Kleyna, J.T., Wilkinson, M.I., Evans, N.W. & Gilmore, G.F. 2004, *MNRAS* 354, L66
Lee, M.G. *et al.* 2003, *AJ* 126, 2840
Tolstoy, E. 2004, *ApJ* 617, L119
Wilkinson, M.I., Kleyna, J.T., Evans, N.W., Gilmore, G.F., Irwin, M.J. & Grebel, E.K. 2004, *ApJ* 611, L21

Discussion

TOLSTOY: The abundance differences (e.g. deep mixing patterns) between "globular cluster" stars and "field" stars in dSphs might suggest that it is difficult to build an entire dSph from globular clusters.

WILKINSON: The star clusters which contributed to the bulk population of the dSphs were probably more akin to Milky Way open clusters in terms of their densities and survival times. Thus, one would not necessarily expect to observe deep mixing in dSphs.

LIN: How would the external perturbation produced by the Milky Way tidal field affect the survival of the substructure in Ursa Minor?

WILKINSON: We performed a number of simulations in which we included the tidal field of the Milky Way (modelled as an isothermal sphere with $10^{12} M_\odot$ inside 50 kpc) for a range of Ursa Minor orbits with periods ~ 1.6 Gyr. In no instance did this tidal force significantly affect the survival of substructure either in a cored or cusped halo.

Near-Field Cosmology with Dwarf Elliptical Galaxies
Proceedings IAU Colloquium No. 198, 2005
H. Jerjen and B. Binggeli, eds.

doi:10.1017/S1743921305003820

Detailed kinematics and morphological features in tidally heated disks

C. Mastropietro[1], B. Moore[1], L. Mayer[1], V. P. Debattista[2,3], R. Piffaretti[1,4] and J. Stadel[1]

[1]Institute for Theoretical Physics, University of Zürich, CH-8057 Zürich, Switzerland
[2] Astronomy Dept., University of Washington Seattle WA 98195-1580
[3]Institut für Astronomie, ETH Zürich, CH-8093 Zürich, Switzerland
[4]Institut für Astrophysik, Leopold-Franzens Universität Innsbruck, A-6020 Innsbruck, Austria

Abstract. The recent discovery of hidden non-axysimmetric and disk-like structures in bright Virgo dwarf elliptical and lenticular galaxies (dE/dSph/dS0) indicates that they may have late-type progenitors. Using N-body simulations we follow the evolution of disk galaxies within a ΛCDM cluster simulated with 10^7 particles, where the hierarchical growth and galaxy harassment are modeled self-consistently. Most of the galaxies undergo significant morphological transformation, even at the outskirts of the cluster, and move through the Hubble sequence from late type disks to dwarf spheroidals. The time evolution of the simulated galaxies is compared with unsharp-masked images obtained from VLT data and the projected kinematics of our models with the latest high resolution spectroscopic studies from the Keck and Palomar telescopes.

Keywords. methods: N-body simulations – galaxies: clusters: general – galaxies: evolution

1. Introduction

Early type dwarfs (dE and dSO) are the most common type of galaxies in the nearby universe, yet their origin and nature are still unknown. These galaxies are found nearly exclusively near bright galaxies and in groups or clusters. Formation scenarios include mainly two different hypotheses: 1) dEs and dSOs are primordial galaxies, 2) they are the result of a morphological transformation of spiral and irregular galaxies accreting into the cluster. Moore, Lake & Katz (1998) have shown that galaxy harassment in clusters can transform spirals into spheroidals. According to this evolutionary scenario we expect to find nearby cluster galaxies that are currently undergoing morphological transformation and retain part of their disk nature. Recent observations confirm that early type galaxies have a broad range of photometric and kinematical characteristics. Jerjen, Kalnajs & Binggeli (2000) and Barazza, Binggeli & Jerjen (2002) discovered hidden spiral structures and bar features in bright dEs in the Virgo cluster. Spirals and disks were also observed within dwarf spheroidal galaxies in the Coma (Graham, Jerjen & Guzmán 2003 and Fornax clusters (De Rijcke *et al.* 2001). Furthermore, the degree of rotational support is found to vary from zero to a value close to one expected for a galaxy flattened by rotation (De Rijcke *et al.* 2001; Simien & Prugniel 2002; Pedraz *et al.* 2002; Geha, Guhathakurta & van der Marel 2002; Geha, Guhathakurta & van der Marel 2003; van Zee, Skillman & Haynes 2004). The aim of this work is to follow the evolution of disk galaxies orbiting in a cluster environment using high resolution N-body simulations and to compare the final harassed remnants with the latest photometric and spectroscopic data.

Gal	Apo-Peri	r_t	r_e	M_h	M_s	b/a	c/a	Gal	Apo-Peri	r_t	r_e	M_h	M_s	b/a	c/a
G1	200-90	8.5	2.0	2.6	2.3	0.79	0.39	G11	...-460	17.2	2.0	8.2	3.6	0.71	0.18
G2	150-100	10.1	1.3	2.8	2.4	0.57	0.40	G12	1130-120	16.9	2.6	14.6	4.1	0.66	0.18
G3	650-60	16.6	2.7	13.0	4.2	0.42	0.18	G13	...-900	17.4	2.6	19.2	4.2	0.41	0.28
G4	300-50	11.6	1.8	5.2	3.2	0.44	0.34	G14	...-...	16.3	2.5	24.4	4.2	0.79	0.29
G5	630-100	19.4	1.9	7.6	3.3	0.60	0.27	G15	2480-...	17.0	2.9	21.1	4.2	0.43	0.19
G6	90-40	3.6	0.7	0.2	0.5	0.81	0.80	G16	720-210	14.7	2.6	18.8	4.1	0.45	0.23
G7	75-210	11.0	1.8	4.2	3.3	0.61	0.38	G17	...-...	15.6	2.6	21.0	4.2	0.38	0.22
G8	100-60	4.6	1.1	1.0	1.5	0.87	0.86	G18	2290-...	15.2	2.7	19.5	4.2	0.41	0.25
G9	120-70	7.9	1.1	1.5	1.8	0.66	0.54	G19	2810-...	15.0	3.0	19.6	4.2	0.77	0.17
G10	270-80	9.0	1.2	1.8	1.8	0.62	0.49	G20	1720-450	16.8	2.9	21.6	4.2	0.34	0.22

Table 1. Final state of the remnants. The second column indicates, when available, the apocentric and pericentric distances of the last orbit. For each galaxy we list three dimensional tidal r_t and effective r_e radius (in kpc), dark matter M_h and stellar M_s mass (in units of $10^9 M_\odot$). The last two columns indicate the axial ratios (respectively intermediate and short axis to major axis) measured within $2r_e$.

2. Galaxy evolution

All the simulations have been carried out using PKDGRAV (Stadel 2001). The galaxy model is a multi–component system with an exponential stellar disk embedded in a NFW (Navarro, Frenk & White 1997) spherical dark matter halo and was constructed using the technique described by Hernquist (1993). The structural parameters of the disk and the halo are chosen so that the resulting rotation curve resembles that of a typical bulgeless late-type (Sc/Sd) disk galaxy. The mass within the virial radius was set equal to 7×10^{10} M$_\odot$ and the fraction of mass in the disk is $\sim 6\%$. The model has a central surface brightness $\mu_B = 22\,\mathrm{mag\,arcsec}^{-2}$. Each galaxy is modeled with 10^6 particles, 10^5 of which are in the disk. The galaxies were evolved for ~ 8.4 Gyrs in a ΛCDM cluster with a mass at $z = 0$ comparable to that of Virgo ($M_{vir} = 3.1 \times 10^{14}$ M$_\odot$) and a virial radius R_{vir} of 1.8 Mpc (Diemand, Moore & Stadel 2004). The number of particles within the virial radius is $\sim 6 \times 10^6$.

The evolution of many of the galaxies within the cluster is quite violent and due to a sequence of strong gravitational encounters with substructures and with the global cluster potential. Most of the central galaxies lose a significant fraction of stars and undergo a complete morphological transformation from disks to spheroidal systems (Fig. 1, left panel). Depending on the orbit and on the number of close encounters that the galaxy suffers, the morphological transformation of the stellar component can be complete and produce a pure spheroidal system, or can lead to the formation of a more elongated structure, which still retains disk and bar features. For a few galaxies orbiting closest to the cluster center the tidal effects from the cluster potential become dominant and the evolution is very rapid. Table 1 summarises the main properties of the remnant galaxies. The initial model has a virial radius of 85 kpc and the effective radius of the stellar distribution is 2.5 kpc. In a few cases the galaxy loses much more dark matter than stars and becomes baryon dominated in the central region. The loss of stellar mass is not a simple function of the mean orbital radius or apocentric/pericentric distance from the cluster center, confirming the importance of harassment. There is a clear correlation between the flattening of the stellar component (c/a) and the final stellar mass: massive stellar remnants have smaller c/a values and retain more of their initial disk nature, while galaxies that have lost most of their stars tend to be more spheroidal.

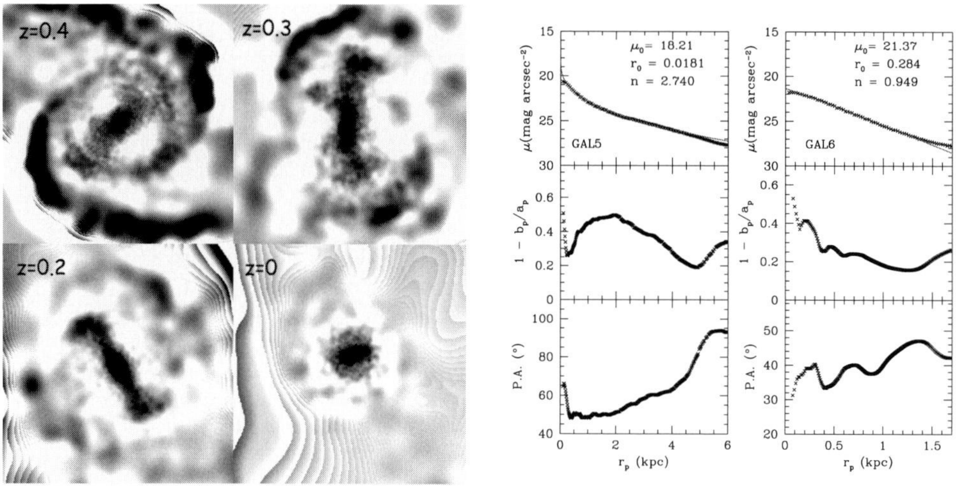

Figure 1. Left panel: morphological transformation of a disk galaxy into a dwarf elliptical. Right panel: *B*-band surface brightness, ellipticity and position angle profile for two galaxies of our sample. The coordinate r_p is the equivalent radius. A Sérsic law is superimposed on the surface brightness profile (red solid line) and the corresponding best fit parameters are given in the top right of the panel. Note that r_0 is in kpc units.

3. Photometric analysis and kinematics

In order to compare the results of our simulations with observations, we projected the remnants along a random line of sight and calculated surface brightness, ellipticity and position angle profiles (see Fig. 2, right panel). Even though there is not a well defined trend in the ellipticity profiles, galaxies which preserve a bar component are generally characterised by large ellipticities in the central 2-3 kpc and by a drop in the outer regions. Often the inner 0.5 kpc exhibits a very low ellipticity corresponding to a small nuclear component produced by buckling instabilities, whereas the external drop is more likely due to disk heating. The surface brightness profiles were fitted with a Sérsic law. The range of surface brightness parameters agrees quite well with the observations of dwarf galaxies in Virgo.

We measure the kinematics of the our projected galaxies using a slit lenght which allows us to observe in most of the galaxies the turnover of the rotation curve. In most of the cases the remnants lose a large fraction of rotation velocity (see Fig. 2, left panel), but only for one galaxy (GAL6) we do not observe any significative rotation along the major axis. In general we find non-rotating and slowly rotating dwarfs to have internal and environmental properties substantially different from the rotating ones. In particular the former are relatively low massive objects with high c/a ratios and less apparent disk features. They seem to be the product of a violent morphological transformation accelerated by the proximity of the cluster center. Although the galaxies lose significant amounts of mass, the velocity dispersion can rise up to $50\,\mathrm{km\,s^{-1}}$ in the central 1-2 kpc. This is due to the bar formation and the transformation of circular to more radial orbits. The velocity dispersion profiles are typically decreasing at larger radii but they rise again in those galaxies for which the slit semi-length approaches or exceeds the tidal radius. As the ratio between rotational velocity and velocity dispersion decreases, the galaxy loses rotational support and the flattening becomes dominated by anisotropic pressure. We plotted in Fig. 2 (right panel) the ratio of the maximum rotational velocity v_{max} to the mean velocity dispersion $\bar{\sigma}$ within a radius of 6 kpc versus the mean isophotal ellipticity

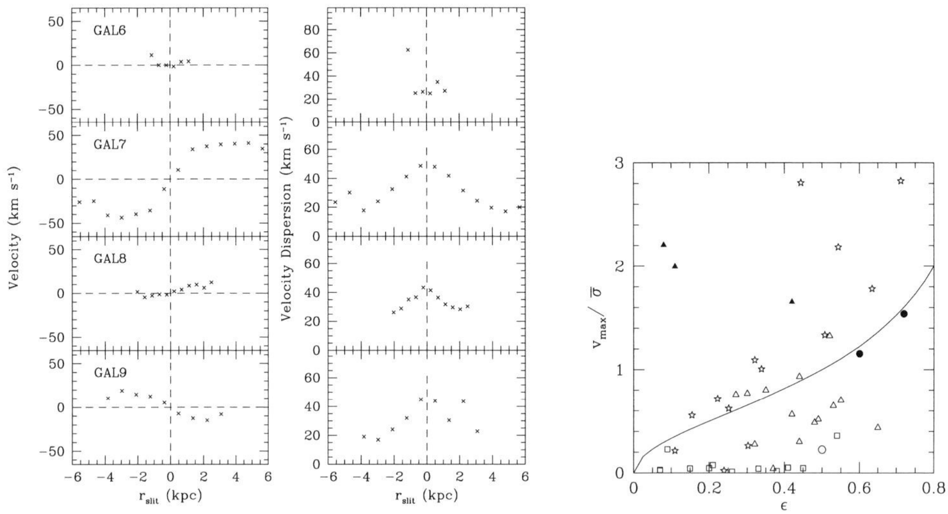

Figure 2. Left panel: projected kinematic profiles of the remnants within the cluster virial radius at $z = 0$. The line of sight velocity (left panels) and the velocity dispersion (right panels) are plotted as a function of the radial distance along the major axis. Right panel: the ratio of the maximum rotational velocity v_{max} to the mean velocity dispersion $\bar{\sigma}$ plotted versus the mean isophotal ellipticity. Stars represent the simulated sample, while the other points are observational data.

calculated in the same radial range. In the same plot we show the observational results for most of the dwarf galaxies in Virgo with published kinematics. We also plot the data for dwarf galaxies in the Fornax cluster by De Rijcke *et al.* (2003). Most of the simulated remnants have a similar range of ellipticities and $v_{max}/\bar{\sigma}$ as the observed dwarfs, even if it is clear that we have some difficulties in reproducing galaxies with a degree of rotational support close to zero. On the other hand it is possible that the low $v_{max}/\bar{\sigma}$ ratio of some galaxies is due to the small radial range covered by the observations.

References

Barazza, F.D., Binggeli, B. & Jerjen, H. 2002, *A&A* 391, 823

Bender, R. & Nieto, J.L. 1990, *A&A* 239, 97

Conselice, C.J., Gallagher, J.S. & Wyse, R.F.G. 2001, *ApJ* 559, 791

De Rijcke, S., Dejonghe, H., Zeilinger, W.W., & Hau, G.K.T. 2001, *ApJ* 559, L21

De Rijcke, S., Dejonghe, H., Zeilinger, W.W. & Hau, G.K.T. 2003, *A&A* 400, 119

Diemand, J., Moore, B. & Stadel, J. 2004, *MNRAS* 353, 624

Geha, M., Guhathakurta, P. & van der Marel, R.P. 2002, *AJ* 124, 3073

Geha, M., Guhathakurta, P. & van der Marel, R.P. 2003, *AJ* 126, 1794

Graham, A.W., Jerjen, H. & Guzmán, R. 2003, *AJ* 126, 1787

Hernquist, L. 1993, *ApJ* 86, 389

Jerjen, H., Kalnajs, A. & Binggeli, B. 2000, *A&A* 358, 845

Moore, B., Lake, G. & Katz, N. 1998, *ApJ* 495, 139

Navarro J.F., Frenk C.S. & White S.D.M. 1997, *ApJ*, 490, 493

Pedraz, S., Gorgas, J., Cardiel, N., Sánchez-Blázquez, P. & Guzmán, R. 2002, *MNRAS* 332, L59

Simien, F. & Prugniel, P. 2002, *A&A* 384, 371

Stadel, J. 2001, *Ph.D. Thesis*

van Zee, L., Skillman, E. D. & Haynes, M. P. 2004, *AJ* 128, 121

Discussion

CONSELICE: How much stellar mass is lost in these simulated disks?

MASTROPIETRO: Up to 90%.

CALDWELL: The models you created that are not purely exponential have support in the real universe, because in nearby clusters the more massive dwarf ellipticals are more centrally concentrated than pure exponentials, and look very much like your models. Also, in addition to the weak disk structure seen in some dwarf ellipticals, many have ages of around 2 - 3 Gyr. Thus if we image clusters at z $\approx$ 0.2, we should see many more dwarfs with disk-like structures shown in your models.

MASTROPIETRO: Yes, most of them should appear like disk galaxies at z $\approx$ 0.4.

Near-Field Cosmology with Dwarf Elliptical Galaxies
Proceedings IAU Colloquium No. 198, 2005
H. Jerjen and B. Binggeli, eds.

doi:10.1017/S1743921305003832

Dwarf "transition" types: searching for the holy grail(s)

Patricia M. Knezek[1], Kate E. Dellenbusch[2], John S. Gallager, III[2], and Kenneth R. Sembach[3]

[1]WIYN Consortium, Inc., 950 N. Cherry Avenue, Tucson, AZ 85719
email: knezek@noao.edu

[2]University of Wisconsin, Dept. of Astronomy, 475 N. Charter Street, Madison, WI, 53706

[3]Space Telescope Science Institute, 3700 San Martin Dr., Baltimore, MD 21218

Abstract. The evolution of dwarf galaxies, particularly any potential morphological evolution among the different types of dwarf galaxies, remains poorly understood. The interstellar medium (ISM) in dwarf galaxies is thought to be more strongly disturbed by star formation episodes in these low mass galaxies, including supernovae-driven winds during starbursts and star formation induced by galaxy-galaxy interactions. Loss of the ISM has been postulated to result in the evolution of dIrrs to dE/dSphs. However, disentangling the relative importance of environmental (galaxy harassment) versus local stocastic processes on star formation, and their effects on subsequent morphological evolution is still a very active area of research. So-called "transition dwarf" galaxies are galaxies that have properties that fall somewhere between gas-rich dIrrs and gas-poor dE/dSphs, and may provide clues to the overall evolutionary history of dwarf galaxies. We report on some recent work on a sample of vigorously star-forming dwarf "transition" galaxies, and suggest that they may be examples of the bridge between dIrrs and dEs/dSphs.

1. Introduction

Dwarf galaxies dominate the population of galaxies by number. As a result, they form a critical component of our overall understanding of how galaxies form and evolve in the universe. Yet many questions about the formation and evolution of dwarf galaxies remain. In particular, one continuing puzzle in galaxy evolution is whether there is an evolutionary connection between the various morphological classes of dwarf galaxies. In particular, are dIrrs and dEs/dSphs related? An extended body of work exists on this topic, both theoretical and observational, and the results indicate that the answer is complex. For example, Knezek, Sembach, & Gallagher (1999), studied three dwarf galaxies proposed as a sequence of dIrr to "transition" dwarf to dE by Sandage & Hoffman (1991). They concluded based on the very low star formation rates and the red optical colors of the two galaxies with gas in this sample that they would not evolve into a dE/dSph within a Hubble time. Grebel, Gallagher, & Harbeck (2003) concentrated on Local Group (LG) dwarfs and determined that while LG dIrrs are unlikely to evolve into LG dSphs based on the fact that dSphs have higher mean stellar metallicities for fixed optical luminosity, the LG "transition dwarfs," whose locus in the luminosity-metallicty diagram is similar to dSphs, *are* likely progenitors of LG dSphs. Meanwhile, van Zee, Skillman, & Haynes (2004) observed a sample of 16 dEs in the Virgo cluster and found that rotation is a significant component of the stellar dynamics of some Virgo dwarfs. Furthermore, based on evidence of a relationship between rotation amplitude and galaxy luminosity that agrees with the Tully-Fisher relation, they propose that the dE population in Virgo arises from stripped dIrrs.

It is likely that the diverse results indicate that there are multiple competing effects. The critical aspect of all proposed evolutionary scenarios for creating gas-poor systems from gas-rich systems, however, is how the gas is removed (e.g., Lin & Faber 1983, Thuan 1985). This might occur as a result of supernova driven winds during starbursts (Dekel & Silk 1986, Babul & Rees 1992, De Young & Heckman 1994, Spaans & Norman 1997), or as a result of an external interaction, such as ram pressure stripping (Kormendy 1985, Binggeli 1986), tidal interactions (e.g. Sofue 1994, Mayer *et al.* 2001 a, b), or galaxy collisions (e.g. Henriksen & Byrd 1996, Moore, Katz, & Lake 1996). Disentangling the physical origins of the two main families of low luminosity galaxies, and searching for paths between them remains a difficult task. The potential key to success is the class of "transition" dwarf galaxies, those systems that show a mixture of early and late-type dwarf galaxy characteristics.

2. What is a "transition" dwarf galaxy?

Dwarf galaxies come in many flavors, and the classification of dwarf galaxies, like the classification of more massive galaxies, can depend how the author defines the sample. Thus, one interpretation of "transition galaxies" was described by Sandage & Hoffman (1991) based on their optical and neutral gas composition. The optical morphology is mixed between that of a dwarf irregular and a dwarf spheroidal, while the gas mass to blue luminosity ratio is low for irregular dwarfs, but high for dwarf spheroidals. There is no criteria based on the current star formation rate of the galaxy. Under this scenario, the expectation is that an irregular galaxy undergoes star formation and either exhausts its fuel supply or loses it through supergalactic winds. The galaxy will then slowly fade away. Potential difficulties with this general definition include the fact that dIrrs generally occupy a different place on the metallicity-luminosity relation from dEs, that dIrrs tend to be rotationally supported, while dEs may not be, and that dIrrs generally have low star formation rates, leading to very long gas depletion timescales (often many Hubble times, e.g. Knezek, Sembach, & Gallagher 1999; van Zee, Haynes, & Salzer 1997).

A related, but separate interpretation of "transition galaxies" is that of a subsample of blue compact dwarf (BCD) galaxies which have central starbursts but smooth outer optical isophotes, and are relatively gas-poor (van Zee, Salzer, & Skillman 2001). The idea here is that the galaxies rapidly use and/or lose their gas through the starburst, and will then slowly fade away to a dSph. Potential difficulties with this scenario include the fact that even the lowest M_{HI}/L_B systems seem to be rotationally supported, and the central starburst remnant will need to fade.

For both interpretations above, the discovery of rotationally supported dEs by van Zee, Skillman, & Haynes (2004), along with the work of Mayer *et al.* (2001a,b) showing that repeated tidal shocks experienced by dwarf galaxies that are satellites can both remove the satellites gas and transform it into a dynamically hot system, suggest that what has traditionally been known as the 'angular momentum problem' may not be an issue in all cases. Furthermore, there is a known starburst galaxy, NGC 5253, which is a member of the Centaurus A group of galaxies, and does *not* appear to be rotationally supported (Caldwell & Phillips 1989). This suggests, at least for dwarf galaxies in moderate to dense environments, the environmental effects of galaxy harrassment may in fact be enough to turn a dIrr into a dE/dSph. Are there other galaxies like NGC 5253?

In this work, we present a sample of five dwarf galaxies selected to be "transition galaxies" based an optical mix of early and late-type morphological characteristics. The sample and observations are presented in Section 3, the results are discussed in Section 4, and our conclusions are in Section 5.

3. Sample and Observations

We selected six dwarf galaxies that appeared to have a mix of early and late-type morphologies based on their appearance on POSS plates, and that had not been observed in Hα. One dwarf, NGC 5798, turned out to be a gas-rich normal dwarf irregular, and will not be discussed further here. It is important to note that environment was *not* a selection criteria, nor was the rate of ongoing star formation.

Direct imaging observations were made at Michigan-Dartmouth-MIT Observatory on the 1.3 m McGraw–Hill telescope. The data were taken on 10 - 12 April 1996 with a 1024x1024 thinned CCD. The pixel scale is $\sim$0.50 $''$/pix for this detector and telescope with the f/7.5 secondary, giving a field of view of $\sim$8.5$'$x8.5$'$. Broadband B and R images were made using the MDM Schombert filter set. These filters quite closely approximate the wavelength coverage of the Kitt Peak National Observatory "Harris Set" of broadband filters, although the transmission efficiencies are lower. Narrowband Hα imaging was done using a set of extragalactic Hα interference filters. Typical integrations times were 20-30 minutes in B, 10-15 minutes in R, and 40-60 minutes in Hα. Several Landolt standard fields (Landolt 1992) were observed each night in the broadband filters, and the spectroscopic standard Feige 34 was observed in R and each Hα filter.

Initial data reduction was carried out in the standard way using IRAF†. In order to obtain the Hα flux for each galaxy, the observations of the spectrophotometric standard Feige 34 were used. Feige 34 was observed under photometric conditions in both the R and Hα filters. Full details of the data reduction can be found in Knezek, Sembach, & Gallagher (1999).

4. Results and Discussion

We have determined the broadband and emission morphologies, as well as the broadband colors and surface brightness distributions for all the dwarfs in our initial sample. Figures 1 and 2 show all six galaxies. The top panel is the image of each galaxy in Hα emission. The bottom panel is the corresponding R band data, shown as surface brightness contours. As noted in the previous section, NGC 5798 is a gas-rich dwarf irregular that was inadvertently included in the sample. Of the five remaining dwarfs, all have fairly smooth outer R band isophotes, consistent with their "mixed" morphology and classification as a "transition" dwarf. Four of the five galaxies have very centrally-peak Hα emssion, with only very faint emission outside the central region. These galaxies are very similar to the 'dwarf amorphous nuclear starburst' (DANS) galaxies first described by Salzer, MacAlpine, & Boroson (1989), but with blue luminosities that are 1-2 magnitudes fainter. The exception is NGC 3353, which has strong central emission, but also extended strong emission that shows a bar-like structure.

Table 1 shows a comparison of the derived properties for the five "transition" dwarfs. The four DANS galaxies all have strong Hα emission, relatively blue colors ($B-R$ $\sim$0.95–1.25), are gas poor ($M_{HI}/L_B \lesssim 0.1$), and are forming stars prodigiously, at a rate of $\sim$2 M_{sun}/yr. NGC 3353 is even more extreme, with a very blue color (B – $R$$\sim$0.85), is relatively gas-rich ($M_{HI}/L_B \sim 0.3$), and is forming stars at a rate that is three times higher than the four DANS galaxies. NGC 3353 is a well known BCD, and perhaps it represents an earlier phase of a starburst phenomenon. All five of these systems will exhaust their gas in less than a Hubble time if they continue to form stars at their current

† IRAF is distributed by the National Optical Astronomy Observatories, which are operated by the Association of Universities for Research in Astronomy, Inc., under contract to the National Science Foundation.

Table 1. Derived Properties of the Dwarf "Transition" Galaxies

Galaxy	M_B	B−R	EW(Hα)	M_{HI}/L_B	SFR(Hα)	log
	(mag)	(mag)	(angstrom)	(M_{sun}/L_{sun})	(M_{sun}/yr)	(M_{HI}/SFR)
IC 745	-17.0	0.96	21	…	1.8	8.56
NGC 3265	-17.5	1.24	48	0.1	2.1	8.02
NGC 3353	-17.3	0.85	…	0.3	6.0	7.95
NGC 3773	-17.3	0.93	6.9	0.06	2.1	7.82
NGC 3928	-16.7	1.23	53	0.1	1.6	8.32

rates. Furthermore, if they do exhaust their gas and begin to fade, they would evolve into dSph-like systems with a color gradient, which may be consistent with what is seen in our own, lower mass, LG dSphs, where some dwarfs appear to have significantly different star formation histories in their inner and outer regions (see Mateo 1998, and references therein).

Morphologically, the four DANS galaxies appear very similar to NGC 5253, the BCD that does not appear to be rotationally supported. However, they are forming stars at nearly an order of magnitude higher rate (NGC 5253 is forming stars at $\sim$0.35 M_{sun}/yr, Marlowe *et al.* 1997). The outer regions of the DANS galaxies are currently *not* undergoing active star formation, and their colors are consistent with an intermediate age population ($\sim$2 − 6 Gyr; Dellenbusch *et al.* 2005). Furthermore, all five galaxies also turn out to share an environmental similarity to NGC 5253. They are all members of groups (IC 745, NGC 3265, NGC 3773) or clusters (NGC 3353, NGC 3928). Given that there was no selection criteria based on environment, and that many BCDs are found to be isolated, perhaps the starbursts in these systems have been induced in a fundamentally different way, e.g. through interactions with other galaxies. In particular, the very centralized nature of the star formation in the majority of the sample suggests that the gas may have been driven towards the center and induced a starburst that will quickly remove the gas and transform the systems into dE/dSphs (Moore, Lake, & Katz 1998).

5. Conclusions

We have studied a sample of five "transition" dwarf galaxies with optical morphologies that have a mix of early and late-type dwarfs characteristics. We find that all five galaxies are forming stars at a very prodigious rate (1.5-6 M_{sun}/yr), and four of the five dwarfs have nearly all their star formation occuring in their centers. Their outer regions no longer appear to be forming stars, and have colors that are consistent with an intermediate age stellar population ($\sim$2-6 Gyr). None of the galaxies is isolated, and the centralized nature of the starburst suggests that the starburst may have been induced by interactions with other galaxies, rather than being internally triggered. If these galaxies continue to form stars at their current rates, they will exhaust their gas in much less than a Hubble time. We suggest that perhaps these systems are true "transition" dwarfs, caught in the act of transforming their gas into stars, and progenitors of more massive dEs/dSphs.

Acknowledgements

The Hα filter set used for this research was kindly loaned to us by Deidre Hunter. Rosa Gonzalez provided us with the scripts for the spectrophotometric data reduction, and Ricky Patterson offered many helpful comments on the surface photometry. In addition, we would like to thank the staff at the Michigan-Dartmouth-MIT Observatory for their assistance and support in this project.

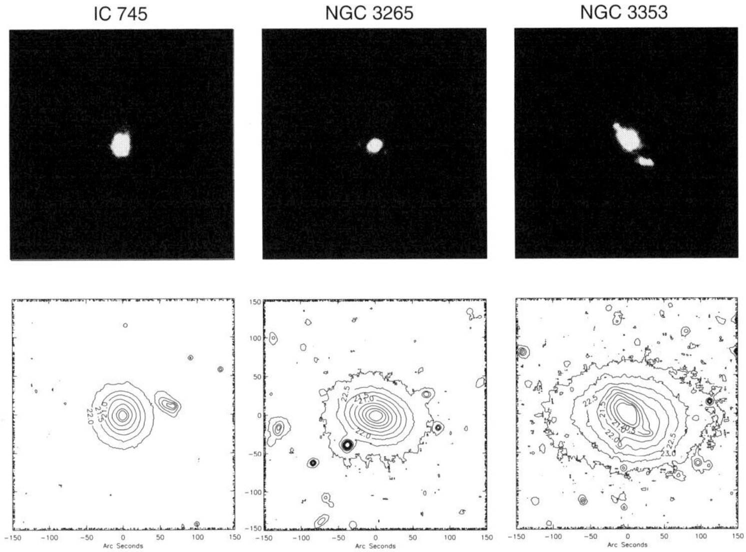

Figure 1. The upper panels show the Hα emission, while the lower panels indicate the R-band surface brightness contours in mag arcsec^{-2}

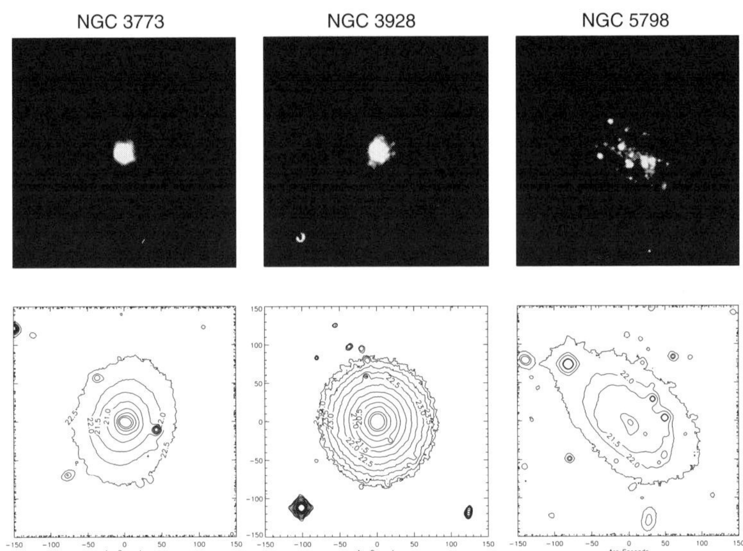

Figure 2. The upper panels show the Hα emission, while the lower panels indicate the R-band surface brightness contours in mag arcsec^{-2}

References

Babul, A. & Rees, M. J. 1992, *MNRAS* 255, 346

Binggeli, B. 1986, in *Star Forming Dwarfs and Related Objects*, eds. D. Kunth, T.X. Thuan, J.T.T. Thuan (Gif sur Yvette: Editiones Frontiéres), p. 52

Caldwell, N. & Phillips, M.M. 1989, *ApJ* 338, 789

De Young, D. & Heckman, T.F. 1994, *ApJ* 431, 598

Dekel, A. & Silk, J. 1986, *ApJ* 303, 39

Dellenbusch, K.D., Knezek, P.M., Gallagher, III, J.S. & Sembach, K.R. 2005, *in prep.*
Grebel, E.K., Gallagher, III, J.S. & Harbeck, D. 2003, *AJ* 125, 1926
Henriksen, M. & Byrd, G. 1996, *ApJ* 459, 82
Knezek, P.M., Sembach, K.R. & Gallagher, III, J.S. 1999, *ApJ* 514, 119
Kormendy, J. 1985, *ApJ* 295, 73
Landolt, A.U. 1992, *AJ* 104, 340
Lin, D.N.C. & Faber, S.M. 1983, *ApJ*(Letters) 266, 21
Marlowe, A.T., Meurer, G.R., Heckman, T.M. & Schommer, R. 1997, *ApJS* 112, 285
Mateo, M.L. 1998, *ARA&A* 36, 435
Mayer, L., Governato, F., Colpi, M., Moore, B., Quinn, T., Wadsley, J., Stadel, J. & Lake, G. 2001a, *ApJ* 559, 754
Mayer, L., Governato, F., Colpi, M., Moore, B., Quinn, T., Wadsley, J., Stadel, J. & Lake, G. 2001b, *ApJ*(Letters) 547, L123
Moore, B., Katz, N. & Lake, G. 1996, *ApJ* 457, 455
Moore, B., Lake, G. & Katz, N. 1998, *ApJ* 495, 139
Salzer, J.J., MacAlpine, G.M. & Boroson, T.A. 1989, *ApJS* 70, 447
Sandage, A. & Hoffman, G.L. 1991, *ApJ*(Letters) 379, L45
Sofue, Y. 1994, *ApJ* 423, 207
Spaans, M. & Norman, C. 1997, *ApJ* 483, 87
Thuan, T. X. 1985, *ApJ* 299, 881
van Zee, L., Haynes, M.P. & Salzer, J.J. 1997, *AJ* 114, 2479
van Zee, L., Salzer, J.J. & Skillman, E.D. 2001, *AJ* 122, 121
van Zee, L., Skillman, E.D. & Haynes, M.P. 2004, *AJ* 128, 121

Discussion

DRINKWATER: Many of the dwarfs morphologically classified as dEs at the edge of the Fornax cluster have significant Hα emission (see poster by Michielsen *et al.*). How would these compare to the transition types you discussed? Their absolute magnitudes are around $M_B = -15$.

KNEZEK: These galaxies may well be members of the same family. The galaxies I showed are brighter ($M_B \approx -17$), but we have not removed emission due to the starburst. If that is done, theirunderlying luminosities may be similar. Our sample might represent one extreme end of a family of galaxies that are modified due to their environment. It might be interesting to examine the relative environmental conditions of the galaxies in our sample (mostly in groups) with these dwarfs on the outskirts of Fornax.

CONSELICE: Does the fact that transition dwarfs are very rare imply a short life-time for this phase? Does this also imply that something must be removing the gas in these systems fairly quickly?

KNEZEK: Yes, the fact that these systems are rare implies that their life-times must be very short.

Near-Field Cosmology with Dwarf Elliptical Galaxies
Proceedings IAU Colloquium No. 198, 2005
H. Jerjen and B. Binggeli, eds.

doi:10.1017/S1743921305003844

Neutral hydrogen in early-type dwarf galaxies of the Sculptor Group

Antoine Bouchard[1,2], Gary DaCosta[1], Helmut Jerjen[1] and Jürgen Ott[2]

[1]Research School of Astronomy & Astrophysics, Mount Stromlo Observatory, Cotter Road, Weston Creek, ACT 2611, Australia
[2]Australia Telescope National Facility, CSIRO, PO Box 76, Epping NSW 1710, Australia

1. Dwarf galaxy evolution and the morphology-density relation

Dwarf galaxies are at the center of a long debated galaxy evolution puzzle. In the Local Group, there seems to be evidence that the exact morphology of a dwarf is strongly correlated with its environment: late-types (dwarf irregulars, dIrr) are further away from major group members than early-types (dwarf ellipticals and dwarf spheroidals, dE). In general, the early-types, with their smooth ellipsoid light distribution, have no detected interstellar medium (ISM), possess low angular momentum and have low current star formation rates. The late-types have an irregular appearance, large quantities of neutral hydrogen (HI), ongoing star formation and higher angular momentum.

The relation between the morphology of a dwarf and the number density of galaxies around it, the so-called morphology-density relation, and the apparent continuous range in dwarf galaxy properties suggests a possible evolutionary connection between the main types, with mixed type (dIrr/dE) galaxies forming the intermediate class. Grebel, Gallagher, & Harbeck (2003) argue against this evolutionary scenario stating two major points: first the fading of a dIrr to the typical luminosity of a dE would produce an object with too low metallicity than what is observed and, second, the removal of the ISM from a dIrr would not produce a non-rotating dE.

The problem with the morphological classification is that it merely describes the behavior of the light distribution around galaxies and does not convey any physical information on the objects it classifies. Studies of the Local Group helped establish empirical relations between morphology and other physical parameters, but have failed to establish a direct quantitative correlation based on solid physical principles.

Consequently, if one wishes to understand dwarf galaxy evolution, a quantitative measure of the environmental impact needs to be established and related to the fundamental properties of galaxies (stellar population, etc.). High density environments maximize the effects of external gas removal mechanisms (e.g. ram pressure, tidal stirring) while a low density environment will minimize such factors, thereby enabling us to differentiate between an environmentally or internally generated effects.

In order to get such a quantitative measure of the environmental impact on the evolution of dwarfs in a low density setting, we have observed five galaxies of the Sculptor group: the dS0/Im ESO294-G010, the two dE/Im ESO540-G030 and ESO540-G032, the dSph ESO410-G005, and the dE Scl-dE1. In the cases of ESO294-G010, ESO540-G030, ESO540-G032, and ESO410-G005, photometric studies seem to indicate the presence of blue stars in the central regions of these galaxies (Karachentsev *et al.* 2000; Jerjen & Rejkuba 2001; Karachentsev *et al.* 2002; Karachentsev *et al.* 2003) although in some cases these may be the result of photometric errors or foreground contamination. In addition

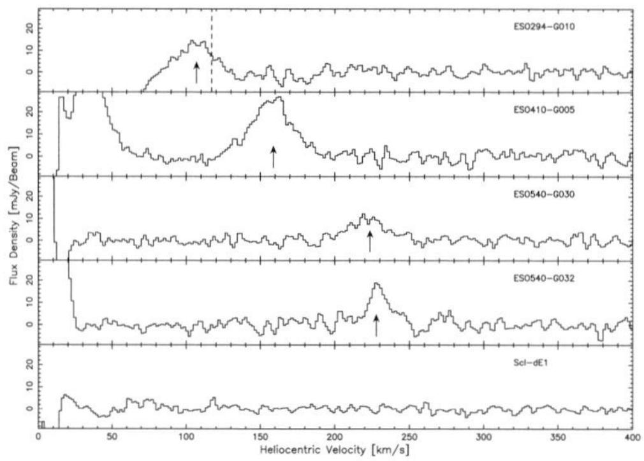

Figure 1. Parkes HI spectra of the 5 observed dwarfs. In the top panel the vertical dashed line shows the optical velocity of the dwarf. The strong features around the 0 km s^{-1} velocity are artifacts from the Galactic HI emission.

to these maybe young blue stars, Jerjen, Freeman & Binggeli (1998) identified a small HII region in ESO294-G010.

The Parkes single dish radio-telescope was used to obtain deep high velocity resolution HI spectra for the five targeted galaxies (Figure 1). The 8 MHz bandpass gave a velocity coverage from -300 to 1300 km s^{-1}, divided in 1024 channels, each 1.6 km s^{-1} wide.

2. Results

Figure 1 shows clear detections of four out of five targets. The objects have $M_{HI} = (2.4 \pm 0.3) \times 10^5\ M_\odot$ for ESO294-G010 (D = 1.7 Mpc), $(7 \pm 2) \times 10^5\ M_\odot$ for ESO410-G005 (D = 1.9 Mpc), $(8 \pm 1) \times 10^5\ M_\odot$ for ESO540-G030 (D = 3.2 Mpc), $(4 \pm 1) \times 10^5\ M_\odot$ for ESO540-G032 (D = 2.2 Mpc), and $M_{HI} < 2 \times 10^5\ M_\odot$ for Scl-dE1 (3σ, D = 4.2 Mpc, assumes a velocity dispersion of 20 km s^{-1}).

In the low density environment of the Sculptor group, most galaxies can be considered as essentially isolated objects. In this case, galaxy interactions will be few, making tidal stirring, disruption and ram pressure stripping very ineffective.

Comparing the case of the dSph ESO410-G005 and of the dE Scl-dE1 is extremely interesting. The former seems to contain some young blue stars (Karachentsev *et al.* 2000) and has significant amount of HI ($M_{HI}/L_B = 0.13$) but is also closer (230 kpc) to its nearest neighbour, NGC55 than Scl-dE1 is to NGC253 (460 kpc). The latter contains no detected HI ($M_{HI}/L_B < 0.08$) and has a predominantly old stellar population. This emphasises that the morphology-density relation is not always applicable in particular cases. The isolated Local Group dE Tucana is another example of this phenomenon.

In conclusion, our findings do support the claim that "*transition-type dwarfs are dSphs that kept their interstellar medium and therefore should replace dSph's in isolated locations where stripping is ineffective*" (Grebel *et al.* 2003), but it fails to readily support a quantitative version of the morphology-density relation

References

Grebel, E.K., Gallagher, J.S. & Harbeck, D. 2003, *AJ* 125, 1926
Jerjen, H., Freeman, K.C. & Binggeli, B. 1998, *AJ* 116, 2873
Jerjen, H. & Rejkuba, M. 2001, *A&A* 371, 487
Karachentsev, I.D. *et al.* 2000, *ApJ* 542, 128
Karachentsev, I.D. *et al.* 2002, *A&A* 389, 812
Karachentsev, I.D. *et al.* 2003, *A&A* 404, 93

Near-Field Cosmology with Dwarf Elliptical Galaxies
Proceedings IAU Colloquium No. 198, 2005
H. Jerjen and B. Binggeli, eds.

doi:10.1017/S1743921305003856

Discovery of remote star clusters in the halo of the Irregular galaxy NGC 6822

Narae Hwang[1], Myung Gyoon Lee[1], Jong Chul Lee[1], Won-Kee Park[1], Hong Soo Park[1], Jang-Hyun Park[2], Sangmo Tony Sohn[2], Sang-Gak Lee[1], Hyung Mok Lee[1], Mun-Suk Chun[3], Young-Wook Lee[3], Young-Jong Sohn[3], In-Soo Yuk[1,2], Sang Chul Kim[1,2], Ho-Il Kim[2] and Wonyong Han[2]

[1]Astronomy Program, SEES, Seoul National University, Seoul 151-742, Korea
email: nhwang@astro.snu.ac.kr

[2]Korea Astronomy & Space Science Institute, 61-1 Whaam Yuseong, Daejeon 305-348, Korea
[3]Center for Space Astrophysics, Yonsei University, Seoul 120-179, Korea

Abstract. We report the discovery of three new star clusters in the halo of the Local Group dwarf irregular galaxy NGC 6822. These clusters were found in the deep images taken with the MegaPrime at the CFHT, covering a total field of 2 deg × 2 deg. The most remote cluster is found to be located as far as 79 arcmin away from the center of NGC 6822. This distance is several times larger than the size of the region in NGC 6822 where star clusters were previously found. Morphological structures of the clusters and color-magnitude diagrams of the resolved stars in the clusters show that at least two of these clusters are proabably old globular clusters.

Keywords. globular clusters: general, galaxies: dwarf, galaxies: individual (NGC 6822)

1. Introduction

Star clusters in the dwarf irregular galaxy NGC 6822 have been first reported by Hubble (1925) and systematically investigated by Hodge (1977). The most recent study by Krienke & Hodge (2004) used the available HST Archive data and cataloged all the star clusters identified including one genuine globular cluster Hubble VII. However, no star cluster survey in the outer halo of NGC 6822 has ever been tried before, although the existence of a rather large stellar halo around the galaxy was suggested by Letarte *et al.* (2002) from their Carbon star survey.

2. Discovery

Visual inspection of the wide field survey data around NGC 6822 has revealed three new star clusters. The locations and morphologies of these clusters are shown in Figure 1. From morphological and photometric studies, two clusters, SC1 and SC2, are regarded genuine old globular clusters with ages more than 3 Gyrs (see Hwang *et al.* 2005 for details). One noteworthy point is that the new star clusters are distributed to the very remote places (*Left* panel of Figure 1). The projected distance from the NGC 6822 center to SC1, the most remote cluster, is about 12 kpc. For comparison, NGC 1841, the outermost star cluster in the LMC, is located at about 13 kpc from the LMC center. Another important point in the images is that all three clusters are extended and are clearly resolved into stars, whereas Hubble VII is not resolved at all in our data (*Right* panel of Figure 1). Further investigation shows that the half-light radii of these new clusters are larger than 10 pc and even larger than 20 pc for SC1.

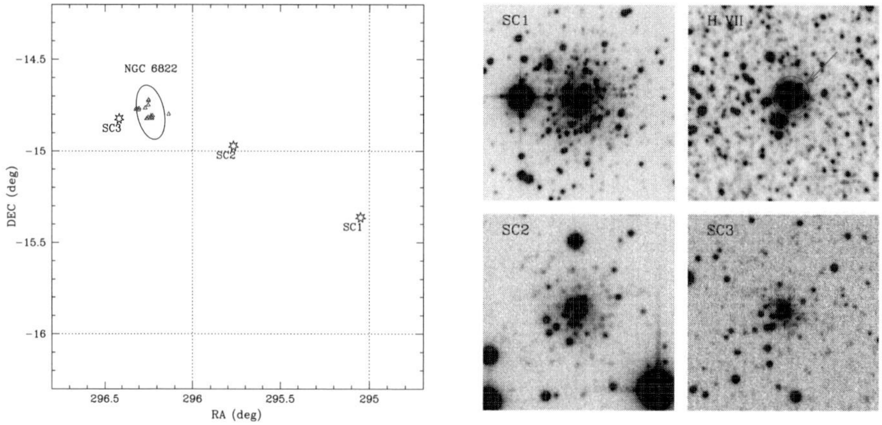

Figure 1. *Left*: Locations of the new star clusters in NGC 6822 (stellar symbols). The SC1 is about 12 kpc away from the galaxy center. Marks (triangles) inside and around the ellipse are previously known star clusters. *Right*: Sloan i band images of the three new star clusters and a known globular cluster Hubble VII in NGC 6822 (upper right; marked by an arrow). Note the resolved member stars of new star clusters. Each image is $37'' \times 37''$ wide.

3. Implications

The existence of the newly discovered star clusters suggests that the underlying halo has a different structure from a giant HI disk-like cloud which is extended along NW-SE direction (de Blok & Walter 2000). These clusters also provide a proof that the halo of NGC 6822 is quite larger than previously expected (see Lee & Hwang 2005 in this volume).

The extended structures of these new clusters are very unusual features. The SC1, among these, is found to be as extended as new star clusters recently discovered in the halo of M31 (Huxor *et al.* 2005; Lee *et al.* 2005). The formation mechanism of these extended star clusters, including the correlation with evolutionary history of the host galaxies, is not clearly understood yet, requiring further studies.

Acknowledgements

N.Hwang was supported in part by the BK21 program. M.G.Lee was supported in part by the ABRL(R14-2002-058-010000-0).

References

de Blok, W.J.G. & Walter, F. 2000, *ApJ* 537, L95

Hodge, P.W. 1977, *ApJS* 33, 69

Hubble, E.P. 1925, *ApJ* 62, 409

Huxor, A.P., Tanvir, N.R., Irwin, M.J., Ibata, R., Collett, J.L., Ferguson, A.M.N., Bridges, T. & Lewis, G.F. 2005, *MNRAS* in press (astro-ph/0412223)

Hwang, N. *et al.* 2005, in preparation

Krienke, K. & Hodge, P. 2004, *PASP* 116, 497

Lee, M.G. & Hwang, N. 2005, *this proceedings*

Lee, M.G. *et al.* 2005, in preparation

Letarte, B., Demers, S., Battinelli, P. & Kunkel, W.E. 2002, *AJ* 123, 832

Near-Field Cosmology with Dwarf Elliptical Galaxies
Proceedings IAU Colloquium No. 198, 2005
H. Jerjen and B. Binggeli, eds.

doi:10.1017/S1743921305003868

Anisotropies of the satellite systems around the Milky Way and Andromeda

Manuel Metz[1], **Pavel Kroupa**[1] **and Helmut Jerjen**[2]

[1]Sternwarte der Universität Bonn, Auf dem Hügel 71, 53332 Bonn, Germany
email: mmetz, pavel@astro.uni-bonn.de

[2]Research School of Astronomy and Astrophysics, Mt. Stromlo Observatory, Australian National University, Cotter Road, Weston ACT 2611, Australia
email: jerjen@mso.anu.edu.au

Abstract. We have studied the distribution of dwarf-galaxy satellite systems around our Milky Way (MW) and the Andromeda (M31) galaxy. The anisotropy is quantified and the form of their distribution is found to be incompatible with that expected if they were cold-dark-matter sub-structures. The origin of these satellites therefore can not be cosmological. Rather the Milky Way and Andromeda satellites probably stem from a local evolutionary mechanism.

Keywords. Galaxy: evolution, Galaxy: halo, galaxies: dwarf, galaxies: formation

1. Introduction

CDM models predict about 500 dark matter dominated subhalos for Milky Way-type systems within a radius of 500 kpc. The distribution of these subhalos follows the density of the dark matter halo of their host galaxy. However, only 13 dwarf galaxies within 500 kpc have been found for the Milky Way and 15 for Andromeda respectivly. The *missing satellite problem* cannot be solved entirely by invoking baryonic processes (Kazantzidis *et al.* 2004). Furthermore, Kroupa *et al.* (2005) argue that the distribution of the MW satellites differs from what one would expect if these are CDM sub-structures. Instead the distribution of the MW satellites is highly unisotropic and can well be described by a planar-like distribution. Here we continue this work by performing a much more robust statistical analysis of both, the Milky Way and Andromeda satellite systems.

2. Algorithm

We have implemented a fitting routine which is an unweighted Algebraic Least-Squares Estimator (see Chojnacki *et al.* 2000)† similar to the method Hartwick (2000) used. Furthermore we now use some more advanced statistical methods: (1) a bootstrap method to investigate the stability of the fitted plane. If the origin of the distribution were NOT disc-like this would show up in a large scatter of the direction of the normal vectors of the fitted planes. (2) For analysis we use some quantities known from statistical analysis of spherical data (Fisher *et al.* 1987): (i) shape- and strength-parameter γ, ζ, (ii) an approximate confidence cone and (iii) a relative RMS-height Δ and $\Delta/R_{\rm cut}$ as in Kroupa *et al.* (2005). We performed tests against random samples drawn from a spherical isotropic

† We used Python, Numeric and SciPy for the implementation. It is important to compare only results derived with identical fitting algorithms, because results may strongly depend on the algorithm used. All routines are available online: `http://www.astro.uni-bonn.de/~mmetz/`.

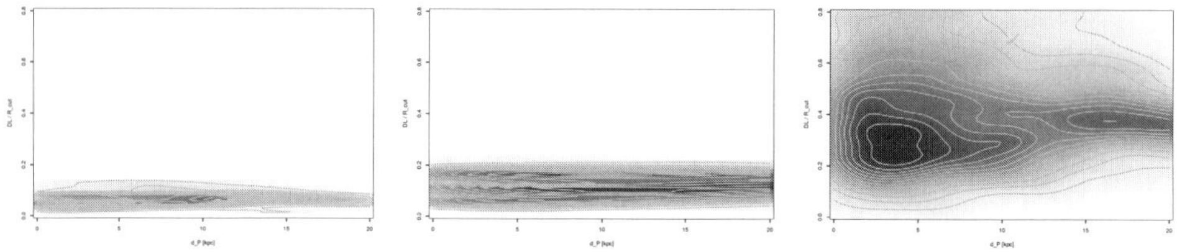

(a) MW (b) M31 (c) random sample

Figure 1. Density plot of $\Delta/R_{\rm cut}$ against $d_{\rm P}$ for for 5 000 bootstrap samples for 11 MW satellites (out to Leo I), 12 M31 satellites (out to LGS 3) and a random sample of 11 satellites

distribution with radial density profiles following a power-law $\propto r^{-p}$, (p =2,2.5) and a Burkert-profile (Burkert 1995).

3. Results and Conclusions

For the analysis we use data as given in Kroupa *et al.* (2005), McConnachie *et al.* (2005) and Grebel *et al.* (2003) and fitted planes to 5 000 bootstrap samples. The principal axes and opening angles of an approx 95% confidence cone are: $(l,b)_{\rm MW} = (158°, -12°)$, (0.3°,0.2°) and $(l,b)_{\rm M31} = (77°, -32°)$, (0.9°,0.3°). *It is remarkable that both satellite planes are highly inclined relative to the plane of the host galaxy.* We derive the following mean thickness ratios for the MW $\langle\Delta/R_{\rm cut}\rangle = 0.07$ and for M31 $\langle\Delta/R_{\rm cut}\rangle = 0.16$. In Fig. 1 $\Delta/R_{\rm cut}$ is plotted against distance $d_{\rm P}$ of the fitted plane from the center of their host. The large scatter in $\Delta/R_{\rm cut}$ is a clear indication for a not planar like origin of the random sample.† We have compared the shape parameters γ and ζ with random samples. We give the combined propability to find larger values for both parameters compared to 10 000 × 5 000 bootstrap samples drawn from random samples with radial density $\propto r^{-2}$. That is the propability for the normals to be more stable than the observed ones: $p_{\rm SP}^{\rm MW} = 0.6\%$ and $p_{\rm SP}^{\rm M31} = 17\%$. Tests performed with random samples with different radial densities ($\propto r^{-2.5}$, Burkert profile) show no significant difference.

With highly robust statistical methods we have shown that the propability of finding a distribution of 11 satellites like those of the MW drawn from a spherical isotropic distribution is less than 0.6%, for M31 less than 17%. The propabilities would be even lower if the MW/M31 dark matter halos are oblate and coplanar to the galactic disks. In addition, the observed distributions are remarkably thin for the MW and M31, confirming our interpretation of the MW and M31 satellite distributions in terms of disk-like systems.

References

Burkert, A. 1995, *ApJ* 447, L25

Chojnacki, W., Brooks, M., van den Hengel, A. & Gawley, D. 2000,*IEEE Transactions on pattern analysis and machine intelligence*, 22, 1294

Fisher, N., Lewis, T. & Embleton, B. 1987, Statistical analysis of spherical data (Cambridge University Press)

Grebel, E.K., Gallagher, J.S. & Harbeck, D. 2003, *Astronomische Nachrichten Suppl.* 324, 90

Hartwick, F.D.A. 2000, *AJ* 119, 2248

Kroupa, P., Theis, C. & Boily, C.M. 2005, *A&A* 431, 517

McConnachie, A.W., Irwin, M.J., Ferguson, A.M.N. *et al.* 2005, *MNRAS* 356, 979

† We note that tests showed that the ALS algorithm is not very sensitive for deriving the distances of planes which to some extend explains the spread in $d_{\rm P}$.

Near-Field Cosmology with Dwarf Elliptical Galaxies
Proceedings IAU Colloquium No. 198, 2005
H. Jerjen and B. Binggeli, eds.

doi:10.1017/S174392130500387X

Are violent events responsible of a galaxy morphological loop?

Cecilia Scannapieco[1], **A.V. Smith Castelli**[1,2] **and P.B. Tissera**[1]

[1]Instituto de Astronomía y Física del Espacio, Conicet – UBA,
CC 67 – Suc 28 (1428) Ciudad de Buenos Aires, Argentina
email: cecilia@iafe.uba.ar, patricia@iafe.uba.ar

[2]Facultad de Ciencias Astronómicas y Geofísicas, Universidad Nacional de La Plata,
Paseo del Bosque s/n (1900) La Plata, Argentina
email: asmith@fcaglp.unlp.edu.ar

Abstract. We use cosmological SPH simulations to investigate the effects of mergers and interactions on the formation of the bulge and disc components of galactic systems. We find that secular evolution during mergers seems to be a key process in the formation of stable disc-bulge systems with observational counterparts and contributes to establish the fundamental relations observed in galaxies. Our findings suggest that the secular evolution phase couples the formation mechanisms of the bulge and disc components. According to our results, depending on the particular stability properties and merger parameters, violents events could drive a morphological loop in which the outcome could be a disc or a spheroid.

Keywords. methods: numerical, galaxies: interactions, galaxies: fundamental parameters

1. Introduction

The understanding of galaxy formation remains a primary goal in astrophysics. Within the current accepted scenario of structure formation, small galaxies are built up first and larger objects are assembled through the merging of less massive building blocks, in a hierarchical way. In this context, mergers and interactions play a crucial role modifying the properties of galaxies such as their mass distribution and star formation activity and affecting the subsequent evolution of the systems.

The formation of disc-like systems is still a matter of debate and several scenarios have been proposed in order to explain the growth of these two components and their interactions. The simplest scenarios currently accepted for the formation of the bulges are two: the first one establishes that the bulge and the disc are formed independently (Andredakis *et al.* 1995), and the second one assumes that the disc forms first and the bulge emerges from it as a consequence of gas inflow during a period of secular evolution (Courteau *et al.* 1996).

Numerical simulations are a powerful tool to tackle galaxy formation since they can account for the joint evolution of baryons and dark matter in a cosmological framework. We use Smooth Particle Hydrodynamical (SPH) simulations to investigate the effects of mergers and interactions on the determination of the structural properties of galaxies.

2. Analysis and Discussion

We analize galactic objects identified from numerical SPH simulations that take into account the gravitational and hydrodynamical evolution of the matter in Cold Dark Matter universes. We run two simulations with $N = 64^3$ (S1) and $N = 2 \times 80^3$ (S2)

total number of particles and the following cosmological parameters: $\Lambda = 0$, $h = 0.5$, $\Omega_b = 0.10$ (S1)and $\Lambda = 0.7$, $h = 0.7$, $\Omega_b = 0.04$ (S2), where ($H = 100h$ km s^{-1} Mpc^{-1}). The simulated boxes represent cubic volumes of $5h^{-1}$ (S1) and $10h^{-1}$ (S2) Mpc length.

We select galaxy-like objects (GLOs) at $z = 0$ by using a density-contrast criterium and follow their evolution with look-back time, constructing their star formation and merger histories (see Scannapieco & Tissera 2003 for details). With the aim at investigating the effects of mergers on the structural properties of the systems we perform bulge-disc decompositions to the projected mass surface density of the GLOs at $z = 0$, and of the progenitors during merger events, assuming an exponential profile for the discs and the Sérsic law for the bulges. Galaxy-like objects identified in our simulations are found to have structural parameters similar to those of observed spiral galaxies (Sáiz *et al.* 2001).

Tissera *et al.* (2002) showed that during merger events, early gas inflows can be triggered depending on the stability properties of the systems. Those with shallower potential wells are more likely to suffer gas inflows since they can be more strongly perturbed from the incoming satellite. Secular evolution phases feed the bulge components which significantly contribute to the stability of the discs. The actual fusion of the baryonic cores increases the star formation activity which in turn affect the subsequent properties of the systems.

Understanding the effects of these two processes (i.e. secular evolution and fusion) on the growth of the disc and bulge components in galaxies can help us to enlighten the picture of galaxy formation. To this end we analize the evolution of the mass distribution of galaxy-like objects during merger events by studying the variation of their structural parameters. We find that secular evolution during the orbital decay phase produces the largest changes in the structural parameters of the systems: bulges get larger with respect to the discs and with more exponential profiles. These changes are found to be more important in unstable systems with shallower potential wells. Fusions introduce further modifications in the opposite direction although they are less important.

We also find that secular evolution phases contribute to establish the observed fundamental relations: after secular evolution, bulges tend to satisfy the observed Fundamental Plane and the disc components the Tully-Fisher relation. Fusions may add some dispersion but do not significantly change them (Smith Castelli *et al.* 2005). Departures from the Fundamental Relations could be indicating a lack of secular phases.

Our findings suggest that interactions play a fundamental role being able to transform discs into spheroids. Discs can also survive violent events, being either partially destroyed or just feeded by the colliding object. The stability properties of the initial system seem to be a fundamental point in the response of the disc to the incoming satellite. Other factors such as orbital parameters and relative masses of the colliding objects would also affect the determination of the structural properties of the outcome systems. The different possible histories of evolution of galactic systems in hierarchical universes could potentially explain the diversity of galaxies in our Universe.

References

Andredakis, Y.C., Peletier, R.F. & Balcells, M. 1995, *MNRAS* 275, 874
Courteau, S., de Jong, R.S. & Broeils, A.H. 1996, *ApJ* 457, L73
Sáiz, A., Domínguez-Tenreiro, R., Tissera, P.B. & Courteau, S. 2001, *MNRAS* 325, 119
Scannapieco, C. & Tissera, P.B. 2003, *MNRAS* 338, 880
Smith Castelli, A.V., Tissera, P.B. & Scannapieco, C. 2005, *MNRAS* submitted
Tissera, P.B., Domínguez-Tenreiro, R., Scannapieco, C. & Sáiz, A. 2002, *MNRAS* 333, 327

Near-Field Cosmology with Dwarf Elliptical Galaxies
Proceedings IAU Colloquium No. 198, 2005
H. Jerjen and B. Binggeli, eds.

doi:10.1017/S1743921305003881

Dark matter or tidal disruption? The case for the Leo I dSph

S.T. Sohn[1,2], S.R. Majewski[2], J.C. Ostheimer[2], W.E. Kunkel[3], R.J. Patterson[2], M.H. Siegel[2,4], P. Guhathakurta[5] and M. Cooper[6]

[1]Korea Astronomy and Space Science Institute, Daejeon 305-348, Korea
email: tonysohn@kasi.re.kr

[2]Department of Astronomy, University of Virginia, Charlottesville, VA 22903
[3]Las Campanas Observatory, Casilla 601, La Serena, Chile
[4]McDonald Observatory, University of Texas, Austin, TX 78712
[5]UCO/Lick Observatory, University of California at Santa Cruz, Santa Cruz, CA 95064
[6]Department of Astronomy, University of California at Berkeley, Berkeley, CA 94720

Abstract. We present results from photometric and spectroscopic survey of giant stars in the Leo I dwarf spheroidal galaxy. We find compelling evidence that this galaxy has been tidally disrupted.

Keywords. Galaxy: halo, galaxies: dwarf, galaxies: kinematics and dynamics, galaxies: Local Group

---

1. Introduction

The high apparent mass-to-light ratios (M/Ls) of Local Group dwarf spheroidal galaxies (dSphs) raises an apparent dilemma in the context of a tidal disruption picture. If dSph galaxies are the visible counterparts of the largest dark matter (DM) lumps (Klypin *et al.* 1999), and the large measured M/L of many dSphs truly reflect large internal DM contents, then the process of tidal disruption should be inhibited. The anticorrelation of dSph's intrinsic brightness with apparent M/L has raised the suggestion that the mass of all dSphs are about the same (Mateo *et al.* 1998). From this standpoint, *no* dSph satellite might be expected to be tidally disrupted. The example of the Sgr dSph (*e.g.*, Majewski *et al.* 2003) is clearly at odds with this, and that it may not be an exception is proposed by studies suggesting tidal disruption in other Mliky Way dSphs. Leo I, at the low end of the dSph M/L scale [$6(M/L)_{\odot,V}$; Mateo *et al.* 1998], provides an interesting contrasting case to the high M/L systems in this DM/tidal disruption debate.

2. Results

We have surveyed a 4.5 degree^2 region centered on Leo I in M, $T2$, + $DDO51$ filter system to explore the extended morphology of this dSph. These photometric data were used to select Leo I giant candidates using the ($M - T2, M - DDO51$) color-color and ($M - T2, M$) color-magnitude diagrams (see *e.g.* Majewski *et al.* 2000 for details). Using Leo I giant candidates, we explore two-dimensional distribution and find compelling evidence that Leo I has been tidally disrupted: (1) we find many Leo I giant candidates outside the King limiting radius, spread out predominantly to the east-west direction (Figure 1–a), (2) our isodensity contour plots show isophotal twists and hint of S-shaped morphology (Figure 1–b), and (3) we find double breaks in the radial density profile (Figure 1–c). We have also measured radial velocities (RVs) of 125 stars using spectra

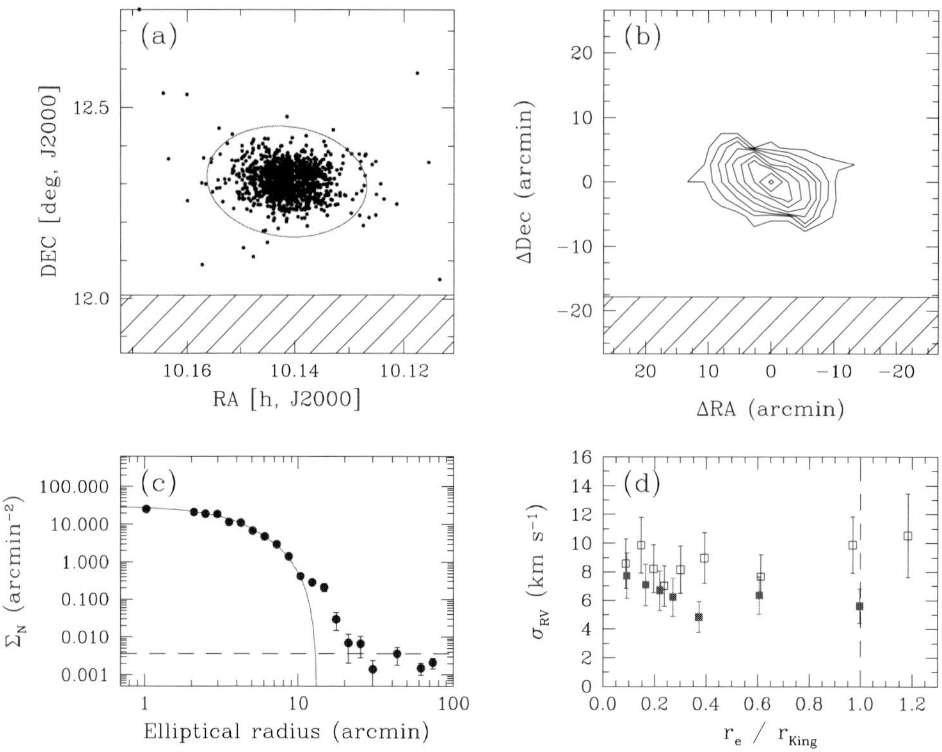

Figure 1. (a) A central region blowup of the distribution on the sky of the Leo I giant candidates. The ellipse shows the King limiting radius. The hatched region is outside of our survey area. (b) Isodensity contour plot of the stars in (a). (c) Fits of the stellar distribution of Leo I giant candidates using a King model. The dased line is the derived level of background, which was subtracted from the data and the model curve shown. (d) Variation of velocity dispersions with elliptical radius from the center of Leo I. The open squares are for the entire sample, while the closed squares are for the sample without the 2σ outliers.

taken with the Keck II + DEIMOS. Out of 105 Leo I members, 28 stars lie at $1.0 < r_e/r_{\rm King} < 1.7$, confirming the existence of the break population found in the radial surface density profile. The RV dispersion profile (Figure 1–d) for the more Gaussian-distributed RV population decrease with distance from the center out to $r_e/r_{\rm King} \sim 0.4$, and flattens beyond that. The profile for the entire sample, however, is flat (or slightly rising) out to the King limiting radius and beyond.

Based on the results found in both photometric and spectrocopic surveys, we conclude that we are seeing a tidal disruption in the Leo I system.

References

Klypin, A., Kravstov, A.V., Valenzuela, O. & Prada, F. 1999, *ApJ* 522, 82

Majewski, S.R., Ostheimer, J.C., Patterson, R.J., Kunkel, W.E., Johnston, K.V. & Geisler, D. 2000, *AJ* 119, 760

Majewski, S.R., Skrutskie, M.F., Weinberg, M.D. & Ostheimer, J.C. 2003, *ApJ* 599, 1082

Mateo, M., Olszewski, E.W., Vogt, S.S. & Keane, M.J. 1998, *AJ* 116, 2315

Near-Field Cosmology with Dwarf Elliptical Galaxies
Proceedings IAU Colloquium No. 198, 2005
H. Jerjen and B. Binggeli, eds.

doi:10.1017/S1743921305003893

dEs and the dI fundamental plane

Ovidiu Vaduvescu and Marshall L. McCall

Department of Physics and Astronomy, York University
128 Petrie Science Building, 4700 Keele St., Toronto, Ontario M3J 1P3, Canada
email: ovidiuv@yorku.ca, mccall@yorku.ca

Abstract. Despite much work, the connection between dwarf elliptical galaxies (dEs) and dwarf irregulars (dIs) remains unclear. Recently, we found that the surface brightness profiles (SBPs) of dIs in the near-infrared can be well fitted with a hyperbolic secant (sech) function. From sech fits, a tight relationship was derived between the absolute magnitude, central surface brightness, and 21cm line widths, which amounts to a fundamental plane for dIs. Here we examine how closely dEs fit into the dI fundamental plane using published data for 22 dEs in the Virgo cluster and the Local Group. Over a 9 mag interval in absolute magnitude the dEs fall in the plane defined by the dIs. The outstanding overlap suggests more than a casual relationship between the two classes.

1. The dI Sample

A sample of 19 field dIs with very accurate distances was observed at CFHT adopting a prudent approach to sample the faint outer regions (Vaduvescu and McCall 2004). We separated resolved and unresolved components using the KILLALL package (Buta and McCall 1999) and then derived the surface brightness profiles (SBPs) by fitting ellipses using the stsdas/ellipse function in IRAF. The hyperbolic secant function (sech) was found to fit accurately all the SBPs in our dI sample (Vaduvescu *et al.* 2005):

$$I = I_0 \ \mathrm{sech}(r/r_0) = \frac{2I_0}{e^{r/r_0} + e^{-r/r_0}} \tag{1.1}$$

where I_0 is the central surface brightness and r_0 the scale length. Note that the sech function becomes exponential at large radius.

2. The dE Sample

The sample of dEs consists of 22 objects with published data in three sub-samples: 11 dEs in the Virgo Cluster with H band images from the GOLDMine database (Gavazzi *et al.* 2002) and velocity dispersions from Geha *et al.* (2003); three dEs/dSphs in the Local Group (M 110, NGC 147, NGC 185) with K images from 2MASS Large Galaxy Atlas (Jarett *et al.* 2003) and velocity dispersions from McElroy (1995) and Held *et al.* (1992); 8 dSphs orbiting our Galaxy (Carina, Draco, Fornax, Leo I, Leo II, Sculptor, Sextans, Ursa Minor) for which star counts in V define surface brightness profiles (Irwin and Hadzidimitriou 1995). We adopted average colors of H-K=0.2 mag and V-K=2.7 mag. In all cases, an exponential model was adopted to fit the outer isophotes: $I = I_0 e^{-r/r_0}$.

3. The Tully-Fisher Relation and the Fundamental Plane

In Figure 1 (left), we plot the Tully-Fisher relation for both the dIs and dEs with W_{20} estimated for dEs from stellar velocity dispersions. It shows considerable scatter.

In Figure 1 (right), we show the fundamental plane defined by the absolute magnitude M_K, the central surface brightness m_0 derived from I_0, and the HI line-width W_{20}. For the dIs, the "absolute magnitude" M_K is the sech magnitude, while for dEs it is the exponential magnitude. The dIs and dEs are marked by crosses and circles, respectively. No corrections for tilt have been applied. The dotted line represents the fit to the dIs alone (defined by the abscissa). The dI fundamental plane shows a remarkable good correlation with a coefficient of correlation 0.91 and a rms scatter of only 0.4 mag. This is comparable to the scatter in the I-band Tully-Fisher relation for spirals adopted in the HST Key Project on the Extragalactic Distance Scale. Over a 9 mag interval in absolute magnitude, the dEs fall in the dI fundamental plane. The outstanding overlap suggests a common scenarion for the early development of the dEs and dIs.

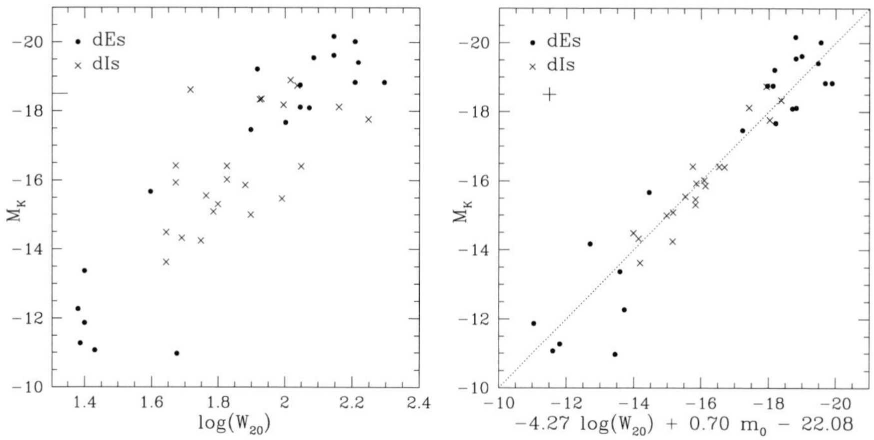

Figure 1. Left: The Tully-Fisher relation appears to hold for both dIs and dEs, although with much scatter. Right: The Fundamental Plane obtained by fitting HI line-widths W_{20}, central brightness m_0, and the sech absolute magnitude M_K for 19 dIs with available data (Vaduvescu *et al.* 2005). The dEs lie on the same plane. A common distance modulus DM=30.62 has been applied for Virgo members. Typical uncertainties are plotted as an error cross.

Acknowledgements

We are grateful to the Natural Sciences and Engineering Research Council of Canada for its continuing support to the PhD Thesis of OV (2005, in preparation). This research has made use of the GOLDMine Database (Gavazzi *et al.* 2002) and data products from the Two Micron All Sky Survey. OV would like to thank to Stefano Zibetti for some feedback related to data reduction of the GOLDMine images.

References

McCall, M.L. & Buta, R.J. 1999, *AJ* 109, 6, 2460
Gavazzi, G. *et al.* 2002, *AA* 400, 451
Geha, M., Guhathakurta, P. & van der Marel, R.P. 2003, *AJ* 103, 3, 851
Held, E.V. *et al.* 1992, *AJ* 103, 3, 851
Irwin, M. & Hadzidimitriou, D. 1995, *MNRAS* 277, 1354
Jarett, T.H. *et al.* 2003, *AJ* 125, 525
McElroy, D.B. 1995, *ApJS* 100, 105
Vaduvescu, O. & McCall, M.L. 2004, *PASP* 116, 820
Vaduvescu, O., McCall, M.L., Richer, M.G. & Fingerhut, R.L. 2005, *AJ*, submitted

Near-Field Cosmology with Dwarf Elliptical Galaxies
Proceedings IAU Colloquium No. 198, 2005
H. Jerjen and B. Binggeli, eds.

doi:10.1017/S174392130500390X

The Tully-Fisher relation of spiral galaxies

I. Yegorova and P. Salucci
SISSA, Via Beirut 2/4 Trieste, Italy
email: yegorova@sissa.it

Abstract. The Tully-Fisher (TF) relation is an empirically establish correlation between the luminosity of a spiral galaxy and its rotational velocity (Tully-Fisher, 1977). We used the Tully-Fisher relation to probe the dark matter (DM) distribution in the optical regions of spiral galaxies. We investigated a sample of rotation curves that includes 957 galaxies. We applied a new technique that takes advantage of the full knowledge of the galaxy rotation curves.

Keywords. galaxies: kinematics and dynamics, dark matter

The Tully-Fisher (TF) relation is an empirically established correlation between the luminosity of a spiral galaxy and its rotational velocity (Tully & Fisher 1977). It has two important properties. First, it is used to obtain cosmological distances. Second, it can be used to study the dynamical properties and evolution of galaxies. We used the TF-relation to probe the dark matter (DM) distribution in the optical regions of spiral galaxies. We investigated a sample of rotation curves (Persic & Salucci 1995) that includes 957 galaxies (see also Mathewson, Ford & Buchhorn 1996). We applied a new technique that takes advantage of the full knowledge of the rotation curves of galaxies. More specifically we have established several new Tully-Fisher like relations, each one related to a different radius, namely at the following fixed fractions of the optical radius R_{opt} (corresponding to about three length scales), i.e. at: 0.2, 0.4, 0.6, 0.8, 1.0, 1.2. The standard TF-relation is evaluated only at just one radius, usually about R_{opt}.

The thorough analysis of our data, which includes 15 000 independent measurements, is done by means of the linear regression models $M_{Bi} = a_i + b_i \log V(R_i)$, where the parameters a_i and b_i are found by least squares. We plot the six TF relations at the above specified radii and in Table 1 we present their main characteristics.

Col(1) – Isophotal radius, Col(2) – Intercept value (a), Col(3) – Standard error of a, Col(4) – Slope value (b), Col(5) – Standard error of b, Col(6) – Scatter, Col(7) – Number of points

In Fig. 1 we show the scatter of these TF-like relations as a function of their reference radius R_i. Notice that at 0.2 R/R_{opt} the scatter is very large because of the influence of the bulge, but, otherwise, it is always very small.

In Fig. 2 we show the slope of the TF-relations. As is apparent the slope steadily increases with increasing distance from the center.

Table 1. The TF-relations at different radii

R/R_opt	*zero point*	*error*	*slope*	*error*	*SD*	*N*
0.2	−11.78	0.103	−4.77	0.054	0.366	739
0.4	−8.207	0.068	−6.141	0.033	0.185	786
0.6	−5.769	0.063	−7.102	0.029	0.146	794
0.8	−4.22	0.09	−7.718	0.042	0.17	657
1.0	−8.2	0.146	−8.197	0.067	0.208	447
1.2	−1.98	0.261	−8.639	0.118	0.253	226

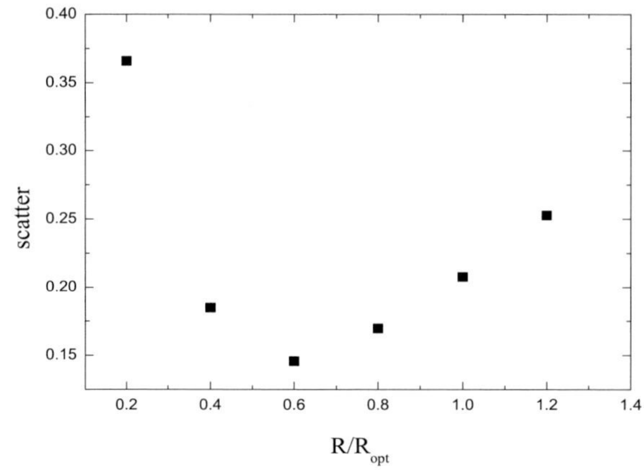

Figure 1. The scatter of the TF-relations as a function of radius

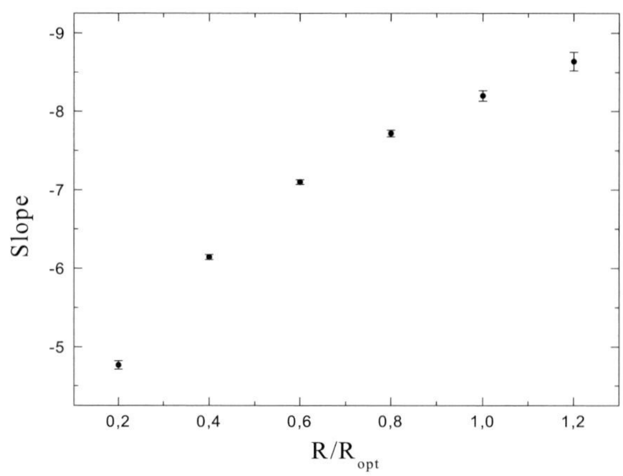

Figure 2. The slope of the TF-relations as a function of radius

The first analysis of this data shows that the dark matter fraction depends on the radius and on the galaxy luminosity. We plan to study quantitatively the behavior of scatter and slope in different models of mass distribution.

References

Mathewson, D.S., Ford, V.L. & Buchhorn, M. 1992, *ApJS* 81, 413
Persic, M. & Salucci, P. 1995, *ApJS* 99, 501
Tully, R.B. & Fisher, J.R. 1977, *A&A* 54, 661

Near-Field Cosmology with Dwarf Elliptical Galaxies
Proceedings IAU Colloquium No. 198, 2005
H. Jerjen and B. Binggeli, eds.

doi:10.1017/S1743921305003911

An HST/ACS survey of early-type galaxies in the Virgo cluster

Patrick Côté

Herzberg Institute of Astrophysics, National Research Council of Canada,
5071 W. Saanich Road, Victoria, BC, V9E 2E7, CANADA
email: Patrick.Cote@nrc-cnrc.gc.ca

Abstract. One hundred early-type members of the Virgo Cluster have been observed with the *Hubble Space Telescope* (*HST*) as part of the ACS Virgo Cluster Survey. The survey and its extensions consists of F475W and F850LP imaging from the *Advanced Camera for Surveys* (ACS), supplemented by additional imaging and spectroscopy from WFPC2/*HST*, *Chandra*, *Spitzer*, Keck, KPNO, and CTIO. In this article, I report briefly on four scientific highlights from the program: (1) the discovery of a population of probable ultra-compact dwarf galaxies; (2) an improved measurement of the frequency of compact stellar nuclei in early-type galaxies; (3) a precise new measurement of Virgo's line-of-sight depth; and (4) new insights into the chemical evolution of galaxies from the color distributions of their globular cluster systems.

Keywords. galaxies: clusters: individual (Virgo)–galaxies: elliptical and lenticular-galaxies: dwarf–galaxies: nuclei–galaxies: structure–galaxies: star clusters

1. Introduction

The Virgo Cluster marks the center of the Local Supercluster (de Vaucouleurs 1956) and, as the attendees of this meeting are well aware, contains by far the largest collection of dwarf galaxies in the local universe. As a result, Virgo's rich population of dwarf galaxies has long played a pivatol role in our understanding of the structure and stellar content of low-mass galaxies, and how they form and evolve in dense cluster environments (e.g., Reaves *et al.* 1983; Caldwell 1983; Binggeli, Sandage & Tammann 1985; Binggeli, Tammann & Sandage 1987; Bothun *et al.* 1985; Ichikawa, Wakamatsu & Okamura 1986; Binggeli & Cameron 1991, 1993; Durrell 1997; Lotz, Miller & Ferguson 2004). Such galaxies have taken on a renewed importance in recent years, following the recognition that their observed numbers, spatial distribution and stellar content impose tight constraints on hierarchical models of structure formation (e.g., Moore *et al.* 1999; Klypin *et al.* 1999; Mayer *et al.* 2002).

With the installation of the *Advanced Camera for Surveys* (ACS) on the *Hubble Space Telescope* (*HST*) in 2002, the already outstanding imaging capabilities of this observing facility improved dramatically (i.e., by roughly a factor of ten in "discovery efficiency"; Ford *et al.* 1998). In an effort to capitalize on this tremendous gain in efficiency, my colleagues and I initiated the *ACS Virgo Cluster Survey* — an ambitious program to image, in two bandpasses (F475W and F850LP), 100 early-type members of the Virgo Cluster. Since then, a number of additional observing programs focussing on the survey galaxies have been undertaken, including imaging and spectroscopy from WFPC2/*HST*, *Chandra*, *Spitzer*, Keck, KPNO, and CTIO. Full details on the observational strategy and science drivers for the original program are given in Côté *et al.* (2004), while scientific results have been presented in Jordán *et al.* (2004ab), Mei *et al.* (2005ab) and Haşegan *et al.* (2005).

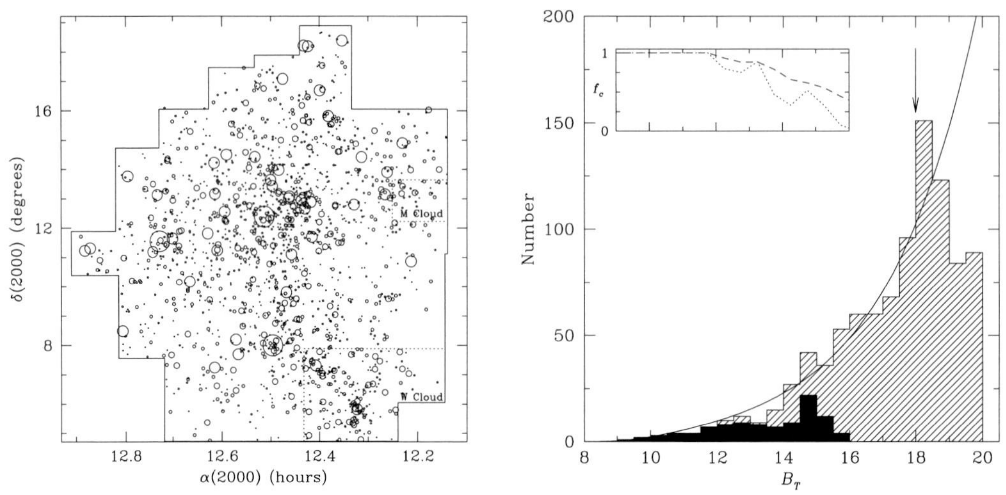

Figure 1. (Left Panel) Distribution of Virgo Cluster galaxies on the plane of the sky, adapted from Binggeli *et al.* (1987). This figure contains a total of 1726 galaxies – with no restriction on morphological type – which are classified as members or possible members of Virgo and have declinations $\delta \gtrsim 5°$. Red symbols denote the full sample of 100 early-type galaxies from the ACS Virgo Cluster Survey (Côté *et al.* 2004). The symbol size in all cases is proportional to blue luminosity. **(Right Panel)** Luminosity function of 956 early-type galaxies classified by Binggeli *et al.* (1987) as members of Virgo (upper, hatched histogram). The arrow shows the completeness limit of the Virgo Cluster Catalog, while the solid curve shows the best-fit Schechter function for E+S0+dE+dS0 galaxies from Sandage *et al.* (1985). The filled lower histogram shows the luminosity function of the 100 early-type galaxies from the ACS Virgo Cluster Survey. (Inset panel) Differential (dotted curve) and cumulative (dashed curve) completeness fractions for the ACS Virgo Cluster Survey over the range $9 \leqslant B_T \leqslant 16$. Figures from Côté *et al.* (2004).

The ACS Virgo Cluster Survey was intended to provide as complete a picture as possible of early-type cluster galaxies in the magnitude range $9.3 \lesssim B_T \lesssim 16$ ($-21.7 \lesssim M_B \lesssim -15$). The distribution of our 100 program galaxies on the plane of the sky is shown in the left panel of Figure 1; the right panel of this figure compares the luminosity distribution of our program galaxies with that of the full sample of early-type members of Virgo from the survey of Binggeli *et al.* (1985). Our sample contains roughly half of all early-type members down to $B_T \leqslant 16$ ($M_B \leqslant -15$) and includes a complete sample of early-types brighter than $B_T = 12$ ($M_B = -19$). In terms of Virgo's dwarf population, 35 of the 100 program galaxies are formally classified as dwarfs according to the morphologies presented in Binggeli *et al.* (1985). If one instead adopts a crude luminosity division between dwarfs and giant galaxies of $M_B \approx -17.5$ (Kormendy 1985), then 53 of the 100 galaxies would qualify as dwarfs.

In this article, I shall focus on a handful of recent scientific highlights from the survey, including the detection of a population of "ultra-compact dwarf" galaxies in Virgo (Haşegan *et al.* 2005), a sharp upward revision in the frequency of compact nuclei in early-type galaxies (Côté *et al.* 2005), a new measurement of the line-of-sight depth of the Virgo Cluster (Mei *et al.* 2005c), and a systematic exploration of the dependence of globular cluster colors on host galaxy properties (Peng *et al.* 2005). More details on each of these topics may be found in the cited papers.

2. Ultra-Compact Dwarf Galaxies

The data reduction pipeline for globular clusters in the ACS Virgo Cluster Survey includes object detection, the identification of cluster candidates, and the measurement of their photometric and structural parameters (Jordán *et al.* 2004a). This final step relies on maximum-likelihood fitting of psf-convolved King models to the two-dimensional light distributions of the detected sources (see the appendix in Jordán *et al.* 2005 for details). While the overwhelming majority of the globular clusters in Virgo have half-light radii, r_h, resembling those of globular clusters in the Milky Way, a handful of objects in our M87 field were found to be much brighter and larger than typical Galactic globulars (with $r_h \sim 0\farcs3 \sim 25$ pc, ten times larger than the Galactic average; see the left panel of Figure 2).

In terms of size and luminosity, these objects are intermediate to globular clusters and dwarf galaxies. Indeed, they resemble the so-called ultra-compact dwarf galaxies reported recently in the Fornax Cluster (e.g., Hilker *et al.* 1999; Drinkwater *et al.* 1999). Follow-up echelle (ESI) spectroscopy with the Keck telescope for six of these objects has been used to measure their mass-to-light ratios (Haşegan *et al.* 2005). The results are plotted in the right panel of Figure 2, which compares the measured mass-to-light ratios to those of globular clusters and the predictions of single-burst stellar population synthesis model from Bruzual & Charlot (2003).

There are two important conclusions to be drawn from this figure. First, our small sample appears to be rather heterogeneous, with three, or possibly four, objects having mass-to-lights that are larger than either globular clusters or the model predictions. These objects are prime candidates for ultra-compact dwarf galaxies. The two remaining objects have mass-to-light ratios which are slightly higher than those of most Galactic globular clusters, but the measured values are nevertheless consistent with the theoretical predictions for globular clusters because of their relatively high metallicities. Second, it is clear that extreme caution must be excercised when attempting to distinguish ultra-compact dwarf galaxies from high-luminosity globular clusters. It is *not* possible to do so on the basis of luminosity alone; structural parameters and (especially) mass-to-light ratios are required.

3. Compact Nuclei in Early-Type Galaxies

HST imaging of spiral galaxies has shown that ≈50–70% of them contain compact stellar nuclei near their photocenters (e.g., Phillips *et al.* 1996; Carollo, Stiavelli & Mack 1999; Böker *et al.* 2004). The origin of these nuclei remains a mystery, with possible formation mechanisms ranging from highly concentrated star formation driven by infall of chemically-enriched gas, to the merger of star clusters driven to the galaxy center by dynamical friction. Discriminating between the various scenarios requires a comparison of the properties of the nuclei across galaxies of differing type and luminosity. Unfortunately, there has been no comparably sensitive search for compact stellar nuclei in the inner regions of early-type galaxies.

Based on the morphological classifications of Binggeli *et al.* (1985, 1987), our sample of 100 early-type galaxies includes 25 nucleated galaxies (24 dE,Ns and one nucleated elliptical). However, this value of $f_n = 25\%$ for the frequency of nuclei in our program galaxies should be considered a firm lower limit on the true frequency. Binggeli *et al.* (1985, 1987) state explicitly that faint nuclei – and those in bright, high-surface brightness galaxies – would have gone undetected in their photographic survey. The left panel of Figure 3 demonstrates the importance of these selection effects by comparing the

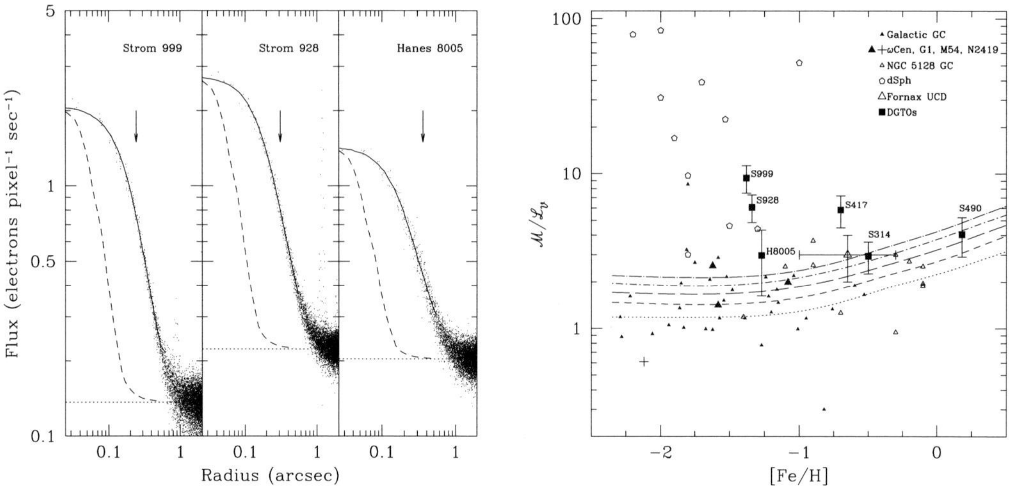

Figure 2. (Left Panel) Radial brightness profiles for three ultra-compact dwarf candidates from the ACS Virgo Cluster Survey. In each panel, the adopted background is indicated by the dotted line. The solid curve shows the best-fit, psf-convolved King model. The dashed curve shows the F475W psf, while the arrow indicates the fitted half-light radius. All three objects are found in our ACS images of M87, and were previously classified as M87 globular clusters on the basis of ground-based photometry and spectroscopy (e.g., Côté *et al.* 2001). **(Right Panel)** Mass-to-light ratio versus metallicity for hot stellar systems. Triangles show globular clusters in the Milky Way and NGC5128. The six squares show candidates ultra-compact dwarfs in M87. From bottom to top, the five curves show the theoretical predictions of the population synthesis models of Bruzual & Charlot (2003) for ages of 7, 9, 11, 13 and 15 Gyr. These models assume a Chabrier (2003) initial mass function. Figures from Haşegan *et al.* (2005).

visual appearance of one dwarf galaxy (the brightest non-nucleated dwarf in our sample according to the original ground-based classifications) with its simuluated appearance from the ground. It should come as no surprise that *HST* is well suited to the detection of faint, compact nuclei in Virgo Cluster galaxies.

The right panel of Figure 3 compares the luminosity distribution of the complete sample of ACS Virgo Cluster Survey galaxies (upper histogram) with the 25 galaxies classified as nucleated in the Virgo Cluster Catalog (filled histogram). Two additional histograms show the distributions for galaxies which we find to have *unambiguous* nuclei (double-hatched histogram) and those which have *certain or possible* nuclei (hatched histogram). These classifications are based on visual inspections of the *HST* images and modeling of the galaxy surface brightness profiles. Depending on the which sample is considered, our estimate for the frequency of nuclei falls in the range 66–82%, which is comparable to *HST* measurements for late-type systems but much higher than previous estimates for early-type galaxies from photographic surveys. The implications of this finding are discussed at length in Côté *et al.* (2005).

4. The Depth of the Virgo Cluster

On the sky, the Virgo Cluster appears complex and irregular, with a diameter of $\approx 10^\circ$. The structure of the cluster along the line of sight is a matter of longstanding debate, with estimates of the back-to-front depth varying wildly. At one extreme, Young & Currie (1995) report a depth of $\pm$6–8 Mpc by employing the shape of the brightness profiles for

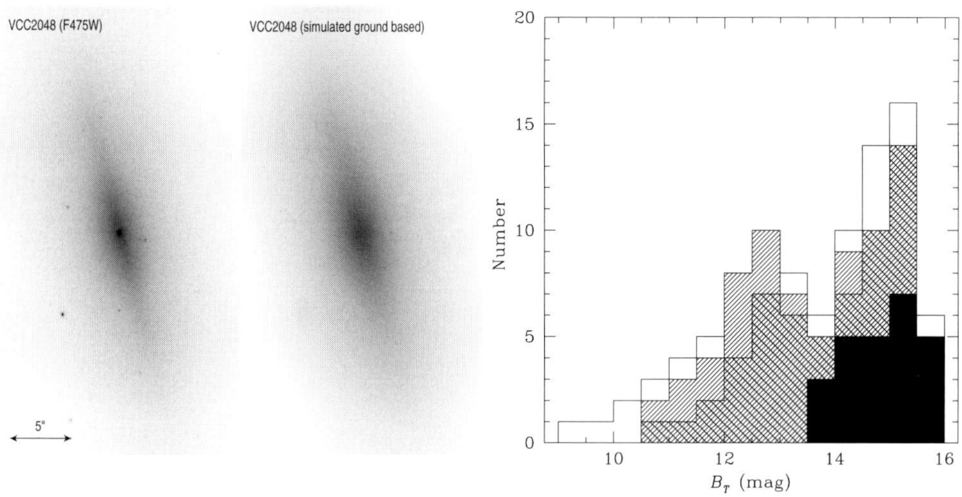

Figure 3. (Left Panel) Two views of the inner region of VCC2048, the brightest dwarf galaxy in our sample according to the morphological classifications of Binggeli, Sandage and Tamman (1985). This galaxy is classified as *non-nucleated* in the Virgo Cluster Catalog, with morphological type d:S0(9). On the left, we show the F475W image from our survey; on the right, we show this same image after binning 4×4 and convolving with a Gaussian of FWMH = $1''.4$ to simulate its appearance in conditions of moderate ground-based seeing. **(Right Panel)** Luminosity distribution of the 94 program galaxies for which a classification as either nucleated or non-nucleated is possible from our ACS imaging (open histogram). The hatched histogram shows the luminosity distribution for the 62 galaxies which we find to show unambiguous evidence of nucleatation. The double-hatched histogram shows this same sample plus 15 additional galaxies which *may* have central nuclei. The solid histogram shows the 25 galaxies in our survey which were classified as nucleated in the Virgo Cluster Catalog (Binggeli *et al.* 1985). Figures from Côté *et al.* (2005).

dwarf elliptical galaxies as a distance indicator, and a similar extent for spiral galaxies in Virgo has been reported by Yasuda, Fukugita, & Okamura (1997) using the Tully-Fisher relation. Meanwhile, measurements of Virgo's depth from the method of surface brightness fluctuations usually find a much smaller depth of a few Mpc (e.g., Neilsen & Tsvetanov 2000).

The ACS Virgo Cluster Survey was designed to provide accurate distances for our program galaxies through the z-band surface brightness fluctuations (see Figure 7 of Côté *et al.* 2004). Details of our implementation of this method are given in Mei *et al.* (2005ab), and in Figure 4 we show our adopted calibration. A thorough analysis of Virgo's three-dimensional structure will be presented in Mei *et al.* (2005c). For the time being, we can use these data to crudely estimate the line-of-sigh depth of Virgo.

The *rms* dispersion in distance modulus of the galaxies shown Figure 4 is 0.12 mag. Based on our simulations, we believe that the average error on these measurements is ~0.03 mag, while the instrinsic stellar population scatter in the z-band fluctuation magnitudes is estimated to be ~0.06 mag. Thus, our estimate for the intrinsic *rms* dispersion in the distance moduli is ≈0.10 mag. For our assumed distance of 16.5 Mpc (Tonry *et al.* 2001), this translates into a 1-σ distance dispersion of 0.8 Mpc, or a 2-σ line-of-sight *depth* of ≈3 Mpc. This value is consistent with previous surface brightness fluctuation measurements based on smaller samples (e.g., Neilsen & Tsvetanov 2000;

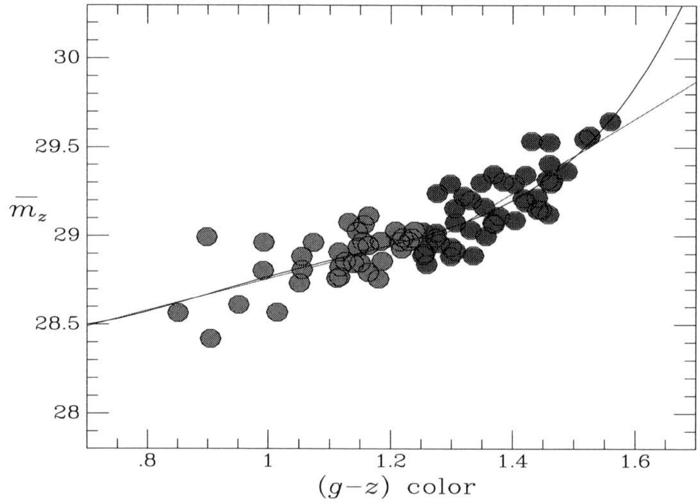

Figure 4. Calibration of the z-band surface brightness fluctuation method for the ACS Virgo Cluster Survey. A total of 82 galaxies are shown in this figure. The broken linear fit is the calibration from Mei *et al.* (2005b), while the continuous curve is a fourth-order polynomial fit. Different symbols are used to denote the galaxies on either side of the dwarf-giant transition region at $(g - z) \approx 1.25$.

West & Blakeslee 2000; Jerjen, Binggeli & Barazza 2004) and effectively rules out the very extended, cigar-like distributions found by some other researchers.

5. Globular Cluster Color Distributions

During the past decade, *HST* has revolutionized our understanding of extragalactic globular cluster systems. Perhaps the single most remarkable development in this field has been the realization that the color distributions of globular cluster belonging to luminous early-type galaxies are usually bimodal in nature (e.g., Gebhardt & Kissler-Patig 1999). The importance of the measured color distributions can scarcely by overestimated as they provide direct, quantitative insights into the processes of mergers and accretions which have shaped their host galaxies (Côté, West & Marzke 2002; Beasley *et al.* 2002; Li, Mac Low & Klessen 2004).

The ACS Virgo Cluster Survey offers a unique opportunity to measure globular cluster color distributions with unprecedented accuracy — given the depth and uniformity of the survey, and the fact that the $(g - z)$ index has twice the metallicity sensitivity of the $(V - I)$ index — and at, the same time, to trace the systematic evolution of the color distributions over a factor of ~450 in host galaxy luminosity. Figure 5 shows the behavior of globular cluster color distributions from the survey (Peng *et al.* 2005). Mean color histogram are shown in bins of host galaxy color; by virtue of the galaxy color-magnitude relation, the panels also reveal the evolution of the color distributions as one moves from bright (red) giants to faint (blue) dwarfs (the upper left and lower right panels, respectively).

The evolution is striking, with two particularly noteworthy features. First, the fractional contribution of the red sub-population shows a dramatic variation with galaxy color and luminosity. Among the giant galaxies, red clusters account for ~60% of the total cluster population; at the faint end of the sample, just ~15% of the clusters are associated with the red component. Second, the peak of the red sub-population shows

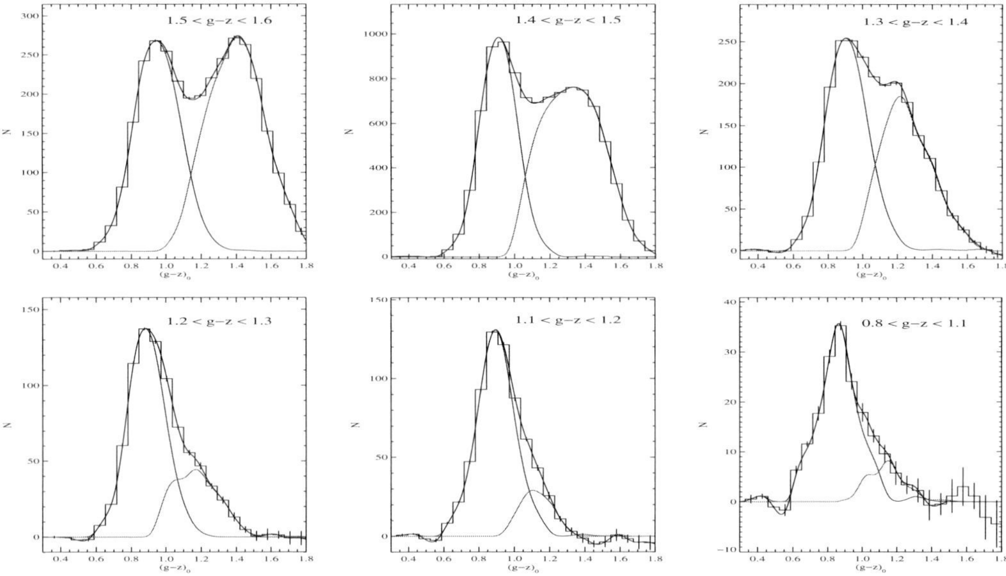

Figure 5. Color histograms and two-gaussian fits to globular clusters from the ACS Virgo Cluster Survey binned by host galaxy color. Note the sharp decrease in the relative number of red clusters — and the blueward shift in the mean color of this population — as one moves from giant to dwarf galaxies. Figure from Peng *et al.* (2005).

an unmistakable migration towards bluer colors as the luminosity of the host galaxy decreases. Peng *et al.* (2005) examines these findings in the context of formation scenarios for galaxies and their globular cluster systems.

Acknowledgements

I thank the members of the ACS Virgo Cluster Survey team for allowing me to highlight some of our findings in this article, and for their hardwork and enthusiasm during the past three years. I also extend my thanks and appreciation to the editors of these proceedings — Helmut Jerjen and Bruno Binggeli — for organizing such an enjoyable and productive meeting. Finally, I would like to acknowledge the heroic efforts of Bruno Binggeli in carrying out a truly remarkable survey of the Virgo Cluster. His tour de force study with Alan Sandage and Gustav Tammann nearly a quarter of a century ago remains the foundation upon which all modern Virgo surveys rest.

References

Beasley, M.A., Baugh, C.M., Forbes, D.A., Sharples, R.M. & Frenk, C S. 2002, *MNRAS* 333, 383

Binggeli, B. & Cameron, L.M. 1991, *A&A* 252, 27

Binggeli, B. & Cameron, L.M. 1993, *A&AS* 98, 297

Binggeli, B., Sandage, A. & Tammann, G.A. 1985, *AJ* 90, 1681

Binggeli, B., Tammann, G.A. & Sandage, A. 1987, *AJ* 94, 251

Böker, T., Sarzi, M., McLaughlin, D.E., van der Marel, R.P., Rix, H.-W., Ho, L.C. & Shields, J.C. 2004, *AJ* 127, 105

Bruzual, G. & Charlot, S. 2003, *MNRAS* 344, 1000

Caldwell, N. 1983, *AJ* 88, 804

Carollo, C.M., Stiavelli, M. & Mack, J. 1998, *AJ* 116, 68

Chabrier, G. 2003, *PASP* 115, 763

Côté, P., West, M.J., & Marzke, R.O. *et al.* 2002, *ApJ* 567, 853

Côté, P., Blakeslee, J.P., Ferrarese, L., Jordán, A., Mei, S., Merritt, D., Milosavljević, M., Peng, E., Tonry, J.L. & West, M.J. 2004, *ApJS* 153, 223

Côté, P. *et al.* 2005, in preparation

de Vaucouleurs, G. 1956, *Vistas Astron*, 2, 1584

Drinkwater, M.J., Phillipps, S., Gregg, M.D., Parker, Q.A., Smith, R.M., Davies, J.I., Jones, J.B. & Sadler, E.M. 1999, *ApJ* 511, L97

Durrell, P.R. 1997, *AJ* 113, 531

Ford, H.C., *et al.* 1998, *Proc. SPIE* 3356, 234

Gebhardt, K. & Kissler-Patig, M. 1999, *AJ* 118, 1526

Haşegan, M., Jordán, A., Côté, P., Djorgovski, S.G., McLaughlin, D.E., Blakeslee, J.P., Mei, S., West, M.J., Peng, E.W., Ferrarese, L., Milosavljević, M., Tonry, J.L. & Merritt, D. 2005, *ApJ* 627, 203

Hilker, M., Infante, L., Vieira, G., Kissler-Patig, M. & Richtler, T. 1999, *A&AS* 134, 75

Ichikawa, S.-I., Wakamatsu, K.-I. & Okamura, S. 1986, *ApJS* 60, 475

Jerjen, H., Binggeli, B. & Barazza F.D. 2004, *AJ* 127, 771

Jordán, A., Blakeslee, J.P., Peng, E., Côté, P., Ferrarese, L., Mei, S., Merritt, D., Milosavljević, M., Tonry, J.L. & West, M.J. 2004a, *ApJS* 154, 509

Jordán, A., Côté, P., Ferrarese, L., Blakeslee, J.P., Mei, S., Merritt, D., Milosavljević, M., Peng, E., Tonry, J.L. & West, M.J. 2004b, *ApJ* 613, 279

Jordán *et al.* 2005, *ApJ* submitted

Klypin, A., Kravtsov, A.V., Valenzuela, O. & Prada, F. 1999, *ApJ* 522, 82

Kormendy, J. 1985, *ApJ* 295, 73

Li, Y., Mac Low, M.-M. & Klessen, R.S. 2004, *ApJ* 614, 29

Lotz, J.M., Miller, B.W., & Ferguson, H.C. 2004, *ApJ* 613, 262

Mayer, L. Moore, B., Quinn, T., Governato, F. & Stadel, J. 2002, *MNRAS* 336, 199

Mei, S., Blakeslee, J.P., Tonry, J.L., Jordán, A., Peng, E.W., Côté, P., Ferrarese, L., Merritt, D., Milosavljevic, M. & West, M.J. 2005a, *ApJS* 156, 113

Mei, S., Blakeslee, J.P., Tonry, J.L., Jordán, A., Peng, E.W., Côté, P., Ferrarese, L., Merritt, D., Milosavljevic, M. & West, M.J. 2005b, *ApJ* 625, 121

Mei, S. *et al.* 2005c, in preparation

Moore, B., Ghigna, S., Governato, F., Lake, G., Quinn, T., Stadel, J. & Tozzi, P. 1999, *ApJ* 524, L19

Neilsen, E.H. & Tsvetanov, Z.I. 2000, *ApJ* 536, 255

Peng, E.W. *et al.* 2005, *ApJ* submitted

Phillips, A.C., Illingworth, G.D., MacKenty, J.W. & Franx, M. 1996, *AJ* 111, 1566

Reaves, G. 1983, *ApJS* 53, 375

Sandage, A., Binggeli, B. & Tammann, G.A. 1985, *AJ* 90, 1759

Tonry, J.L., Dressler, A., Blakeslee, J.P., Ajhar, E.A., Fletcher, A.B., Luppino, G.A., Metzger, M.R. & Moore, C.B. 2001, *ApJ* 546, 681

Yasuda, N., Fukugita, M. & Okamura, S. 1997, *ApJS* 108, 417

Young, C.K. & Currie, M.J. 1995, *MNRAS* 273, 1141

West, M.J. & Blakeslee, J.P. 2000, *ApJ* 543, 27

Near-Field Cosmology with Dwarf Elliptical Galaxies
Proceedings IAU Colloquium No. 198, 2005
H. Jerjen and B. Binggeli, eds.

doi:10.1017/S1743921305003923

Constraining galaxy formation models with dwarf ellipticals in clusters

Christopher J. Conselice
California Institute of Technology, Pasadena CA, USA

Abstract. Recent observations demonstrate that dwarf galaxies in clusters, despite their faintness, may be a critical galaxy type for understanding the physical processes behind galaxy formation. Dwarfs are the most common type of galaxy and are particularly abundant in rich galaxy clusters. The dominate model for explaining the formation of these systems thorough Cold Dark Matter models is that the bulk of their stellar mass formed early, and within their present environments. Recent results however suggest that some dwarfs appear in clusters after the bulk of its members form, a scenario not predicted in standard hierarchical structure formation models. Many of these systems appear to be younger and more metal rich than dwarfs in lower density areas, suggesting they are possibly created by a tidal process induced by the cluster. Several general galaxy cluster observations, including steep luminosity functions and the origin of intracluster light, are natural outcomes of these processes.

1. Introduction

Although dwarf galaxies are the faintest and lowest mass galaxies in the universe, they likely hold important clues for understanding galaxy formation and the nature of dark matter. The reason for this is quite simple: low-mass galaxies, and particularly low-mass galaxies in clusters, are the most common galaxies in the nearby universe (Ferguson & Binggeli 1994). Any ultimate galaxy evolution/formation theory must be able to predict and accurately describe the properties of these objects. In popular galaxy formation models, such as hierarchical assembly (e.g., Cole *et al.* 2000), massive dark halos form by the mergers of lower mass ones early in the universe, and the first galaxies likely have low stellar mass. By understanding dwarf galaxies, we are also thus potentially studying the very first galaxies to form. On the other hand, observations reveal that few low-mass galaxies could have formed all of their stars early in the universe at $z > 7$, with considerable evidence for star formation occurring in the last few Gyrs in dwarf spheroidals (e.g., Mateo 1998).

The traditional approach to studying low-mass galaxies is to examine those in the Local Group (LG). It is now well established that LG dwarf elliptical and dwarf spheroidal galaxies have varying star formation histories, with metal-poor populations as old as classical halo globular galaxies but also with evidence for recent star formation (see e.g., Mateo 1998). There are also some low-mass LG galaxies, such as Sagittarius, that contain surprisingly metal-rich populations given their luminosities (e.g., Ibata, Gilmore & Irwin 1995). Because LG dE galaxies are close we can resolve their stellar populations, and thus we can learn much more about them than we do dwarfs in more dense, but distant environments. Because of this we know a great deal concerning LG dwarf properties including their internal kinematics and star formation histories. Many of the lowest mass galaxies in the LG, such as Draco and Ursa Minor, have very high inferred central M_{tot}/L ratios (Kleyna *et al.* 2002) and apparently contain the densest dark matter halos of all known galaxies, in qualitative agreement with the original CDM predictions (e.g., Lake 1990). Low-mass galaxies in clusters, however, have different kinematic and spatial properties than these LG systems (e.g., Conselice, Gallagher, & Wyse 2001,2003), suggesting they might have a different formation scenario.

Observationally, dwarf galaxies are typically faint ($M_B > -18$), with low surface brightnesses ($\mu_B > 23$ mag arcsec^2). Since dwarfs are so common, they are in every sense normal galaxies. The most common type among dwarfs are dwarf ellipticals/spheroids, which dominate the number density of galaxy clusters down to $M_B = -11$ (Ferguson & Bingelli 1994; Trentham *et al.* 2002). Based on studies of luminosity functions in clusters, there are also more dwarf ellipticals per giant in denser regions than in the field. This implies that clusters of galaxies cannot form through simple mergers of galaxy groups. Some dwarf systems must form within the cluster environment. The nature of this over-density may be the result of initial conditions, or 'non-standard' galaxy formation. That is, dwarfs may have formed after the cluster was in place. There is now evidence for this, the implications of which can explain several galaxy cluster phenomenon. On the other hand, there is also evidence that some dwarf ellipticals/spheroidals in clusters are dominated by old stellar populations. In this paper we review the current observations of dwarfs in clusters and attempt to interpret these systems in terms of known properties of Local Group dwarf ellipticals/spheroidals, and in the milieu of theoretical ideas concerning low mass galaxy formation in a cosmological context.

2. Cluster Dwarf Galaxy Properties

Dwarfs have been studied in detail in nearby rich clusters, such as Virgo, Fornax and Coma, and Perseus. Most of the data we discuss therefore comes from these sources. In particular a significant fraction of the data presented in this article comes from the papers by Conselice, Gallagher & Wyse (2001,2002,2003) and Conselice *et al.* (2003b). We list below our current understanding of dwarf properties in terms of various observational quantities.

Number Densities: The luminosity function (LF) in all galaxy environments are dominated by dwarf galaxies. The over-density of dwarfs in clusters is about 5–10 times that in groups, that is the dwarf to giant ratio is nearly a factor of ten higher. Another way to quantify this is through the faint end slope of the LF, α. The value of α in rich clusters, such as Virgo is typically around $\alpha = -1.2$ to -1.5 (Sandage *et al.* 1985; Trentham *et al.* 2002), with some results suggesting even steeper LFs with $\alpha = -1.6$ (Sabatini *et al.* 2003). This is steeper than field values, such as in the Local Group ($\alpha = -1.1$; van den Bergh 1992), yet flatter than the value predicted by CDM for all environments ($\alpha = -2$). Environment clearly affects the way these systems are produced, which is generally not a prediction in CDM models (cf. Tully *et al.* 2002). This also suggests that clusters of galaxies *cannot* simply form from the mergers of lower density groups of galaxies.

Spatial Positions: While Local Group dwarf galaxies, particularly dwarf ellipticals, are strongly clustered around giant galaxies in the Local Group, the opposite is found for low-mass galaxies in clusters, where most are neither clustered around, nor distributed globally similar to, giant elliptical galaxies (Conselice *et al.* 2001). Both dwarf ellipticals and irregulars also have a broader distribution within clusters, that is they are not clustered towards the center, but are spread throughout (e.g., Conselice *et al.* 2001). This is also the pattern seen for spirals and irregulars in clusters.

Radial Velocities: The radial velocities of low-mass cluster galaxies, including S0s, spirals, dwarf irregulars and dwarf ellipticals have a wider distribution than the ellipticals (see Figure 1a). For example, Virgo cluster elliptical galaxies have a narrow Gaussian

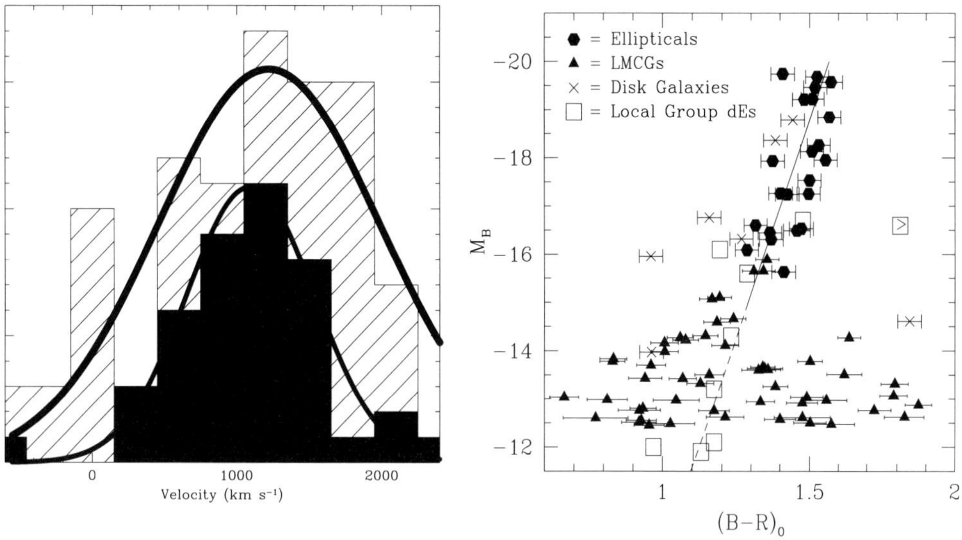

Figure 1. (a) Velocity histograms for giant ellipticals (solid) and dwarf ellipticals (shaded) in the Virgo cluster (Conselice *et al.* 2001). (b) Color magnitude diagram for galaxies in the Perseus cluster, demonstrating the large color scatter for systems with $M_B > -15$. The solid boxes are where Local Group dEs/dSphs would fit on this plot. Dwarf ellipticals are labeled as low mass cluster galaxies (LMCGs).

velocity distribution, with $\sigma = 462$ km s^{-1}, concentrated at the mean radial velocity of the cluster. The other populations, including the over 100 classified dwarf ellipticals in Virgo with radial velocities, have much broader, and non-Gaussian, velocity distributions ($\sigma \sim 700$ km s^{-1}), all with velocity dispersion ratios with the ellipticals consistent with their being accreted (e.g., Conselice *et al.* 2001). There is also significant sub-structure within these velocity distributions, unlike the case for the giant ellipticals.

Stellar Populations: Currently, we know with some certainty that dwarf galaxies have either young/metal rich or old/metal poor stellar populations (e.g., Poggianti *et al.* 2001). This is further seen in complete color-magnitude diagrams in nearby rich clusters, such as Perseus, down to $M_B = -12$ (Conselice *et al.* 2002, 2003a). Fainter dwarfs also tend to be even more heterogeneous, with a large scatter from the color-magnitude relationship (CMR) (Conselice *et al.* 2003a; Rakos *et al.* 2001) (Figure 1b). This trend is found in several nearby clusters, including Fornax, Coma and Perseus, and can be explained by different dwarfs having mixtures of ages and metallicities (e.g., Poggianti *et al.* 2001; Rakos *et al.* 2001; Conselice *et al.* 2003a). Stromgren and broad-band photometry reveal that the redder dwarfs are metal enriched systems (Figure 2). However, old stellar populations do exist in dwarf galaxies (Poggianti *et al.* 2001) and in the globular clusters surrounding dwarfs (e.g., Conselice in prep).

Internal Kinematics: One key observational test for the origin of dwarf ellipticals is whether they rotate or not. Models show that a dwarf elliptical which has been transformed from a spiral should reveal some rotation. Using 8-10 meter class telescopes, the evidence is ambiguous with some dEs showing rotation (Pedraz *et al.* 2002), while others clearly do not (Geha *et al.* 2003). There is no obvious difference between these two populations in terms of morphology or chemical abundances (Geha *et al.* 2003), although

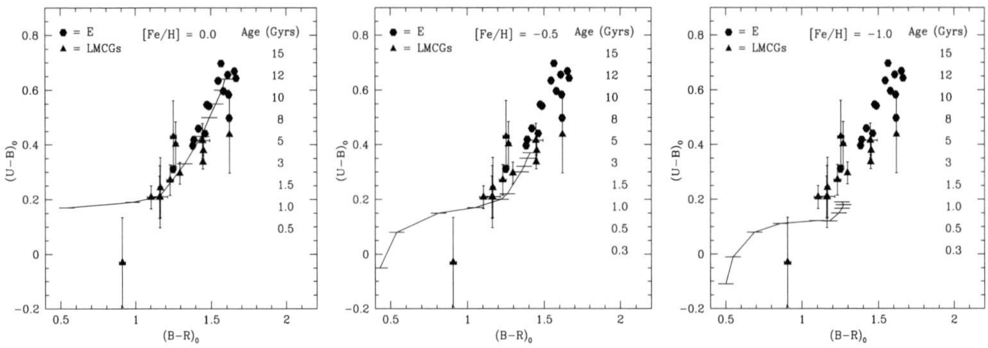

Figure 2. Three stellar synthesis modeled age tracks on a UBR color-color diagram at constant metallicities of solar, [Fe/H] = -0.5, and [Fe/H] $= -1$. The age range is 0.3 Gyrs to 15 Gyrs for the [Fe/H] = -0.5 and -1 models and 0.5 Gyrs to 15 Gyrs for the solar metallicity models. Dwarf ellipticals are labeled as low mass cluster galaxies (LMCGs).

these samples are still very small. The central velocity dispersions of these systems is also quite low, indicating that dark matter does not dominate, at least in their centers.

Furthermore, in addition to predicting low-metallicity systems, Dekel & Silk (1986) show, based on their model assumptions, how the internal velocities of low-mass galaxies should correlate with M_{tot}/L ratios and luminosities, L. The relationship they predict is $M_{tot}/L \sim L^{-0.37}$, and $\sigma \sim L^{0.19}$. These predictions can be tested utilizing Virgo Cluster dE internal velocity measurements made by Pedraz *et al.* (2002) and Geha *et al.* (2003). After fitting a power law to the relationship between σ and L, $\sigma \sim L^{0.31 \pm 0.05}$. The fitted exponent on L is 2σ away from the relationship predicted by Dekel & Silk. This is another indication that cluster dwarf ellipticals are possibly formed through multiple methods.

Gas Content: Surprisingly, some dEs in clusters contain evidence for gas in various phases, both cold HI (Conselice *et al.* 2003b; Buyle *et al.* 2005) and warm gas in the form of Hα (Michielsen *et al.* 2004). These studies have only been performed in the nearest clusters, Virgo and Fornax due to the difficulty of making these observations. Although 15% of dEs have evidence for HI, all of these systems are located in the outer parts of Virgo (Conselice *et al.* 2003b) and Fornax (Buyle *et al.* 2005). Figure 3 shows the location of HI emitters in the Virgo cluster and the fraction of dEs with HI emission as a function of radius from the center of the Virgo cluster.

3. Cluster Dwarf Galaxy Origins

Any successful theory of dwarf galaxies, particularly for explaining how dwarf ellipticals form, must account for the following properties: over-density in relation to massive galaxies in rich environments, mix of stellar populations, ability to survive in dense environments, diffuse spatial and velocity structure, and mixtures of rotation. This theory must also explain why fainter dwarfs are more heterogeneous than brighter ones.

In the simple collapse + feedback scenarios (Dekel & Silk 1986), dwarfs are formed when gas collapses and forms stars. These stars produce winds that expel gas from these systems, halting any future star formation. In this formation scenario dwarfs formed before the cluster ellipticals, or at least formed within groups that later merged to form clusters. Fainter dwarfs however, cannot all be born in groups which were later accreted

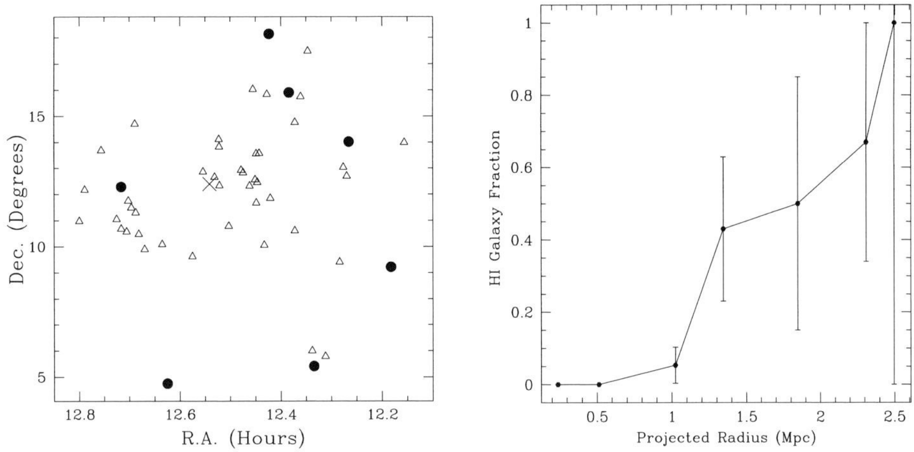

Figure 3. (a) Distribution on the sky of dE galaxies observed at 21 cm in the Virgo Cluster. The open triangles represent non-detections, while the filled circles show the locations of the dE galaxies with detected H I. The cross toward the center shows the location of the giant elliptical M87. (b) Fraction of Virgo classified dwarf elliptical galaxies detected in H I as a function of projected distance from the center of the cluster.

into clusters along with the massive galaxies, due to the high dwarf to giant galaxy ratio found in clusters (Conselice *et al.* 2001, 2003a).

One idea to explain this, as suggested by Tully *et al.* (2002), is that the dark matter halos of dwarfs are 'squelched' in lower density environments due to a large ultraviolet background after the universe was reionized. This explains the differences in α between different environments, but does not explain how within the same environment there is a great diversity in the dwarf population. It also does not easily explain why many dwarfs appear to have recently been accreted into clusters.

The above evidence suggests that simple low-mass galaxy formation scenarios can be safely ruled out for some dwarfs. The first clue that dwarfs are not produced through standard methodologies came from studies that showed dwarfs are generally found in abundance in dense areas such as nearby clusters. In fact, in standard CDM scenarios dwarfs should be more common in lower density environments, but we see the direct opposite (Trentham *et al.* 2002). Dwarfs within dense environments also have a broader distribution, both spatially and in terms of their radial velocities, than giant ellipticals, similar to the pattern of infalling spirals. This is an indication that both are recent additions to clusters (Conselice *et al.* 2001). Internal dynamic evidence also suggests that at least some dwarfs are rotating (e.g., Geha *et al.* 2003), a feature not seen in Local Group dEs. Finally, the stellar populations of some faint $M_B > -15$ dwarf galaxies appear to be metal rich (near solar), implying that the dwarf population itself is inhomogeneous and may have multiple origins. We discuss several scenarios which have been proposed to explain these observations and argue that one method which can reproduce these trends involves the formation of dwarfs from already existing galaxies through a tidal effect.

One alternative idea is that some modern dwarfs formed after the cluster itself was in place by collapsing out of enriched intracluster gas. Another is that the intracluster medium (ICM) is able to retain enriched gas that in the Dekel and Silk (1986) paradigm would be ejected by feedback, but remains due to the confinement pressure of the ICM (Babul & Rees 1992). This scenario would explain the higher metallicities of some of the fainter dwarfs.

An alternative scenario, now gaining in popularity, is that dwarfs form in clusters through a tidal origin. Two main possibilities for this are tidal dwarfs (Duc & Mirabel 1994), and as the remnants of stripped disks or dwarf irregulars (Conselice *et al.* 2003a). The velocity and spatial distributions of dwarfs suggests that they were likely accreted into clusters during the last few Gyrs (Conselice *et al.* 2001). This, combined with the high metallicities of these cluster dwarfs, and the fact that their stellar populations are fundamentally different than field dwarfs (e.g., Conselice *et al.* 2003a; Figure 1b; Figure 2) suggests that the cluster environment has morphologically transformed, or stripped, accreted galaxy material into dwarfs.

There is evidence for this process currently occurring in nearby clusters (e.g., Conselice & Gallagher 1999). If dEs form from infalling spirals then there should also be systems now being transformed which appear morphologically as dEs, but retain some of the gas left over from their precursor. These systems would only exist in the outer parts of clusters, as any that dwarfs venturing towards the core will be rapidly stripped of material. For example, very deep Arecibo observations of Virgo dEs reveal that $\sim$15% of a sample of 56 have HI detections, all of which are located outside the core of the cluster (Conselice *et al.* 2003b). Other detailed morphological investigations of nearby cluster dwarf ellipticals show that they contain a wide diversity of structures, some with tidal features, and others with apparent spiral structures (e.g., Jerjen *et al.* 2000; De Rijcke *et al.* 2003; Barazza *et al.* 2003; Graham *et al.* 2003).

Despite the above, there is considerable evidence that some dEs in clusters are indeed an old population, some with metal poor populations (e.g., Lotz *et al.* 2004). A single cluster dwarf scenerio is unlikely to explain the great diversity we see, thus multiple formation methods are likely necessary.

4. Implications of Late Dwarf Formation in Clusters

Environment likely plays a role in all aspects of galaxy formation and evolution. If indeed any dwarfs form after their host cluster due to a tidal process, there are several cluster features these processes can possibly explain. The first is that the LF of clusters will change after these lower mass galaxies form. The LF of galaxies in the Perseus cluster becomes flatter, with a similar faint end slope (α) as the field, after removing these red galaxies (Figure 4). Figure 4 shows how the luminosity function changes slope from $\alpha = -1.4$ to $\alpha = -1.2$ due to galaxies in the Perseus cluster which are redder than 2σ from the color-magnitude relation, defined by the giant cluster galaxies (Conselice 2002).

There are several implications for this process, beyond possibly steepening the luminosity function. One of these is the origin of intracluser light, which can make up 50% of the light coming from dense clusters (e.g., Adami *et al.* 2005). If cluster dEs originate from tidally disturbed galaxies then the amount of light liberated is enough to account for all of this intracluster light as arising from the debris from tidally disturbed galaxies (Conselice *et al.* 2003a). For example, the total luminosity of intracluster light in the Virgo Cluster within 2° of M87 is 2×10^{11} $L_{\odot}$. If this material originates from tidally striped objects whose remnants are dwarfs, we can compute how much material on average each dE must have lost. There are 170 dEs, 148 dE,Ns, and 14 S0 galaxies within this radius in Virgo. On average, if all the dE and dE,N galaxies are remnants of stripped galaxies then 0.6×10^9 $L_{\odot}$ of light was lost by each object. If we consider only dE galaxies as remnants of this process then 1.1×10^9 $L_{\odot}$ was lost by each dE. Therefore, if we assume that dEs were once as massive as a typical disk galaxy, then the amount of

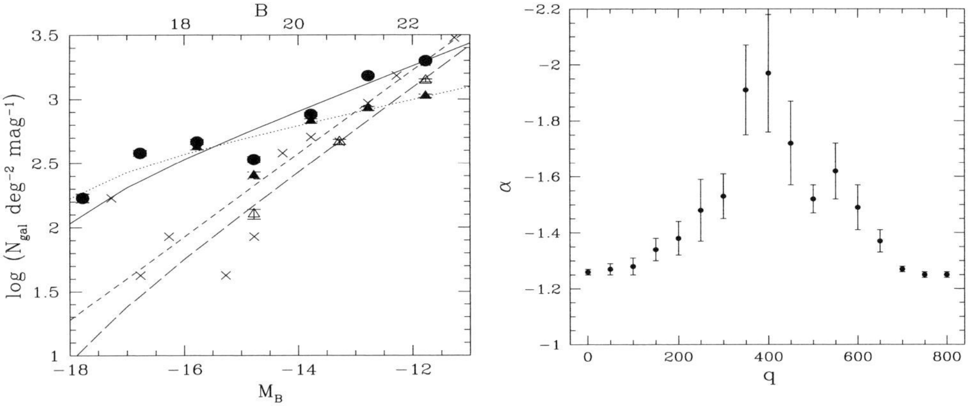

Figure 4. (a) The Perseus cluster luminosity function plotted in various ways down to $M_B = -11$. The solid round points, and fitted solid line, is the total luminosity function of the central region of Perseus. The luminosity function for dEs redder and bluer than the CMR prediction are plotted as open and solid triangles and long dashed and dotted lines. The crosses mark the density of background galaxies. (b) Modeled cluster luminosity function slope, α, as a function of the number of high-speed maximum interactions (q ~time) cluster galaxies undergo during evolution in a Perseus like cluster (see Conselice 2002).

material these systems lost through tidal effects is enough to account for the intracluster light in rich clusters of galaxies.

There are still however many questions that need to be answered by observing dwarfs in clusters in more detail. There has never been a spectroscopic survey of dwarfs fainter than $M_B = -15$ in clusters of galaxies, for example. We are also just learning about the internal properties of dwarfs through high resolution imaging and spectroscopy. Making progress on these topics will require surveys of dwarfs in all environments, and connecting these observations with galaxy evolution in higher redshift clusters.

Acknowledgements

I acknowledge my collaborators for their participation in the projects described here. I also thank the organizers of this symposium for creating an interesting meeting, and for their patience in receiving this review.

References

Adami, C., *et al.* 2005, *A&A* 429, 39
Babul, A. & Rees, M.J. 1992, *MNRAS* 255, 346
Barazza, F.D., Binggeli, B. & Jerjen, H. 2003, *A&A* 407, 121
Buyle, P., De Rijcke, S., Michielsen, D., *et al.* 2005, *astro-ph/0504270*
Cole *et al.* 2000, *MNRAS* 319, 168
Conselice, C.J. & Gallagher, J.S. 1999, *AJ* 117, 75
Conselice, C.J., Gallagher, J.S. & Wyse, R.F.G. 2001, *ApJ* 559, 791
Conselice, C.J., Gallagher, J.S. & Wyse, R.F.G. 2002, *AJ* 123, 2246
Conselice, C.J., Gallagher, J.S. & Wyse, R.F.G. 2003a, *AJ* 125, 66
Conselice, C.J. *et al.* 2003b, *ApJ* 591, 167
Conselice, C.J. 2002, *ApJ* 573, 5L
De Rijcke, S. *et al.* 2003, *A&A* 400, 119
Dekel, A. & Silk, J. 1986, *ApJ* 303, 39
Duc, P.-A. & Mirabel, I.F. 1994, *A&A* 289, 83

Ferguson, H.C. & Binggeli, B. 1994, *A&ARv* 6, 67
Geha, M., Guhathakurta, P. & van der Marel, R.P. 2003, *AJ* 126, 1794
Graham, A.W., Jerjen, H. & Guzman, R. 2003, *AJ* 126, 1787
Ibata, R. *et al.* 1998, *MNRAS* 277, 781
Jerjen, H., Kalnajs, A. & Binggeli, B. 2000, *A&A* 358, 845
Kleyna, J., *et al.* 2002, *MNRAS* 330, 792
Lake, G. 1998, *MNRAS* 244, 701
Mateo, M. 1998, *ARA&A* 36, 435
Michielsen, D. *et al.* 2004, *MNRAS* 353, 1293
Moore, B. *et al.* 1998, *ApJ* 495, 139
Pedraz, S. *et al.* 2002, *MNRAS* 332, 59L
Poggianti, B.M. *et al.* 2001, *ApJ* 562, 689
Rakos, K., *et al.* 2001, *AJ* 121, 1974
Sabatini, S. *et al.* 2003, *MNRAS* 341, 981'
Sandage, A. *et al.* 1985, *AJ* 90, 1759
Trentham, N. *et al.* 2002, *ApJ* 569, 573
Tully, R.B. *et al.* 2002, *ApJ* 569, 573
van den Bergh, S. 1992, *A&A* 264, 75

Discussion

FERGUSON: In the Virgo and Fornax clusters, the spatial distributions of nucleated and non-nucleated dEs brighter than M = -14 appear to be different. the dE,Ns are as concentrated toward the cluster centers as the giant Es. It will be interesting to see if this still holds with the new classifications from the ACS Virgo cluster survey.

CONSELICE: Yes, that is indeed the case, and this probably reflects the fact that there exists an old population of dEs in clusters, as well as a younger population of dEs. The bright nucleated dEs are likely taken from the population of older or original dEs, while pure dEs may be from the newer formed population.

MIESKE: If the red LSB galaxies are stripped bulges, shouldn't then there be huge amounts of galaxies in the process of disruption and also many more that got totally destroyed? Seeing a population of LSB galaxies ≈500 times fainter than possible progenitors seems like a huge coincidence "snapshot". Taking into account that there should be more massive, intermediate ones and also more very faint (possibly entirely disrupted) ones, can you get such a high number of luminous progenitors?

CONSELICE: The galaxies we se today as red dEs at $M_B > -15$ may still be loosing mass and over time could disappear entirely. There are examples in clusters of red more luminous galaxies, such as amorphous or disk-like galaxies, but these aren't dEs yet. In the past, clusters had more of these infalling galaxies, as seen in the Butcher-Oemler effect, and these may be the progenitors of today's dEs.

KROUPA: It could be conceivable that the dE populations in clusters are tidal-dwarf galaxies (Okazaki & Taniguchi 2000). This would explain their wide distribution and different chemical properties as they would have formed from pre-enriched material in merging parents.

CONSELICE: Yes, the red dEs could be produced through tidal dwarfs, we have a discussion of this possibility in the paper Conselice *et al.* (2003). It does not matter, in a sense, if the new dEs are produced in tidal dwarfs or from the stripped remains of a larger galaxy. Both processes will produce red dEs with large spatial and velocity ditributions.

Near-Field Cosmology with Dwarf Elliptical Galaxies
Proceedings IAU Colloquium No. 198, 2005
H. Jerjen and B. Binggeli, eds.

doi:10.1017/S1743921305003935

Low luminosity and ultra low surface brightness galaxies in the Virgo cluster

Nelson Caldwell

Smithsonian Astrophysical Observatory, 60 Garden Street, Cambridge, MA, USA
email: caldwell@cfa.harvard.edu

Abstract. A CCD survey of selected areas in the Virgo cluster was undertaken with the goal of finding dwarf ellipticals fainter in surface brightness than known in the Local Group. This was successful, and galaxies as much as 2 magnitudes fainter than e.g., Umi, have been found. These tend to be larger in radius, due to the detection methods, and thus more luminous than the LG as well. Two fields containing such galaxies were imaged with ACS/HST, with the result that individual RGB stars have been resolved in these galaxies, which are 15 Mpc distant.

Keywords. galaxies: dwarf, galaxies: luminosity function, galaxies: stellar content

1. Introduction

What are the limits of dwarf galaxy sizes and densities? Is the star formation history different at the extremes of surface density and luminosity?

These are two questions that I investigated using a CCD survey of several areas within the Virgo cluster. Using the KPNO Mosaic camera on the Mayall 4m telescope, Taft Armandroff & I took images totaling 6000s in B, and 2900s in R. The combined images of these fields (7 in the cluster, 2 outside for background measurements), were then searched for dwarf elliptical galaxies via two techniques:

1.1. *Visual inspection*

Images were displayed on a computer monitor at moderate scale. Galaxies were identified as dEs if they were smooth, and had appropriate sizes and surface brightnesses. At extremely low SB, there is a possibility that the objects are either artifacts, or background clusters of galaxies. The two colors helped both in the sense of requiring multiple detections and appropriate colors (see figure 1). This method is of course tedious and prone to miss obvious objects, but it does find the large, low SB objects which automated techniques tend to miss.

1.2. *Automatic detection*

The same images were analyzed using a combination of SExtractor (Bertin & Arnouts 1996) and GIM2d (Simard 1998). The former program was used to identify pixels belonging to distinct galaxies (and set the local background), while the latter program was used to fit a model to each galaxy, thus providing structural parameters that can be filtered to exclude objects not conforming to those of a dE galaxy. This technique is also tedious! But it does allow one to measure the small, low luminosity objects at least in a statistical manner, which the visual technique cannot do.

2. Virgo Extremes

Figure 2 shows the resultant distribution of data in the magnitude-surface brightness plane from the automatic method. Objects below the straight line are in the area

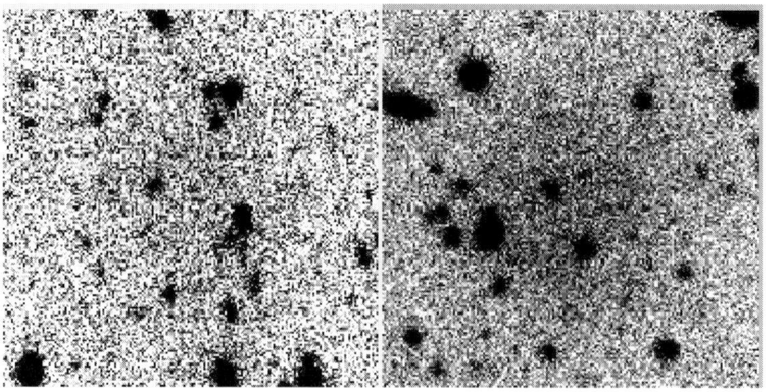

Figure 4. Two Ultra Low LSB targets

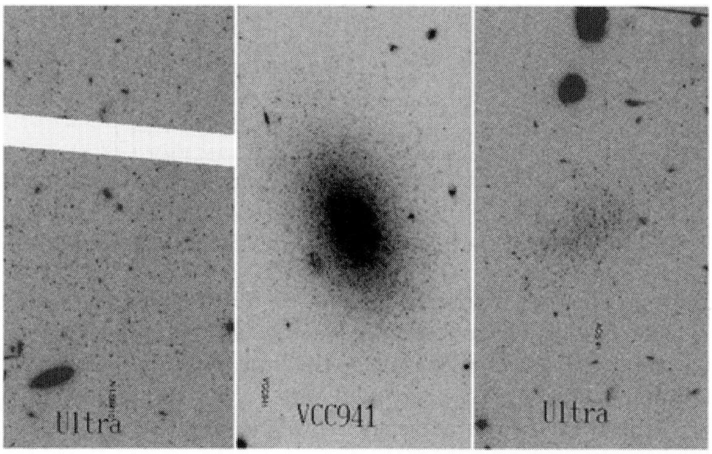

Figure 5. HST/ACS I band images

fainter dE galaxies listed in Binggeli *et al.* (1985, we thus find $M_B = -12.1$, but is also bright in surface brightness with $B_o = 23.2$). It too is clearly resolved, and is thus also a member. At the right is a bonus galaxy, one picked up by the automated method above, but discounted by me as a background object on inspection. It is in fact also a member, with $M_B = -8.2$, making it the one of the least luminous galaxies studied.

The second ACS field shows much the same thing: the candidate dE galaxy is well-resolved and thus a cluster member. I am forming stellar color-magnitude diagrams, but difficulties with aperture corrections and zero-points prevents me from publishing them here at this time. In general however, it appears that the metal abundances are low, perhaps $[Fe/H] \sim -1.7$, but not extreme. As well, the dE galaxies do not appear to have large amount of upper AGB stars, so ages are not as young as found in M81-F8D1, for comparison.

4. Conclusions

Clearly, extremely low stellar density galaxies do exist, and are even found in the densest regions of the Virgo cluster. One must think that these objects are dark matter dominated, lest they be disrupted easily by the frequent passages of larger galaxies in the core. The large observed range in optical radii among galaxies (a factor of 5) at low surface brightness is intriguing. Is this due to star formation processes, or rather DM halo density distribution variations at low mass?

Acknowledgements

This work has been done in collaboration with Dan McIntosh, T.A. Armandroff, and G.S. Da Costa.

References

Binggeli, B., Sandage, A. & Tammann, G. 1985, *AJ* 90, 1681
Bertin, E. & Arnouts, S. 1996, *A&AS* 117, 393
Simard, L. 1998, *ASPC* 145, 108

Discussion

MIESKE: If you restrict your sample to those galaxies with $\mu <$ that of the Local Group, is the average size still larger than that of the LG dwarfs?

CALDWELL: If we consider the "candidate" dEs found automatically, via SExtractor and GIM2d, then I think the answer is that the mean Virgo sizes are similar to those in the LG. What is apparent however, is that there are dEs in Virgo with much larger sizes than found in the LG, and currently I don't have a technique to find those automatically in Virgo (they were found by inspection of the images)

.

GALLAGHER: Comment: Great project and spectacular results! Question: Can you be sure that these are independent objects and not part of larger debris structures that are common in Virgo (e.g., Mihos CWRU study)?

CALDWELL: The two "ultra-low" surface brightness objects targeted have a definite radial falloff in stellar density from a well-defined center, and, as well as I can measure, are elliptical on the sky. In other words, even with their low central surface brightnesses, they still stand out from their surroundings.

FERGUSON: Would you have been able to detect AGB stars in your V band images?

CALDWELL: Not the reddest ones of course, whose V-I colors would mean that the V magnitudes would be below my detection limit. The behaviour of AGB stars in the M81 dwarf F8D1 though indicates we would expect to see a large number of AGB stars just above the RGB tip, which aren't evident in these Virgo galaxies.

Near-Field Cosmology with Dwarf Elliptical Galaxies
Proceedings IAU Colloquium No. 198, 2005
H. Jerjen and B. Binggeli, eds.

doi:10.1017/S1743921305003947

The counterparts of Local Group dwarf spheroidals in nearby clusters

Michael Hilker[1], Steffen Mieske[1] and L. Infante[2]

[1]Sternwarte der Universität Bonn, Auf dem Hügel 71, 53121 Bonn, Germany
email: mhilker@astro.uni-bonn.de, smieske@astro.uni-bonn.de

[2]Departamento de Astronomía y Astrofísica, P. Universidad Católica,
Casilla 104, Santiago 22, Chile
email: linfante@astro.puc.cl

Abstract. In this contribution, we report on the discovery of dwarf spheroidals (dSphs) in the core of the Fornax cluster. Their photometric properties – like magnitude, colour, surface brightness – are very similar to those of Local Group dSphs. However, at a given total magnitude, dSphs in Fornax seem to be more extended than their Local Group counterparts. The membership of several dwarf galaxy candidates in Fornax has been confirmed by surface brightness fluctuation measurements on deep wide-field images taken with the Magellan telescope. The analysis of these images also confirms the flat faint end slope of the luminosity function for dSphs in Fornax which contradicts the expected large number of small dark matter halos connected to dwarf galaxies in ΛCDM theory. Dwarf spheroidals have also been detected in the Hydra I and Centaurus cluster. A preliminary analysis of their photometric properties shows that they obey similar scaling relations as their counterparts in Fornax and the Local Group.

Keywords. Galaxies: clusters: individual (Fornax, Hydra I, Centaurus), galaxies: dwarf, fundamental parameters, luminosity function, photometry

1. Introduction

The identification of dwarf galaxies in different environments plays an important role for the verification of cosmological models. Cold dark matter models predict a large number of small dark matter (DM) halos as sub-structures around Milky Way-sized halos as well as galaxy cluster-sized halos (e.g. Moore *et al.* 1999). If every small DM halo would contain luminous matter, many dwarf galaxies are expected in basically all environments. However, already in the Local Group there exists a strong discrepancy, the so-called "missing satellite problem" (e.g. Klypin *et al.* 1999). The faint end of its galaxy luminosity function is quite flat ($\alpha = -1.1$, Pritchet & van den Bergh 1999) compared with CDM model predictions. Several studies of galaxy clusters show rising numbers of low-luminosity dEs down to about $M_V \simeq -11$ mag with very different faint-end slopes in the range $-1.1 > \alpha > -2.2$ (e.g. Trentham & Tully 2002). However, it is not known whether the luminosity function in clusters continues to $M_V = -8.5$, as it does in the Local Group. Do dSphs actually exist in galaxy clusters? And how abundant are they?

Here, we present the identification and analysis of dwarf spheroidals in nearby clusters, namely the Fornax, Hydra I and Centaurus cluster.

2. WFCCD survey of dwarf spheroidals in the Fornax cluster core

End of 1999, we observed the core of the Fornax cluster through the Johnson VI filters with the wide-field CCD (WFCCD) at the 100-inch du Pont telescope at Las Campanas,

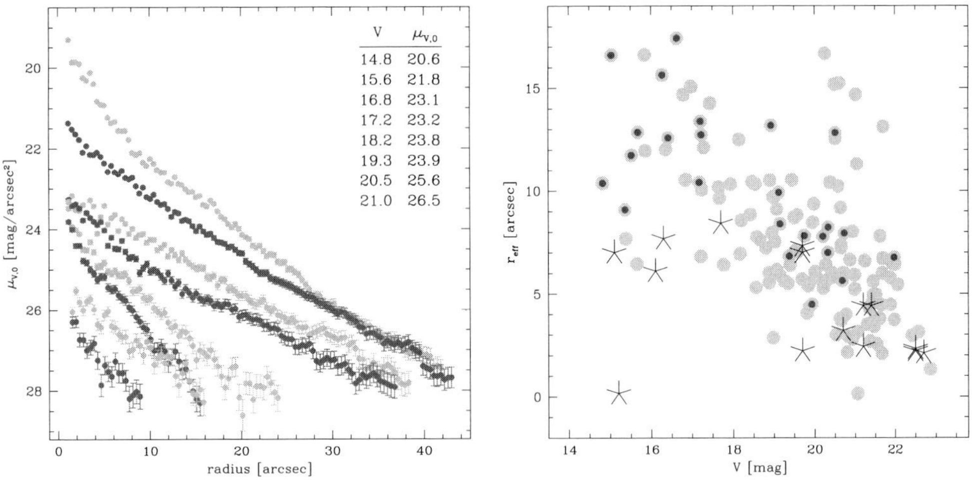

Figure 1. Left: Surface brightness profiles of several dEs and dSphs in the Fornax cluster. The total apparent V magnitudes and central surface brightnesses (exponential law) are indicated. **Right:** Scale lengths of exponential fits to the outer parts of the surface brightness profiles for dSphs/dEs in Fornax (grey circles) and the Local Group (asterisks) versus V magnitude.

Chile. The pixel scale of 0.77 arcsec/pixel favours the detection of very low surface brightness galaxies. The photometric parameters of the dEs/dSphs have been derived from the analysis of their surface brightness profiles (see Fig. 1): the total magnitude by a curve of growth analysis, the color within an aperture of 8 arcsec diameter, and the central surface brightness from an exponential fit to the outer part of the profile.

First results of our study were presented in Hilker *et al.* (2003). About 70 new dSph candidates have been discovered beyond the limits of the Fornax Cluster Catalog (FCC, Ferguson 1989), the faintest one with an absolute magnitude of $M_V \simeq -8.8$ mag and a central surface brightness of $\mu_V \simeq 27$ mag/arcsec^2. The dSphs follow similar magnitude-surface brightness and also colour-magnitude relations as their counterparts in the Local Group (see Figs. 1 and 2 in Hilker *et al.* 2003). The faint-end slope of the luminosity function of the dSphs is flat ($\alpha \simeq -1.1 \pm 0.1$, see Fig. 3 in Hilker *et al.* 2003).

The only difference between the Local Group dSphs and those in the Fornax cluster are their sizes. Fig. 2 shows that at a given total magnitude, the dEs/dSphs in Fornax are on average larger than their Local Group counterparts (data from Grebel *et al.* 2003). This is also reflected in the magnitude-surface brightness diagram (Fig. 3) where the Local Group dwarfs have on average a higher central surface brightness than the Fornax dSphs. This finding might be explained by environmental effects, i.e. the strong tidal forces in a cluster that lead to more extended, lower surface brightness dwarf galaxies, and finally to a destruction of dSphs close to the cluster center (e.g. Hilker *et al.* 1999).

3. SBF measurements in the Fornax Deep Field

The WFCCD study has shown that the combination of deep multi-color photometry in a wide field with a sufficient resolution is crucial in order to identify dwarf spheroidal candidates in nearby clusters. However, one still can not be sure that all dSph candiates are real cluster members, as long as one has not measured their distances.

Therefore, to improve the membership assignment, we have re-imaged the central Fornax cluster with substantially better spatial resolution (0.2 arcsec/pixel) using IMACS and Magellan at LCO. From the first part of our survey, we directly determined the

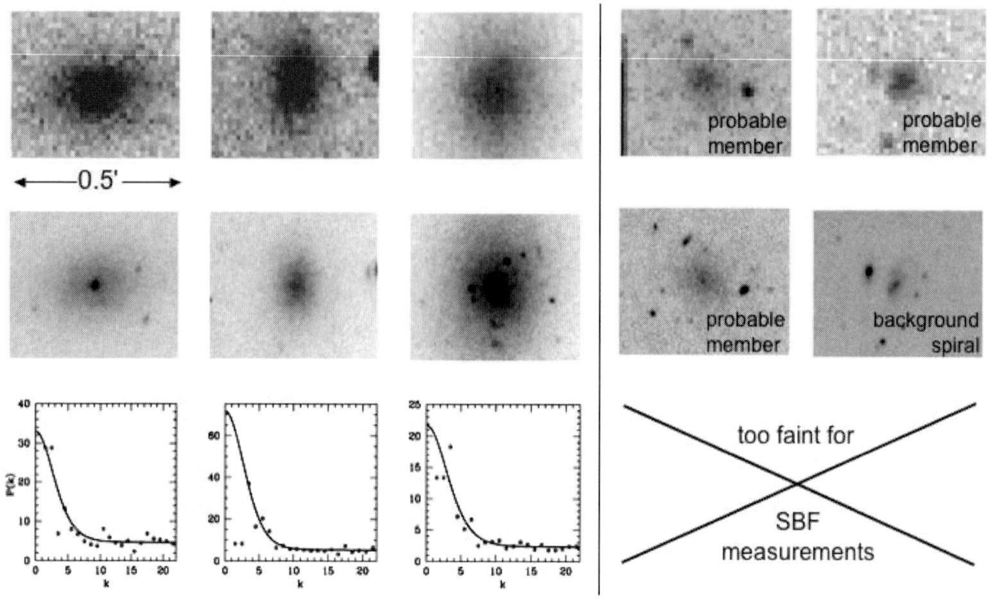

Figure 2. Left: three examples for SBF measurements in dEs. The top and middle row show thumbnail images galaxies from the WFCCD and IMACS data sets, respectively. In the bottom row the azimuthally averaged power spectra of the SBF measurement and the corresponding fits are plotted. **Right:** four galaxies are shown that are too faint or small for SBF measurements.

cluster membership for 10 previously unconfirmed candidate dEs in the magnitude range $-14.2 < M_V < -11$ mag using the SBF method (Fig. 2, see Mieske *et al.* 2003 for details of this method). This extends the magnitude range of confirmed cluster members far into the regime where the faint end slope α dominates the galaxy luminosity function.

Furthermore, we improved the morphological cluster membership assignments for fainter galaxies ($M_V < -10$ mag). For the vast majority of dE candidates we confirm the probable cluster membership, such that α changes by less than 0.02 compared to the previously determined value. Only very few dSph candidates turn out te be likely background galaxies due to their better resolved morphology (see Fig. 2). We find two new dSph candidates from our IMACS imaging (see Fig. 3). Including them does not change α, either. This further confirms the strong discrepancy between the number of low mass dark matter halos expected in a ΛCDM universe and the number of low luminosity galaxies.

The new membership determinations from the first part of the "Fornax Deep Field" survey are summarised in the magnitude-surface brightness plot (see Fig. 3). Except for a few barely resolved galaxies (close to the dashed line), the morphological membership classification seems to work very reliable.

4. Dwarf spheroidals in the Hydra I and Centaurus cluster

The central parts of the Hydra I and Centaurus cluster have been imaged through Johnson VI with FORS1 at the VLT in dark time and under excellent seeing conditions (for first results, see Mieske & Hilker 2003). A first visual inspection of the images revealed a wealth of faint, spatially resolved dSph candidates. The surface brightness profiles of a small random sample has been measured. Their structural and photometric parameters are such that they fall on top of the colour-magnitude and magnitude-surface brightness relation defined by the Fornax dSphs. The publication of these results is in preparation.

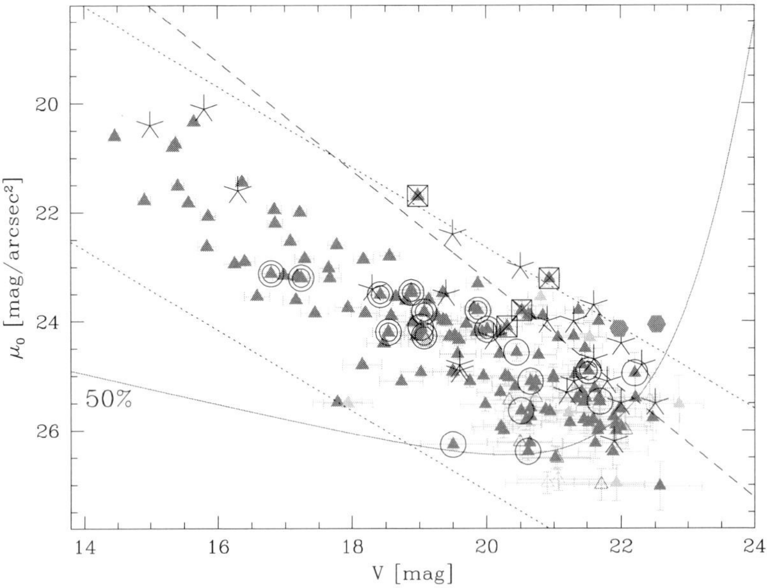

Figure 3. Magnitude-surface brightness diagram of dEs and dSphs in the Fornax cluster (triangles) and the Local Group (asterisks). Light grey triangles mark galaxies that lie outside 3σ of the colour-magnitude relation, open triangles those galaxies that were identified only by eye. Double circles indicate cluster membership confirmed with SBF, single circles probable membership due to morphology. Probable background galaxies (due to morphology) are crossed out. The two filled hexagons mark two additional cluster member candidates found from the IMACS imaging. The dashed line indicates a scale length of 2.5 arcsec for an exponential profile. The dotted lines are the 3σ deviations from the magnitude-surface brightness relation as defined by the Fornax dEs. The solid curve shows the completeness limit of 50%.

5. Conclusions

1. The counterparts of Local Group dSphs do exist in Fornax. They follow similar colour-magnitude and magnitude-surface brightness relations, except that the Fornax dSphs seem to be in average larger at a given total magnitude than the Local Group dSphs.
2. SBF-cluster memberships and improved morphological classifications do not change the shallow faint end slope of $\alpha \simeq -1.1$ derived from the WFCCD-data by more than 0.02, posing a challenge to ΛCDM theory.
3. Preliminary analysis of deep VLT images in Hydra I and Centaurus show that dSph candidates exist in these clusters, and that they follow similar scaling relations as their counterparts in Fornax. More on this in forthcoming publications.

References

Ferguson, H.C. 1989, *AJ* 98, 367
Grebel, E.K., Gallagher, J.S. & Harbeck, D. 2003, *AJ* 125, 1926
Hilker, M., Infante, L. & Richtler, T. 1999, *A&AS* 138, 55
Hilker M., Mieske S. & Infante L. 2003, *A&A* (Letters) 397, L9
Klypin, A., Kravtsov, A.V., Valenzuela, O. & Prada, F. 1999, *ApJ* 522, 82
Mieske, S. & Hilker, M. 2003, *A&A* 410, 445
Mieske, S., Hilker, M. & Infante, L. 2003, *A&A* 403, 43
Moore, B., Gighna, S., Governato, F., *et al.* 1999, *ApJ* (Letters) 524, L19
Pritchet, C.J. & van den Bergh, S. 1999, *AJ* 118, 883
Trentham, N. & Tully, R.B. 2002, *MNRAS* 335, 712

Discussion

Moore: The luminosity function you find is similar to the Local Group, suggesting a comparable "missing satellite problem". In a scenario where reionisation cuts off small galaxy formation you would expect more dwarf galaxies – comments?

Hilker: Right, if there originally existed a large number of small galaxies in Fornax, the nowadays flat luminosity function of dEs/dSphs requires evolutionary processes that reduced the number of small galaxies in the cluster center, like for example tidal disruption. The decline in number counts of dEs in the very center of the cluster might be a hint for this. However, maybe the reionisation scenario can be adjusted in such a way that no/less small galaxies have formed (and the nowadays existing dSphs have a different origin?).

Near–Field Cosmology with Dwarf Elliptical Galaxies
Proceedings IAU Colloquium No. 198, 2005
H.Jerjen and B. Binggeli, eds.

doi:10.1017/S1743921305003959

Dwarf spheroidal galaxies in nearby groups imaged with HST

I.D. Karachentsev[1], V.E. Karachentseva[2] and M.E. Sharina[1]

[1]Special Astrophysical Observatory, Russian Academy of Sciences, N.Arkhyz, Russia
email:ikar@sao.ru

[2]Astronomical Observatory of Kiev University, Kiev, Ukraine

Abstract. An all-sky list of 88 nearby dwarf spheroidal (dSph) galaxies with distances $D < 10$ Mpc is considered. Most of the objects have recently been found by Karachentseva & Karachentsev based on POSSII/ESO-SERC survey. A hundred more dSph galaxies are expected in this volume, being missed so far because of their low luminosity and low surface brightness. Apart from 22 dSph members of the Local Group, there are 33 dSphs in other nearby groups that have been resolved recently into stars with HST. Only 5% of the local dSphs are situated outside the known groups.

We discuss observational correlation between basic parameters of the dwarf spheroidal galaxies, in particular, absolute magnitude, surface brightness, metallicity, and so-called "tidal index". The observed number of dSphs in group increases with luminosity of its brightest galaxy. In a "synthetic" nearby group, dwarf spheroidal galaxies are distributed in depth quite symmetrically about the principal galaxy, having an rms distance scatter of 200 kpc. Projected radial distribution of dSphs in the synthetic group follows the profile $N(R) \sim \exp(-R/200 \text{ kpc})$.

Keywords. dwarf, fundamental parameters

1. Introduction

According to the entire sky catalog of 1500 low surface brightness dwarf galaxies with angular diameters larger than $0.4'$ (Karachentseva & Sharina 1988), dwarf spheroidal galaxies (dSphs) are strongly concentrated towards the nearest groups and clusters. Distributions of dSphs in Virgo and Fornax cluster were studied in details by Binggeli *et al.* 1987 and Ferguson & Sandage 1989. A comprehensive review of basic properties of dwarf spheroidal galaxies in the Local Group was outlined by Mateo 1998. Recent mass measurements of accurate distances to nearby galaxies summarized in Catalog of 451 neighboring galaxies (Karachentsev *et al.* 2004) allow us to study distribution and properties of dSphs in the Local ($D < 10$ Mpc) Volume.

2. DSph population within 10 Mpc

Among 451 galaxies of the Local Volume (LV), 88 objects are classified as dSphs because of their smooth shape, low surface brightness, absence of H_α emission and undetectably low HI flux. Nine of them (LGS-3, Phoenix, KK 27, KDG 61, Antlia, HS 117, DDO 113, KKR 25, and DDO 216) can be classified rather as dwarf galaxies of transition (dSph/dIr) type since the presence of a fraction of young stars or noticable HI content. Among 66 dSphs in the LV situated outside the Local Group, 50% of the sample have being imaged so far with WFPC2 or ACS at Hubble Space Telescope. No nucleated galaxies have been found as yet.

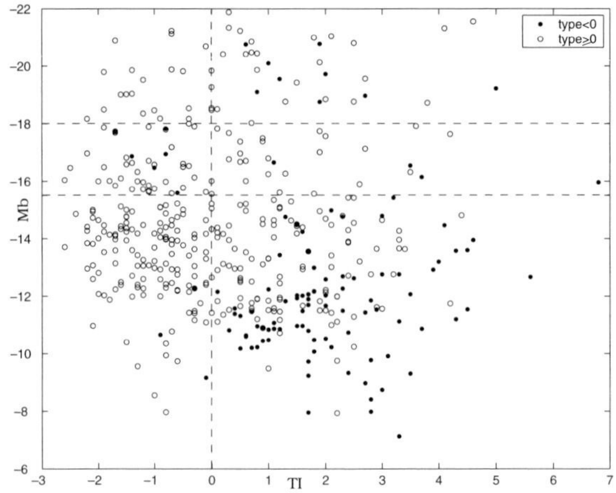

Figure 1. Absolute blue magnitude vs. tidal index for the LV galaxies

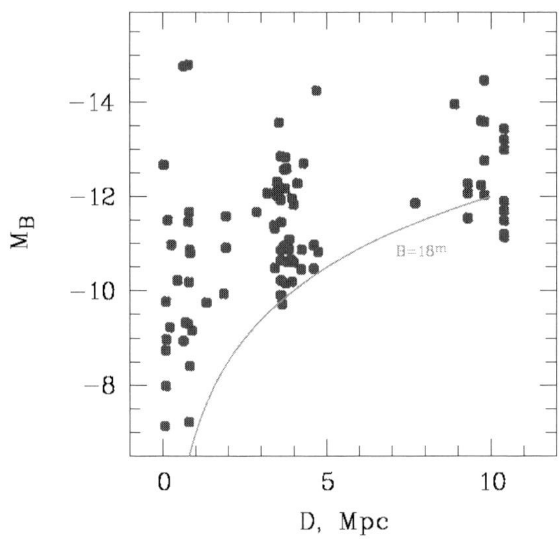

Figure 2. Absolute magnitude vs. distance for dSph galaxies

To quantify the environment of a LV galaxy, the so-called "tidal index" (or isolateness index) was used: $TI_i = \max[\log(M_k/D^3_{ik}] + C, \quad k = 1, 2, ...$ where M_k is the mass of a neighboring galaxy located at a distance D_{ik} from the one considered; a constant C is choosen so that the negative values of TI correspond to isolated galaxies of the general field, while the positive values to group members. Fig. 1 shows the distribution of 451 LV galaxies in blue absolute magnitude and tidal index. The filled circles indicate early-type objects ($T < 0$). Among the high-luminosity galaxies with $M_B < -18^m$, all E and SO galaxies are within the range of positive TI, i.e. in groups. On the bottom side, among dwarf galaxies with $M_B > -15.5^m$, the overwhelming majority of early- type (dSph) objects are in the dense environment ($TI > 0$) too.

Fig. 2 presents the distribution of dSph galaxies according to their distances and absolute blue magnitudes corrected for the Galactic extinction. The line corresponds to a limiting apparent magnitude 18^m. Lack of the faintest galaxies towards the LV edge is clearly seen. Assuming our sample within 2 Mpc to be 90% complete, we derive about 100 dSphs missed within 10 Mpc because of their faintness.

The pair-wise distributions of the LV dSphs according to their absolute magnitude M_B, mean blue surface brightness Σ_B, apparent axial ratio b/a, and tidal index TI are given in Fig. 3. Objects of the transition dSph/dIr type are indicated by triangles, and members of the Local Group are shown by filled symbols. As seen from these data, dwarf spheroidal galaxies outside the LG occupy about the same range of the considered parameters as the LG members themselves. There is no significant difference between galaxies of the dSph/dIr type and pure dSph objects either.

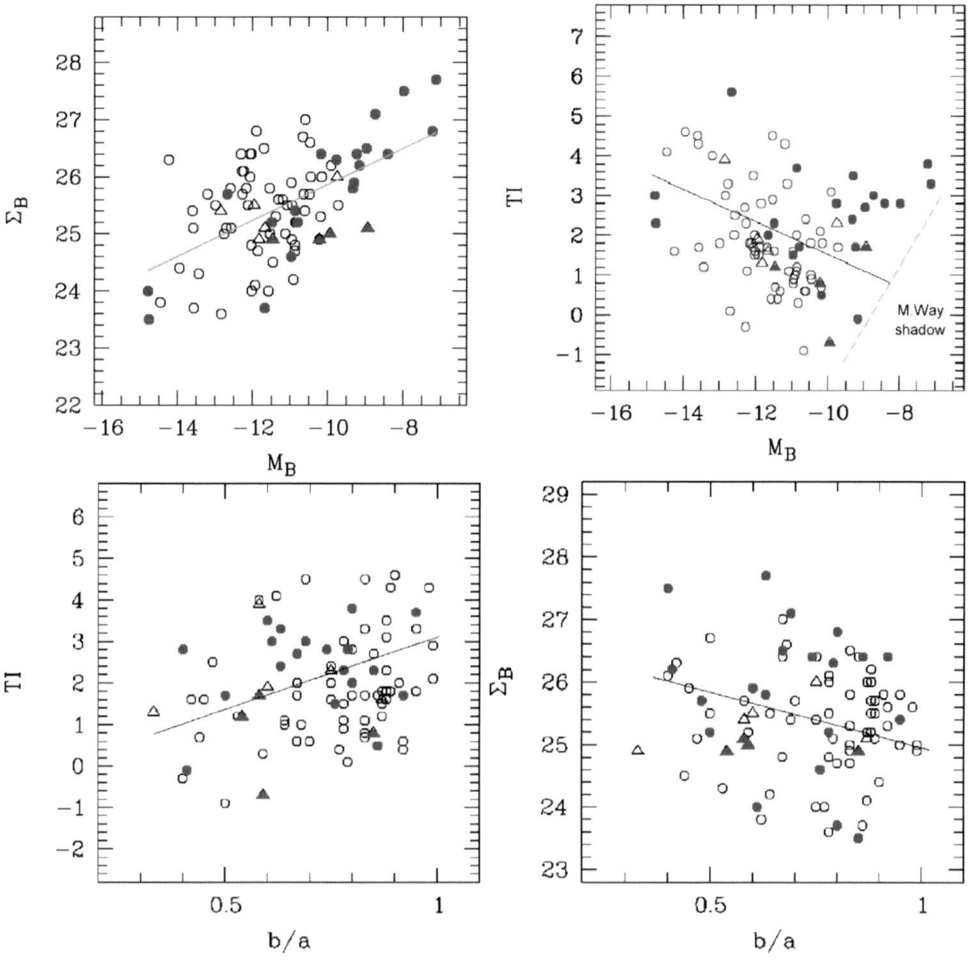

Figure 3. Correlations between absolute magnitude, average surface brightness, axial ratio and tidal index for dSphs. Galaxies of transition type are shown by triangles and members of the LG are indicated by filled symbols

3. Dwarf spheroidal galaxies imaged with ACS HST

Over the past few years, 33 dSph galaxies outside the LG have been observed with WFPC2 and ACS at HST. This number exceeds the total number of dSphs in the Local Group itself (N = 22). 600-second exposures were taken in the F606W and F814W filters for each object. The photometric reduction was carried out using the HSTphot stellar photometry package described by Dolphin 2000. The ACS I- images of the dSph galaxy

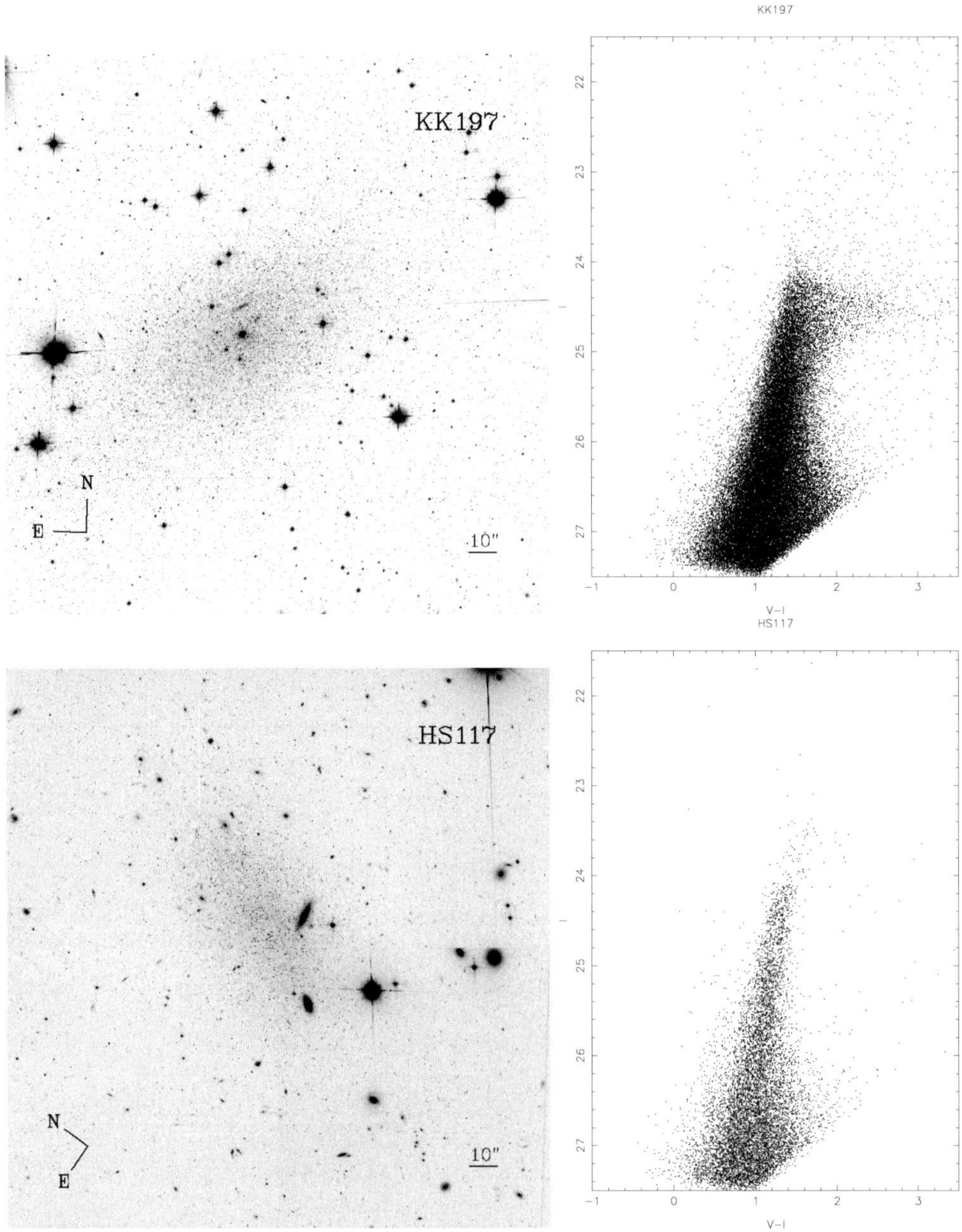

Figure 4. I-band ACS images and CM diagrams for dSph galaxy KK 197 and dSph/dIr galaxy HS 117

KK 197 in the Centaurus A group and the dSph/dIr galaxy HS117 in the M81 group are shown on left panel of Fig. 4. Their $(V-I)$ color-magnitude diagrams (right panel of Fig. 4) exhibit a lot of red giant branch (RGB) stars. Based on the tip of RGB, the galaxy distances are 3.73 Mpc and 3.93 Mpc, respectively, confirming their membership in corresponding groups. The galaxy KK 197 has a blue absolute magnitude of −12.83 and a 50 kpc projected separation from the giant galaxy Cen A. The galaxy HS 117 has a lower blue absolute magnitude −11.96 and a projected separation of 190 kpc from the

M81. The superior quality of ACS led to detection of about 44 000 stars in KK 197 and about 9 000 stars in HS 117.

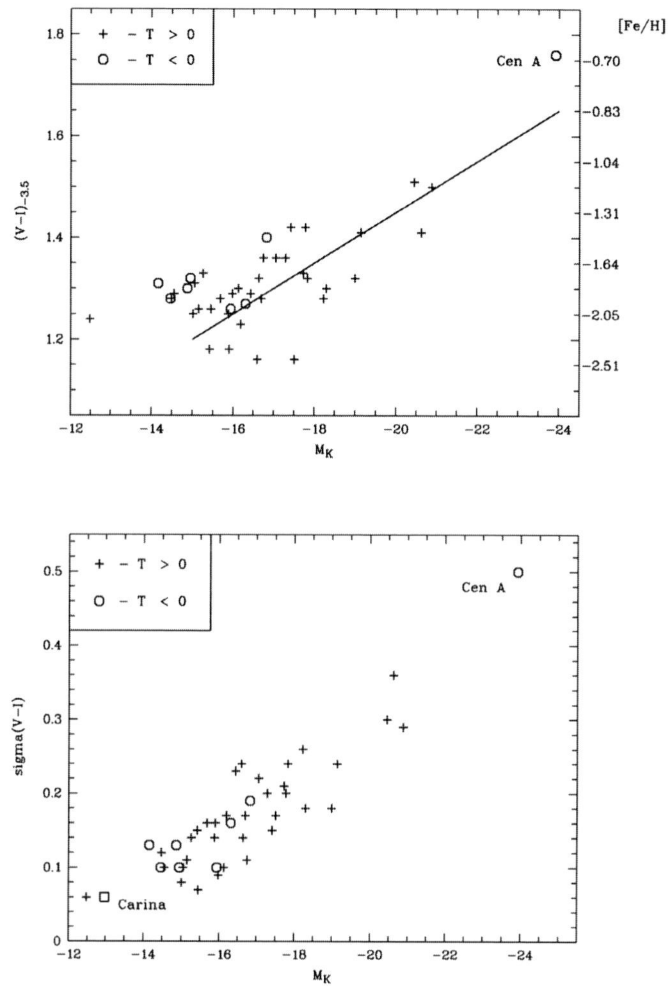

Figure 5. RGB color (upper panel) and dispersion of RGB colors vs K-band absolute magnitude for dSph (circles) and dIr (crosses) galaxies

4. RGB color – luminosity relation

As it was found by Da Costa & Armandroff 1990 and Lee *et al.* 1993, the reddening-corrected $(V-I)$ color of the RGB at $M_I = -3.5$ is a monotonic sequence with respect to galaxy metallicity [Fe/H]. According to Grebel *et al.* 2003, the metallicity of dwarf galaxies in the LG correlates with their luminosity, and dSphs are less luminous at equal metallicity than dIrs. We measured $(V-I)_{-3.5}$ colors for 42 nearby galaxies observed with ACS and compared them with the absolute magnitude of the galaxies in the K-band derived from 2MASS (Karachentsev & Kut'kin 2005). A plot of the mean $(V-I)_{-3.5}$ color vs. absolute K-magnitude is shown on upper panel of Fig. 5. Late-type and early-type galaxies are indicated by crosses and open circles, respectively. We put on the plot also the giant elliptical galaxy Cen A taking its metallicity from Rejkuba 2004. Here, the right scale shows the metallicity conversed from $(V-I)$ color via the relation by Lee *et al.* 1993. The known color — luminosity relation, as well as the offset of dSphs with respect to dIrs is seen distinctly.

Apart from the mean RGB color, the scatter of RGB colors measured at the same $M_I = -3.5$ level exhibits a correlation with K-luminosity of the galaxies too (lower panel of Fig. 5). The existence of such a kind of relation can be suspected from the data on the LG galaxies collected by Mateo 1998. The LG galaxy Carina dSph having a dispersion of [Fe/H] = 0.25 dex (Koch *et al.* 2004) is indicated in Fig. 5 by open square. Surprisingly, we did not recognize any significant difference between dSphs and dIrs on the diagram.

5. DSphs in a synthetic nearby group

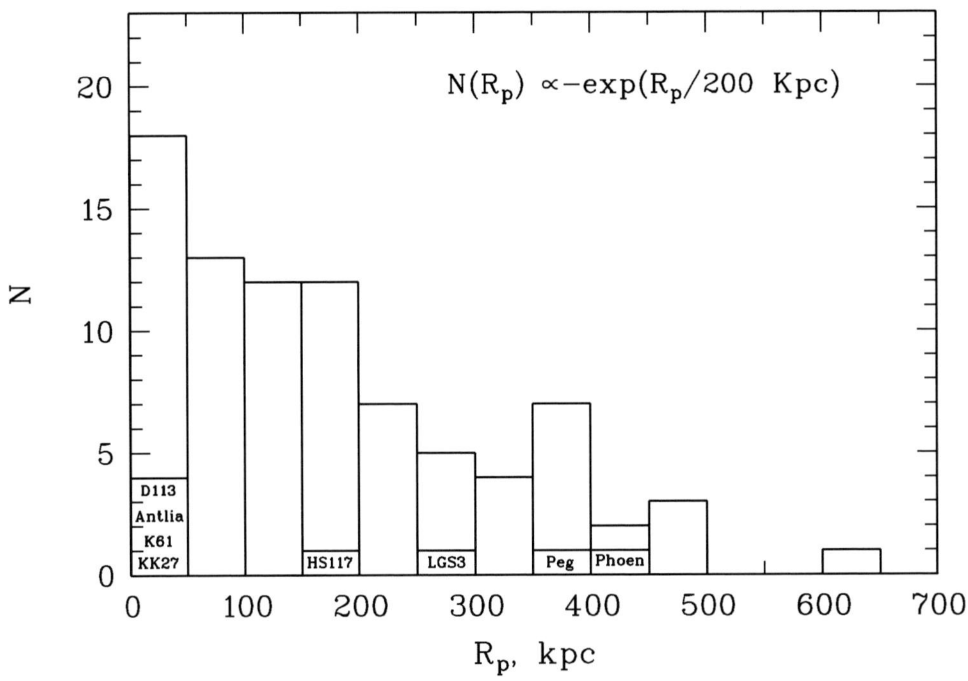

Figure 6. Number distribution of dSphs according to their projected distance from the brightest galaxy in nearby groups

Comparing numbers of dSph and dIr galaxies in nearby groups, Karachentsev 2005 noted that the fraction of dSphs tends to increase with luminosity of the bulge of the main group member and to decrease with crossing time of the group, which has a quite obvious evolutionary interpretation. Table 1 presents parameters of the brightest galaxy in nearby groups and numbers of dSph companions. Its columns give the following data: (1) name of the galaxies ranked according to their infrared luminosity, (2) morphological type of the brightest galaxy in each group, (3) mean distance to the group, (4) absolute K- magnitude of the main galaxy, (5) tidal index for the brightest galaxy (showing presence of another bright enough galaxy in the group), (6) total number of the known

Table 1. Parameters of the brightest galaxy in a group and number of dSphs

Main galaxy	T_1	D Mpc	$(M_K)_1$	$(TI)_1$	N_{tot}	N_{sph}	N^c_{sph}	$< R_p >_{sph}$ kpc	$< R_p >_{Ir}$ kpc
Sombrero	1	9.33	−24.91	0.3	6	4	4	102	336
NGC 253	5	3.94	−24.22	0.3	6	3	2	414	512
NGC 3115	−3	9.68	−24.06	1.9	6	2	2	146	255
M 81	3	3.63	−24.00	2.2	28	12	9	151	256
CentaurusA	−2	3.66	−23.92	0.6	24	14	4	254	329
M 101	6	7.38	−23.83	0.6	5	0	0	–	286
NGC 3368	3	10.4	−23.78	0.6	21	6	6	191	279
NGC 2784	−2	9.84	−23.72	2.0	5	4	4	62	223
NGC 2903	4	8.9	−23.72	1.8	2	1	1	24	366
NGC 5236	5	4.47	−23.66	0.8	13	4	1	220	139
M 31	3	0.77	−23.47	4.6	17	11	3	176	394
NGC 4945	6	3.6	−23.36	0.7	3	2	2	145	305
NGC 4736	2	4.66	−23.24	−0.5	5	1	0	448	360
Milky Way	4	0.01	−23.16	2.5	14	10	2	148	326
NGC 2683	3	7.73	−23.12	0.2	2	1	1	78	29
NGC 3489	1	10.4	−22.72	1.5	6	1	1	25	258
NGC 3412	−1	10.4	−22.43	1.9	8	2	2	70	150
NGC 2403	6	3.30	−21.42	−0.0	2	1	1	77	212
NGC 1313	7	4.27	−20.62	−1.6	1	1	1	24	-
NGC 55	8	2.12	−20.38	−0.4	2	2	1	165	-
NGC 4214	10	2.94	−19.44	−0.7	1	1	1	9	-
NGC 3109	10	1.32	−16.73	2.3	1	1	0	28	-

companions, (7) total number of dSph companions, (8) number of dSph companions brighter than $M_B = -11^m$ (this quantity reduces the incompleteness of distant groups with respect to nearby ones seen in Fig. 2), (9, 10) mean projected separations of the dSphs and of late-type companions from the principal galaxy.

These data show that 84 of 88 LV dSphs are situated in the known groups. Among the rest four field objects, two (KKs3, KK 258) need to be still confirmed as true dwarf speroidal galaxies. The highest numbers of dSphs are seen around the most luminous giant galaxies of types $T < 4$. The presence of another bright galaxy in group favours higher number of dSphs companions too (the case of M 81 and M 82). Note that the giant, but bulgeless galaxy M 101, which dominates in its luminosity over the rest of the group members, has no dSph companions at all. Comparing the mean projected separation of dSphs, 174 kpc, with the mean separation of the late-type companions, 276 kpc, we find that in the nearby groups dSphs occupy a volume, which is only a quater of the volume populated by the late-type members.

The distribution of dSphs according to their projected separations from the brightest group galaxy is presented in Fig. 6. Eight transition objects are indicated by their abbreviations. The observed distribution can be fitted by an exponent $N(R_p) \sim exp(-R_p/200$ kpc). No significant offset of dSph/dIr galaxies is seen with respect to the pure dSphs on this diagram. We also considered the distribution of distances of 39 dSph companions along the line of sight with respect to the principal galaxy in nearby groups. The distribution looks to be quite symmetric showing the mean difference $< D(dSph) - D_1 >= +24 \pm 33$ kpc and dispersion 202 kpc, which is roughly consistent with their distribution according to projected linear distances.

6. Brief conclusions

We presented arguments that basic properties of dwarf spheroidal galaxies situated in nearby groups are quite similar to parameters of the Local Group dSphs that have been studied in much more details.

There are 88 known dSphs (and dSph/dIrs) within 10 Mpc, and a hundred dSphs are still missing in the same volume. Almost all dSphs in the LV (~95%) occur in galaxy groups. No "nucleated" galaxies fainter than $M_B = -15.5^m$ were found in the LV. This threshold corresponds (accidentally?!) to the "specific frequency" =1 for globular clusters.

Distributions of the LV dSphs according to their absolute magnitude, surface brightness, apparent axial ratio and tidal index show that a) on the $\{\Sigma_B$ vs. $M_B\}$- diagram dSphs follow the line of constant spatial brightness, $\Sigma_B \sim (1/3) \times M_B$, b) the brighter dSphs, the higher their tidal index is (survival effect?), c) round dSphs tend to be closer to giant galaxies, d) elongated dSphs tend to have lower Σ_B (inclination effect?).

The RGB color – luminosity relation, known for the LG dSphs, is also valid for the dSphs in other nearby groups. The LV dSphs are found to be $\sim 1^m$ fainter than dIrs of the same metallicity, like dSphs in the Local Group. Surprisingly, the dispersion of metallicity (defined as the dispersion of RGB colors) decreases strongly (8 times!) from giant galaxies towards dwarfs.

The distribution of dSphs in nearby groups with respect to the principal galaxy has a characteristic scale of 200 kpc. The number of dSphs in group increases with bulge luminosity of the brightest group member. In the nearby groups, Sphs occupy only a quater of the volume populated by the late-type members of the groups.

Acknowledgements

This work was supported by RFFI grant 04–02–16115 and DFG-RFBR grant 02–02–04012. This search has made use of the 2MASS survey.

References

Binggeli B., Tammann G.A. & Sandage A. 1987, *AJ* 94, 251
Da Costa G.S. & Armandroff T.E. 1990, *AJ* 100 162
Dolphin A.E. 2000, *ApJ* 550, 554
Ferguson H.C. & Sandage A. 1989, *ApJ* 346, 53
Karachentsev I.D. 2005, *AJ* 129, 178
Karachentsev I.D., Karachentseva V.E., Huchtmeier W.K. & Makarov D.I. 2004 *AJ* 127, 2031
Karachentsev I.D. & Kut'kin A.M 2005 *Astronomy Letters* 31, 299
Karachentseva V.E. & Sharina M.E. 1988, *Comm. Special Astrophys. Obs.* 57, 5
Koch A., Grebel E.K., Harbeck D., *et al.* 2004 *astro-ph/0410435*
Lee M.G., Freedman W.L. & Madore B.F. 1993, *ApJ* 417 553
Mateo M. 1998,*ARAA*. 36, 435
Rejkuba M. 2004, *astro-ph/0407636*

Discussion

SHARINA: Can one make conclusions about metallicities of RGB stars without reliable transformations from the ACS instrumental to the standard Johnson-Cousins system available? The transformations are nonlinear and strongly depend on colour.

KARACHENTSEV: We made such conclusions using uniform data for all galaxies. Since there are no reliable transformations at the moment published, these conclusions are very preliminary.

Near-Field Cosmology with Dwarf Elliptical Galaxies
Proceedings IAU Colloquium No. 198, 2005
H. Jerjen and B. Binggeli, eds.

doi:10.1017/S1743921305003960

Structural properties of dwarf ellipticals and the connection with (ordinary) elliptical galaxies

Alister W. Graham

Mount Stromlo and Siding Spring Observatories, Australian National University, Private Bag, Weston Creek PO, ACT 2611, Australia.
email: Graham@mso.anu.edu.au

Abstract. This article reviews the popular reasons for the belief that dwarf elliptical galaxies and (ordinary) elliptical galaxies are distinct and separate species. They include: light–profile shape (or similarly image concentration); the magnitude–central surface brightness diagram; the magnitude–effective surface brightness diagram (or similarly the magnitude–effective radius diagram); and the Fundamental Plane. It is shown how a continuous trend between luminosity and a) light–profile shape, and b) central surface brightness (until the onest of core formation at $M_B \sim -20.5$ mag), results in a unification of the dwarf elliptical and (ordinary) elliptical galaxies. Neither the above four reasons, nor the luminosity function (at least in the Virgo cluster) provide evidence for a division at $M_B = -18$ mag between the dwarf elliptical and (ordinary) elliptical galaxies. Instead, they appear to be continuous extensions of each other.

Keywords. galaxies: dwarf, galaxies: elliptical and lenticular, cD, galaxies: formation, galaxies: fundamental parameters, galaxies: luminosity function, galaxies: structure

1. Introduction

For at least the past two decades, astronomers have believed that dwarf elliptical (dE) galaxies (not to be confused with dwarf spheroidal galaxies, which typically have absolute magnitudes fainter than $M_B = -13$ mag) are distinct objects from the more luminous (ordinary) elliptical (E) galaxies. The divide between these allegedly different classes is supposed to occur at an absolute magnitude of $M_B = -18$ mag (e.g., Kormendy 1985).

In this talk I will review the most oftenly quoted reasons why the dEs are thought to be different from the Es, and then go on to explain why I think these past claims for a discontinuity at $M_B = -18$ mag should be reconsidered. I will argue that a continuity in structural properties exists across this alleged boundary.

Much, but not all, of this talk is based on Sections 1 and 4 from Graham & Guzmán (2003), where further details may be found.

2. Past evidence of continuity

Before addressing the issue of galaxy structure, which is where past claims of discontinuity have their roots, it is relevant to review various other physical aspects of the dE and E population. Several physical quantities vary smoothly across the alleged division at −18 B–mag.

The average globular cluster metallicity is known to vary continuously, as a function of host galaxy magnitude, across the alleged dE/E divide (e.g., Forbes *et al.* 1996, their figure 14). Côté (2005, these proceedings) has shown how this arises from a systematic variation in the relative numbers of red and blue globular clusters, as opposed to a single population changing its mean metallicity.

Furthermore, the host galaxies themselves follow a luminosity–metallicity relation that shows no sign of a division or break at $M_B = -18$ mag (e.g., Caldwell & Bothun 1987). Related to this, the colour–magnitude relation shows no sign of a division at $M_B = -18$ mag (e.g., Caldwell 1983). Now, on average, dEs do have a different metallicity from the brighter Es, but the key point is that there is a continuous change in chemical composition between the two. If one was to measure the average height and weight of a sample of adults, and then do the same for a sample of children, one would of course conclude that these populations have different physical properties. But this obviously tells us little as to whether or not children are the same specie as adults.

Aside from the stellar populations not knowing about the supposed dE/E transition, there is also no dynamical evidence. Including dozens of dEs, Matković & Guzmán (2005, and also these proceedings) and De Rijcke *et al.* (2005, see also these proceedings) have both shown that the luminosity–velocity dispersion ($L - \sigma$) relation continues linearly across the alleged boundary between the dEs and Es. Interestingly, both studies found that the slope of this relation is $\sim$2, jumping to $\sim$4 and thus recovering the Faber–Jackson (1976) relation at magnitudes greater than ~ -20.5 B–mag.

What then, are the popular reasons which have led people to believe that the dEs form a uniquely different class of object to the Es?

3. Past evidence of discontinuity

This section lists the three most often quoted reasons why dwarf and giant elliptical galaxies are thought to be different species of object. The reasons pertain to galaxy structure. In the light of new data, however, the following section will explain why these three arguments are no longer valid. The following section will also counter two further claims of support for a dE/E discontinuity at $M_B = -18$ mag.

(i) Dwarf elliptical galaxies have exponential light–profiles (Faber & Lin 1983; Binggeli, Sandage & Tarenghi 1984), while elliptical galaxies have $R^{1/4}$ profiles (de Vaucouleurs 1948, 1959; de Vaucouleurs & Capaccioli 1979).

(ii) Dwarf elliptical galaxies define a distribution that is almost at right angles to that traced by the brighter elliptical galaxies in the diagram of central surface brightness, μ_0, versus absolute magnitude, M, (Kormendy 1985, his figure 3). Such behaviour has been taken as evidence that different formation processes must have shaped the dEs and the Es. The divide is said to occur at an absolute magnitude of -18 B–mag.

(iii) In diagrams which plot a galaxy's effective surface brightness, $\mu_{\rm e}$, dwarf elliptical galaxies follow a distribution that is clearly offset to that defined by the brighter elliptical galaxies. This is the case for the $M - \mu_{\rm e}$ and $R_{\rm e} - \mu_{\rm e}$ diagram (Capaccioli, Caon & D'Onofrio 1992, their figure 4; Binggeli & Ferguson 1994; their section 2.2.2), and also for the Fundamental Plane (Djorgovski & Davis 1987) when constructed using effective surface brightnesses (e.g., Geha, Guhathakurta & van der Marel 2002).

4. A unified picture

In the Author's opinion, there have been two key advances since the above three observations were established. The first has come from a closer scrutiny of optical light–profiles. Studies such as those by Schombert (1986) revealed that the $R^{1/4}$–law was not as universal as previously thought. Similarly, analyses by Caldwell & Bothun (1987) and Binggeli & Cameron (1991) revealed that the dEs have magnitude–dependent departures

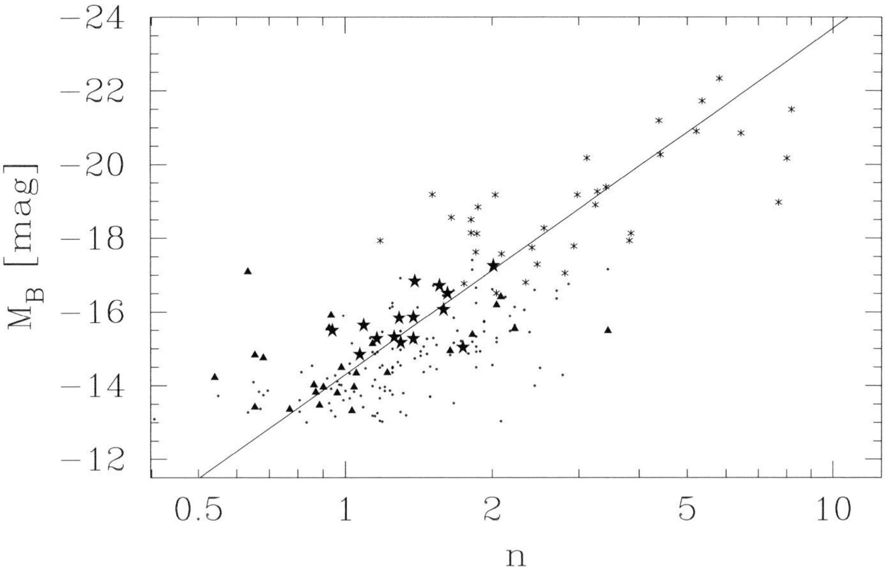

Figure 1. Absolute B–band galaxy magnitude versus the logarithm of the Sérsic shape index n. Due to biasing from the magnitude cutoff at $M_B \sim -13$ mag, the line $M_B = -9.4\log(n) - 14.3$ has been estimated by eye. Dots represent dE galaxies from Binggeli & Jerjen (1998), triangles are dE galaxies from Stiavelli *et al.* (2001), large stars are dE galaxies from Graham & Guzmán (2003), asterix are intermediate to bright E galaxies from Caon *et al.* (1993) and D'Onofrio *et al.* (1994). Lenticular galaxies have been excluded.

from the exponential model. The subsequent use of high–quality CCD imagery no doubt helped to refine this picture. Second, the superb resolution provided by the Hubble Space Telescope has allowed us to probe the centers of galaxies on scales previously washed out by atmospheric seeing. The impact of these two things on the above three points is given below.

(i) There is no question that de Vaucouleurs' $R^{1/4}$ model fits many elliptical galaxies remarkably well. It is a clever and useful empirical function. However, as noted in Djorgovski & Kormendy (1989), it fits best elliptical galaxies with $M_B \sim -21$ mag; brighter and fainter galaxies having different curvature than described by the $R^{1/4}$ model. Thanks to the pioneering work of Capaccioli and collaborators, most notably Caon, Capaccioli & D'Onofrio (1993), numerous studies have now shown that Sérsic's (1963, 1968) $R^{1/n}$ generalisation† accommodates for the real (Trujillo, Graham & Caon 2001; their section 2) and systematic changes in light–profile shape n with absolute magnitude (e.g., D'Onofrio, Capaccioli, & Caon 1994; Graham *et al.* 1996; Vennik *et al.* 1996). Similarly, the stellar distribution in dwarf elliptical galaxies is also a function of absolute magnitude (e.g., Davies *et al.* 1988; Cellone, Forte, & Geisler 1994; Vennik & Richter 1994; Young & Currie 1994, 1995; Karachentseva *et al.* 1996; Jerjen & Binggeli 1997). This is shown in Figure 1 using a heterogeneous sample of literature data. Presumably the actual level of scatter is less than seen here. One may alternatively plot concentration, rather than profile shape, as done in Graham, Trujillo & Caon (2001).

(ii) Using *HST* data from the Nuker team (Faber *et al.* 1997), the M–μ_0 relation is now much better defined than it was twenty years ago, especially at the bright end. Furthermore, the increased number of galaxies around the alleged transition magnitude of −18 B–mag has also improved our understanding of what is going on. What is immediately

†A useful compilation of various Sérsic expressions can be found in Graham & Driver (2005).

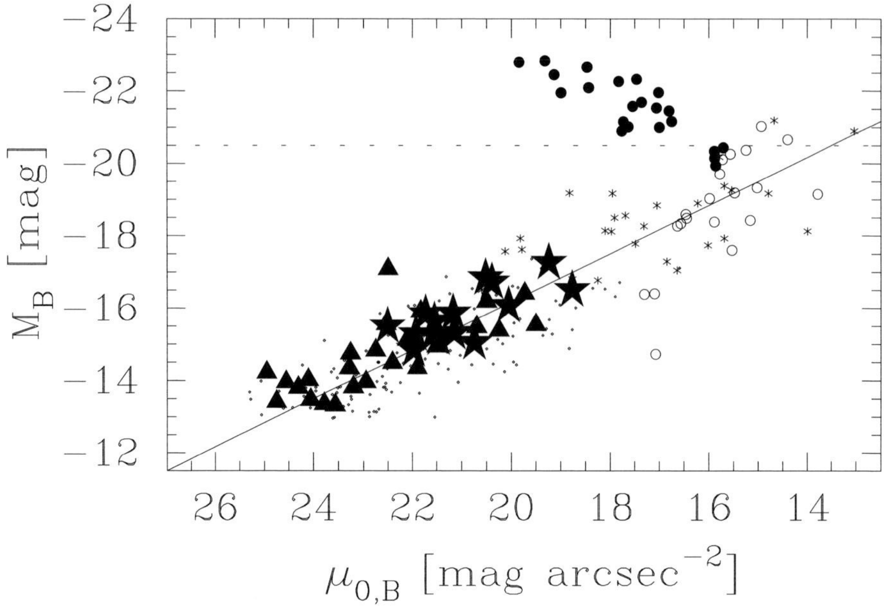

Figure 2. Galaxy magnitude versus (host galaxy) central surface brightness. Due to biasing from the magnitude cutoff at $M_B \sim -13$ mag, the line $M_B = (2/3)\mu_0 - 29.5$ has been estimated by eye rather than from a linear regression routine. The central surface brightness values obtained from Sérsic models fitted to luminous ($M_B \lesssim -20.5$ mag) E galaxies follow this same relation (Jerjen & Binggeli 1997), although they were not used to define it. The symbols are the same as in Figure 1, with the additional open circles representing the so–called "power–law" E galaxies from Faber *et al.* (1997), and the filled circles representing the "core" E galaxies from these same authors.

clear from Figure 2 is that there is no break at $M_B = -18$ mag. Instead, a linear trend exists across this supposed divide, and extends from $M_B = -13$ mag to $M_B \sim -20.5$ mag. This diagram does not imply that the dEs ($M_B > -18$ mag) experienced a different formation process to the Es.

The behaviour of the elliptical galaxies in Figure 2 is similar to that seen in the $L - \sigma$ diagram, in which galaxies more luminous than $M_B \sim -20.5$ mag depart from the linear distribution defined by the less luminous Es and dEs (see Matković & Guzmán 2005, these proceedings, their figure 1).

The deviant galaxies at the high–luminosity end in Figure 2 have all been identified by the Nuker team as "core" galaxies: their central light–profiles display a clear break and downward flattening relative to the inward extrapolation of their outer (Sérsic) stellar distribution (Graham *et al.* 2003; Trujillo *et al.* 2004). Such "core" galaxies may have formed from the dissipationless merger of two or more lesser Es, in which the progenitor galaxy's supermassive black holes have coalesced at the center of the new galaxy via the gravitational slingshot, and thus evacuation, of the central stars (Begelman, Blandford, & Rees 1980; Ebisuzaki, Makino, & Okumura 1991). This local, rather than global, physical process, reduces the central flux by ∼0.1% of the total galaxy flux (Graham 2004).

It is worth noting that the central surface brightness one would obtain from the inward extrapolation of the undisturbed outer light–profile of the "core" galaxies yields values that fall on the linear relation shown in Figure 2 (Jerjen & Binggeli 1997).

(iii) The difference between the central surface brightness, μ_0, and the effective surface brightness, μ_e (the surface brightness at the effective half–light radius R_e), is the same

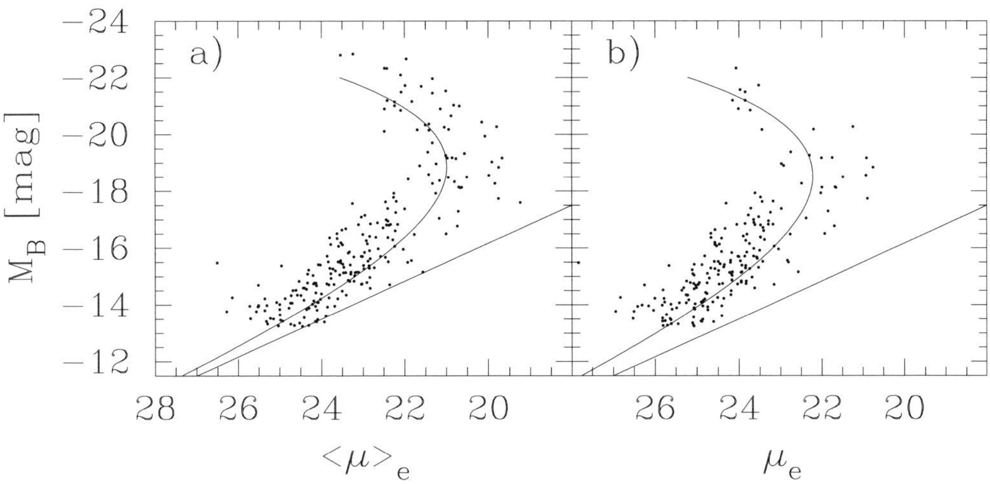

Figure 3. The curved lines show the expected behavior of galaxies in the (a) magnitude–mean effective surface brightness and (b) magnitude–effective surface brightness diagrams given the linear relations seen in Figures 1 and 2. The straight line shown here is from Figure 2.

for every $R^{1/4}$ profile. This difference, however, varies as a function of profile shape n, and is given by the expression $\mu_e = \mu_0 + 1.086b$, where $b \sim (2n - 1/3)$. So, as the galaxy magnitudes brighten in Figure 2, the value of n increases (Figure 1) and thus the value of $\mu_e - \mu_0$ increases. This is shown in Figure 3b. Related to this, the mean surface brightness within the effective radius, $\langle\mu\rangle_e$, can be derived from the expression $\langle\mu\rangle_e = \mu_e - 2.5\log[e^b n\Gamma(2n)/b^{2n}]$, where Γ is the gamma function (e.g., Graham & Driver 2005).

The different slopes for the dE and E galaxy distributions in Figure 3 is merely a consequence of a continuously varying profile shape with galaxy luminosity, it does not imply distinctly different galaxy formation processes for dEs and Es.

4.1. *Other issues*

(a) Kormendy (2004) has claimed that the decline in the dE galaxy luminosity function at bright magnitudes, and the associated decline in the E galaxy luminosity function at faint magnitudes is evidence that they are distinct populations. This argument of course falls down if one assigns a galaxy to the dE or E population based on its luminosity. Similarly, if one assigns an object to a class based on its appearance (i.e., using image concentration or light–profile shape), the situation remains unchanged (see Figure 1).

Figure 4 shows the luminosity function of the Virgo dE + E galaxies. There is obviously nothing special about $M_B = -18$ mag, and certainly no evidence for a deficit of galaxies at this magnitude, nor is it the overlap region of two distributions peaking at fainter and brighter magnitudes.

(b) From the beginning it has been known that galaxies with central velocity dispersions less than about 100 km s^{-1} do not follow the "Fundamental Plane" (Djorgovski & Davis 1987) as defined by luminous elliptical galaxies.

Using $L = 2\pi R_e^2 \langle I\rangle_e$, it can be shown that the same mechanism that causes the curved relations in Figure 3 is also behind the different slopes in the $\langle\mu\rangle_e$–$\log R_e$ and $M - \log R_e$ diagram for the dEs and Es. It therefore comes as no surprise that the dEs are not distributed on the Fundamental Plane. As was noted in Graham & Guzmán (2004), the Fundamental Plane is simply the tangent sheet to the high–luminosity end of a curved surface in the three–space of $\langle\mu\rangle_e, R_e$, and σ. Only bright Es that follow the

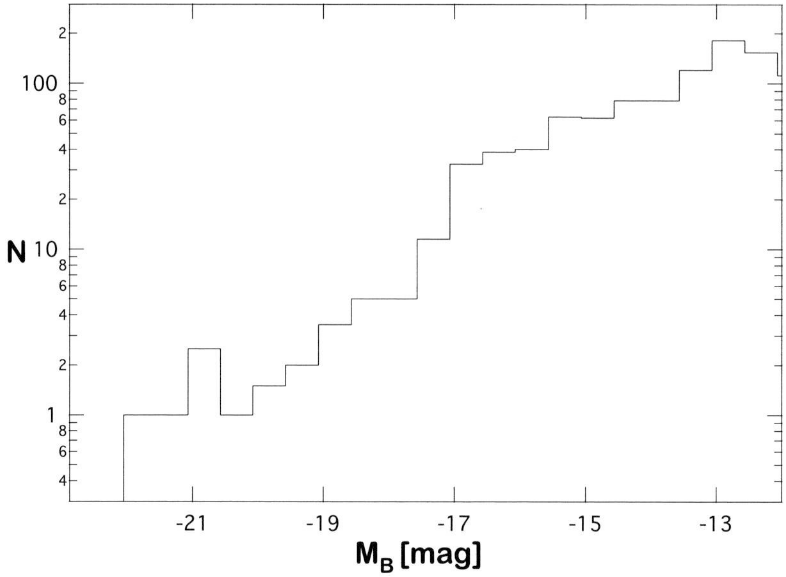

Figure 4. Virgo cluster dE + E galaxy luminosity function. There is no evidence for a division at $M_B = -18$ mag. Likely (50% chance) cluster members are weighted by 0.5. Figure courtesy of H. Jerjen.

Kormendy (1977) relation (see Capaccioli *et al.* 1992, their figure 4) will reside on typical constructions of the Fundamental Plane. The use of central surface brightness mitigates the apparent different behaviour of the dEs and Es (Guzmán *et al.* 2005, in prep).

5. Conclusions

Chemically and dynamically, the dEs appear to form a continuous extension with the Es. Both the colour–magnitude relation (but see Lisker *et al.*, these proceedings) and the L–σ relation show a continuous linear trend across the alleged boundary at -18 B–mag. Similarly, the dE + E galaxy luminosity function shows no evidence of a divide at this magnitude.

In terms of galaxy structure, dEs and Es display a continuous range of properties that vary smoothly across the alleged dE/E divide. The shape of their (projected) stellar distribution, quantified through Sérsic's index n, increases steadily with luminosity from values around 0.5 for the fainter dwarfs, to values greater than 4 for the more luminous ellipticals. The central (host galaxy) surface brightness brightens linearly (from $M_B \sim -13$ mag) as the host galaxy magnitude brightens. There is no discontinuity at $M_B = -18$ mag in the M–μ_0 diagram.

The above two relations account for the near–orthogonal distribution of dEs and Es in diagrams of effective surface brightness versus (i) magnitude and (ii) effective radius. They also explain why the dEs depart from the Es in the traditional construction of the Fundamental Plane, which uses effective surface brightness.

We report a clear break in the M–μ_0 relation at $M_B \sim -20.5$ mag, such that more luminous galaxies have fainter central surface brightnesses. Thanks to the resolution of *HST*, this departure of the brightest elliptical galaxies from the main M–μ_0 relation is understood to be associated with the presence of partially depleted cores.

Acknowledgements

I wish to thank my collaborators Rafael Guzmán, Ana Matković and Ileana Vass who have contributed to various aspects of this paper. I am also grateful to Helmut Jerjen who provided the luminosity function shown in Figure 4. This research was supported in part by NASA grant HST-AR-09927.01-A from the Space Telescope Science Institute, which is operated by the Association of Universities for Research in Astronomy, Inc., under NASA contract NAS5-26555.

References

Begelman, M.C., Blandford, R.D. & Rees, M.J. 1980, *Nature* 287, 307

Binggeli, B. & Cameron, L.M. 1991, *A&A* 252, 27

Binggeli, B. & Jerjen, H. 1998, *A&A* 333, 17

Binggeli, B., Sandage, A. & Tarenghi, M. 1984, *AJ* 89, 64

Caldwell, N. 1983, *AJ* 88, 804

Caldwell, N. & Bothun, G.D. 1987, *AJ* 94, 1126

Caon, N., Capaccioli, M. & D'Onofrio, M. 1993, *MNRAS* 265, 1013

Capaccioli, M., Caon, N. & D'Onofrio, M. 1992, *MNRAS* 259, 323

Cellone, S.A., Forte, J.C., & Geisler, D. 1994, *ApJS* 93, 397

Davies, J.I., Phillipps, S., Cawson, M.G.M., Disney, M.J. & Kibblewhite, E.J. 1988, *MNRAS* 232, 239

De Rijcke, S., Michielsen, D., Dejonghe, H., Zeilinger, W.W. & Hau, G.K.T. 2005, *A&A* in press (astro-ph/0412553)

de Vaucouleurs, G. 1948, *Annales d'Astrophysique* 11, 247

de Vaucouleurs, G. 1959, in: S. Flügge (ed.) *Handbuch der Physik* (Berlin: Springer) p. 275

de Vaucouleurs, G. & Capaccioli, M. 1979, *ApJS* 40, 669

Djorgovski, S. & Davis, M. 1987, *ApJ* 313, 59

Djorgovski, S. & Kormendy, J. 1989, *ARA&A* 27, 235

D'Onofrio, M., Capaccioli, M. & Caon, N. 1994, *MNRAS* 271, 523

Ebisuzaki, T., Makino, J. & Okumura, S.K. 1991, *Nature* 354, 212

Faber, S.M. & Jackson, R.E. 1976, *ApJ* 204, 668

Faber, S.M. & Lin, D.M.C. 1983, *ApJ* 266, L17

Faber, S.M., *et al.* 1997, *AJ* 114, 1771

Ferguson, H.C. & Binggeli, B. 1994, *A&ARv* 6, 67

Forbes, D.A., Franx, M., Illingworth, G.D. & Carollo, C.M. 1996, *ApJ* 467, 126

Geha, M., Guhathakurta, P. & van der Marel, R.P. 2002, *AJ*, 124, 3073

Graham, A.W. 2004, *ApJ* 613, L33

Graham, A.W. & Driver, S.P. 2005, *PASA* 22(2), 118

Graham, A.W, Erwin, P., Trujillo, I. & Asensio–Ramos, A. 2003, *AJ* 125, 2951

Graham, A.W. & Guzmán, R. 2003, *AJ* 125, 2936

Graham, A.W. & Guzmán, R. 2004, in: D.L. Block, I. Puerari, K.C. Freeman, R. Groess & E.K. Block (eds.), *Penetrating bars through masks of cosmic dust*, (Dordrecht: Springer), vol. 319, p. 723

Graham, A.W., Lauer, T., Colless, M.M. & Postman, M. 1996, *ApJ* 465, 534

Jerjen, H., & Binggeli, B. 1997, in: *The Nature of Elliptical Galaxies; The Second Stromlo Symposium*, ASP Conf. Ser., vol. 116, p. 239

Karachentseva, V.E., Prugniel, P., Vennik, J., Richter, G.M., Thuan, T.X. & Martin, J.M. 1996, *A&ASS* 117, 343

Kormendy, J. 2004, in: D.L. Block, I. Puerari, K.C. Freeman, R. Groess & E.K. Block (eds.), *Penetrating bars through masks of cosmic dust*, (Dordrecht: Springer), vol. 319, p. 728

Kormendy, J. 1977, *ApJ* 218, 333

Kormendy, J. 1985, *ApJ* 295, 73

Matković, A. & Guzmán, R. 2005, *MNRAS*, submitted

Schombert, J.M. 1986, *ApJS* 60, 603

Sérsic, J.-L. 1963, *Boletin de la Asociacion Argentina de Astronomia* vol. 6, p. 41
Sérsic, J.-L. 1968, Atlas de Galaxias Australes (Cordoba: Observatorio Astronomico)
Stiavelli, M., Miller, B.W., Ferguson, H.C., Mack, J., Whitmore, B.C. & Lotz, J.M. 2001, *AJ* 121, 1385
Trujillo, I., Erwin, P., Asensio Ramos, A. & Graham, A.W. 2004, *AJ* 127, 1917
Trujillo, I., Graham, A.W. & Caon, N. 2001, *MNRAS* 326, 869
Vennik, J., Hopp, U., Kovachev, B., Kuhn, B. & Elsässer, H. 1996, *A&ASS* 117, 261
Vennik, J. & Richter, G.M. 1994, *Astron. Nachr.* 315, H3, 245
Young, C.K. & Currie, M.J. 1994, *MNRAS* 268, L11
Young, C.K. & Currie, M.J. 1995, *MNRAS* 273, 1141

Discussion

TAYLOR: What are the smallest black holes you need in this picture to scour out the cored galaxies?

GRAHAM: Bearing in mind that it's not simply the black hole masses, but whether the galaxy merger is dissipationless or not — as gas can facilitate black hole coalescence and also help repopulate the loss cone — the smallest black holes in galaxies with cores have masses around $10^8 M_\odot$.

CONSELICE: The low–luminosity ellipticals also have black holes, and probably also form through mergers. If this is the case, as seems likely, then why don't the low–luminosity ellipticals also have depleted cores? Does this suggest that giant and fainter ellipticals really do have two formation methods?

GRAHAM: A good question. Giant elliptical galaxies with partially depleted cores tend to be brighter than ~ -20.5 B–mag, while "power–law" galaxies tend to be fainter. It has been suggested that the "core" galaxies formed via a dissipationless merger event while the "power–law" galaxies formed from a relatively gas rich merger. No division at -18 B–mag is apparent between the "power–law" ellipticals and the dwarf ellipticals. In passing I note that the boxy/disky isophote divide also occurs at ~ -20.5 B–mag, as does the change in slope of the L–σ relation (Matković & Guzmán and de Rijcke *et al.* these proceedings).

Near-Field Cosmology with Dwarf Elliptical Galaxies
Proceedings IAU Colloquium No. 198, 2005
H. Jerjen and B. Binggeli, eds.

doi:10.1017/S1743921305003972

The colours of Virgo dEs as seen by SDSS

Thorsten Lisker, Eva K. Grebel and Bruno Binggeli

Astronomical Institute, University of Basel, Switzerland; email: tlisker@astro.unibas.ch

Abstract. A colour analysis of a carefully selected sample of over 200 Virgo cluster dEs and 8 giant ellipticals (Es) in the SDSS yields the following results: a) the nucleated dEs (dENs) follow a tight colour-magnitude relation (CMR) which does not broaden significantly at fainter ($M_r \geqslant -15$) magnitudes, b) the CMR of dENs smoothly transitions to the CMR of Es, but the latter changes its slope at about $M_r \approx -20$, c) the brighter dENs are consistent with ages of $t \geqslant 5\,\mathrm{Gyr}$ and a decrease of metallicity towards fainter magnitudes, but a possible additional luminosity-age relation is not ruled out, d) it is crucial to treat dS0s separately from dEs since their different colour properties would otherwise bias the comparison of nucleated and non-nucleated dEs (dEnoNs), e) the dEnoNs show a weak trend towards bluer colours than dENs.

Keywords. galaxies: dwarf, galaxies: clusters: general, galaxies: fundamental parameters

1. Introduction and sample selection

Numerous studies of dwarf ellipticals (dEs) have put the homogeneity of this class of galaxy in question. In particular, the two main subclasses, nucleated and non-nucleated dEs (dENs and dEnoNs), might have a different history: dENs are more strongly clustered, are brighter, and have been claimed to show redder colours (Binggeli & Cameron 1991; Rakos *et al.* 2004). To explore the differences and similarities of these galaxies in greater detail, we study here the fundamental relation of dEs – the colour-magnitude relation (CMR) – separately for dENs and dEnoNs. This requires a large and homogeneous dE sample which ideally should also enable us to address the issue of a possible transition from dwarf to giant ellipticals (Es). Contrary to the earlier belief, it has been shown recently that the structural parameters of dEs and Es exhibit a smooth transition (Graham & Guzman 2003), raising the question whether this is also the case for their colour properties.

The SDSS Data Release 3 provides reduced and calibrated images in the u, g, r, i, and z band as well as photometric measurements for Virgo cluster galaxies. However, low surface brightness objects like the dEs can be 'shredded', i.e. parts of them are assigned to multiple separate source detections by the detection algorithm, resulting in unreliable photometry. A comparison with the work of Gavazzi *et al.* (2005) showed that we must restrict ourselves to dEs brighter than $m_r \lesssim 18$ that are already listed in the Virgo Cluster Catalog (VCC; Binggeli *et al.* 1985) and carefully exclude misaligned or shredded source detections visually. Moreover, the SDSS photometric pipeline overestimates the local sky flux due to the large apparent object sizes, for which we apply an appropriate correction. Our final sample comprises 220 dEs+dS0s; the sample of 8 giant ellipticals was selected in the same manner. Colours for the dwarfs were derived from aperture photometry on the SDSS images, excluding the inner $r \leqslant 2''$ to avoid nucleus light entering the aperture. Since this does not require any photometric measurements from the SDSS pipeline, our sample is extended to 349 dEs+dS0s in diagrams where only colours are required. Errors were estimated from the signal-to-noise ratio and the uncertainty of the sky flux correction. Galactic extinction corrections are provided by the SDSS database.

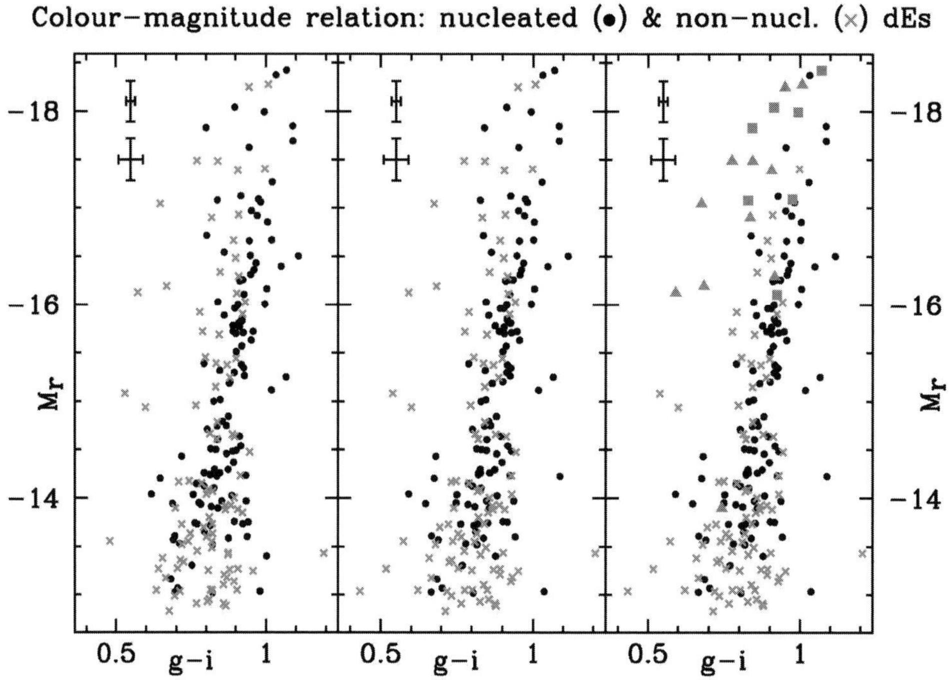

Figure 1. CMR for dENs (filled circles), dEnoNs (crosses), nucleated dS0s (squares) and non-nucleated dS0s (triangles). Upper and lower error bars indicate mean magnitude uncertainties for $M_r \lesssim -15.5$ and $M_r \gtrsim -15.5$, respectively. A distance modulus of $m - M = 31$ is adopted. *Left:* Absolute r-band magnitude versus fixed aperture ($r \approx 8''$) $g - i$ colour, including dS0s into the dE class. *Middle:* CMR with $g - i$ colour measured from half-light radius apertures, including dS0s into the dE class. *Right:* CMR with $g - i$ colour measured from half-light radius apertures, now showing dS0s separately.

2. Results

The classification of several dwarfs as dS0 in the VCC had been based on the two-dimensional appearance of the galaxies rather than on their azimuthally averaged profiles – the latter are indistinguishable from those of the dEs (Binggeli & Cameron 1991). Consequently, the existence of a separate class of dS0 galaxies has been questioned: Ryden *et al.* (1999) conclude that there are no compelling arguments for it, and e.g. Barazza *et al.* (2003) subsume dEs and dS0s into one class. We demonstrate the danger of this approach in Fig. 1, where we show the CMR separately for dENs and dEnoNs, treating dS0s as dEs in the middle panel but separating them in the right panel. From the middle panel one would conclude that the CMR of dEnoNs shows a steeper slope than the one of dENs, whereas the right panel reveals only a slight tendency towards bluer colours of dEnoNs compared to dENs, also given the paucity of bright dEnoNs. Here and in the following figures colour is measured within the half-light radius; the left panel of Fig. 1 confirms that no significant difference appears when using fixed apertures instead. The CMR of dENs is tight and does not broaden towards fainter magnitudes, as opposed to the findings of previous studies of dEs (e.g. Conselice *et al.* 2003). Note that the increase in colour errors by more than a factor of two would require any significant broadening to be larger than that, which is not the case.

We extend the CMR to faint Es in Fig. 2, illustrating that there is a smooth transition from dEs to Es which is mainly populated by dENs. The left panel shows the positions of

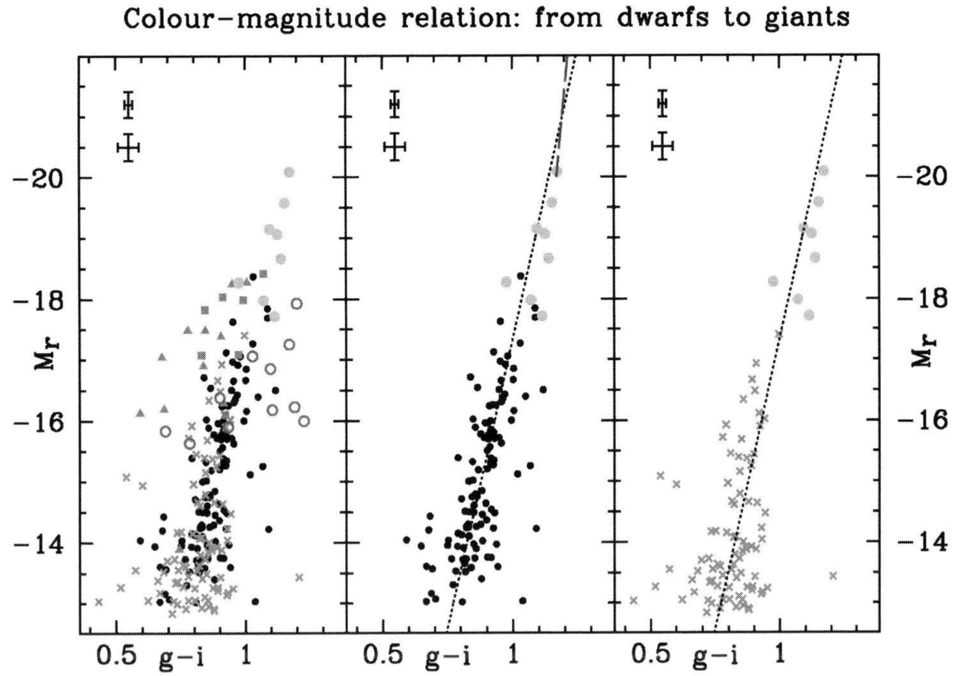

Figure 2. *Left:* CMR for dENs (filled circles), dEnoNs (crosses), nucleated dS0s (squares), non-nucleated dS0s (triangles), compact faint (M32-like) ellipticals (open circles), and Es (large filled circles). *Middle:* CMR for dENs and Es only. The dotted line shows a least-squares fit to the dENs, the dashed line is the slope of the CMR of brighter giant ellipticals from Chang *et al.* 2005. *Right:* CMR for dEnoNs and Es only. The dotted line is the same as in the middle panel.

various types of galaxy and demonstrates that compact faint 'M32-like' ellipticals neither belong to the E nor to the dE class but define their own colour-magnitude region instead. In the middle panel we show that the Es basically follow the linear extension of the CMR of dENs as derived from a least-squares fit. However, a study of nearby Es in the SDSS by Chang *et al.* (2005) finds a steeper slope which is well constrained for magnitudes $M_r \lesssim -20$ and indicates a slope change at about this magnitude. In the right panel we focus on the CMR of dEnoNs in order to see whether or not they follow the relation defined by dENs and Es. Unfortunately the small number of bright dEnoNs prevents any definite conclusion – despite a weak tendency towards bluer colours the dEnoNs are not inconsistent with the slope defined by the dENs.

A rough age and metallicity distribution of our sample can be obtained through a multicolour comparison of the observations with population synthesis models from Bruzual & Charlot (2003), as shown in Fig. 3. Since the range of values for $i - z$ is rather small, we select only objects brighter than $m_r \leqslant 16$ to avoid large errors that would dilute the comparison. Brighter objects tend to be redder because of the CMR (left panel of Fig. 3), therefore the colours can only be explained with a decrease in metallicity towards fainter magnitudes (right panel), consistent with previous studies. Young ages ($t \lesssim 2\,\mathrm{Gyr}$) are clearly ruled out, and for the dENs the data is consistent with ages $t \gtrsim 5\,\mathrm{Gyr}$. However, we cannot exclude a possible relation of age and luminosity *on top* of the dominant luminosity-metallicity relation; moreover, the trend of dEnoNs showing bluer colours appears to be strongest in $u - g$, possibly indicating younger ages than for the dENs.

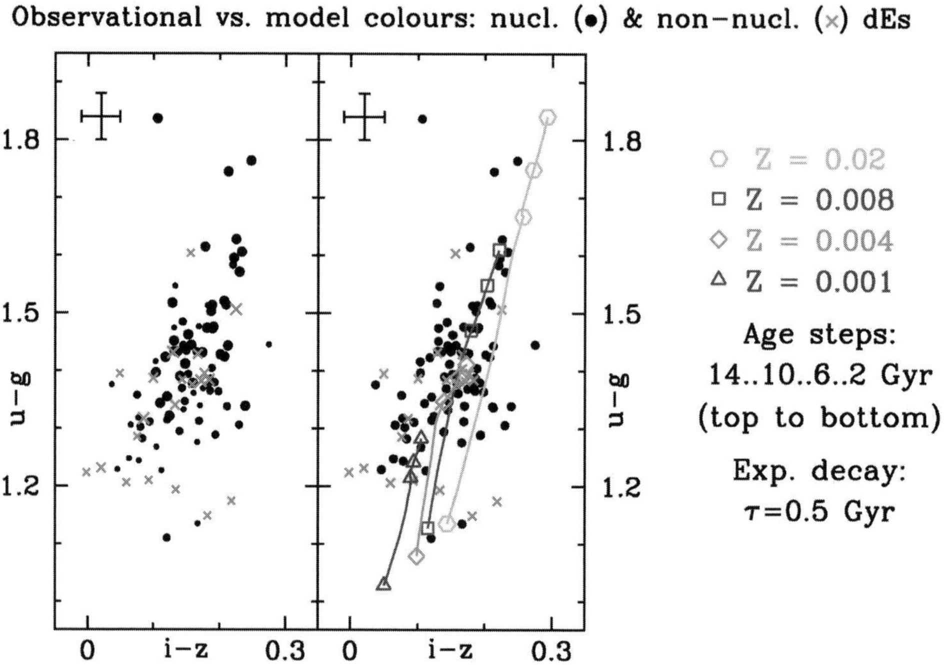

Figure 3. Colour-colour plot for dENs (filled circles) and dEnoNs (crosses). Only objects brighter than $m_r \leqslant 16$ were selected to ensure small errors in the colour. *Left:* The effect of the CMR is illustrated by the symbol sizes: larger for brighter objects. *Right:* Comparison of observed and model colours. Each track from Bruzual & Charlot (2003) shows the colours of a single-burst stellar population with a decay time of $\tau = 0.5$ Gyr and an age of 2 (bottom) to 14 (top) Gyr using Padova 2000 isochrones and a Chabrier IMF.

3. Conclusions

Our large sample of Virgo cluster dEs+dS0s from the SDSS shows that dENs follow a tight CMR down to $M_r \approx -13$ without any significant broadening in colour. The faint Es follow the CMR defined by dENs, but a change in slope occurs at $M_r \approx -20$. After excluding dS0 galaxies, only a weak trend towards bluer colours of dEnoNs compared to dENs is found. The colours of the brighter dEs rule out young ages ($t \lesssim 2$ Gyr) and the dENs are consistent with ages $t \gtrsim 5$ Gyr.

Acknowledgements

We thank the SDSS collaboration for the wealth of DR3 data. The SDSS Web site is http://www.sdss.org/.

References

Barazza, F.D., Binggeli, B. & Jerjen, H. 2003, *A&A* 407, 121
Binggeli, B., Sandage, A. & Tammann, G.A. 1985, *AJ* 90, 1681
Binggeli, B. & Cameron, L.M. 1991, *A&A* 252, 27
Bruzual, G. & Charlot, S. 2003, *MNRAS* 344, 1000
Chang, R., Gallazzi, A., Kauffmann, G. *et al.* 2005, *astro-ph/0502117*
Gavazzi, G., Donati, A., Cucciati, O. *et al.* 2005, *A&A* 430, 411
Graham, A.W. & Guzmán, R. 2003, *AJ* 125, 2936
Rakos, K. & Schombert, J. 2004, *AJ* 127, 1502
Ryden, B.S., Terndrup, D.M., Pogge, R.W. & Lauer, T.R. 1999, *ApJ* 517, 650

Discussion

FERGUSON: Quite a few galaxies in your colour-colour diagram were inconsistent with any of the models. Are you confident that there are not systematic uncertainties in the models or data? How far do you think you could shift the model colours, e.g. using different isochrones?

LISKER: I was using Padova 2000 isochrones against the recommendation of Bruzual & Charlot (2003), since the Padova 1994 isochrones were offset towards redder $i-z$ colours by a noticeable amount – so indeed, we are facing a 'model error' in addition to the observational error.

RAKOS: Would you agree that the difference in the slope you have shown between E and dE galaxies is introduced by different metallicity?

LISKER: Since both the shallower slope of the CMR at $M_r \lesssim -20$ and the steeper slope at $M_r \gtrsim -20$ are luminosity-metallicity relations, I am not sure whether the *change* in the slope can be caused by metallicity itself.

CONSELICE: If you were to include all of the fainter dEs from the VCC that are not included in your sample, you would probably see a larger scatter in the CMR at faint magnitudes.

LISKER: We cannot test this so far, since the SDSS photometry catalogs are not useful for the faintest dEs, mainly due to galaxy shredding. However, we already go down to $M_r \approx -13$, clearly fainter than where a significant increase in the colour spread has been claimed to appear.

Near-Field Cosmology With Dwarf Elliptical Galaxies
Proceedings IAU Colloquium No. 198, 2005
H. Jerjen and B. Binggeli, eds.

doi:10.1017/S1743921305003984

Formation and evolution of dwarf elliptical galaxies: Structural and kinematical properties

S. De Rijcke[1]†, D. Michielsen[1], H. Dejonghe[1], W.W. Zeilinger[2] and G.K.T. Hau[3]

[1]Sterrenkundig Observatorium, Ghent University, Krijgslaan 281, S9, B-9000 Gent, Belgium
email: sven.derijcke@UGent.be

[2]Institut für Astronomie, Universität Wien, Türkenschanzstraße 17, A-1180 Wien, Austria
[3]Department of Physics, South Road, Durham, DH1 3LE, UK

Abstract. We confront predictions of the currently popular theories for dE formation and evolution with the observed position of dEs in $\log L_B$ vs. $\log \sigma$, $\log L_B$ vs. $\log R_e$, $\log L_B$ vs. $\log I_e$, and $\log R_e$ vs. $\log I_e$ diagrams and in the $(\log \sigma, \log R_e, \log I_e)$ parameter space in which bright and intermediate-luminosity elliptical galaxies and bulges of spirals define a Fundamental Plane (FP). We show that the dE sequences in the various univariate diagrams are disjunct from those traced by bright and intermediate-luminosity elliptical galaxies and bulges of spirals. It appears that semi-analytical models (SAMs) that incorporate quiescent star formation with an essentially z-independent star-formation efficiency, combined with post-merger starbursts and the dynamical response after supernova-driven gas-loss, are able to reproduce the position of the dEs in the various univariate and bivariate diagrams.

Keywords. galaxies: dwarf – galaxies: fundamental parameters – galaxies: kinematics and dynamics

1. Introduction

For almost two decades now it has been known that dynamically hot galaxies (elliptical galaxies and bulges of spiral galaxies) are not scattered randomly in the three-dimensional space spanned by B-band luminosity ($\log L_B$, expressed in solar B-band luminosities), half-light radius ($\log R_e$, expressed in kiloparsecs), and velocity dispersion ($\log \sigma$, expressed in km s^{-1}), but that instead they occupy a slender plane: the Fundamental Plane (FP) (Djorgovski & Davis 1987, Dressler *et al.* 1987, Bender *et al.* 1992). Projections of the FP onto the coordinate planes, in combination with the particular way in which galaxies are distributed within the FP, produce the univariate relations between (i) luminosity and velocity dispersion ($\log L_B = \text{const.} + 3.74 \log \sigma$, Faber & Jackson 1976), (ii) luminosity and half-light radius ($\log L_B = \text{const.} + 1.19 \log R_e$, Fish 1963), (iii) surface brightness (expressed in solar B-band luminosities per square parsec) and half-light radius ($\log I_e = \text{const.} - 0.81 \log R_e$, Kormendy 1977), (iv) and luminosity and surface brightness ($\log L_B = \text{const.} - 1.46 \log I_e$, Binggeli *et al.* 1984). The numerical values of these slopes and zero-points were taken from Guzmán *et al.* 1993, who made a compilation of these so-called fundamental relations for a sample of 66 Coma bright ellipticals.

The locus of the dEs and dS0s in the $(\log \sigma, \log R_e, \log I_e)$ space and in the univariate diagrams that require kinematical information, such as the $\sigma - L_B$ relation, was up to now rather uncertain. After the early work by Nieto *et al.* 1990, Bender *et al.* 1992, and

† Postdoctoral Fellow of the Fund for Scientific Research – Flanders (Belgium)(F.W.O.)

Guzmán *et al.* 1993, relatively little attention has been paid to the relation between the internal kinematics and the structural parameters of these faint, small elliptical galaxies and, particularly, to what we can learn from such relations regarding the origin and evolution of dEs. Still, explaining the scaling relations among the structural parameters of dwarf galaxies, which are believed to be the building blocks of more massive galaxies, is a crucial test for theories of cosmological structure formation. A comparison of theoretical predictions with observations can put strong constraints on the cosmological star-formation history and help to refine prescriptions for, e.g., star formation and energy feedback from supernova explosions, which play a crucial role in low-mass dwarf galaxies.

2. Observations and data reduction

Within the framework of an ESO Large Program, we observed deep major and minor axis spectra with unprecedented spatial and spectral resolution of a sample of 15 dEs and dS0s, both in group (NGC5044, NGC5898, and NGC3258 groups) and cluster environments (Fornax cluster). We also collected Bessel VRI-band images of 22 dEs and dS0s, including the 15 dEs/dS0s of the spectroscopic sample. The data were obtained with the FORS2 imaging spectrograph mounted on the unit telescopes Kueyen and Yepun of the VLT.

From our images, we measured the surface-brightness profile, position angle, and ellipticity as a function of the geometric mean of major and minor axis distance denoted by a and b respectively, using our own software. The spectra, with typical exposure times of 5–8 h per position angle and a seeing in the range $0.3''$–$1.0''$ FWHM, cover the wavelength region around the strong CaII triplet absorption lines (∼8600 Å). We extracted the stellar kinematical information by fitting a weighted mix of late G to late K giant stars, broadened with a parameterised line-of-sight velocity distribution (LOSVD) to the galaxy spectra. We approximated the LOSVD by a fourth-order Gauss-Hermite series (Gerhard 1993, van der Marel & Franx 1993). The strong CaII lines, which contain most of the kinematical information, are rather insensitive to the age and metallicity of an old stellar population (see e.g. Michielsen *et al.* 2003, Falcón-Barroso *et al.* 2003, Saglia *et al.* 2002), so template mismatch does not significantly affect our results. The spectra contain useful kinematical information out to 1.5-2 R_e.

3. Univariate relations

In the following, we confront the predictions of models for galaxy formation with the relations observed between the structural and kinematical parameters of dynamically hot galaxies. Semi-analytical models (SAMs) (Somerville *et al.* 2001, Nagashima & Yoshii 2004) include the hierarchical merger tree that leads up to the formation of a galaxy of a given mass in a ΛCDM cosmology. SAMs make use of prescriptions for star-formation, energy feedback from supernova explosions, gas cooling, tidal stripping, dust extinction, and the dynamical response to starburst-induced gas ejection. Despite their simplicity, the models presented by Yoshii & Arimoto 1987 already capture a lot of the physics of galaxy formation, at least in the mass-regime of the dEs, which apparently were assembled from what were still largely gaseous progenitors. *In all figures, we will use them as a simple and instructive proxy for the physically more motivated SAMs.* Alternatively, dEs could stem from late-type disk galaxies that entered the clusters and groups of galaxies about 5 Gyr ago ($z \sim 0.5$). N-body simulations show that gravitational interactions trigger bar-formation in any small late-type disk galaxy (Scd-Irr) orbiting in a cluster (Moore *et al.* 1996, Moore *et al* 1998) or around a massive galaxy in a group environment

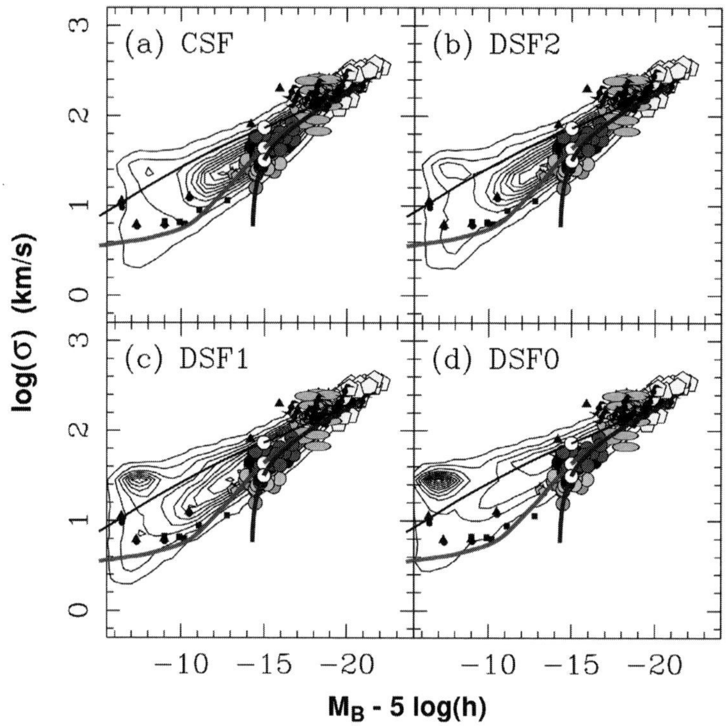

Figure 1. The luminosity L_B vs. velocity dispersion σ relation. The bright galaxies follow a trend $L_B \propto \sigma^4$ whereas the dwarf ellipticals follow a sequence defined by $L_B \propto \sigma^{1.6}$. This can be reproduced by the SAMs (contours) by allowing protogalaxies to expand after supernova-induced galactic winds following each major merger.

(Mayer *et al.* 2001) and strip large amounts of stars, gas, and dark matter from it by tidal forces. Internal dynamical processes, such as the buckling of the bar, subsequently transform a disk galaxy into a dynamically hot spheroidal dE within a timespan of a few Gyr (Mastropietro *et al.* 2004).

3.1. *The $\sigma - L_{\rm B}$ relation*

Bright and intermediate-luminosity ellipticals and bulges of spiral galaxies adhere closely to the $\sigma - L_{\rm B}$ relation: $L_B \propto \sigma^\alpha$, with $\alpha \sim 4$ (Faber & Jackson 1976). As is obvious from Fig. 1, the $\sigma - L_{\rm B}$ relation becomes noticeably flatter below $M_B \sim -18.3$ mag. A straight-line fit to the dE data yields

$$\log L_B = 6.02^{\pm 0.31} + 1.57^{\pm 0.19} \log \sigma. \tag{3.1}$$

A dE $\sigma - L_{\rm B}$ relation that is steeper than that of bright ellipticals is definitely excluded by these data. These results contradict the universality of the $\sigma - L_{\rm B}$ relation, which was suggested by Guzmán *et al.* 1993. A currently running project on the properties of faint early-type galaxies ($-22 < M_R < -17.5$) in the central 1deg of the Coma cluster yields a $\sigma - L_{\rm B}$ relation $L \propto \sigma^{2.01}$ at the faint end, which is consistent with our results (Matković & Guzmán, private communication).

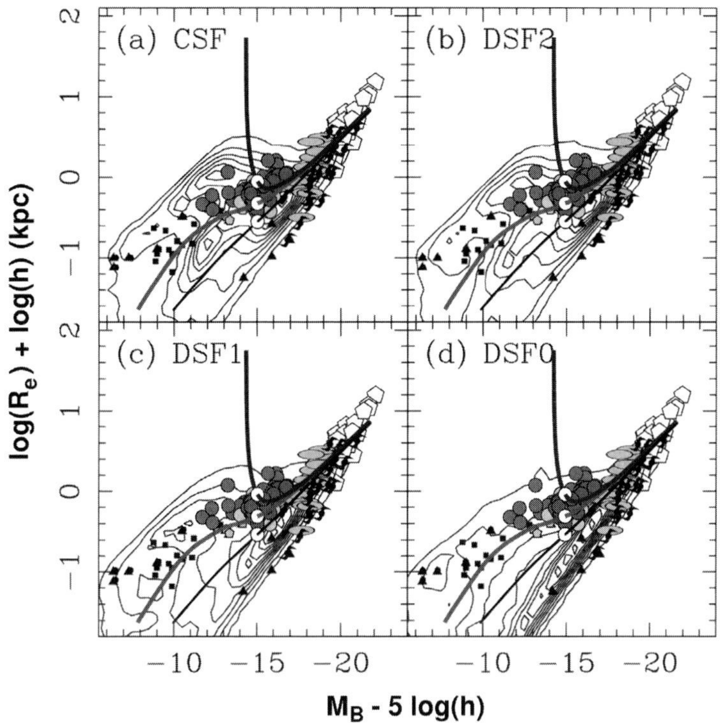

Figure 2. The luminosity L_B vs. half-light radius R_e relation. Bright galaxies follow a well defined relation whereas dEs and dSphs are much more diffuse.

3.2. *The* $R_e - L_B$ *relation*

The simple fact that diffuse, low surface-brightness dEs exist puts strong constraints on the redshift dependence of the cosmic star-formation rate. Nagashima & Yoshii 2004 have convincingly shown that structure-formation models in a ΛCDM universe with a short star-formation timescale at high z fail to produce such inflated dEs. In such a universe, dEs are assembled from progenitors that have already converted most of their gas into stars, and they are expected to trace the same sequences as the giant ellipticals, which are formed further down the merger tree from almost purely stellar progenitors, independent of the cosmic star-formation rate. Only models that have long enough star-formation timescales at high z, such that dEs can be formed by the mergers of gaseous progenitors, agree with these observations, e.g. the models of Nagashima & Yoshii 2004 with a redshift independent star-formation time scale $\tau_* = 1.3$ Gyr, where the star-formation rate is given by $M_{\rm gas}/\tau_*$ with $M_{\rm gas}$ the HI mass. The starburst triggered by each merger and the ensuing supernova-explosions eject gas, and thus lead to a population of diffuse dwarf galaxies with low velocity dispersions, as observed.

4. Harassment

These findings do not necessarily falsify the harassment scenario. The dEs observed so far overlap in κ-space with the present-day analogs of possible dE progenitors (the Scd and Irr galaxies). This overlap leaves open the possibility that we have observed both dEs that formed via hierarchical merging *and* dEs that formed via harassment.

The dEs formed in these simulations are expected to rotate quite rapidly and some should still display some memory of their former state. Some of the dEs observed during the Large Program provide us with very strong evidence that harassment has indeed played an important role in their past evolution; this evidence includes embedded stellar disks (Barazza *et al.* 2002, De Rijcke *et al.* 2003a, Graham *et al.* 2003) or kinematically decoupled cores (De Rijcke *et al.* 2004).

5. Conclusions

Models for the evolution of dwarf and intermediate-luminosity elliptical galaxies, based on the idea that these stellar systems grow from collapsing primordial density fluctuations, are able to reproduce the observed relations between parameters that quantify their structure (L_B, R_e, I_e) and internal dynamics (σ) quite well. While dEs follow well-defined sequences in the various univariate diagrams, the correlations are not as tight as in the case of bright ellipticals. This cannot be due solely to measurement uncertainties, e.g. very deep photometry is now available that allows to determine L_B and R_e with very small errors, still the scatter on the dE $\log R_e$-$\log I_e$ relation is large. This cosmic scatter may be a consequence of the sensitivity of these low-mass systems to both internal (supernova explosions, feedback efficiency, the details of galactic winds, etc.) and external processes (gravitational interactions, tidal stripping of stars and ram-pressure stripping of gas, etc.) in group and cluster environments. Hence, these objects are ideal laboratories for studying these physical processes to which bright ellipticals seem to be quite insensitive.

Acknowledgements

Based on observations made at the European Southern Observatory, Chile (ESO Large Programme Nr. 165.N-0115). SDR wishes to thank Philippe Prugniel for fruitful discussions while visiting the CRAL-Observatoire de Lyon. WWZ acknowledges the support of the Austrian Science Fund (Project P14783). DM acknowledges the support of the Bijzonder OnderzoeksFonds (BOF Universiteit Gent)

References

Barazza, F.D., Binggeli, B. & Jerjen, H. 2002, *A&A* 391, 823
Bender, R., Burstein, D. & Faber, S.M. 1992, *ApJ* 399, 462
Binggeli, B., Sandage, A. & Tarenghi, M. 1984, *AJ* 89, 64
De Rijcke, S., Dejonghe, H., Zeilinger, W.W. & Hau, G.K.T. 2003, *A&A* 400, 119
De Rijcke, S., Dejonghe, H., Zeilinger, W.W. & Hau, G.K.T. 2004, *A&A* 426, 53
De Rijcke, S., Michielsen, D., Dejonghe, H., Zeilinger, W.W. & Hau, G.K.T. 2005, *A&A*, accepted
Djorgovski, S. & Davis, M. 1987, *ApJ* 313, 59
Dressler, A., Faber, S.M., Burstein, D., Davies, R.L., Lynden-Bell, D., Terlevich, R. & Wegner, G. 1987, *ApJ* 313, L37
Faber S.M. & Jackson R.E. 1976, *ApJ* 204, 668
Falcón-Barroso, J., Peletier, R.F., Vazdekis, A. & Balcells, M. 2003, *ApJ* 588, L17
Fish R.A. 1963, *AJ* 68, 72
Geha, M., Guhathakurta, P. & van der Marel, R.P. 2003, *AJ* 126, 1794
Gerhard, O.E. 1993, *MNRAS* 265, 213
Graham, A.W., Jerjen, H. & Guzmán, R. 2003, *AJ* 126, 1787
Guzmán, R., Lucey, J.R. & Bower, R.G., 1993, *MNRAS* 265, 731
Kormendy, J., 1977, *ApJ* 218, 333
Mastropietro, C., Moore, B., Mayer, L., Debattista, V.P., Piffaretti, R. & Stadel, J. 2004, *MNRAS* submitted, astro-ph/0411648

Mayer, L., Governato, F., Colpi, M., Moore, B., Quinn, T., Wadsley, J., Stadel, J. & Lake, G. 2001, *ApJ* 559, 754
Michielsen, D., De Rijcke, S., Dejonghe, H., Zeilinger, W.W. & Hau, G.K.T. 2003, *ApJ* 597, L21
Moore B., Katz, N., Lake, G., Dressler, A. & Oemler Jr., A. 1996, *Nature* 379, 613
Moore, B., Lake, G. & Katz, N. 1998, *ApJ* 495, 139
Nagashima, M. & Yoshii, Y. 2004, *ApJ* 610, 23
Nieto J.-L., Bender, R., Davoust, E. & Prugniel, P. 1990, *A&A* 230, L17
Saglia, R.P., Maraston, C., Thomas, D., Bender, R. & Colless, M. 2002, *ApJ* 579, L13
Somerville, R.S., Primack, J.R. & Faber, S.M. 2001, *MNRAS* 320, 504
van der Marel, R.P. & Franx, M. 1993, *ApJ* 407, 525
van Zee, L., Skillman, E.D. & Haynes, M.P. 2004, *AJ* 128, 121
van Zee, L., Barton, E.J. & Skillman, E.D. 2004, *AJ* 128, 2797
Yoshii, Y. & Arimoto, N. 1987, *A&A* 188, 13

Discussion

PRUGNIEL: Why do you say that disks are evidence for harassment?

DE RIJCKE: As was pointed out by Moore *et al.* (1998) and Mastropietro at al. (2004), harassment not always completely transforms a late-type galaxy into a dwarf elliptical. In fact, most of the simulated dEs end up having embedded disks, bars, spiral arms, etc. So our observations corroborate this idea.

FONT: Regarding the $\sigma - L$ relation: it seems that the theoretical models you presented cannot fit the low-σ (hence low-mass) regime well. One would expect that for low-mass galaxies winds play an important role, yet neither of the mass ejection models seem to fit the data. Are these theoretical models reliable? Do they take into account environmental effects?

DE RIJCKE: These models rely on an empirical relation between total mass and binding energy. Since gas is ejected when it is heated above the escape velocity, this is clearly an important ingredient of the models. The empirical relation between mass and binding energy does, however, not provide a good description of dwarf spheroidals, so this may be the reason the models fail to reproduce the properties of the dSphs. More recent semi-analytic models for galaxy formation through hierarchical merging give a much better description of the dSph properties. None of these models take into account environmental effects.

Near-Field Cosmology with Dwarf Elliptical Galaxies
Proceedings IAU Colloquium No. 198, 2005
H. Jerjen and B. Binggeli, eds.

doi:10.1017/S1743921305003996

The impact of accurate distances to dwarf elliptical galaxies on cosmology

Helmut Jerjen

Research School of Astronomy and Astrophysics, Australian National University, Australia
email: jerjen@mso.anu.edu.au

Abstract. The Surface Brightness Fluctuation method has been shown to be a powerful distance indicator for dwarf elliptical galaxies to very low surface brightness levels. It is applicable to stellar systems that are out of reach for classical indicators requiring resolved stellar populations such as the tip magnitude of the red giant branch. I briefly discuss a few results from recent SBF studies of dEs to demonstrate the significance of the SBF method to address long-standing issues related to cosmography, dark matter in galaxy groups, substructures in clusters, and the discrepancy between the mass function of collapsed objects and the faint end of the galaxy luminosity function. For the analysis of the large number of galaxy images that need to be processed as part of such SBF studies we are currently developing a fast, semi-automatic reduction pipeline that will be made readily available to the astronomical community.

Keywords. galaxies: clusters: individual (Virgo)–galaxies: elliptical and lenticular-galaxies: dwarf galaxies: distances and redshifts, galaxies: luminosity function, mass function

---

1. Motivation

The ability to spatial locate newly detected dwarf galaxy candidates is of fundamental importance as it enables us among others to allocate them to an environment and to measure physical quantities such as their luminosity, mass, and star formation history. For instance any dwarf galaxy that can be found either in or close to the Local Group is of greatest interest. Such stellar systems are relatively rare and due to their proximity represent prime targets for detailed stellar population and dark matter studies. That explains why the scientific interest in Andromeda IV, discovered during van den Bergh's (1972) search for dwarf spheroidal companions to M31, was initially high but depreciated quickly when a radial velocity and a tentative distance measurement revealed 28 years later (Ferguson, Gallagher & Wyse 2000) that it actually is a background irregular galaxy well outside the confines of the Local Group.

But before a distance can be measured the dwarf galaxy needs to be detected first. A modern and probably the most promising way to find gas-rich *dwarf irregular* (dIrr) galaxies in the local Universe is via deep, blind HI surveys using newly developed multi-beam instruments at radio telescopes. Because low luminous and low surface brightness dIrrs have increasingly large HI mass fractions (e.g. Warren *et al.* 2004), simultaneous detection and distance measurement via redshift remains possible at 21 cm even in extreme cases like DDO154 (Krumm & Burstein 1984; Carignan & Freeman 1988) or ESO215-G?009 (Warren *et al.* 2004) where only an elusive stellar component is present, embedded in an extended disk of neutral hydrogen that has not been involved in the galaxy's star-formation processes.

Evidence for a high detection completeness with this search method is provided by the HI Parkes All-Sky Survey (HIPASS, Barnes *et al.* 2001). Perhaps somewhat surprising, HIPASS found *no genuine dark galaxies, ie. extragalactic HI clouds or dark matter halos filled with HI but no stars* above the detection limit of $M_{HI} > 10^7\,M_\odot$. All but a very small number of the ∼6000 HIPASS detections in the surveyed velocity range $-700\,\mathrm{km\,s^{-1}} < v_\odot < 12,700\,\mathrm{km\,s^{-1}}$ could be associated with stars (Koribalski *et al.* 2004, Doyle *et al.* 2005). The few exceptions are tidal features (e.g. Ryder *et al.* 2001, Oosterloo *et al.* 2004) or high velocity clouds (Kilborn *et al.* 2000).

Finding *dwarf elliptical* (dE) galaxies is by nature a more difficult task. Karachentsev with collaborators and many others identified large numbers of dE candidates in the direction of nearby groups and clusters over the last decade using primarily wide-field photographic plates. The challenge remains to measure distances to all of them to separate the wheat from the chaff. Unlike in the case of the dIrrs, the absence of gas in dEs prevents a relative simple distance estimate via redshift. Moreover, the large scatter in well-known empirical scaling relations based on integral galaxy parameters such as the surface brightness–luminosity (Binggeli & Cameron 1991) or the shape parameter–luminosity relation (Binggeli & Jerjen 1998) makes them unsuitable for individual distance measurements.

That is why in the last decade, a formidable observational effort has been devoted to derive distances to dEs by using the tip magnitude of the red giant branch (TRGB), a powerful Pop II distance indicator (Karachentsev *et al.* 2004 and references therein). But the requirement of accurate photometry for individual stars 1–2 magnitudes below the tip sets a practical application limit at ≈5 Mpc. TRGB measurements for galaxies beyond that limit become quickly very costly and time consuming as was demonstrated by Harris *et al.* (1998) who spent 9 hours with the HST to obtain the TRGB distance for one single dE (IC3388) in the Virgo cluster.

Because of this limitation, a relatively new distance indicator the so-called surface brightness fluctuation (SBF) method has received increasing attention in recent years. The method quantifies the statistical pixel-to-pixel variation of star counts across a galaxy image with the major technical advantage of working on unresolved stellar populations. Ground-based CCD images with relative short integration times can be used instead of high resolution space-based observations which makes the SBF method extremely efficient. Distance measurements for early-type galaxies far beyond the reach of any other distance indicator become possible.

The theoretical framework of the SBF method was developed by Tonry & Schneider (1988) with one of the main aims to study the distribution and kinematical properties of large-scale structure outlined by luminous giant ellipticals in the local Universe (e.g. Tonry *et al.* 2001). While SBF applications were most exclusively focussed on these high surface brightness systems, Jerjen and collaborators (Jerjen, Freeman & Binggeli 1998, 2000; Jerjen *et al.* 2001, Rekola, Jerjen & Flynn 2005) demonstrated that the method works equally well for low surface brightness dEs as faint as $\mu_{\mathrm{B,eff}} = 26\,\mathrm{mag\,arcsec^{-2}}$ and $M_B = -10\,\mathrm{mag}$. With this possibility to measure distances to large numbers of dEs easily, including the newly identified dE candidates, a special opportunity arises to make progress in areas most relevant to near-field cosmology including

- Mapping the gravitational centers in the Supergalactic plane and Local supercluster.
- Substructures and dark matter contents of galaxy clusters.
- dE density in the local Universe.
- Shape of the galaxy luminosity function to the lowest luminosities.

2. SBF calibration

As the strength of the SBF signal and the derived fluctuation magnitude $\overline{m}$ depend not only on the distance of the galaxy but also on its stellar composition, the calibration of

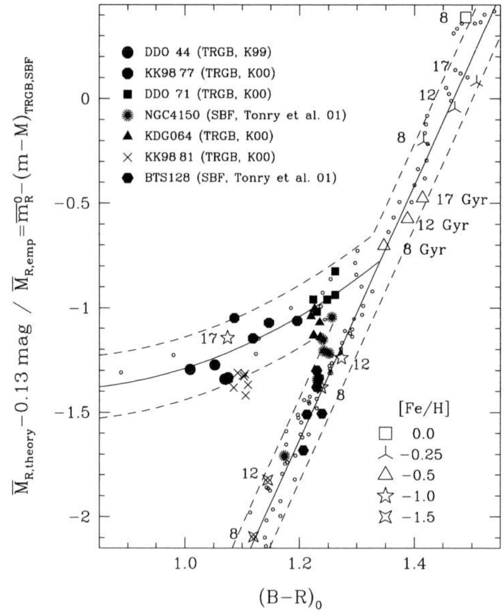

Figure 1. The SBF calibration diagram for the R-band. The open symbols are the $(B-R, \overline{M}_R)$ values for 170 synthetic stellar populations based on Worthey (1994) models and Padova isochrones (Bertelli *et al.* 1994). An offset correction of 0.13 mag is applied to the theoretical values to match the empirical results from 36 SBF fields measured in 7 nearby dwarf galaxies with independent TRGB and SBF distances (filled symbols). The solid lines represent the best analytical functions describing the loci of the models while the dashed lines indicate the ± 0.15 mag uncertainty bands.

the method must take into account the basic properties of the stars. For a first theoretical calibration of the SBF method in the R-band, Jerjen *et al.* (1998, 2000) studied the effect of different mixtures of ages and metallicity as well as the starformation history on the observable quantities, fluctuation magnitude $\overline{M}_R$ and $B-R$ colour. For that purpose, Worthey's (1994) models combined with Padova isochrones (Bertelli *et al.* 1994) were employed to simulate a range of synthesis stellar populations with two starformation bursts to make allowance for the variety of starformation histories observed among Local Group dEs/dSphs (Grebel 2001). The age of the first burst was set at 8, 12, or 17 Gyr with a metallicity in the range from -2.0 to 0. We added a second population, 5 Gyr old and of solar metallicity, that contributed 0, 10, 20, or 30% in mass to the entire population. In all cases we assumed a Salpeter IMF. The $(B-R, \overline{M}_R)$ values for a total of 170 synthetic stellar populations were calculated.

The theoretical predictions were then tested with 36 SBF measurements in seven nearby galaxies (Jerjen *et al.* 2001, Rekola, Jerjen & Flynn 2005) for which independent distances were available (Karachentsev *et al.* 1999, 2000; Tonry *et al.* 2001). Good agreement was found after a systematic offset correction of 0.13 mag was applied to the theoretical values.These corrected $(B-R, \overline{M}_R)$ data points are shown in Fig. 1, indicated by the various open symbols. The two lines are best analytical fits of the empirically anchored theoretical data and represent the calibration equations for the SBF method in the R-band currently available (Jerjen *et al.* 2001). The typical overall uncertainty in a SBF distance measurement is $\approx$10% which includes the intrinsic scatter of the Worthey+Padova model values and an uncertainty of 0.1 mag in the TRGB zero point.

Calibrations for other filter sets have been established by various authors (e.g. V, I filters, Tonry *et al.* 1997, Mieske & Hilker 2003; SDSS g, z filters, Mei *et al.* 2005).

3. Some SBF applications of cosmological relevance

3.1. *3D-mapping of cluster cores*

There is a growing interest in the structural analysis of galaxy clusters to evaluate the importance of external mechanisms driving galaxy evolution in different kinds of environments, from cluster cores to their outskirts. In this context the characterization of the physical conditions at the immediate location of a cluster galaxy (distance) and the galaxy's dynamical status (velocity) are essential quantities.

Methods for detecting cluster substructure classically resort to projected galaxy positions and redshifts. But as redshifts are generally available only for high surface brightness galaxies, Es, S0s, and spirals where the latter are no genuine cluster galaxies and the number of Es+S0s in a cluster is relatively small, samples tend to be dominated by late-type galaxies that are infalling at a late stage of the cluster evolution and thus are kinematically distinct from genuine cluster galaxies. Consequently, these studies preferencially probe substructures in the cluster outskirts which are due to directional, inhomogenous galaxy aggretion along the larger-scale filaments a cluster is embedded in.

The detection of substructures/group remnants in the densest cluster regions where early-type galaxies dominate is more difficult as this is hampered by the small number of available giant ellipticals and the general lack of redshift information for dEs, the largest galaxy population that resides long enough in the cluster to trace the gravitational potential. Accurate distances of a sufficient number of early-type galaxies are indispensable for such studies.

In a pilot project, Jerjen, Binggeli & Barazza (2004) measured SBF distances to 16 luminous dEs and combined them with SBF distances to giant ellipticals from the literature (Neilsen & Tsevtanov 2000, Tonry *et al.* 2001) to trace the gravitational centers of the Virgo cluster along the line-of sight. The sample dEs showed a substantial spread in distances ranging from 14.9 to 21.3 Mpc (see Fig. 2) which confirmed that the Virgo cluster core has an elongated shape, consistent with the previous findings of Neilsen & Tsvetanov (2000), West & Blakeslee (2000) and Arnaboldi *et al.* (2000). However, somewhat unexpected was that this seemingly dynamically relaxed system of early-type galaxies were clumped in two major concentrations at $(m - M) = 31.02$ mag (15.8 Mpc) and at 31.33 mag (18.5 Mpc) with a number ratio of 3:1 (Fig. 2). An adaptive kernel analysis of the distribution found that these clumps are significance at the 99% and 89% level, respectively, and that there might be even a minor third clump of galaxies at $\approx$31.55 mag.

By combining the newly obtained SBF distances with redshift information from the literature the long-standing mystery around the four highly blueshifted dE galaxies VCC810, 815, 846, and 928 could be solved. The distance-velocity map (Fig. 2) revealed that these systems are part of the second clump approximately 2.7 Mpc at the back of the major galaxy concentration at 15.8 Mpc. Together with M86, these dEs constitute a dynamically independent group infalling into the cluster from the backside for the first time. These dwarf systems might have been late-type disk galaxies which, by gravitational interactions with the ICM and other cluster members, lost their gas and transformed into the spheroids we see today (Mao & Mo 1998, Moore *et al.* 1998, Mastropietro *et al.* 2005). Therefore, they present prime targets for follow-up studies to search for HI cloud remnants and structural features as have been already found in other Virgo dEs (Jerjen,

panels of Fig. 3. The fluctuation signal remains above the noise level out to wave number $k \sim 15$. Indeed, the SBF detection with HST/ACS of a dwarf with $M_I = -11.5$ mag will be better than the SBF detection with the current Subaru data of the galaxy with $M_R = -16.5$ mag. Based on these simulation predictions we will comfortably be able to measure SBF signals and distances for our faintest candidates.

The results from this SBF study will enable us to establish a distance-based galaxy luminosity function for that particular group environment down to $M_R = -11$ mag. Moreover, the feedback from the NGC1407 Group will then be combined with spectroscopic and neutral hydrogen data measured for the few late-type group member candidates to allocate accurate membership probabilities to the morphology-based five-staged rating scheme employed by Trentham & Tully (2002). Transferring the calibrated rating scheme to other galaxy groups under study will then help to much better constrain the luminosity functions found in other environments, permitting progress on the important question of whether there are differences with environment.

4. Conclusion

With the brief discussion of some recent projects and results based on SBF distances of early-type dwarf galaxies I hope I was able to let readers discover the great potential of this powerful dE distance indicator. Due to its simplicity and efficiency there is little doubt that this method will play an important role in the future 3-D surveying of the local universe, particularly taking advantage of the deep, high quality CCD imaging data produced by the steadily increasing number of wide-field cameras on present and new generation telescopes.

Acknowledgements

HJ would like to express his gratitude to the IAU for the financial support of this most fruitful and enjoyable meeting. Part of my research on dwarf galaxies is supported by the ARC Discovery Grant DP0343156.

References

Arnaboldi, M., *et al.* 2002, *AJ* 123, 760
Barazza, F.D., Binggeli, B. & Jerjen, H. 2002, *A&A* 391, 823
Barnes, D.G., *et al.* 2001, *MNRAS* 322, 486
Bertelli, G., *et al.* 1994, *A&AS* 106, 275
Binggeli, B. & Cameron, L.M. 1991, *A&A* 252, 27
Binggeli, B. & Jerjen, H. 1998, *A&A* 333, 17
Binggeli, B., Sandage, A. & Tammann, G.A. 1988, *ARAA* 26, 509
Bode, P., Ostriker, J.P. & Turok, M. 2001, *ApJ* 556, 93
Carignan, C. & Freeman, K.C. 1988, *ApJ* 332, 33
Côté, P. *et al. this volume*
Dekel, A. & Silk, J. 1986, *ApJ* 303, 39
Doyle, M. *et al.* 2005, *MNRAS*, 361, 34
Dunn, L. & Jerjen, H. *this volume*
Ferguson, A., Gallagher, J. & Wyse, R. 2000, *AJ* 120, 821
Gnedin, N.Y. 2000, *ApJ* 542, 535
Grebel, E.K. 2001, *Astrophysics and Space Science Supplement* 277, 231
Haiman, Z., Thoul, A.A. & Loeb, A. 1996, *ApJ* 464, 523
Harris, W.E., Durrell, P.R., Pierce, M.J. & Secker, J. 1998, *Nature* 395, 45
Hilker, M., Mieske, S. & Infante, L. 2003, *A&A* 397, L9

Jerjen, H., Binggeli, B. & Barazza, F. 2004, *AJ* 127, 771
Jerjen, H., Freeman, K.C. & Binggeli, B. 1998, *AJ* 116, 2873
Jerjen, H., Freeman, K.C. & Binggeli, B. 2000, *AJ* 119, 166
Jerjen, H., Kalnajs, A. & Binggeli, B. 2000, *A&A* 358, 845
Jerjen, H. *et al.* 2001, *A&A* 380,90
Kambas, A. *et al.* 2000, *AJ* 120, 1316
Karachentsev, I.D. *et al.* 1999, *A&A* 352, 399
Karachentsev, I.D. *et al.* 2000, *A&A* 363, 117
Karachentsev, *et al.* 2004, *AJ* 127, 2031
Kilborn., V. *et al.* 2000, *AJ* 120, 1342
Klypin, A. *et al.* 1999, *ApJ* 522, 82
Koribalski, B.S. *et al.* 2004, *AJ* 128, 16
Krumm, N. & Burstein, D. 1984, *AJ* 89, 1319
Mao, S. & Mo, H.J. 1998, *MNRAS* 296, 847
Mastropietro, C. *et al.* 2005, astro-ph/0411648
Mei, S. *et al.* 2005, *ApJ* 625, 121
Mieske, S. & Hilker, M. 2003, *A&A* 410, 445
Moore, B., Lake, G. & Katz, N. 1998, *ApJ* 495, 139
Moore, B. *et al.* 1999, *ApJ* 524, L19
Neilsen, E.H., Jr. & Tsvetanov, Z.I. 2000, *ApJ* 536, 255
Oosterloo, T. *et al.* *IAU Symposium No. 217* (San Francisco: Astronomical Society of the Pacific), p. 48
Phillipps, S. *et al.* 1998, ApJ, 493, L59
Rekola, R., Jerjen, H. & Flynn, C. 2005, *A&A* 437, 823
Ryder, S.D. *et al.* 2001, *ApJ* 555, 232
Spergel, D.N. & Steinhardt, P.J. 2000, *Phys Rev Lett* 84, 3760
Thoul, A.A. & Weinberg, D.H. 1996, *ApJ* 465, 608
Tonry, J.L. *et al.* 1997, *ApJ* 475, 399
Tonry, J.L. *et al.* 2001, *ApJ* 546, 681
Tonry, J.L. & Schneider, D.P. 1988, *AJ* 96, 807
Trentham, N. & Hodgkins, S. 2002, *MNRAS* 333, 423
Trentham, N. & Tully, R.B. 2002, *MNRAS* 335, 712
Tully, R.B. 2005, *ApJ* 618, 214
van den Bergh, S. 1972 *ApJ* 171, L31
Warren, B. E., Jerjen, H. & Koribalski, B. S. 2004, *AJ* 128, 1152
West, M.J. & Blakeslee, J.P. 2000, *ApJ* 543, L27
Worthey, G. 1994, *ApJS* 95, 107

Discussion

NELSON CALDWELL: It's interesting that the dwarfs you have observed so far fall either on the linear or parabolic part of the $\overline{m}$ vs. B-R relation. Naively, one might expect a range of star formation histories in galaxies and so a scattering of points between the two curves. With more objects observed, do you think the dEs will fill in the area between the curves?

HELMUT JERJEN: A possible explanation for this empty region is that the colour of the RGB is largely independent of age but is strongly dependent on metallicity (Fig. 1 of Da Costa 1997) and that for a fixed metallicity the colour of the HB has a discontinuity from blue for the oldest populations ($\sim$15 Gyr) to red at younger ages ($\leqslant$10 Gyr; Fig. 4 of Da Costa 1997). The latter effect is more marked for lower metallicities. These two trends describe the behavior of $(B-R)$ and $\overline{M}_R$ in Fig. 1. In the models, only the very old and metal-poor populations are located on the parabolic curve. With an increasing fraction of younger stars the total population gets redder and the model values reach quickly the linear curve.

STEFFEN MIESKE: How do you know that the models you are using for the SBF calibration are correct?

HELMUT JERJEN: In Jerjen, Freeman & Binggeli (1998) we compared the Worthey+Yale and Worthey+Padua results with empirical data and found much better agreements with the latter models. This good agreement still exists after using the latest SBF data for seven calibration galaxies shown in Fig. 1. But ultimately we are aiming for a model-independent SBF calibration based solely on nearby dEs.

IGOR CHILINGARIAN: For the color-metallicity relation you assume old ages (>8 Gyr), but in the paper of Geha *et al.* who observed several tens of dEs in Virgo the mean age is estimated 3 Gyr (using Lick indices). How will this affect the relation?

HELMUT JERJEN: The $(B-R,\overline{M})$ data points for stellar populations of younger ages ($<$ 8 Gyr) would move down the linear branch in Fig. 1 blending with the results for older stellar populations but lower metallicities. Caldwell, Rose & Concannon (2003) also tried to disentangle the effects of age and metallicity in a sample of Virgo early-type galaxies. The Balmer lines in these dwarfs were interpreted within the Lick index system as being primarily caused by young age, rather than by low metallicity leading to galaxy ages ranged from 1–16 Gyr. This interpretation stands in contrast to Maraston & Thomas (2000, 2001), who showed that a mix of an old metal-rich and an old metal-poor component can produce strong Balmer and metal lines without invoking a young population. I also would like to recall another result from age-sensitive narrow-band photometry in the modified Strömgren filter system by Rakos *et al.* (2001). These authors derived a mean age of 10 ± 1 Gyr for a sample of 27 Fornax dEs.

Near-Field Cosmology with Dwarf Elliptical Galaxies
Proceedings IAU Colloquium No. 198, 2005
H. Jerjen and B. Binggeli, eds.

doi:10.1017/S174392130500400X

Environmental dependencies in the luminosity function of galaxies

R. Brent Tully

Institute for Astronomy, University of Hawaii, Honolulu, HI 96822, USA
email: tully@ifa.hawaii.edu

Abstract. The evidence is becoming strong that the luminosity function of galaxies varies with environment. Higher density, more dynamically evolved regions appear to have more dwarfs per giant. The situation is becoming clearer as a result of wide field imaging surveys with the Canada-France-Hawaii and Subaru telescopes and spectroscopy of faint dwarfs with the Keck telescope. We report here on extensive observations of the small but dense NGC 5846 Group. The faint end of the luminosity function rises relatively steeply in this case.

Keywords. galaxies: luminosity function, galaxies: dwarf

1. Introduction

At the inception of this program, circa 1997, the following thoughts provided motivation. The properties of the luminosity function of galaxies at the time were poorly constrained by observations at the faint end. Most of the observational attention had been directed toward rich clusters of galaxies because they could provide good statistics and contrast against interlopers. However rich clusters are relatively distant and at faint levels the luminosity function is only defined after statistical corrections for contamination. Experience in and near the Local Group teaches us that the vast majority of dwarf galaxies are low surface brightness, a property that distinguishes them from background objects. If we confine a study to groups that are clean from the perspective of confusion from nearby structures then we might reliably identify group members without need for statistical corrections. By looking at nearby groups, we can observe down to faint magnitudes, over a variety of environments including some where the density is low.

We began with a study of the Ursa Major (UMa) Cluster because we were familiar with the region from work on the bright members (Tully *et al.* 1996; Verheijen & Sancisi 2001). The UMa Cluster almost qualifies as Abell richness class 0 but it is different from most clusters to receive attention because it is irregular, low density, and spiral rich. The cluster is at roughly the same distance as the Virgo Cluster and subtends 100 sq. deg. We sampled 15% of the area with strips along the major and minor axes of the elongated cluster with observations using the 12k imager at the Canada-France-Hawaii Telescope and the neutral hydrogen receiver at the Very Large Array. We conducted a double blind search for dwarf galaxies with Neil Trentham conducting the optical search and Marc Verheijen conducting the HI search.

The HI survey turned up all the previously known galaxies including a couple that had excaped detection in HI but only uncovered 11 new objects. The HI mass flunction in UMa was demonstrated to be flat at low masses (Verheijen *et al.* 2000). The optical survey turned up 3 times as many candidates, still not a large number. Probably many, if not most, of these candidates would have been detected in HI if the sensitivity achieved with the VLA could have been pushed a factor 10 below our $10^7 M_\odot$ limit. In any event,

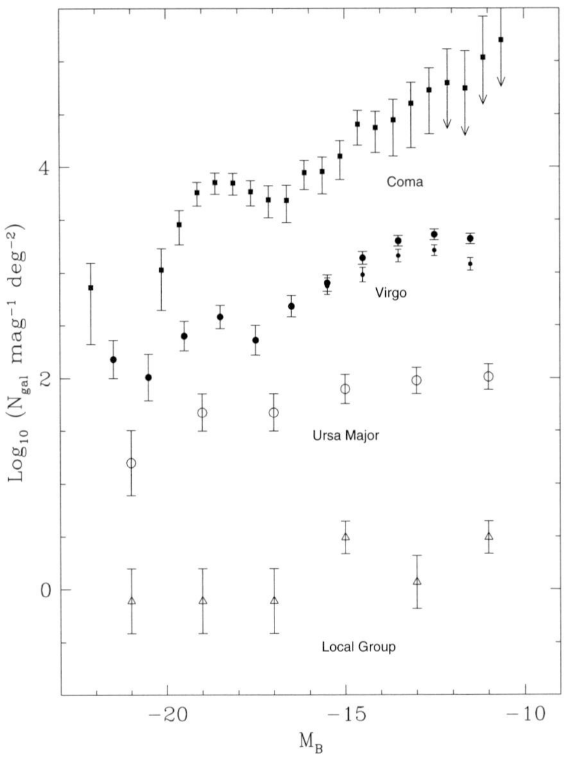

Figure 1. *B*-band luminosity functions for the Coma, Virgo, Ursa Major clusters and the Local Group. Vertical scales are shifted for clarity. Downward arrows in the case of the Coma Cluster reflect background contamination uncertainties.

the optical luminosity function in UMa is almost flat at the faint end (see Figure 1 extracted from Trentham *et al.* 2001).

2. Environmental Dependencies?

The luminosity function in the UMa Cluster is flatter than others have found in denser, more dynamically evolved places, reminiscent of the situation found in the Local Group (Klypin *et al.* 1999; Moore *et al.* 1999). The luminosity function in these low density environments is *much flatter* than the mass function anticipated by ΛCDM hierarchical clustering theory (Sheth & Tormen 1999). The theory predicts that the mass spectrum is *steeper* in lower density environments, not flatter (Sigad *et al.* 2000). What could explain an increased departure of the luminosity function from the mass function in lower density environments?

Thoul & Weinberg (1996), Gnedin (2000) and others have discussed the way reionization could have suppressed the accumulation of gas in small dark matter halos. Hot intergalactic gas has too much energy to fall into a small dark matter halo that collapses after the epoch of reionization. By contrast, a small halo that collapses before reionization will be accompanied by cold gas. Early forming low mass halos would contain gas and form stars while late forming low mass halos would not contain gas, hence not be the sites of star formation. This scenario could explain the trends seen in the properties

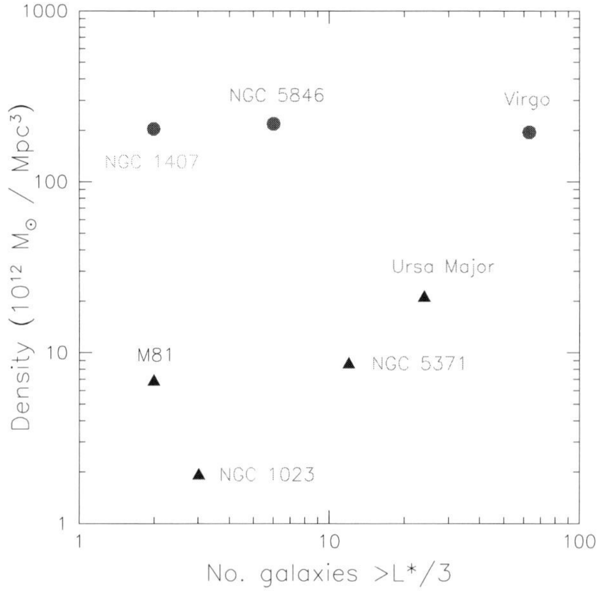

Figure 2. Seven cluster or group environments will be surveyed over the course of this program. The environments are characterized by a density parameter (virial mass / volume) and a richness parameter (number of members more luminous than $L^\star/3$). Groups of predominantly early-type and late-type galaxies are identified by circles and triangles, respectively. The present paper summarizes results obtained for the dense but relatively poor NGC 5846 Group.

of the luminosity function because low mass halos collapse earlier in environments that become rich clusters than in low density environments like those that become the Local Group or the Ursa Major Cluster (Tully *et al.* 2002). Semi-analytic models suggest that gas retention is strongly favored in places that become dense, massive clusters because some halos form very early. Even in these densest places, though, *most* halos will form after reionization; ie, gas will not be pulled in with the collapse of most halos.

3. Pilot Program to Survey a Variety of Environments

Our next step was to take a quick look at a wide variety of environments within the Local Supercluster. Subaru SuprimeCam was used to image the inner parts of 5 additional groups, which added to UMa gave us 6 samples (Trentham & Tully 2002). Trends were seen in the suspected sense: more dynamically evolved regions have higher dwarf/giant ratios. The mean slope at the faint end of the luminosity function is characterized by a Schechter (1976) slope parameter $\alpha = -1.19 \pm 0.03$. However, not enough sky was surveyed in each group for sufficient statistics on individual groups.

The problem required much more sky coverage. Entire groups should be surveyed and groups with very different properties should be considered. We began a program with the wide field imagers on the Canada-France-Hawaii Telescope, culminating in the use of the 1 sq. deg. field MegaCam. Fields were chosen to maximize the range of group properties from numerically rich to poor in bright galaxies and from dense and dynamically evolved to sparse and dynamically young. The seven groups identified in Figure 2 have been or will be observed. We report here on the case of the NGC 5846 Group.

4. The Small, Dense NGC 5846 Group

The number of luminous galaxies in the NGC 5846 Group is modest, with only 3 $L^{\star}$ galaxies, but the group is dynamically evolved, with almost all members of early type. It is a dumbbell system: members congregate around the two bright ellipticals NGC 5813 and NGC 5846. Extended X-ray emission is centered on these two major galaxies. The distance to the group is 26 Mpc. The velocity dispersion, from 86 redshifts, is 322 km/s, resulting in a virial mass estimate of $8 \times 10^{13} M_{\odot}$ and mass to light ratio of 320 in solar units.

The definition of the luminosity function to faint levels depends on the ability to identify group members free of contamination. We first select candidates based on quantitative surface brightness criteria that would assure inclusion of dwarfs seen in the local neighborhood and redshift-confirmed members of the UMa and Virgo clusters. These low surface brightness candidates were then evaluated on morphological grounds and given qualitative group membership ratings (1) probable, (2) possible, (3) plausible, and (4) implausible.

Our rating scheme was then tested in two ways. The first and most rigorous test was to obtain new velocities. The Sloan Digital Sky Survey (SDSS) now provides redshifts for essentially all our galaxies with $R < 17$ ($M_R < -15$). We have obtained additional spectra for galaxies up to 2 mag fainter with Keck Telescope. It was found that (a) all 24 galaxies with new velocities that were rated 1 and 2 are members, (b) 16 of 23 galaxies rated 3 are members, (c) only 1 galaxy rated 4 has a new velocity that shows it to be a member, and (d) 5 high surface brightness galaxies have turned out to be members. The second test involves intercomparisons of spatial correlations. Galaxies rated 1 and 2 are as highly correlated with the group as galaxies confirmed as members on the basis of velocities, indicating that essentially 100% of these objects belong to the group. By contrast, the correlation with the galaxies rated 3 indicates that $50\% \pm 10\%$ of these candidates are in the group. It is concluded that 255 ± 15 of our candidates are associated with the group. Of the dwarfs, 80% are early types.

There turns out to be a clean separation between high and low surface brightness objects in plots of surface brightness vs. luminosity. There are 3 very compact ellipticals; one rather like M32. It turns out that these three unusual objects are the *nearest neighbors* in projection to the dominant NGC 5846 galaxy. We suggest that these 3 galaxies have been tidally truncated.

The identification of group members allows us to build the luminosity function seen in Figure 3. There is a saddle point in the observed distribution at $M_R \sim -20$, similar to a feature seen in other samples, and not something that can be described by a Schechter function. Following our procedure in the Subaru pilot program (Trentham & Tully 2002), we fit the luminous, high surface brightness galaxies with a Gaussian function ($M_g = -21.0$, $\sigma_g = 1.1^m$) and fit the faint, low surface brightness component with a Schechter function ($M_d^{\star} = -19.5$, $\alpha_d = -1.36$). The luminosity function is well defined to a completion limit of $M_R = -12$.

The luminosity function for the NGC 5846 Group is compared with what was found from other environments in Figure 4. The faint end slope of $\alpha = -1.36 \pm 0.04$ is significantly steeper than the mean slope for the 5 Subaru groups of $\alpha = -1.19 \pm 0.03$, and steeper still than the slope $\alpha \sim -1.0$ found for the UMa sample. However it is shallower than the Press-Schechter mass function slope. The NGC 5846 Group environment is dense and dynamically evolved, even though the number of bright galaxies is modest and the total mass is not extreme. Our information is still limited but we are seeing a

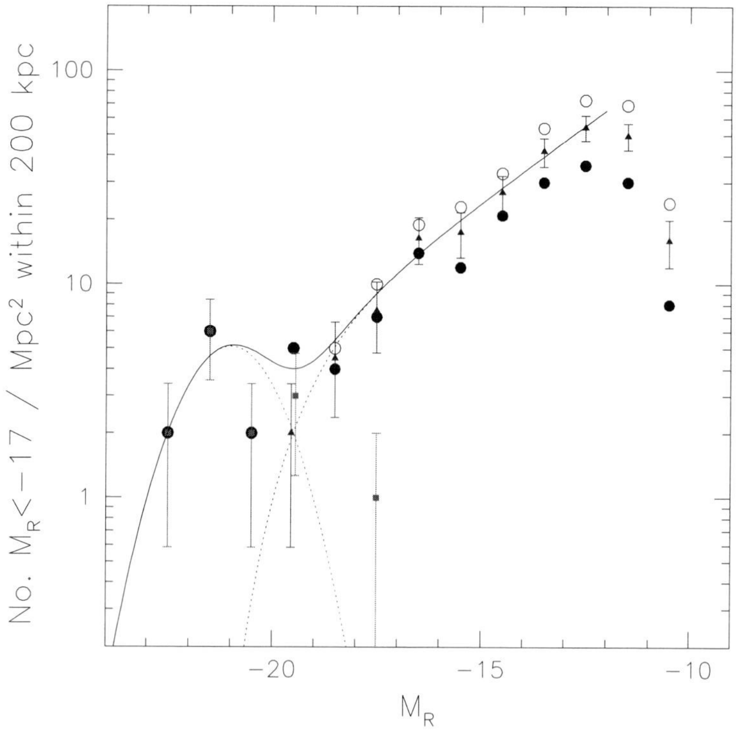

Figure 3. NGC 5846 Group luminosity function normalized by the surface density of galaxies with $M_R < -17$ at 200 kpc radius. Large filled circles: number of galaxies per half-magnitude bin either confirmed as a group member with a redshift or almost a certain member with a morphological rating 1 or 2. Large open circles: same except also including morphological rating 3. Small filled squares: high surface brightness only. Small filled triangles: low surface brightness only, splitting the difference between the values represented by the filled and open large circles. Dotted curves: separate fits to the high and low surface brightness distributions. Solid curve: sum of the two component fits. Parameters of the fit: $M_g = -21.0 \pm 0.3$, $\sigma_g = 1.1 \pm 0.2$, $N_g = 5.1 \pm 1.2$, $M_d = -19.5 \pm 0.6$, $\alpha_d = 1.36 \pm 0.04$, $N_d = 5.9 \pm 0.4$.

pattern that appears consistent with the reionization model (Tully *et al.* 2002). Details of the NGC 5846 study are presented by Mahdavi *et al.* (2005).

Acknowledgements

My collaborators in this work are Neil Trentham, Andisheh Mahdavi, Rachel Somerville, and Marc Verheijen. Support is provided by the US National Science Foundation award AST-03-07706.

References

Gnedin, N.Y. 2000, *ApJ* 542, 535

Klypin, A., Kratsov, A.V., Valenzuela, O. & Prada, F. 1999, *ApJ* 522, 82

Mahdavi, A., Trentham, N. & Tully, R.B. 2005, *AJ* to be submitted

Moore, B., Ghigna, S., Governato, F., Lake, G., Quinn, T., Stadel, J. & Tozzi, P. 1999, *ApJ* 524, L19

Schechter, P.L. 1976, *ApJ* 203, 297

Sheth, R.K. & Tormen, G. 1999, *MNRAS* 308, 119

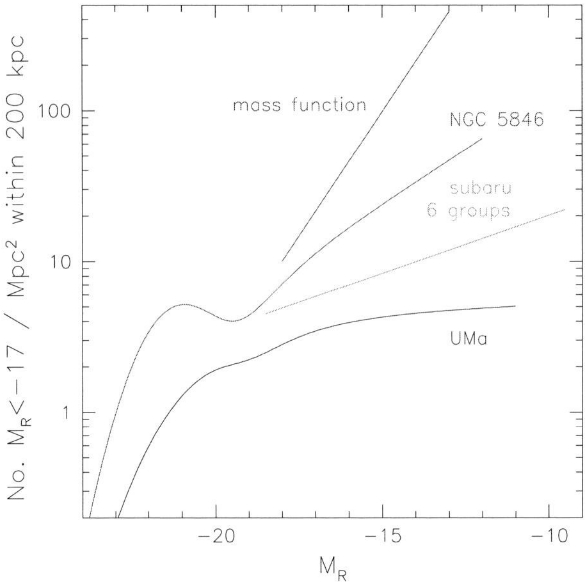

Figure 4. Schematic comparison of the NGC 5846 Group luminosity function with other cases. The curve representing the NGC 5846 Group is the solid curve in Fig. 3. The curves labeled 'subaru 6 groups' and 'UMa' are taken from Trentham & Tully 2002. The 6 groups curve represents an average found for the dwarf regime in 6 groups discussed by Trentham & Tully 2002. The density normalizations are carried out in a consistent way in these separate cases. The steeper line labeled 'mass function' indicated the low mass slope of a modified Press-Schechter distribution with arbitrary vertical scaling.

Sigad, Y. *et al.* 2000, *astro-ph/0010222*
Thoul, A.A. & Weinberg, D. 1996, *ApJ* 465, 608
Trentham, N. & Tully, R.B. 2002 *MNRAS* 335, 712
Trentham, N., Tully, R.B. & Verheijen, M.A.W. 2001 *MNRAS* 325, 385
Tully, R.B., Somerville, R.S., Trentham, N. & Verheijen, M.A.W. 2002, *ApJ* 569, 573
Tully, R.B., Verheijen, M.A.W., Pierce, M.J., Huang, J.S., & Wainscoat, R.J. 1996, *AJ* 112, 2471
Verheijen, M.A.W. & Sancisi, R. 2001, *A&A* 370, 765
Verheijen, M.A.W., Trentham, N., Tully, R.B. & Zwaan, M.A. 2000 2000, *ASP Conf. Ser.* 218, 263

Discussion

GALLAGHER: (Great project!) Have you found a way to combine your model associating cluster dwarfs with primordial fluctuations with the observations of intermediate ages (velocity distributions, stellar pop ages) in cluster dE/dSph galaxies?

TULLY: Archaic populations are found in almost all dwarfs where an appropriate observation has been made. As you say, *some* dE/dSph manifest intermediate age populations as well. Our model suggests why some dwarf halos would start out with gas and have the capability of forming stars. It says nothing concerning the *rate* of conversion of this gas into stars.

Near-Field Cosmology with Dwarf Elliptical Galaxies
Proceedings IAU Colloquium No. 198, 2005
H. Jerjen and B. Binggeli, eds.

doi:10.1017/S1743921305004011

Diffuse light in $z \sim 0.25$ galaxy clusters: constraining tidal damage and the faint end of the luminosity function

Stefano Zibetti[1] and Simon D.M. White[2]

[1]Max-Planck-Institut für Extraterrestrische Physik
Gießenbachstraße, D-85748, Garching bei München, Germany
email: szibetti@mpe.mpg.de

[2]Max-Planck-Institut für Astrophysik – Garching bei München, Germany

Abstract. The starlight coming from the intergalactic space in galaxy clusters and groups witnesses the violent tidal interactions that galaxies experience in these dense environments. Such interactions may be (at least partly) responsible for the transformation of normal star-forming galaxies into passive dwarf ellipticals (dEs).

In this contribution we present the first systematic study of the IntraCluster Light (ICL) for a statistically representative sample (Zibetti *et al.* 2005), which comprises 683 clusters selected between $z = 0.2$ and 0.3 from $\sim$1500 deg^2 in the SDSS. Their ICL is studied by stacking the images in the g-, r-, and i-band after masking out all galaxies and polluting sources. In this way a very uniform background illumination is obtained, that allows us to measure surface brightnesses as faint as 31 mag arcsec^{-2} and to trace the ICL out to 700 kpc from the central galaxy. We find that the local fraction of light contributed by intracluster stars rapidly decreases as a function of the clustercentric distance, from $\sim$40% at 100 kpc to $\sim$5% at 500 kpc. By comparing the distribution and colours of the ICL and of the clusters galaxies, we find indication that the main source of ICL are the stars stripped from galaxies that plunge deeply into the cluster potential well along radial orbits. Thus, if dEs are the remnants of these stripped progenitors we should expect similar radial orbital anisotropies and correlations between the dE luminosity function and the amount of ICL in different clusters.

The diffuse emission we measure is contaminated by faint unresolved galaxies: this makes our flux estimate depend to some extent on the assumed luminosity function, but, on the other hand, allows us to constrain the number of faint galaxies. Our present results disfavour steep ($\alpha < -1.35$) faint-end powerlaw slopes.

Keywords. galaxies: interactions, clusters: general, intergalactic medium, dwarf, elliptical and lenticular, cD, evolution, formation

1. Introduction

The diffuse intracluster light (ICL) is produced by stars which are not bound to any individual galaxy in a cluster, but orbit freely in the global gravitational potential of the cluster. The ICL has now been detected in a number of nearby groups and clusters both by means of deep surface photometry (e.g. Bernstein *et al.* 1995; Feldmeier *et al.* 2002; Gonzalez, Zabludoff & Zaritsky 2005) and by resolving red giant stars (Durrell *et al.* 2002) and planetary nebulae (e.g. Arnaboldi *et al.* 1996). As for the origin of this stellar component, the presence of tidal features and dynamical substructure, that has emerged from a number of studies, strongly supports the hypothesis that intergalactic stars are created via tidal stripping from galaxies and during merger events, rather than *in situ* from isolated extragalactic star forming regions.

Tidal stripping and repeated tidal perturbations in a dense environment (harassment) may be also invoked as the main mechanism responsible for transforming low mass star-forming discs into dwarf elliptical galaxies (dEs) and compact dwarves. The investigation of the systematic properties of the ICL puts quantitative constraints on the tidal damage that galaxies suffer in clusters and, therefore, can provide invaluable clues to understand the formation of dEs.

Zibetti *et al.* (2005) have recently analysed the systematic properties of the diffuse stellar emission in a sample of 683 clusters by stacking their images from the Sloan Digital Sky Survey (SDSS, York *et al.* 2000). In this contribution we will review their results, focusing on the implications for the stripping scenario for the formation of dEs.

2. Sample and photometric technique

The stacking technique aims at obtaining high photometric sensitivity, while strongly reducing the systematic defects that arise in individual observations from flat field inhomogeneities and scattered light from "polluting" (fore- and background) sources. This is achieved by averaging a large number of images of different targets, where the unwanted sources are efficiently masked. Typical required number of images is several hundreds, hence the need for a large imaging database as the SDSS. The results obtained from the stacking of a large number of clusters grant a high statistical significance, but have the disadvantage of lacking information about substructure and cluster-to-cluster variations, that must be derived from the analysis of different subsamples.

For the present study a sample of 683 clusters of galaxies has been drawn from ~1,500 $\deg^2$ of the first data release of the SDSS (Abazajian *et al.* 2003). We have used the maxBCG cluster finder algorithm developed by J. Annis to select clusters in the same richness range as the Abell catalogue, between $z = 0.2$ and 0.3. The redshift constraint has been introduced to limit the size of the stacked images to a single SDSS frame, thus avoiding to mosaic images with different background properties for a given cluster. Moreover, at the median z of the sample the $g - r$ colour conveniently maps the 4000Å break, providing valuable information about the intracluster stellar population.

Lacking any kinematic characterisation, we *define* as ICL all the flux beyond the isophote 25 mag arcsec^{-2} (observed r-band) of any galaxy. All galaxies, except the central brightest cluster galaxy (BCG), are masked to this extent. Bright stars are masked out to 3 times the isophotal radius, to prevent their bright haloes being included in the stacking. Images are then centred on the BCG, size-rescaled to the same metric radius, and intensity-rescaled to the same photometric calibration, corrected for Galactic extinction and cosmological SB dimming. A simple average of the masked images is then computed (stacked image). A stacking of images where only the brightest stars were masked has been performed as well, to obtain a map of the total luminosity of the cluster to be compared to the ICL.

We have tested the efficiency of our masking algorithm and quantified the fraction of galaxy light which fails to be masked. Realistic SB distribution have been used to simulate galaxies in the same observed conditions as for real clusters, with different luminosity functions (LFs) parametrised á la Schechter. Depending on the adopted parameters (and particularly on the powerlaw slope of the faint end α) we find that 5 to 20–25% of the light in galaxies can be missed. As we will show below, this effect can be used along with other observational constraints to put limits to the faint end slope of the LF.

The following *results* are derived assuming that the fraction of unmasked galaxy light is 15%, as obtained for the parameters of the Coma cluster LF ($M_r^* = -21.37$, $\alpha = -1.18$, Mobasher *et al.* 1998), passively evolved to $z = 0.25$.

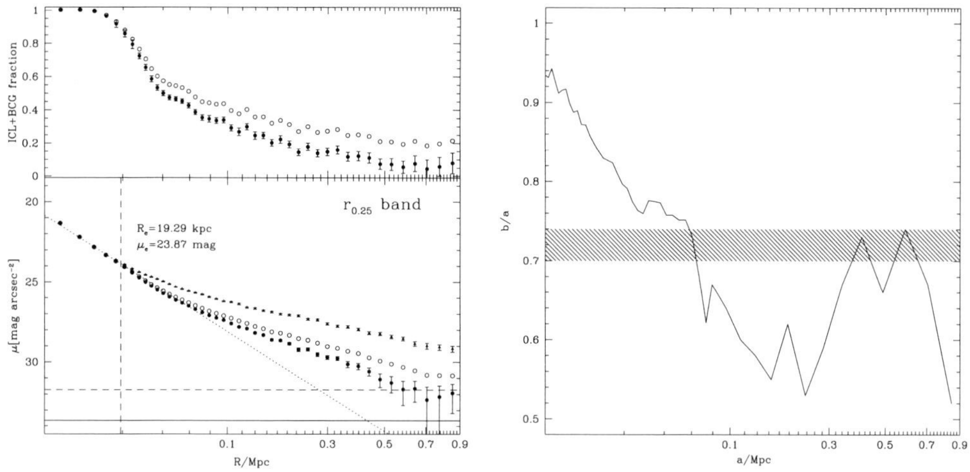

Figure 1. Panel a): r-band surface photometry. The lower section shows the surface brightness profiles for all components (triangles with error bars), for the diffuse light as measured (open circles) and for the corrected ICL (filled circles with error bars). The dotted line represents the best fitting de Vaucouleurs profile to the BCG core, with the vertical dashed line marking R_e. The two horizontal lines show 1σ background uncertainty for the total (dashed) and for the diffuse light. The upper section displays the local relative amount of diffuse light (open circles) and of the corrected ICL+BCG component (filled circles with error bars). Panel b): Isophotal shapes for clusters with a flattened BCG ($b/a < 0.7$ from 2-D de Vaucouleurs fit). The solid line displays the axial ratio of the (uncorrected) diffuse light + BCG, with typical errors of 0.05 in the central regions to 0.1 at 500 kpc. The shaded area is the 1σ interval for the distribution of the total light.

3. Results

By extracting azimuthally averaged SB profiles we can trace the diffuse light out to ~700 kpc from the central galaxy at levels of 31 mag arcsec^{-2}. In the lower panel of fig. 1a) the combined contribution of the BCG and of the ICL is shown by filled circles with error bars. Starting from 80 kpc, the ICL clearly appears as an excess with respect to the inner de Vaucouleurs profile that characterises the BCG. Compared to the total light of all stellar components (triangles), the ICL is remarkably more centrally concentrated, providing a local relative contribution that decreases from 40% of the total SB at 100 kpc, to less than 5% at 600 kpc and beyond. This clearly indicates that the creation/accumulation of intracluster stars is much more efficient in the centre of the cluster. The colour profiles of the ICL ($g - r$ and $r - i$, obtained from the g-, r- and i-band stacked images) are consistent with $(g - r)$ or just marginally redder than $(r - i)$ the average colours of galaxies, thus suggesting that the intracluster stellar population is not dissimilar from that in galaxies. On a subsample of clusters with BCG having a significant flattening ($b/a < 0.7$ from 2D de Vaucouleurs SB fitting), the stacking has been performed after aligning the images along the major axis of the BCG. The isophotal analysis (fig. 1b) shows that the ICL is strongly aligned with the BCG and displays even a higher degree of flattening. In fact, there is a monotonic decrease of b/a from the BCG core out to 300 kpc, where $b/a \sim 0.5$. It is remarkable that the distribution of galaxies is less aligned with the BCG and/or less flattened than the ICL. This may indicate that the intracluster stars are stripped more efficiently from galaxies that move along radial orbits aligned with the main axis of the BCG.

We have studied the dependence of the ICL SB and luminosity on cluster richness and BCG luminosity. The SB of the ICL is enhanced in rich clusters and in those having a luminous BCG, and *vice versa*. The clusters with the faintest BCGs appear particularly deficient in ICL. However, the relative contribution of ICL to the total luminosity within 500 kpc is roughly constant 11% for all clusters.

4. dEs as stripped galaxies?

N-body hydrodynamical simulations (also Mastropietro in this Colloquium) have demonstrated that dynamical harassment, tidal and ram-pressure stripping can transform small star-forming disc galaxies into passive dEs (see also Merritt 1984, Moore *et al.* 1996). It is therefore conceivable that dEs and the ICL share the same origin. The broadband colours are consistent with this scenario, and the fact that the total luminosity of dEs is similar to that of the ICL is very intriguing too. On the other hand, if dEs are the remnants of the galaxies from which a large number of stars have been stripped into the intergalactic space, we should expect (i) that in anisotropic clusters dEs display a similar flat and aligned distribution as the ICL, in addition to some degree of radial orbital anisotropy; and (ii) that the luminosity function of dEs correlates with the amount of ICL. The study of these correlations between ICL and dEs may therefore provide invaluable clues on the origins of these two components and on the violent history of galaxy clusters.

5. Constraints to the faint end of the LF

In all the analysis conducted above, the measured diffuse light is converted into "real" ICL assuming that the fraction of galaxy light that is left unmasked is 15%, according to our simulations for a LF as in Coma. As already mentioned, by varying the parameters of the LF values between 5 and 20–25% are found. However we have a number of constraints that limit the range of variation to <5% around the assumed 15%. M^* is constrained within a few hundredths of mag by directly observed number counts. On the other hand, if a too steep LF is chosen, then the predicted unmasked flux becomes larger than the actual flux in the diffuse component. In particular, if we assume that the LF is independent of radius, then the minimum value of the local fraction of diffuse light represents the maximum fraction of light that might be missed by our masks. Since this fraction is 20% at $R \sim 600$ kpc (see fig. 1a), we calculate that the faint-end powerlaw of a simple Schechter LF cannot be steeper than $\alpha = -1.35$.

References

Abazajian K., *et al.*, 2003, *AJ* 126, 2081

Arnaboldi M., Freeman K.C., Mendez R.H., Capaccioli M., Ciardullo R., Ford H., Gerhard O., Hui X., Jacoby G.H., Kudritzki R.P. & Quinn P.J., 1996, *ApJ* 472, 145

Bernstein G.M., Nichol R.C., Tyson J.A., Ulmer M.P. & Wittman D., 1995, *AJ* 110, 1507

Durrell P.R., Ciardullo R., Feldmeier J.J., Jacoby G.H. & Sigurdsson S., 2002, *ApJ* 570, 119

Feldmeier J.J., Mihos J.C., Morrison H.L., Rodney S.A. & Harding P., 2002, *ApJ* 575, 779

Gonzalez A.H., Zabludoff A.I. & Zaritsky D., 2005, *ApJ* 618, 195

Merritt D., 1984, *ApJ* 276, 26

Moore B., Katz N., Lake G., Dressler A. & Oemler A., 1996, *Nature* 379, 613

York D.G., *et al.*, 2000, *AJ* 120, 1579

Zibetti S., White S.D.M., Schneider D.P. & Brinkmann J., 2005, *MNRAS* 358, 949

Discussion

CONSELICE: Does the fact that the luminosity function of clusters becomes flatter in their cores suggest that the intracluster light is produced by galaxy destruction?

ZIBETTI: It may be the case. Galaxy destruction may remove preferentially faint galaxies, so that a flatter LF can be associated to larger amounts of ICL in the cluster cores.

GALLAGHER: Could a large fraction of the light come from massive galaxies? These could be more vulnerable to mass stripping than dwarfs.

ZIBETTI: Yes, it is possible. Recent simulations (Willman *et al.* 2004) in fact show that roughly half of the expected ICL can originate from galaxies with $L > L^*$. However, since simulations still have many limitations, the hypothesis of dE progenitors being the primary contributors of ICL cannot be ruled out, although I don't want to push too much for this hypothesis.

Near-Field Cosmology with Dwarf Elliptical Galaxies
Proceedings IAU Colloquium No. 198, 2005
H. Jerjen and B. Binggeli, eds.

doi:10.1017/S1743921305004023

Optimizing the tip of the red giant branch distance estimator

L. Rizzi[1], R.B. Tully[1], D. Makarov[1,2], L. Makarova[1,2], S. Sakai[3], E.J. Shaya[4]

[1]Institute for Astronomy, University of Hawai'i, 2680 Woodlawn Drive, HI 96822, USA
email: rizzi@ifa.hawaii.edu

[2]Russian Academy of Sciences, Special Astrophysical Observatory

[3]University of California,Division of Astronomy and Astrophysics, Los Angeles

[4]GSFC, RITSS & U. of Maryland

Abstract. Until very recently, our knowledge of the local peculiar velocity field has been severely hampered by the lack of reliable distance measurements. HST has dramatically changed this situation, allowing astronomers to obtain accurate distances to more than 150 nearby galaxies. This number could easily reach 400 if enough observing time would be dedicated to snapshot observation of the objects in the catalog of Karachentsev *et al.* (2004). Such a dense grid of objects correctly placed in their 3D position would provide key information on the amplitude of peculiar motions, the radial domain of bound groups, the clustering and morphological segregation properties of galaxies, and the incidence of extreme dwarfs galaxies. The key instrument to measure distances with HST is the Tip of the Red Giant Branch technique. The full exploitation of this powerful distance estimator requires a deeper understanding of the possible sources of errors and biases, such as the absolute calibration of the I-band magnitude of the tip and its dependency on age and metallicity of the underlying population, the possible contamination by AGB stars, the breakdown of the methodology in sparsely populated colour-magnitude diagrams and when the tip is close to the photometric limit.

Keywords. galaxies: distances and redshifts, stars: distances

1. Introduction

Until the year 2000 very little data had been available to map the peculiar velocity field of galaxies in the vicinity of the Local Group. Noticeably, this was due to the lack of information on distances, while extensive compilations of radial velocities do exist that include information for thousands of galaxies. The knowledge of the peculiar velocity field plays a key role in putting reliable constraints on the current cosmological models. The Local Group itself is in a highly nonlinear collapse regime, and in a larger volume deviations from the Hubble flow can be expected due to the gravitational influence of nearby groups as well as by Virgo-centric and Great Attractor flow. Earlier hints that local expansion flow is very quiet (Tamman & Kraan 1978,Tully 1982) have been recently confirmed (Karachentsev *et al.* 2002). It is now appreciated that this cold flow is the signature of the Dark Energy (Baryshev *et al.* 2001) and that the earlier the transition from Dark Matter to Dark Energy dominance, the lower the amplitude of local peculiar velocities.

HST has dramatically improved our knowledge of the local peculiar velocity field thanks to its superior angular resolution. Using the Tip of the Red Giant Branch (TRGB) distance estimator, it's now possible to measure distances out to 5 to 7 Mpc with a single orbit, thus making an extremely efficient use of telescope time. For this reason, our group has started a long-term observational campaign aimed at collecting single-orbit images of the largest possible number of galaxies of the local universe. The project consists of two major parts. The first is an archive proposal aimed at collecting all the useful observations already obtained in the past with WFPC2 . All these observations will be reduced again using a uniform recipe. The second part of the project is based on several snapshot proposal submitted and accepted from 1999 to now, to get both WFPC2 and ACS images of the additional galaxies. An important step of this project is the optimization of the TRGB distance estimator. Indeed, it's important to understand what is the best method to identify the tip, to assess the uncertainties related to the absolute calibration of the tip, to measure the possible errors related to the presence of an AGB population, and to understand what happens when we reach the photometric limit. In this contribution, I will focus mainly on the different methods for detecting the tip, and I will discuss the possible biases related to AGB contamination, poorly populated Color-Magnitude Diagrams (CMD), and the detection of the tip near the photometric limit.

2. Testing the TRGB detection methods

2.1. *Crowding, Photometric Errors, AGB contamination*

Several different methods have been developed to detect the level of the RGB tip. With few exceptions, they can be grouped in two main categories, the first based on edge detector (ED) algorithms, and the second based on maximum-likelihood (ML) analysis. The ED algorithm we tested is the Sobel filter with the kernel $[-1, 0, +1]$ (Lee, Freedman, & Madore 1993). We tested this method applying it to both the linear and the logarithmic I-band luminosity function of our objects. We also tested two different ML algorithms. The first was presented by Méndez *et al.* (2002). The second has been recently developed by Makarov *et al.* (2005) and it is derived from the previous one, with the additional possibility to take into account the asymmetric distribution of photometric errors known as *bias* and the completeness function, both measured directly on artificial stars experiments.

All the methods were applied to synthetic color-magnitude diagrams (CMD's), constructed using the ZVAR code (Bertelli *et al.* 1992). The photometric errors and the completeness effects were applied using a suitable set of artificial star experiments. To avoid biases, no attempt at modeling the errors or the completeness function was made, and the artificial stars table was actually used as a look-up table (see Rizzi *et al.* 2002 for details). Two different star formation histories were adopted, the first consisting of an old single episode about 14 Gyr ago (closely reproducing the case of a globular cluster), the second based on a continuous star formation activity from 14 Gyr ago until now. The model galaxies were put at distance moduli ranging from 27 to 30, the latter being the extreme case in which the TRGB is at the detection limit of the photometry.

The results of our simulations are shown in Figure 1. Left and right panels show the contribution to the total distance error due to the TRGB detection alone, for all the galaxies in the sample (distance moduli up to 30, or 10 Mpc). Results for globular cluster-like and continuous star formation histories are shown in the two panels, respectively. ED methods, applied both to linear and logarithmic luminosity function, and ML methods, both in the version of Méndez *et al.* (2002) and Makarov *et al.* (2005), are shown. Left and right panel of Figure 1 clearly show that is a significant difference in the behavior of the methods we tested. In particular, it is easily appreciated that for the nearest cases all the methods give very good results. Moving to more distant galaxies, ED methods seem

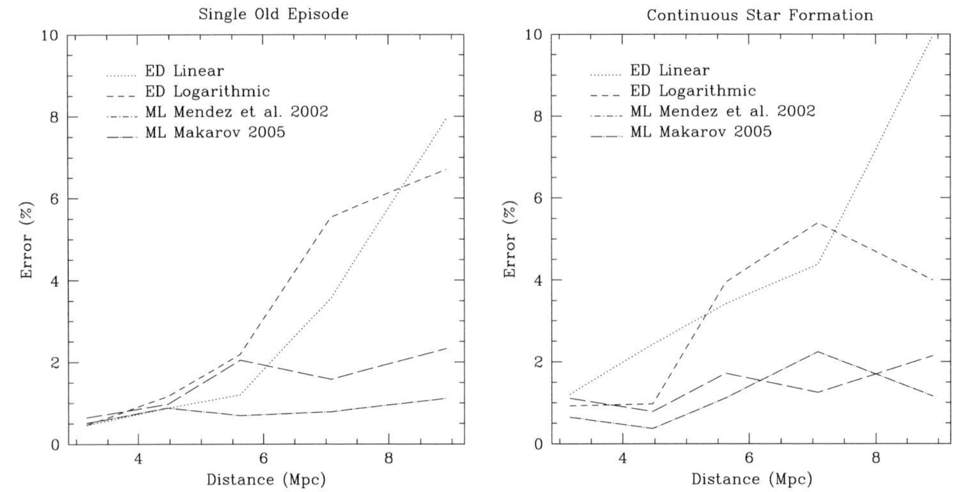

Figure 1. Contribution to the total distance error due to the TRGB detection alone. Left panel: Globular cluster-like galaxies. Right panel: Galaxies with continuous star formation.

to be more affected by the problems connected with the degraded photometric quality and by the increased crowding.

Right panel of Figure 1 shows what happens in a galaxy with a very strong AGB contamination, obtained by using a continuous star formation history. In this case, the RGBT is superimposed on an extended and bright AGB component. Comparison of left and right panel of this Figure clearly shows that ML methods are basically insensitive to the presence of AGB stars, while ED starts to produce significant errors as soon as the galaxies are more distant than 4 Mpc.

2.2. *Poorly populated CM diagrams*

To simulate the effect of poorly populated CM diagrams, we investigated the case of a globular cluster-like galaxy, and an intermediate distance of $(m - M)_0 = 28$. A rather common way to quantify the star population in a CM diagram is the number of stars in the first magnitude bin after the tip (hereafter, $N^{(-1)}$). We generated galaxies with $N^{(-1)}$ between 300 and 30 and applied the detection methods to all of them. Results are shown in Figure 2

Left panel of Figure 2 shows that there is a general trend for a large scatter in the TRGB detection when $N^{(-1)}$ falls below 100. This is consistent with previous similar studies (e.g., Freedman & Madore 1995). Careful scrutiny of the this plot also shows that the average dispersion of measurements is significantly lower at any $N^{(-1)}$ for ML methods, compared to ED methods. This is further demonstrated by the right panel of Figure 2, that shows the r.m.s. of TRGB detections against $N^{(-1)}$. Not only the r.m.s. of ED methods is always higher, but ED methods also tend to break down at a $N^{(-1)}$ of about 100 stars, while ML methods seems to be able to produce reasonable results at least down to 50 stars. Incidentally, we also note that we don't find in our simulations the large systematic deviations pointed out by Madore & Freedman (1995). The difference is most likely due to the different type of simulated galaxies used for the simulations. Madore & Freedman (1995), indeed, use fiducial lines from globular clusters populated with power-laws distributions, so they don't have stars brighter than the TRGB. Removing stars from these simulated galaxies can only produce fainter TRGB detection, while the use of

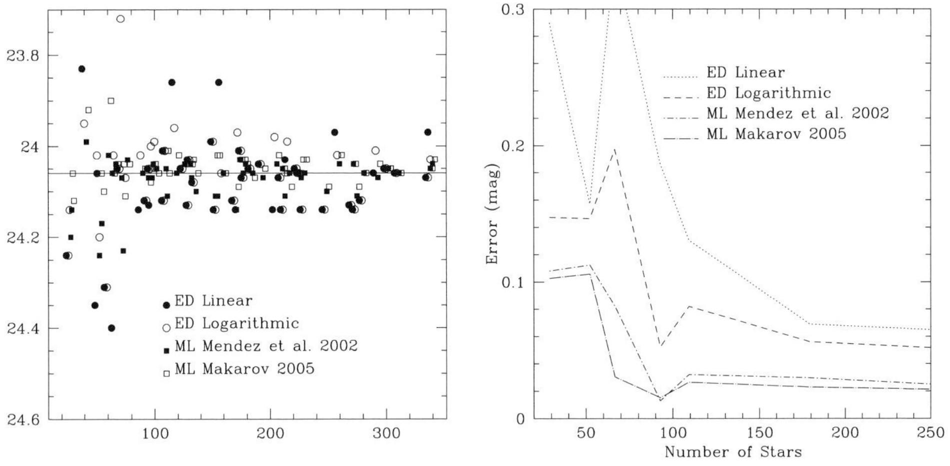

Figure 2. TRGB detection against the number of stars in the first magnitude bin. Left panel: Deviations from the expected TRGB position at 24.06. Right panel: R.m.s. of the different methods against number of stars in the first magnitude bin.

synthetic stellar populations produced with stellar models results in deviations in both directions.

3. Conclusions

We applied different TRGB detection methods to a set of simulated galaxy. We verified that ML methods are less sensitive to crowding and photometric errors than ED methods. We also verified that even a prominent AGB component does not significantly affect the detection. Finally, we applied the methods to poorly populated diagrams, and found that ED methods produce significantly higher errors for each level of $N^{(-1)}$. ML methods seem to be able to detect TRGB down to a level of only $N^{-1} = 50$.

References

Baryshev, Y.V., Chernin, A.D. & Teerikorpi, P. 2001, *A&A* 378, 729

Bertelli, G., Mateo, M., Chiosi, C. & Bressan, A. 1992, *ApJ* 388, 400

Madore, B.F. & Freedman, W.L. 1995, *AJ* 109, 1645

Lee, M.G., Freedman, W.L. & Madore, B.F. 1993, *ApJ* 417, 553

Makarov, D., Makarova, L., Rizzi, L., Tully, B., Sakai, S. & Shaya, E., 2005, *in preparation*

Méndez, B., Davis, M., Moustakas, J., Newman, J., Madore, B.F. & Freedman, W.L. 2002, *AJ* 124, 213

Karachentsev, I.D., Karachentseva, V.E., Huchtmeier, W.K. & Makarov, D.I. 2004, *AJ* 127, 2031

Rizzi, L., Held, E.V., Bertelli, G., Nasi, E., Saviane, I. & Vallenari, A. 2002, *ASP Conf. Ser. 274*, Observed HR Diagrams and Stellar Evolution, 490

Tammann, G.A. & Kraan, R. 1978, *IAU Symposium 79*, Large Scale Structures in the Universe, 71

Tully, R.B. 1982, *ApJ* 257, 389

Near-Field Cosmology with Dwarf Elliptical Galaxies
Proceedings IAU Colloquium No. 198, 2005
H. Jerjen and B. Binggeli, eds.

doi:10.1017/S1743921305004035

RASS-SDSS galaxy cluster survey. A ubiquitous dwarf galaxy population in clusters

P. Popesso[1], A. Biviano[2], H. Böhringer[3] and M. Romaniello[1]

[1] European Southern Observatory, Karl Scharzschild Strasse 2, D-85748
email:ppopesso@eso.org

[2] Osservatorio Astronomico di Trieste, via G. B. Tiepolo 11, I-34131, Trieste, Italy

[3] Max-Planck-Institut fur extraterrestrische Physik, 85748 Garching, Germany

Abstract. We analyze the Luminosity Functions (LFs) of a subsample of 69 clusters from the RASS-SDSS galaxy cluster catalog. The cluster LFs are computed within the physical sizes of the galaxy systems, r_{500} and r_{200}. When calculated within the cluster physical sizes, all the cluster LFs appear to have the same shape, well fitted by a composite of two Schechter functions with a marked upturn and a steepening at the faint-end.

We determine the composite LF for early- and late-type galaxies, where the typing is based on the galaxy $u - r$ colors. The late-type LF is well fitted by a single Schechter function with a steep slope. The early-type LF instead cannot be fitted by a single Schechter function, and a composite of two Schechter functions is needed. The faint-end upturn of the global cluster LF is due to the early-type cluster galaxies. The shape of the bright-end tail of the early-type LF does not seem to depend upon the local galaxy density or the distance from the cluster center. The late-type LF shows a significant variation only very near the cluster center. On the other hand, the faint-end tail of the early-type LF shows a significant and continuous variation with the environment.

Keywords. galaxies: fundamental parameters, galaxies: clusters: general, galaxies: dwarf

1. Introduction

The galaxy Luminosity Function (LF) is a fundamental tool for understanding galaxy evolution and faint galaxy populations. The shape of the cluster LF provides information on the initial formation and subsequent evolution of galaxies in clusters while the slope of the faint-end indicates how steeply the dwarf number counts rise as a function of magnitude.

Much work has been done on the cluster LF, with various groups finding differences in its shape and the faint-end slope. Whether the LF of cluster galaxies is universal or not, and whether it is different from the LF of field galaxies are still debated issues. Moreover, the LF of cluster galaxies is typically observed to steepen faint-ward of $M_g \sim -18$, with power-law slopes $\alpha \sim -1.8 \pm 0.4$. This corresponds to the debated upturn of the cluster LF due to an excess of dwarf galaxies relative to the field LF.

In this work we use SDSS data (Abazajian *et al.* 2003) to show that, when measured within the physical size of the system, given by either r_{200} or r_{500}, the LF is invariant for all clusters, both at the bright and at the faint end. We also investigate the nature of the dwarf galaxies in clusters by studying their color distribution. The analysis is performed

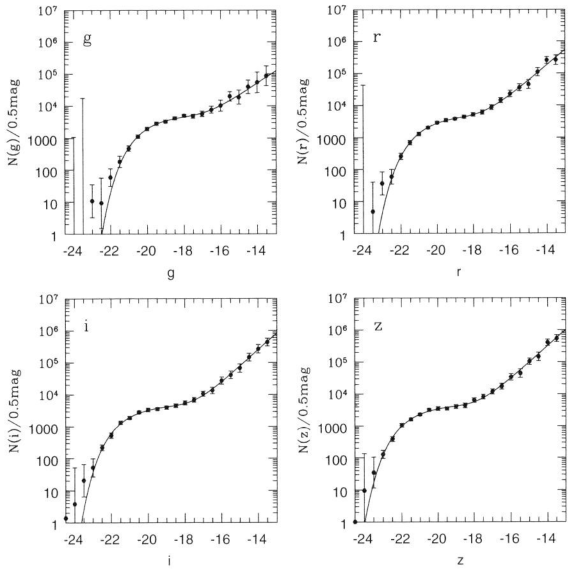

Figure 1. The 4 panels show the composite LFs in the 4 Sloan bands. The individual LFs used to calculate the composite LFs are measured within the physical sizes of the clusters, as given by r_{200}.

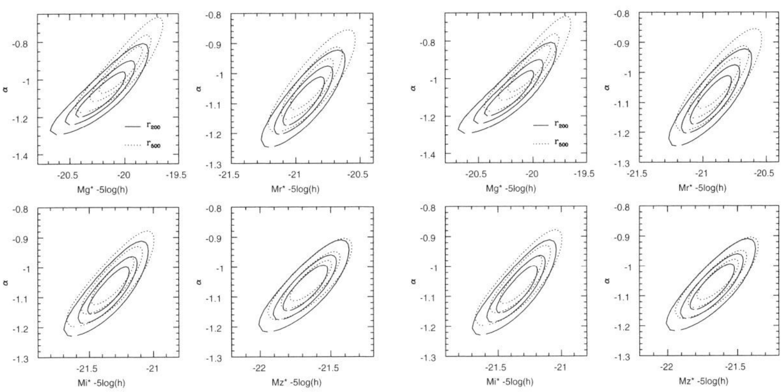

Figure 2. Contour plots of the 68%, 95%, and 99% confidence levels of the parameters of the bright-end component (4 panels on the left-side) and faint-end component (4 panels on the right-side) of the double-Schechter function fit to the 4 SDSS bands composite LFs. Solid (dotted) contours show the results for the composite LF computed within r_{200} (respectively r_{500}).

on a subsample of 69 clusters taken from RASS-SDSS galaxy cluster sample (Popesso *et al.* 2004, 2005).

2. The Cluster Luminosity Function

The composite LF is obtained by stacking the individual cluster LFs estimated within r_{200}, according to the prescription of Colless (1989). Fig. 1 shows the cluster LF in four SDSS photometric bands. In all the bands, a single Schechter function does not provide

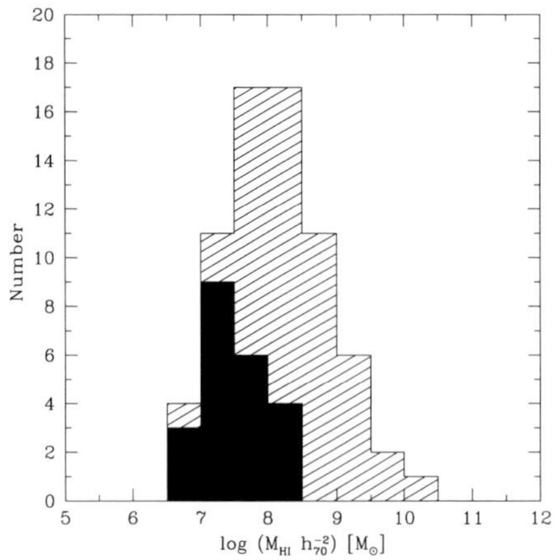

Figure 1. Number of detections per HI mass bin. The shaded histogram represents the distribution of the whole sample of 69 detections, the filled histogram represents the distribution of 22 detections observed for the first time in HI.

2. The CVn blind survey

We designed an HI survey which is extremely sensitive for objects with HI masses below 10^8 $M_{\odot}$. The main difficulty for the HI observations are the long integration times. To be able to detect a statistically meaningful sample of objects with low HI masses in a reasonable amount of time, we have chosen to observe in a slightly over-dense region instead of surveying the field. The Canis Venatici (CVn) groups of galaxies are an excellent region for this purpose. The CVn galaxies extend along the line of sight up to a distance determined by $cz \approx 1200$ km s^{-1}. This region contains a large number of the dwarf galaxies, rich in HI, as detected in the previous surveys (in the optical by Binggeli et al. 1990, in the HI by Kraan-Korteweg *et al.* 1999).

For the survey, we observed 60 × 12 hr using the Westerbork Synthesis Radio Telescope (WSRT) in mosaic mode. The observed area on the sky is ≈86 deg^2. It is covered by observing 60 strips of different declination, separated by 15 arcmin. Each strip of constant declination has 24 pointings separated by 15 arcmin in Right Ascension. The total integration time per pointing was 80.1 min. Only data with $-350 < cz < 1400$ km s^{-1} were used in further data reduction and analysis process. Given that the size of the WSRT primary beam is 34 arcmin, for the used observational grid we obtained nearly uniform sensitivity over the whole observed area. This makes the CVn survey extremely suitable to detect objects with small HI masses.

The data were calibrated and analysed using scripts written by Oosterloo and Kovač based on the the MIRIAD (Sault at al. 1995) routines. The output of this process is ∼1400 3-dimensional data-cubes, each cube being created by extracting the central part of the combined 9 nearest pointings. The spatial resolution of the produced data-cubes is ∼30 × 60 arcsec. The velocity resolution is ∼33 km s^{-1}. The noise in the cubes is just below 0.9 mJy Beam^{-1}. The data-cubes were automatically searched for signal, using algorithms implemented by Kovač. The performance of the algorithms has been tested by inserting simulated objects into the real data-cubes. The HI mass of the detections was estimated using $M_{HI} = 236 D^2 \int S dv [M_{\odot}]$, where D is the distance to the object in Mpc, S is the flux in mJy integrated over the velocity range with significant detection, and the width of the channels is dv given in km s^{-1}.

We used the minimum detectable integrated flux to estimate the maximum distance up to which the object with a certain HI mass can be detected in our survey. Using the real and the simulated detections we calculated that $S_{\rm min} = 0.2$ Jy km s^{-1} is the minimum detectable integrated flux to classify a detection in our survey as a real object. The minimal HI mass which can have an object detected in our survey is then given by $\log M_{\rm HI,min} = 4.67 + 2\log D[M_\odot]$. Objects with HI masses $10^6\ h_{70}^{-2}\ M_\odot$ can be detected up to a distance 4.6 h_{70}^{-1} Mpc and with HI masses $10^7\ h_{70}^{-2}\ M_\odot$ up to 14.6 h_{70}^{-1} Mpc.

In total, we detect 69 objects with HI emission. The cross-correlation with objects from the existing catalogues has been done using the astronomical data bases NED (http://nedwww.ipac.caltech.edu) and HyperLeda (http://leda.univ-lyon1.fr). We found that 4 of our HI detections have not been previously catalogued and 1 of the objects does not have an optical counterpart. The last detection is most probably a tidal satellite. It is important to point out that 22 of our 69 detections have not been previously detected in HI. The number distribution of detections per HI mass bin is shown in Figure 1.

At this moment the largest uncertainty in the estimated HI masses of the detected objects is their distance. We obtain the distances from measuring recessional velocities and correcting them for the motion of the Sun in the Galaxy, motion of the Galaxy in the Local group and for the infall of the Local Group towards Virgo, using the infall velocity of the Local Group according to Theureau *et al.* (1998).

3. The HI Mass Function

The HI mass function (HIMF) is defined analogously to the optical luminosity function, so one can use identical methods for its derivation. We used the $\Sigma(1/V_{max})$ method introduced by Schmidt (1986). This method consists of summing reciprocal values of the maximum surveyed volume in which the objects of a certain HI mass can be detected.

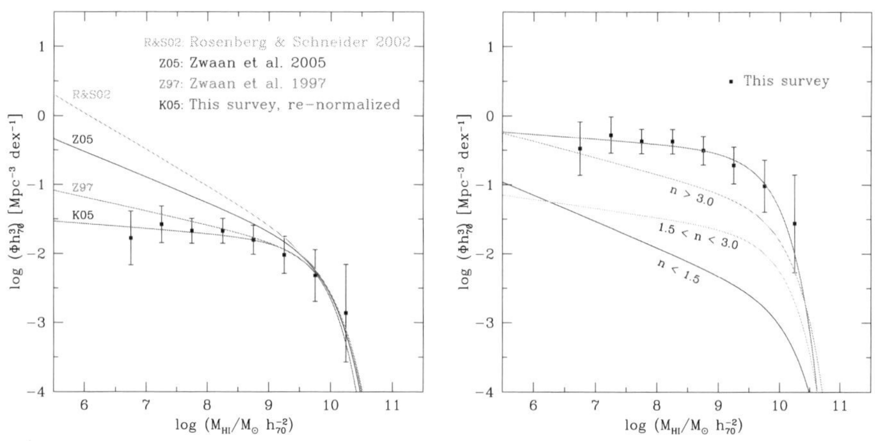

Figure 2. Left panel: The HIMF from our survey compared to the HIMFs derived from three recent blind HI surveys. Our function is re-normalized to match the high mass end of the HIMF determined from the plotted surveys. **Right panel:** Comparison of the HIMF from our survey to the HIMF derived for regions of three different local matter densities. The higher values of n correspond to the region of higher density (Springob *et al.* 2005).

We fitted the binned data with a Schechter function. The best-fit values are $\alpha = -1.07$, $\log(M^*/M_\odot) = 9.73$ and $\phi = 0.129 h_{70}^{-3}$ Mpc^3. The calculated HIMF is presented in Figure 2. In the left panel of this Figure, the HIMF from this survey is re-normalized to match the the high mass end of the HIMF derived from three recent blind HI surveys. The inferred number density of the HI objects of different mass is 20 × higher then plotted. For the surveyed volume, the number density of the objects with HI masses spanning ∼2 decades (10^7–10^9 in $M_\odot$) is approximately constant. This result is not consistent with the slopes of the HIMF as found by Zwaan *et al* (2005) and Rosenberg & Schneider (2002). The disagreement is even larger when comparing this slope to the analytically derived slope ~ -1.8 of the low mass end of the CDM mass function (Press & Schechter 1974).

Recently, a lot of effort was made to explain the effect of the environment on the properties of galaxies and their distributions. The HI surveys cover a much smaller area than the optical surveys. The only way to study the environmental effects on the HIMF is by direct comparison of the estimated HIMF from the surveys carried out in different environments (clusters, groups, voids). The main conclusion from this comparison is that HI galaxies detected in a denser environment have HIMF characterised with the flatter end compared to the HIMF derived from the field. Springob *et al.* (2005) investigated the HIMF of optically selected HI galaxies. Their sample was large enough to investigate the dependence of the HIMF on the environment. Since their subsamples of galaxies in regions of different densities had the same fractions of the morphological types, the difference in the derived HIMF must be due to the density of the environment. Although the obtained results had statistically small differences, the conclusion was that the HIMF in the higher-density region shows a flattening in the faint end and lower values of M_*. Thus, the result of the HIMF from our survey is in agreement with this finding (Figure 2, right panel).

Acknowledgements

The Westerbork Synthesis Radio Telescope is operated by the ASTRON (Netherlands Foundation for Research in Astronomy) with support from the Netherlands Foundation for Scientific Research (NWO). K.K. is grateful to the IAU for the received grant.

References

Binggeli, B., Tarenghi, M. & Sandage A. 1990, *A&A* 228, 42

Briggs, F.H. 2004, *IAUS* 217, 26

de Heij, V., Braun, R. & Burton, W.B. 2002, *A&A* 392, 417

Kraan-Korteweg, R.C., van Driel, W., Briggs, F., Binggeli, B. & Mostefaoui, T. I. 1999, *A&AS* 135, 255

Press, W.H. & Schechter, P. 1974, *ApJ* 187, 425

Rosenberg, J.L. & Schneider, S.E. 2002, *ApJ* 567, 247

Sault, R.J., Teuben, P.J. & Wright, M.C.H. 1995, in:*ASP Conf.Ser. 77: Astronomical Data Analysis Software and Systems IV.* ,433

Schmidt, M. 1986, *ApJ* 151, 393

Springob, C.M., Haynes, M.P. & Giovanelli, R. 2005, *ApJ* 621, 215

Theureau, G., Rauzy, S., Bottinelli, L. & Gouguenheim, L. 1998, *A&A* 340, 21

Verde, L., Oh, S.P. & Jimenez, R. 2002, *MNRAS* 336, 541

Zwaan, M.A., Briggs, F. H., Sprayberry, D. & Sorar, E. 1997, *ApJ* 490, 173

Zwaan, M.A., Meyer, M.J., Staveley-Smith, L. & Webster, R.L. 2005, *MNRAS* 359, L30

Discussion

KARACHENTSEV: About 50 % of the area surveyed by you lies within the Arecibo zone. This area will be surveyed soon with much deeper detection limit. It would be reasonable to shift your strip of survey at higher declination, i.e. outside the Arecibo zone.

Near-Field Cosmology with Dwarf Elliptical Galaxies
Proceedings IAU Colloquium No. 198, 2005
H. Jerjen and B. Binggeli, eds.

doi:10.1017/S1743921305004059

Kinematic properties and stellar populations of dwarf and faint early–type galaxies in the Coma cluster

Ana Matković and Rafael Guzmán

Department of Astronomy, University of Florida, PO Box 112055, Gainesville, FL 32611-2055, USA
email: matkovic@astro.ufl.edu and guzman@astro.ufl.edu

Abstract. We obtained velocity dispersion and line strength index measurements for 69 faint early–type galaxies in the core of the Coma cluster. The full sample spanning $-22.0 < M_R < -17.5$ mag includes 36 dE/dS0 galaxies. We examine the L–σ relation and compare it to that of bright ellipticals from the literature. The slope defined by the faint early–type galaxies including dEs/dS0s is significantly shallower, $L \propto \sigma^{2.01 \pm 0.36}$, than that defined by luminous elliptical galaxies, $L \propto \sigma^4$. Moreover, there is no evidence for a change of slope at the typical division between dwarfs and ordinary elliptical galaxies at $M_B = -18$ mag; rather, the change in slope appears at $M_B = -20.5$ mag. We show that rotation in the fainter galaxies cannot be solely responsible for the change of slope in the L–σ relation. We also plot metallicity and age-dependent line strength indices in diagnostic diagrams and find that our galaxies span a wide range in both age and metallicity.

Keywords. kinematics, clusters: individual (Coma), dwarf, elliptical and lenticular, formation, fundamental parameters.

---

1. Introduction

The kinematic properties of dwarf elliptical galaxies (dEs) offer crucial insight into galaxy formation scenarios. Unfortunately, their low surface brightness makes observations, particularly those useful to kinematic studies, challenging. Despite this difficulty, several studies measuring the velocity dispersions of dEs have recently emerged, the latest two being De Rijcke *et al.* (2004) and Smith *et al.* (2004). To further this research, we present a statistically representative sample of faint early–type galaxies in different environments within one cluster, an essential tool for testing different formation mechanisms of dE galaxies.

The primary goal of this study is to characterize kinematic and stellar populations of dwarf and other low–luminosity early–type galaxies as a function of their environment. For this purpose we observed galaxies in the center and outskirts of the Coma cluster. Here we present our preliminary results, which concentrate on the central region.

2. Observations and Data Reduction

We observed the Coma cluster faint early–type galaxies with the HYDRA multi-fiber spectrograph on the 3.5 m WIYN telescope at Kitt Peak National Observatory. We obtained galaxies in two 45 arcmin diameter regions in the center and outskirts of the Coma cluster. The selection of the faint early–type galaxies was done using the $C-M$, and $C-C$ diagrams containing the U, B, and R bands (see Matković & Guzmán 2005 for more detail). In our sample, we define faint early–type galaxies to have $-15.8 \geqslant M_B \geqslant -20.5$ mag† where 36 out of 69 of these objects are dwarf galaxies ($M_B > -18$, Ferguson & Binggeli 1994). We define the galaxies brighter than $M_B \leqslant -20.5$ as bright ellipticals.

The wavelength range of 4120 – 5600 Å was optimal for discerning a few of the most prominent features of early–type galaxies: G-band, Hβ, Mg_2, and Feλ5350. The FWHM of 1.91 Å allowed us to detect velocity dispersions down to 35 km s^{-1}. Due to the low throughput of the fibers and to reach $S/N > 15$, our galaxies had long exposure times, ranging from 4 to 18 hours.

We reduced all the spectra using the 'dohydra' task within IRAF, while we measured the velocity dispersions (σ) with the REDUCEME software (Cardiel 1999). The software implements the Fourier Quotient method. We determined the uncertainty in the σ measurements using the boot-strapping technique.

3. $L - \sigma$ Relation

Faber & Jackson (1976) showed that the bright early–type galaxies exhibit a relation between their luminosity and velocity dispersion such that $L \propto \sigma^n$, where n was thought to be ≈4 (Faber & Jackson 1976; Sargent *et al.* 1977; Schechter & Gunn 1979; Schechter 1980; Tonry & Davis 1981; Terlevich *et al.* 1981). In 1981, Tonry (1981) observed that the faint early–type galaxies followed a slightly shallower relation than the giant ellipticals, $L \propto \sigma^3$. This result was confirmed by Davies *et al.* (1983) and Held *et al.* (1992) whose samples yielded $n \approx 2.5$. Both Tonry (1981) and Davies *et al.* (1983) discussed rotation as a possible cause for the change of slope in the $L - \sigma$ relation, while Held *et al.* (1992) found it consistent with the supernova driven galactic winds models. The data samples of these authors only included approximately a dozen dEs.

Here we show the $L - \sigma$ relation for our sample of 69 faint early–type galaxies in the central 45 arcmin diamteter of the Coma cluster. This sample contains 36 dEs/dS0s. We also include velocity dispersion measurements from the literature (see Figure 1). The $L - \sigma$ relation exhibits a change of slope where $L \propto \sigma^{2.01\pm0.36}$ for faint early–type galaxies, while the bright early–type galaxies seem to follow the FJ relation.

One of the possible reasons for the change of slope in the $L - \sigma$ relation may be the presence of galaxies other than elliptical. We used the bulge-to-total (B/T) ratios from Gutiérrez *et al.* (2004) to classify the galaxies in this combined sample as bulge-dominated for $B/T = 1$, bulge+exponential component for $0.5 < B/T < 1.0$, and single exponential component galaxies for $B/T < 0.5$. When we fit a $L - \sigma$ relation for each of these classes of galaxies, we observe a slight difference in slope. However, this difference is not sufficient to explain the change of slope in the $L - \sigma$ relation from bright to faint early–type galaxies. Furthermore, it is difficult to classify galaxies at the distance of the Coma cluster without resolved photometry.

Next, we investigate the possibility of rotation as the cause of the change of slope in the $L - \sigma$ relation. Assuming a universal relation between L, kinetic energy per unit mass, V^2, one can derive a relation between L, σ, and the anisotropy parameter, V_{rot}/σ. For bright elliptical galaxies $L = b\sigma^4$ implies $V_{rot} = 0$, as is observed. However, for the faint early–type galaxies, $L = c\sigma^2$ implies a systematic increase in the anisotropy parameter:

$$\frac{V_{rot}^2}{\sigma^2} = \sqrt{\frac{c}{b}\frac{1}{\sigma}} - 3$$

† Throughout this paper we use $H_0 = 70$ km s^{-1} Mpc^{-1}, and a distance modulus for Coma cluster of 35.078, ($d = 99$ Mpc).

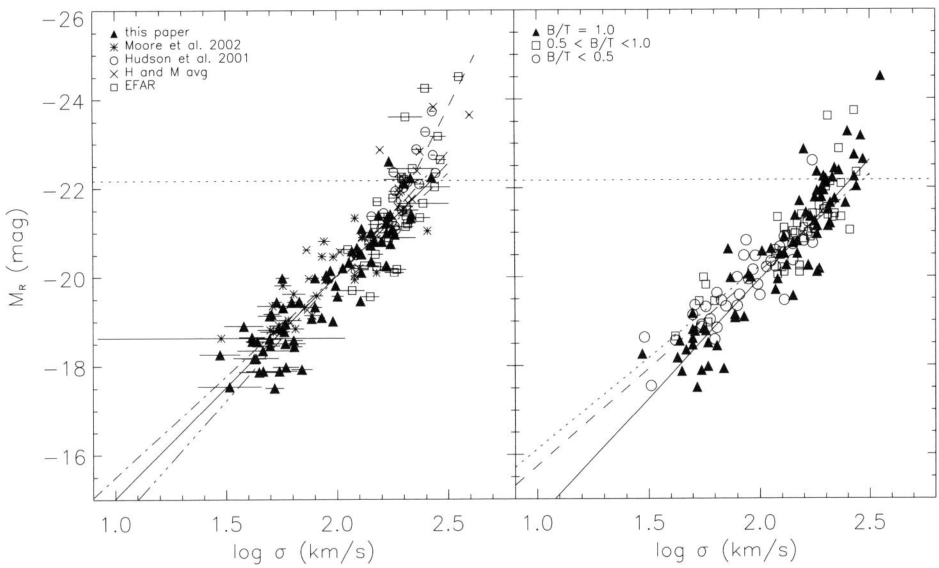

Figure 1. $M_R - \sigma$ plots. (Left) Coma galaxies in our sample and from the literature, as indicated in the legend. The *dotted* line separates the bright from faint early–type galaxies at $M_R = -22.17$ mag, while the *dashed* line is the most recent Faber-Jackson relation from the literature (Forbes & Ponman, 1999). The *solid* line is the least squares bisector fit to all faint early–type galaxies. (Right) Coma cluster galaxies for which bulge-to-total (B/T) ratios were available (Gutiérrez *et al.* 2004). The *solid* line is the bisector fit to the single exponential component($B/T < 0.5$) galaxies, the *dashed* line for bulge+single component ($0.5 < B/T < 1.0$), and the *dotted* line bulge-dominated ($B/T = 1$) galaxies.

In Figure 2 the solid line shows this relation. It represents the change in the anisotropy parameter with velocity dispersion, assuming that rotation was solely responsible for the change of slope in $L - \sigma$. It is crucial to note that previously observed galaxies (Geha, Guhathakurta & van der Marel 2003; Simien & Prugniel 2002; Pedraz *et al.* 2002; Davies *et al.* 1983) do not follow this relation. Therefore, *rotation cannot explain the change of slope in* $L - \sigma$ for these faint early–type galaxies.

De Rijcke *et al.* (2004) also find that $L \propto \sigma^2$ for their sample of dEs and show that this is consistent with galactic wind models.

4. Ages and Metallicities

We also measured line strength indices for our sample of faint early–type galaxies. Preliminary investigation of relations between different combinations of age and metallicity sensitive indices with the velocity dispersion hint that the faintest galaxies, dEs/dS0s, may exhibit a wider range in metallicity and age than the remaining early–type galaxies.

In Figure 3, we plot the age sensitive index, Hβ, vs. the metallicity sensitive index, [MgFe]. We represent dEs/dS0s as triangles, while asterisks are other faint early–type galaxies. Overall, the galaxies show a wide range in age and metallicity. The dwarf galaxies span a wider range in both age and metallicity with some lying outside the stellar synthesis model grids. These are potentially very old, low metallicity dwarf galaxies whose stellar populations may be similar to those of globular clusters.

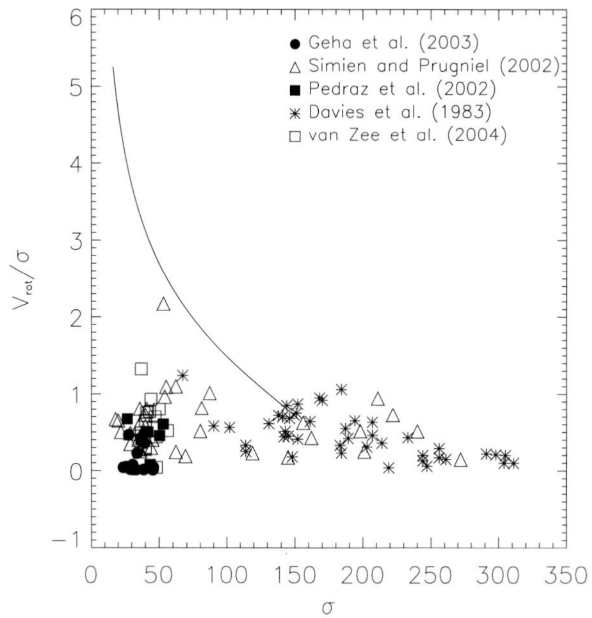

Figure 2. V_{rot}/σ vs. σ plot. The solid line represents the expected systematic increase in the anisotropy parameter assuming that rotation is responsible for the change of slope in the $L - \sigma$ relation. The different symbols are observations from the literature as indicated in the legend.

5. Summary

- We show preliminary results of a larger environmental study intended to characterize internal kinematics and stellar populations of faint early–type galaxies in the Coma cluster.

- We find that $L \propto \sigma^{2.01\pm0.36}$ for faint early–type galaxies ($-15.8 > M_B > -20.5$), while the bright elliptical galaxies follow the FJ relation well. After investigating possible causes of this change of slope, we conclude that rotation cannot explain it.

- The Hβ vs. [MgFe] plot indicates that all the faint early–type galaxies in our sample exhibit a wide range in metallicity and age, with dEs/dS0s showing a wider range in these properties. A few galaxies in this sample fall outside the stellar synthesis grids and may be old, low metallicity systems reminiscent of stellar populations of globular clusters.

Acknowledgements

We would like to acknowledge Patricia Sánchez-Blázquez, Javier Gorgas, and Nicolás Cardiel for help with the velocity dispersion and line strength index measurements. We also thank Alister Graham for many insightful discussions.

References

Cardiel, N. 1999, Dissertation, *Star formation in Central Cluster Galaxies*, Univ. Complutense, Madrid

De Rijcke, S., Michielsen, D., Dejonghe, H., Zeilinger, W.W. & Hau, G.K.T. 2004, *astro-ph/0412553, v2*

Davies, R.L., Efstathiou, G., Fall, M., Illingworth, G. & Schechter, R.L. 1983, *AJ* 266, 41

Faber, S.M. & Jackson, R.E. 1976, *ApJ* 204, 668

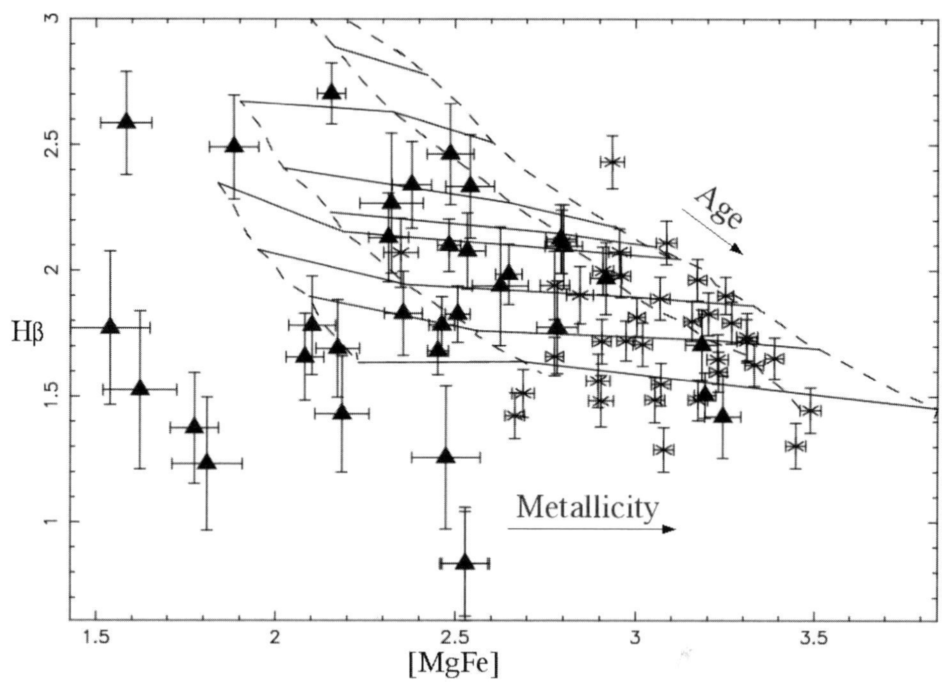

Figure 3. Hβ vs. [MgFe] plot. The solid triangles are galaxies classified as dwarf galaxies ($M_B > -18$), while asterisks are the remaining galaxies in our sample. The ages represented by solid lines, top-to-bottom, are: 1.78, 2.51, 3.16, 3.98, 5.62, 7.98, 12.2, 15.85 Gyr; while the metallicity lines are dashed, left-to-right, [Fe/H] = −0.68, −0.38, 0.0, +0.2 dex.

Ferguson, H.C. & Binggeli, B. 1994, *A&A Rev.* 6, 67
Forbes, D.A. & Ponman, T.J. 1999, *MNRAS* 309, 623
Geha, M., Guhathakurta, P. & van der Marel, R. 2003, *AJ* 126, 1794
Gutiérrez, C.M., Trujillo, I., Aguerri, J.A., Graham, A.W. & Caon, N. 2004, *ApJ* 602, 644
Held, E.V., de Zeeuw, T., Mould, J. & Picard, A., 1992, *AJ* 103,851
Matković, A. & Guzmán, R. 2005, *MNRAS* submitted
Pedraz, S., Gorgas, J., Cariel, N., Sanchez-Blazquez, P. & Guzmán, R. 2002, *MNRAS* 332, L59
Sargent, W.L.W., Schechter, P.L., Boksenberg, A. & Shortridge, K. 1977, *ApJ* 212, 326
Schechter, P.L. & Gunn, J. E. 1979, *ApJ* 229, 472
Schechter, P.L. 1980, *AJ* 85, 801
Simien, F. & Prugniel, P. 2002, *A&A* 384, 371
Smith, R.J. *et al.* 2004, *AJ* 128, 1558
Terlevich, A.I., Davies, R.L., Faber, S.M. & Burstein, D. 1981, *MNRAS* 196, 381
Tonry, J.L. 1981, *ApJ* 251, 1
Tonry, J.L. & Davis, M. 1981, *ApJ* 246, 680
van Zee, L., Skillman, E.D. & Haynes, M.P. 2004, *AJ* 128, 121

Discussion

N. Caldwell: Your result that the lower velocity dispersion galaxies have a large spread in age is also found for Virgo and field galaxies, using similar techniques (Caldwell *et al.* 2003). Also, the galaxies which fall off the grid would appear to lie in the area occupied by globular clusters. Though this is a shortcoming of the models, it may mean that those galaxies have populations similar to those of globular clusters.

MATKOVIC: Yes, I completely agree with your findings in Virgo, and I dosee the possibility of some of these dwarf galaxies having stellar populations similar to those of globular clusters as I mentioned in the talk.

C. CONSELICE: At what magnitude to you begin to see a spread in ages and metallicities? Is it a range at all magnitudes?

MATKOVIC: The spread in both age and metallicity is seen over all the magnitudes. Keep in mind that our sample does not include galaxies fainter than $M_B \sim -15.8$.

HELD: Your results on the $L-\sigma$ relation for dwarf elliptical galaxies interestingly confirm – as does De Rijcke's talk – the result obtained by Held *et al.* (1992, AJ) based on the much smaller dataset available at that time.

MATKOVIC: Thank you very much for pointing out this reference.

SILVA: Did you correct for Hβ emission?

MATKOVIC: Yes, that is important, however in the Coma cluster we observed, we do not see any emission.

Near-Field Cosmology with Dwarf Elliptical Galaxies
Proceedings IAU Colloquium No. 198, 2005
H. Jerjen and B. Binggeli, eds.

doi:10.1017/S1743921305004060

Resolving stellar populations using broad band photometry

Maren Hempel[1] & Markus Kissler-Patig[2]

[1]Department of Physics and Astronomy, Michigan State University, East Lansing, MI 48824, USA
email: hempel@pa.msu.edu
[2]European Southern Observatory, Karl-Schwarzschild Str.2, 85748 Garching, Germany
email: mkissler@eso.org

Abstract. Setting constraints on the evolutionary history of a galaxy requires informations about the age and metallicity of its stellar content. However, outside the Local Group of galaxies representative surveys on the age and metallicity of stellar populations in galaxies of different types (e.g. luminosity, environment) can only be performed with photometry. The latter requires special care in order to overcome the age- metallicity degeneracy of broad band colors. Combining optical and near-infrared observations of globular clusters is a powerful tool in this respect, using the different sensitivity of broad band colors regarding both parameters. Based on Single Stellar Population (SSP) models we derive the cumulative age distribution for various globular cluster systems and compare it to simulated systems with a known age distribution. So far this method has been tested successfully on various early-type galaxies, revealing a second generation of globular clusters in two galaxies in a sample of eight with convincing reliability.

Keywords. galaxies: formation, galaxies: evolution, star clusters, techniques: photometry

1. Introduction

To determine the age and metallicity of stellar populations in galaxies is crucial for our understanding of the underlying formation scenario. Fortunately, a specific formation scenario, e.g. hierarchical merging (e.g. Toomre & Toomre 1972; Kauffmann, White & Guiderdoni 1993), monolithic collapse (e.g. Chiosi & Carraro 2000), or accretion (e.g. Hilker, Infante & Richtler 1999) leaves its imprint in the age structure of the galaxy. The detection of different age populations within a galaxy will therefore allow us to set some constraints on its formation. This, however, requires the ability to disentangle stellar populations of different ages. Globular clusters represent Single Stellar Populations almost perfectly. They are found in almost every galaxy and can be observed at distances as far as 100 Mpc (e.g. Coma galaxy cluster, Woodworth & Harris 2000). Consequently, they are ideal tools to trace the major star formation events in their host galaxy, and we are now left with the task to detect globular cluster populations and to derive their relative ages. The method we present here is based on photometric data, allowing to observe representative globular cluster samples in each of the galaxies.

One of the features of globular cluster systems, which has to be explained by the formation/evolution of the host galaxy, is the bimodal color distribution, seemingly rather common in early-type galaxies (e.g. Gebhardt & Kissler-Patig 1999; Kundu & Whitmore 2001a; 2001b). Before we attempt to argue about the origin of the color distribution and its dependency of the galactocentric distance (see Figure 1) we need to know whether the different colors are due to age or metallicity differences. In order to lift the age-metallicity

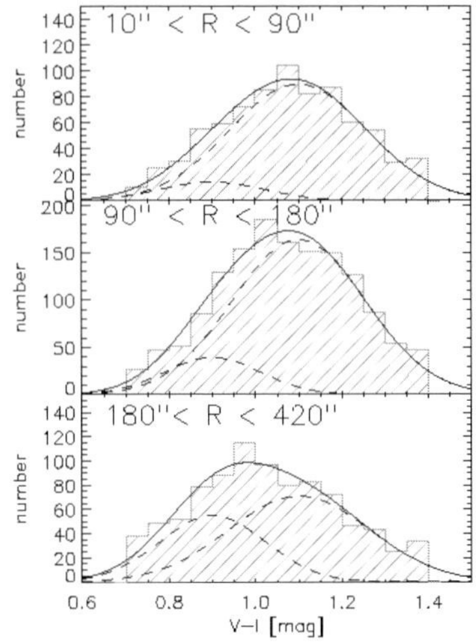

Figure 1. Radial color distribution in NGC 3311, the central cluster galaxy in the Hydra I cluster (see also Hilker 2003).The dashed lines in each plot refer to the two color populations, whereas the solid line represents their sum. The plots indicate the stronger concentration of the red population towards the center of the galaxy.

degeneracy of optical broad band colors (Worthey 1994) we combine optical, e.g. $(V-I)$, with infrared colors such as $(V-K)$ or $(V-H)$, taking advantage of the fact that the former are in comparison more age sensitive than infrared colors, which are dominated by the metallicity of the stellar populations.

2. Data and Models

To derive the relative ages of globular cluster sub-populations we combine optical and near -infrared photometric data. So far these were obtained with HST/WFPC2 (optical), and either VLT/ISAAC or HST/NICMOS2 in the near infrared. In order to probe the effect of external parameters, i.e. environment, or galaxy luminosity our galaxy sample contains cD galaxies (e.g. NGC 3311, M87), galaxies in groups (e.g. NGC 4365, NGC 5846, IC 4051) as well as rather isolated galaxies (e.g. NGC 7192, NGC 3115). The link to the galaxy formation is set via the probability for galaxy mergers (see Section 1), a formation scenario which not only allows but predicts bimodal color distributions in globular cluster systems. In Figure 2a we show an example of the final data set used for our analysis. The different isochrones (2 and 15 Gyr) show clearly the advantage of combined optical/near-infrared photometry with respect to the above mentioned age-metallicity degeneracy. However, this plot tells us also that the color of an individual globular cluster is not accurate enough to allow the determination of individual ages. Keeping in mind that at this point of the project we are mostly interested in detecting age sub-populations and to derive their relative ages we do not consider this to be a major drawback. A detailed description of how we infer the existence of globular cluster sub-populations differing in age by several Gyr can be found in Hempel & Kissler-Patig (2004) and we will only describe the major steps here.

• Compare the optical color of the observed GCs with the predictions of Single Stellar Populations (e.g. Maraston 2005; Bruzual & Charlot 2003; Vazdekis 1999) and select each GC into the corresponding age bin (see Figure 2b).

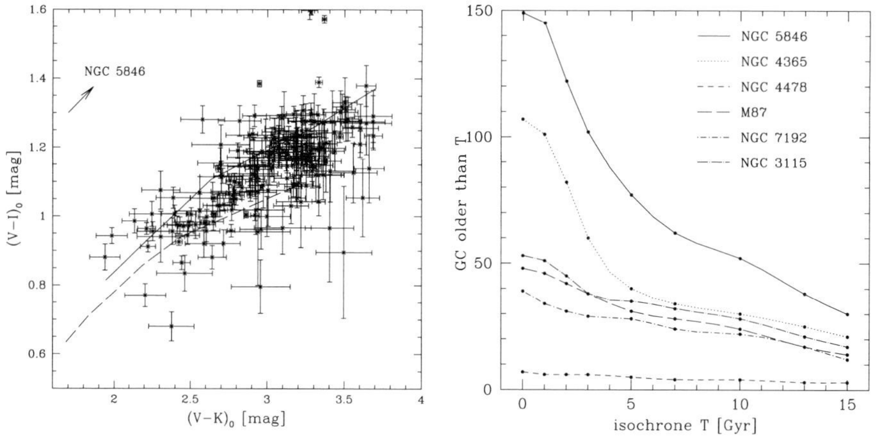

Figure 2. *Left*:$(V-I) vs. (V-K)$-color-color diagram for the NGC 5846 globular cluster system. The solid and dashed line mark the 15 Gyr and 2 Gyr isochrone, following the Bruzual & Charlot SSP models (Bruzual 2000). *Right*: Cumulative age distribution in various galaxies derived with the above described procedure (see also Hempel & Kissler-Patig (2004)).

- Simulate the color-color distributions for GCSs with a given age structure, e.g. a composition of 13 Gyr and 5 Gyr old objects. The number ratio between both populations may vary in 10 % increments between a purely old or young population. In dependence on the availability of SSP isochrones we assume GC ages of 1, 1.5, 2, 3, 5, 7, 10, and 13 Gyr, respectively.
- Derive the cumulative age distribution of the simulated color distributions in the same way as for the observed systems.
- Compare the cumulative age distribution in the observed GCS with the complete set of simulated systems (different age and size for sub-populations) using a reduced χ^2-test. The identification of a "best fitting" model allows us to set constraints on the age structure in the observed systems.

3. Results

This project aims at the detection of globular cluster sub-populations in early-type galaxies, considering various external parameters. For two selected galaxies, NGC 5846 and NGC 7192 the results are presented in Figure 3. The former is a member of a group of galaxies, whereas NGC 7192 represents isolated galaxies, accompanied by only 2 close neighbors. In case of NGC 5846 the derived age distribution is best matched by a mixed population of old (13 Gyr) and ~10 Gyr younger objects. Hereby we note that for our purpose, the detection of age populations, an accuracy of 3 Gyr is more than sufficient. However, we need to discuss the surprisingly large fraction of intermediate age globular clusters. As described in Hempel & Kissler-Patig (2004) photometric globular cluster surveys so far have been centered on the host galaxy, which leads to considerable spatial bias (see also Schweizer & Seitzer 1998; Mihos & Hibbard 1995). A more accurate determination of the relative sizes of sub-populations requires therefore spatially extended observations. Whereas in a larger group of galaxies interactions between its members are not unlikely, isolated galaxies like NGC 7192 will correspond much better to the picture of early-type galaxies as old and passively evolving galaxies. In agreement to that we find the cumulative age distribution dominated by an old population. At this point we do not distinguish between the possibility of a small fraction of intermediate age globular

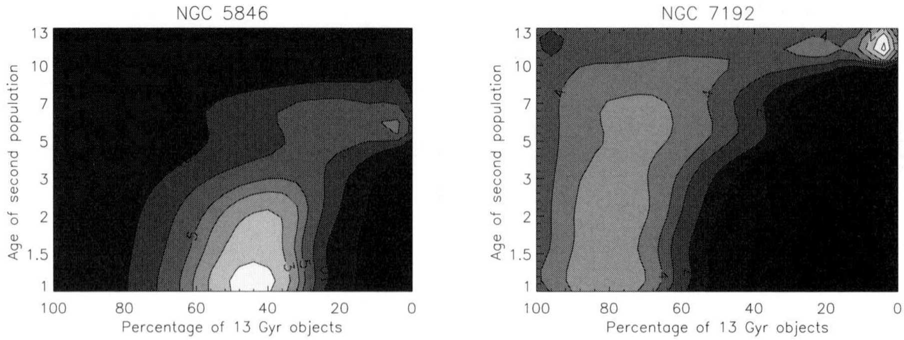

Figure 3. Result of the χ^2-test for NGC 5846 (left) and NGC 7192 (right). The contours represent the χ^2 values of the test. The best fitting model to the cumulative age distribution in NGC 5846 is clearly dominated by a population of intermediate age (≈1–2 Gyr) globular clusters, whereas in NGC 7192 highest agreement between observation and model is obtained assuming either a purely 13 Gyr old population or a small contribution of second generation globular clusters with a highly uncertain age.

clusters or a system which is dominated by a second generation of globular clusters which is only slightly younger than the first generation (between 10 and 13 Gyr).

4. Conclusions

With the semi-numerical approach for age determination in globular cluster systems we are able to detect age sub-populations in early-type galaxies. Especially in case of the giant elliptical galaxies and their extremely rich globular cluster systems this will lead to a better understanding about their formation. For instance, the knowledge whether the observed blue colors of globular clusters are caused by very low metallicities or young/intermediate ages gives evidence for a possible accretion or stripping from dwarf galaxies or merger induced star formation. Although the data set needs to be expanded considerably we can already say that the major star formation epochs of early-type galaxies are far more diverse than previously thought.

References

Ashman, K.M. & Zepf, S.E. 1998, *Globular Cluster Systems*, Cambridge University Press
Beasley, M.A., Sharples, R.M., Bridges, T.J. *et al.* 2000, *MNRAS* 318, 1249
Bruzual, A.G. 2000, *private communication*
Bruzual, A.G. & Charlot, S. 2003, *MNRAS* 344, 1000
Chiosi, C. & Carraro, G., 2000, *MNRAS* 335, 335
Gebhardt, K. & Kissler-Patig, M. 1999, *AJ* 118, 1525
Hempel, M. & Kissler-Patig, M. 2004, *A&A* 419, 863
Hilker, M., Infante, L. & Richtler, T. 1999, *AAS* 138, 55
Hilker, M. 2003, *Extragalactic Globular Cluster Systems*, ESO Astrophysics Symp., eds. Kissler-Patig, 173
Kauffman, G., White, S.D.M. & Guiderdoni, B. 1993, *MNRAS* 264, 201
Kundu, A. & Whitmore, B.C. 2001, *AJ* 121, 2950
Kundu, A. & Whitmore, B.C. 2001, *AJ* 122, 1251
Maraston, C. 2005, *MNRAS* submitted
Vazdekis, A. 1999, *ApJ* 513, 224
Woodworth, S.C. & Harris, W.E. 2000, *AJ* 119, 2699
Worthey, G. 1994, *ApJS* 95, 107

Near-Field Cosmology with Dwarf Elliptical Galaxies
Proceedings IAU Colloquium No. 198, 2005
H. Jerjen and B. Binggeli, eds.

doi:10.1017/S1743921305004072

SAPAC: a SBF analysis pipeline for the astronomical community

Laura P. Dunn and Helmut Jerjen

RSAA, Mt Stromlo Observatory, ANU, Cotter Road, Weston ACT 2611 Australia
email: laurad@mso.anu.edu.au, jerjen@mso.anu.edu.au

Abstract. Large volumes of CCD imaging data that will become available from wide-field cameras at telescopes such as the CFHT, SUBARU, VST, or VISTA in the near future are highly suitable for systematic *distance surveys of early-type galaxies* using the Surface Brightness Fluctuation (SBF) method. For the efficient processing of such large data sets, we are developing the first semi-automatic SBF analysis pipeline named SAPAC. After a brief description of the SBF method we discuss the image quality needed for a successful distance measurement and give some background information on SAPAC.

Keywords. Methods: data analysis, galaxies: distances and redshifts

1. The SBF method in a nutshell

Employing the surface brightness fluctuation signal of unresolved stars in distant galaxies is an effective and inexpensive new way to measure accurate distances to early-type (dwarf) galaxies. Unlike other extragalactic distance indicators (e.g. TRGB, RR Lyrae stars), this method does *not* require resolved stars therefore allowing distance measurements for early-type galaxies far beyond the practical limits of any of the classical distance indicators ($\sim$5 Mpc). With Fourier analysis techniques, the SBF method quantifies the mean stellar flux per CCD pixel and rms variation due to Poisson noise across a designated area in a dwarf galaxy.

Initially the SBF method was almost exclusively applied on nearby giant ellipticals and MW globular clusters (e.g. Tonry *et al.* 1989, 1994) but was found to work equally well with dwarf elliptical (dE) galaxies (e.g. Jerjen *et al.* 1998, 2000, 2001, 2004, and Rekola *et al.* 2005). As dE galaxies are by far the most numerous galaxy type at the current cosmological epoch, the SBF method in combination with wide-field CCD imaging offers the opportunity for the first time to spatially locate dEs in vast numbers and thereby to map in 3D the densest environments of the local Universe (for first results see contributions by Côté *et al.*, Jerjen, Jordan *et al.*, and Rekola *et al.* in this volume). First SBF distances are published for dEs as distant as 15Mpc (using 2m ground-based telescopes) and 25Mpc (using 8m VLT+FORS and HST & ACS).

2. Analysis prerequisites

The *minimal requirements* for the SBF analysis of an early-type galaxy are:

• Galaxy morphology: the light distribution of the stellar system must be radially symmetric and have minimal structure. An overall elliptical shape of the galaxy is crucial as this is modelled and subtracted as part of the SBF analysis.

• Photometry: calibrated CCD images are required in two photometric bands, e.g. (B, R) or (g_{475}, z_{850}), as the fluctuation magnitude shows a colour dependency.

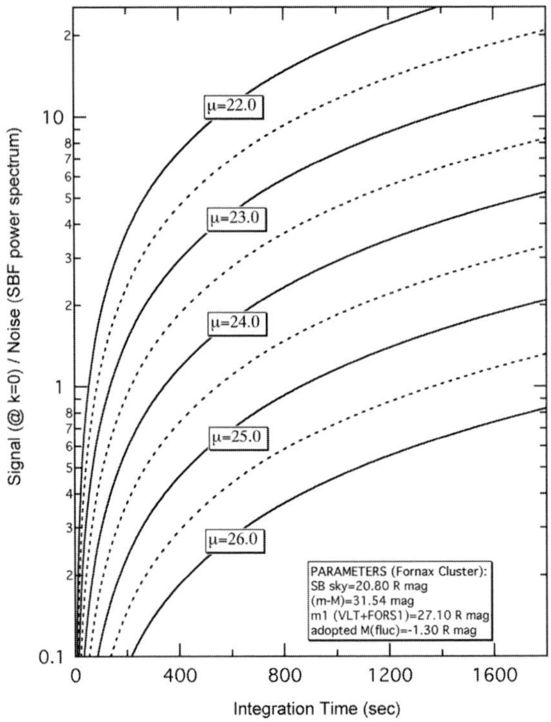

Figure 1. An illustration how the signal-to-noise in the SBF power spectrum increases with length of exposure time and galaxy surface brightness at the distance of the Fornax Cluster.

- Image quality: FWHM $\leqslant r_{\rm eff}['']/20$, where $r_{\rm eff}$ is the half-light radius of the galaxy.
- Integration time: $t = {\rm S/N}\cdot 10^{0.4\cdot(\mu_{\rm gal}-\mu_{\rm sky}+DM+\overline{M}-m_1)}$, where $\mu_{\rm gal}$ is the mean surface brightness of the galaxy, $\mu_{\rm sky}$ the surface brightness of the sky background, DM the estimates distance modulus of the galaxy, $\overline{M}$ the fluctuation luminosity of the underlying stellar population, and m_1 the magnitude of a star providing 1 count/sec on the CCD detector at the telescope.

To give a general idea of these constraints, Fig. 1 illustrates the depth required for an image of a dE at the distance of the Fornax cluster observed with VLT+FORS1. The SBF amplitude above the shot noise level (signal-to-noise) in the power spectrum is shown as a function of integration time and mean effective surface brightness of the galaxy. A SBF distance can be determined when the S/N is approximately 0.5, (see Fig. 8 in Rekola *et al.* 2005), but that depends largely on the image quality i.e. seeing. For example, to achieve a S/N$\sim$2 in the galaxy power spectrum, the minimum exposure time required for a dE with a mean surface brightness of 25 mag arcsec^{-2} is 1600s.

It is interesting to note that this exposure time is by a factor of 20 shorter than the 32,000s of HST time spent by Harris *et al.* (1998) to measure the TRGB distance of a dwarf elliptical at a similar distance.

3. SBF Reduction Pipeline

Previous SBF work has entailed individuals hand selecting regions in galaxy images for the analysis. To make the results as impartial as possible and data reduction more efficient we are developing a rapid, semi-automatic SBF analysis package named SAPAC that can process large numbers of galaxies. SAPAC is a software package that carries out a semi-automatic SBF analysis of any early-type galaxy for which CCD data meets the requirements as discussed above. For a detailed description of the fluctuation magnitude calibration and the individual reduction steps such as the modelling of the galaxy, foreground star removal, selection of SBF fields etc. we refer the reader to Jerjen (2003). SAPAC consists of Perl scripts using and IRAF module and uses a sophisticated graphical user interface, also written in Perl. The average processing time for 10 SBF fields in a galaxy and measuring a distance is approximately 20 minutes. Initially we have concentrated the pipeline on B, R images, but the implementation of calibration information for a wider range of commonly used filter sets for SBF work like J, H, K of the SDSS g, z filters (Mei *et al.* 2005) is in process.

Potential users of SAPAC who are interested in testing this package for calculating accurate distances of early-type dwarfs are welcome to contact Laura Dunn. This software package will be made available to the astronomical community soon.

Acknowledgements

L.P.D would like to acknowledge partial financial support from the Astronomical Society of Australia, the International Astronomical Union, and Alex Rodgers Travel Scholarship.

References

Harris, W. E., Durrell, P. R., Pierce, M. J. & Secker, J. 1998, *Nature* 395, 45
Jerjen, H. 2003, *A&A* 398, 63
Jerjen, H., Bingeli, B. & Barazza, F. 2004, *AJ* 127, 771
Jerjen, H., Freeman, K.C. & Binggeli, B. 1998, *AJ* 116, 2873
Jerjen, H., Freeman, K.C. & Binggeli, B. 2000, *AJ* 119, 166
Jerjen, H., Rekola, R., Takalo, L., Coleman, M. & Valtonen, M. 2001, *A&A* 380,90
Mei, S., Blakeslee, J.P., Tonry, J.L., Jordán, A., Peng, E.W., Côté, P., Ferrarese, L., West, M., Merritt, D. & Milosavlijevic, M. 2005, *ApJ* 625, 121
Rekola, R., Jerjen, H. & Flynn, C. 2005, *A&A* 437, 823
Tonry, J.L., AJhar, E.A. & Luppino, G.A. 1989, *ApJ* 346, L57
Tonry, J.L. & Schneider, D.P. 1988, *AJ* 96, 807

Near-Field Cosmology with Dwarf Elliptical Galaxies
Proceedings IAU Colloquium No. 198, 2005
H. Jerjen and B. Binggeli, eds.

doi:10.1017/S1743921305004084

The ACS Fornax cluster survey

Andrés Jordán[1,2] and the ACSFCS Team†

[1] European Southern Observatory, Karl-Schwarzschild-Str. 2, 85748 Garching bei München, Germany; email: ajordan@eso.org

[2] Astrophysics, Denys Wilkinson Building, University of Oxford, 1 Keble Road, OX1 3RH, UK

Abstract. We are conducting a survey of 44 early-type galaxies in the Fornax Cluster using the *Advanced Camera for Surveys* on the *Hubble Space Telecope*: the ACS Fornax Cluster Survey. We briefly describe the survey, the selection of the target galaxies and some of the scientific objectives.

Keywords. Surveys, galaxies: general, galaxies: clusters: individual (Fornax), galaxies: star clusters

---

The ACS Fornax Cluster Survey is a Cycle 13 program to image, in the F475W (g_{475}) and F850LP (z_{850}) bandpasses, 44 early-type galaxies in the Fornax Cluster using the *Advanced Camera for Surveys* (ACS) on the *Hubble Space Telescope* (HST). Observations are still being carried out and the program is expected to be completed during 2005.

Selection of target galaxies was based on the Fornax Cluster Catalog (FCC; Ferguson 1989). This survey, the most complete and homogeneous available for Fornax, covers $\sim 40^\circ$ squared centered at $\alpha \sim 3^h 35^m$ and $\delta \sim -35.7^\circ$. Within the survey area, the catalog contains 340 likely cluster members. Their classification was done mainly based on galaxy morphology, supplemented where possible by a small set of radial velocities.

The ACS Fornax Cluster Survey sample was built by selecting all members of the FCC with $B \leqslant 15.5$ mag that were classified as early-types (E, S0, dE, dS0, dE,N, dS0,N) by the FCC *and* by the NASA Extragalactic Distance Database (NED). This sample is comprised of 42 galaxies, *12 of which are classified as dwarfs.* Additionally we included NGC 1340 and IC 2006 in the sample. While not included in the FCC, these galaxies were selected because of their usefulness for calibrating the surface brightness fluctuations (SBF) distance method, one of the main science objectives of this survey.

Combined with a similar survey of Virgo (Côté *et al.* 2004), this represents the most comprehensive imaging survey to date of early-type galaxies in cluster environments in terms of depth, spatial resolution, sample size and homogeneity. The ACS Fornax cluster survey is *complete* down to its limiting magnitude, as shown in Figure 1.

The scientific objectives of the survey include:

• A homogeneous study of the globular cluster (GC) systems of the whole sample, including determination of photometrical and structural parameters for thousands of GCs.

• Morphologies, luminosity profiles and dust content. We will carry out a comprehensive study of the nuclear morphologies, luminosity profiles and dust properties of the target galaxies, and probe their dependence on cluster environment.

• Extragalactic Distance Scale via the SBF Method. The compact nature of Fornax combined with the high SBF signal-to-noise at *HST* resolution will allow us to determine

† J.P. Blakeslee (JHU), P. Côté (HIA/Rutgers), L. Ferrarese (HIA/Rutgers), L. Infante (PUC), S. Mei (JHU), D. Merritt (RIT), M. Milosavljević (Caltech), E.W. Peng (HIA), J.L. Tonry (IfA, Hawaii), M.J. West (Hawaii)

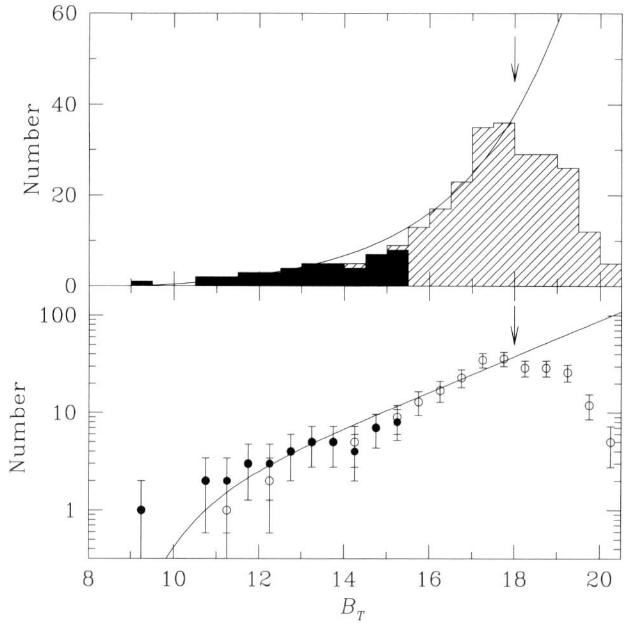

Figure 1. *Upper panel)* Luminosity function of 269 early-type galaxies that are classified by Ferguson (1989) as members of the Fornax cluster. (*upper, hatched histogram*). The arrow shows the FCC completeness limit, while the solid curve shows the best-fit Schechter function for E+S0+dE+dS0 galaxies from Ferguson & Sandage (1991). The filled lower histogram shows the luminosity function for the 44 early-type galaxies in the ACS Fornax Cluster Survey. (*Lower panel*) Same as above, expect in logarithmic form. The open circles show the luminosity function of 269 early-type members of the Fornax cluster according to Ferguson (1989). The arrow shows the completeness limit of the FCC, while the filled circles show the luminosity function of galaxies in the ACS Fornax Cluster Survey. The solid curve is the same as that shown in the upper panel.

the best calibration of the z_{850}-band SBF method. We will determine the zeropoints to 0.05 mag and measure its stellar population dependence as inferred from $(g_{475} - z_{850})$ colors.

• We will perform a detailed study of the properties of dwarf nuclei, the most massive GCs and Ultra-Compact Dwarfs, and the relation between them (e.g., Haşegan *et al.* 2005).

• Using archival *Chandra* data, we will study the properties of low-mass X-ray binaries and their connection with GCs for our program galaxies (e.g., Jordán *et al.* 2004).

References

Côté, P., Blakeslee, J.P., Ferrarese, L., *et al.* 2004, ApJS, 153, 223
Ferguson, H.C. 1989, *AJ* 98, 367
Ferguson, H.C. & Sandage, A. 1991, *AJ* 101, 765
Haşegan, M., Jordán, A., Côté, P. *et al.* 2005, *ApJ* in press (astro-ph/0503566)
Jordán, A., Côté, P., Ferrarese, L. *et al.* 2004, *ApJ* 613, 279

Near-Field Cosmology with Dwarf Elliptical Galaxies
Proceedings IAU Colloquium No. 198, 2005
H. Jerjen and B. Binggeli, eds.

doi:10.1017/S1743921305004096

A multicolour view of the nuclei of dEs

Thorsten Lisker, Eva K. Grebel and Bruno Binggeli

Astronomical Institute, University of Basel, Switzerland; email: tlisker@astro.unibas.ch

Abstract. Using a sample of over 100 nucleated Virgo cluster dEs from the SDSS we show that interchanging nucleus and galaxy colours does not alter the colour-magnitude relation. No significant difference is found between nuclei and their parent galaxies in a multicolour comparison except for a weak tendency that some nuclei are slightly redder than their hosts.

Keywords. galaxies: nuclei, galaxies: dwarf, galaxies: clusters: general

---

Nuclei of dwarf ellipticals presumably are compact central star clusters that form a bright extension of globular clusters (GCs) in their sizes and luminosities (Lotz *et al.* 2004). Their formation process is poorly understood – possible scenarios include early in-situ formation, late star formation out of central gas, or infall and merging of massive GCs via dynamical friction (Oh & Lin 2000). Photometric studies have been contradictory so far: while Lotz *et al.* (2004) find from HST/WFPC2 data the nuclei to be slightly bluer than their host galaxies and to be consistent with GC colours, no colour difference was seen in previous ground-based studies.

For 187 Virgo dENs from the SDSS Data Release 3 we performed aperture photometry for a central ($r < 1.2''$) as well as a half-light radius aperture, excluding the inner $r < 2''$ for the latter. Although the central aperture is intended to represent the nucleus, in most cases the underlying galaxy still dominates the flux as illustrated in the left panel of Fig. 1. We would therefore not be able to see very small differences in nucleus and galaxy colour, but the large sample size and wide colour basis compensate for this to some extent and should unveil any significant systematic differences. Figure 1 (right panel) shows that nucleus and galaxy u–z colours agree within the errors – still, a tiny offset towards redder nuclei seems to be present. This offset does not become stronger for larger nucleus-to-galaxy ratios (larger symbols), indicating that it might at best apply to a fraction of objects rather than to the full sample. In the colour-magnitude relation (Fig. 2, left panel, using 113 dENs) nucleus and galaxy colours can be interchanged without altering its shape; only at the very faint end where measurement errors are large, there seems to be a number of slightly redder nuclei. Moreover, several nuclei show redder u-g colours in the colour-colour diagram (Fig. 2, right panel). However, the measurement errors are large compared to the relatively small range of colours and thus prevent any definite statement on the number of these nuclei or on a possible systematic effect applying to all objects. Clearly, high-quality spectroscopy would be the proper way to decide whether or not nuclei and their host galaxies do consist of the same stellar population.

Acknowledgements

We thank the SDSS collaboration (http://www.sdss.org/) for the wealth of DR3 data.

References

Lotz, J.M., Miller, B.W. & Ferguson, H.C. 2004, *ApJ* 613, 262
Oh, K.S. & Lin, D.N.C. 2000, *ApJ* 543, 620

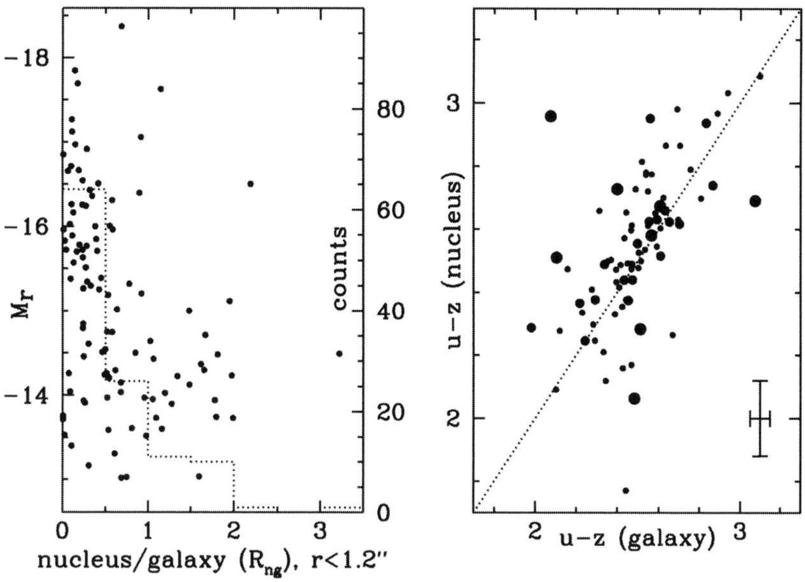

Figure 1. *Left:* Absolute galaxy magnitude in the r band (adopting $m - M = 31$) versus the flux ratio of nucleus to galaxy within $r < 1.2''$. The dotted line shows a histogram with number counts given on the right y-axis. *Right:* Comparison of nucleus and galaxy u-z colours. Large symbols represent objects with a nucleus-to-galaxy ratio $R_{ng} \geqslant 1$, intermediate symbols stand for $0.5 \leqslant R_{ng} < 1$, and small symbols mean $R_{ng} < 0.5$. Only objects brighter than $m_r \leqslant 16$ were selected to minimize measurement uncertainties. The dotted line stands for equality.

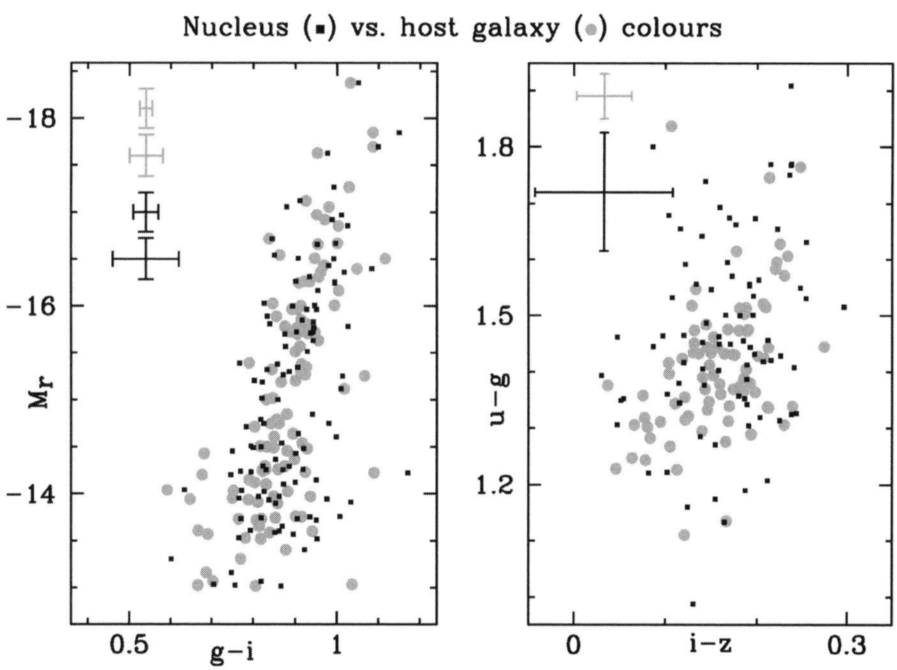

Figure 2. *Left:* Absolute galaxy magnitude in the r band ($m - M = 31$) versus galaxy colours (circles) and nucleus colours (squares). Colours were measured from half-light radius and central ($r < 1.2''$) apertures, respectively. Upper error bars are for the brighter part of the sample, lower error bars for the fainter part. *Right:* Colour-colour plot; symbols as in the left panel. Only objects brighter than $m_r \leqslant 16$ were selected to minimize measurement uncertainties.

Near-Field Cosmology with Dwarf Elliptical Galaxies
Proceedings IAU Colloquium No. 198, 2005
H. Jerjen and B. Binggeli, eds.

doi:10.1017/S1743921305004102

N-body/SPH study of dEs in a cluster environment

Dolf Michielsen, S. De Rijcke and H. Dejonghe

Sterrenkundig Observatorium, Ghent University, Krijgslaan 281, B-9000 Ghent, Belgium
email: dolf.michielsen@UGent.be

Abstract. Until recently, dwarf elliptical galaxies (dEs) were mostly believed to be gas-less systems, having lost their interstellar medium (ISM) through a galactic wind or through ram pressure stripping in the dense cluster environment. However, recent Hα and H I surveys indicate that some dEs in cluster environments are able to retain part of their gas (Drinkwater *et al.* (2001), Conselice *et al.* (2001)). In a detailed observational study of the ISM of a sample of Fornax dEs, we detected genuine dEs containing warm and cold gas (De Rijcke *et al.* (2003), Michielsen *et al.* (2004), Buyle *et al.* (2005)).

Using an N-body/SPH code, we explore the two gas-removing scenarios in order to assess their efficiency in removing the ISM from dEs. We simulated dEs subjected to ram pressure stripping by the intracluster medium (ICM) and found that, while small dEs are instantaneously stripped, more massive dEs are able to retain part of their ISM over long time-scales. Isolated dEs on the other hand are expected to lose their gas through a galactic wind. Our simulations show that supernova explosions can blow (super)bubbles in the ISM, thereby ejecting part, though not all, of the ISM.

Keywords. methods: n-body simulations, galaxies: dwarf, galaxies: ISM

1. Ram pressure stripping

We used a 3D SPH code to model ram pressure stripping from dEs that initially contain an ionized ISM in hydrostatic equilibrium with the Burkert dark matter (DM) potential (Burkert, 1995). For DM halo masses ranging from 10^6 to $10^{10}\,\mathrm{M}_\odot$, two different sets of ISM initial density and pressure profiles were adopted. First the ISM is set up to be density confined within the ICM (referred to as DC), with a temperature proportional to the DM halo mass ($T_{\mathrm{ISM}} \propto M_{\mathrm{DM}}^{4/7}$). This results in a one-parameter family of models (Mori & Burkert, 2000). Second, the ISM is considered to be pressure confined within the surrounding ICM (referred to as PC), and we assume a fixed temperature $T_{\mathrm{ISM}} = 10^4\,\mathrm{K}$. The PC conditions have much smaller ISM boundary radius than the DC conditions. The ICM properties chosen are typical for a moderate cluster like the Virgo Cluster ($T_{\mathrm{ICM}} = 10^7\,\mathrm{K}$, $n_{\mathrm{ICM}} = 10^{-4}\,\mathrm{cm}^{-3}$), while the relative velocity of the galaxy through the cluster was fixed at $1000\,\mathrm{km\,s}^{-1}$.

When subjected to the ram pressure, the ISM of small dEs ($M_{\mathrm{DM}} < 10^8\,\mathrm{M}_\odot$) is instantaneously stripped. This was to be expected since the central thermal pressure of the ISM is lower than the ram pressure of the ICM. However, the central pressure in the more massive dwarfs ($M_{\mathrm{DM}} \geqslant 10^9\,\mathrm{M}_\odot$) is considerably higher in the PC, than in the DC simulations, due to the higher central densities. Therefore, these dEs are able to retain their ISM over several dynamical time-scales, during which the gas may cool and form stars. Star formation may be even enhanced by the increased density due to the ram pressure of the ICM.

2. Galactic winds

We used a 3D N-body/SPH code including radiative cooling, star formation and feedback to model the formation of an isolated dE. The $9 \times 10^8\,M_\odot$ DM halo is set up in a King distribution with concentration parameter $c = 2$. Initially, the $10^8\,M_\odot$ of gas is distributed homogeneously within the DM tidal radius. Star formation occurs in collapsing, cold and dense gas particles (Katz, 1992). Each newly formed star particle represents a simple stellar population. The most massive stars explode as SNe, each injecting 10^{50} ergs of thermal energy into the surrounding gas.

Different tidal radii are adopted, giving different initial density distributions. The star formation rates (SFRs) differ significantly for models with different initial densities (Carraro *et al.*, 2001). The densest model shows an intense starburst of ~100 Myr, whereafter the remaining ISM is blown out in a galactic wind. Models with lower initial density have lower SFRs spanning longer time-scales. The SNe exploding in these low-density systems blow (super)bubbles in the ISM, and part of the hot, enriched gas escapes the galaxy, while part of the gas is able to fall back. The increased density caused by shocks enhances the star-formation. After an initial prolonged starburst phase, the low-density models keep on forming stars at a slow rate for several Gyr.

3. Conclusions and outlook

Our simulations show that the gas-removal processes in dEs are not always efficient. The ISM of more massive dEs can survive ram pressure stripping for long timescales, while galactic winds blow out only part of the ISM. These results corroborate our observations of genuine dEs containing a warm and cold ISM.

Until now, simulations of dE formation including star formation and feedback concentrated on isolated dEs. However, if ram pressure stripping is not instantaneous, the thermal pressure of a surrounding ICM could confine the galactic wind inside the dE so that more gas is available for star formation and enriched gas cannot easily escape the galaxy. Furthermore, the increased density induced by ram pressure is assumed to increase the SFR. We now have the tools to investigate this scenario. Simulating a variety of ICM and dE properties will shed more light on the formation and evolution of dEs in a dense environment.

Acknowledgements

D. M. acknowledges the financial support of the Bijzonder OnderzoeksFonds (Ghent University). S. D. R. acknowledges the financial support of the Belgian Fund for Scientific Research (F.W.O.).

References

Burkert, A. 1995, *ApJ* 447, 25

Buyle, P., De Rijcke, S., Michielsen, D., Baes, M. & Dejonghe, H. 2005, *MNRAS* accepted (astro-ph/0504270)

Carraro, G., Chiosi C., Girardi, L. & Lia, C., 2001, *MNRAS* 327, 69

Conselice, C.J., Gallagher, J.S. & Wyse, R.F.G. 2003, *AJ* 125, 66

De Rijcke, S., Zeilinger, W.W., Dejonghe, H. & Hau G.T.K. 2003b, *MNRAS* 339, 225

Drinkwater M.J., Gregg M.D., Holman B.A. & Brown M.J.I. 2001, *MNRAS* 326, 1076

Katz, N. 1992, *ApJ* 391, 502

Michielsen, D., De Rijcke, S., Zeilinger, W.W., Prugniel, P., Dejonghe, H. & Roberts, S. 2004, *MNRAS* 353, 1293

Mori, M. & Burkert, A. 2000, *ApJ* 538, 339

Near-Field Cosmology with Dwarf Elliptical Galaxies
Proceedings IAU Colloquium No. 198, 2005
H. Jerjen and B. Binggeli, eds.

doi:10.1017/S1743921305004114

Dwarf galaxy candidates in the NGC 3665 galaxy group

Bettina Ogbuagu-Poledna
and Werner W. Zeilinger

Institute of Astronomy, Türkenschanzstr. 17, A-1180 Vienna, Austria
email: poledna@astro.univie.ac.at

Abstract. We present preliminary results of a search for candidate dwarf galaxies in the NGC 3665 galaxy group. A 0.1×0.1 Mpc^2 field around the dominant early-type galaxy NGC 3665 was investigated using BRI images from the 2.5m INT WFC archive. A search with the SExtractor yielded 148 dwarf galaxy candidates in the range $-12 > M_B > -16$. The resulting luminosity function for the group can be described by a standard Schechter function.

Keywords. galaxies: dwarf, galaxies: evolution

1. Introduction

Present-day galaxy groups contain a significant fraction of the galaxy population of the local universe. It is therefore of crucial importance to understand the dynamical and chemical galaxy evolution in such environments. Galaxy groups are an ideal laboratory to study not only galaxy-galaxy interactions but also the evolution of cosmic structures at a "cellular" level. The detection of a diffuse hot intra-group medium in a number of groups ascertained their physical reality; e.g. Mulchaey *et al.* (1996), Ponman *et al.* (1996). Dwarf galaxies are an important indicator for the dynamical evolution of a group. Zabludoff & Mulchaey (1998) detected a significant dwarf galaxy population down to $M_B \simeq -14$ in X-ray bright groups which is apparently absent in X-ray non-detected groups of the same optical morphology. The importance of galaxy-galaxy interactions in group environments is illustrated by the fact that bulge-dominated systems are more common and that disk systems with apparent morphological signatures of interactions show higher star formation rates than in the field, e.g. Tran *et al.* (2001).

2. Characteristic Properties of the NGC 3665 Group

The NGC 3665 galaxy group was first mentioned by Geller & Huchra (1983) and subsequently described by Tully (1988) and Garcia (1993). The group is located at a distance of 27.6 Mpc ($H_0 = 75$ km s^{-1} Mpc^{-1}) and appears as a loose system of mostly late-type galaxies around the dominant early-type galaxy NGC 3665. A search of an area of 1 Mpc^2 around NGC 3665 resulted in only 10 luminous group members yielding a velocity dispersion of 65 km s^{-1} for the group. However, velocity dispersions of at least 100-200 km/s are expected for a virialized group by relating the minimum density of a group to the density of the universe (Mamon 1994). Therefore the derived velocity dispersion of the NGC 3665 group remains suspect. The presence of a hot diffuse intra-group medium in the group has been confirmed and it belongs to the faintest and coolest systems (Helsdon *et al.* 2005). BRI images obtained from the WFC archive of the 2.5m INT were analyzed in order to search for further group member candidates. The images were centered upon NGC 3665 yielding a field of view of $\approx 34' \times 34'$ which corresponds to about 0.1 Mpc^2.

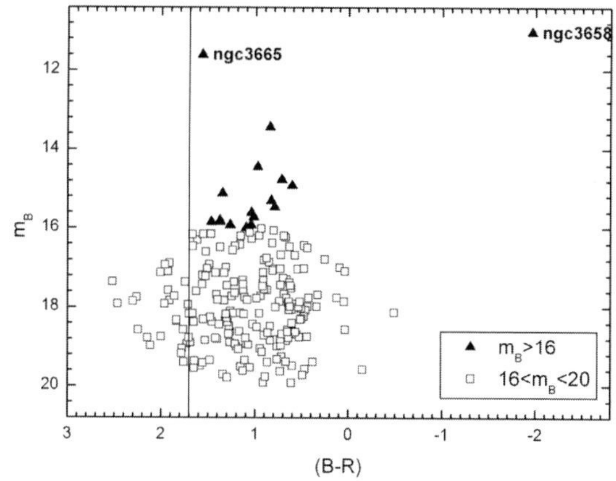

Figure 1. Colour-magnitude diagram of all group member candidates.

Standard procedures were applied to calibrate the images photometrically. The candidate search was performed using the SExtractor (Bertin & Arnouts (1996)). The search based upon a stellaricity indix of < 0.9, $m_B < 16$ for bright galaxies and $16 < m_B < 20$ for dwarf galaxies. As a minimum size for dwarf galaxies a value of $1.5''$ was adopted based upon the seeing measurements.

3. Preliminary Results

A (B-R)-diagramm for all likely members was constructed (see Figure 1). After applying a cut-off at (B-R) = 1.7 (as it is applied in Miles *et al.* (2004)), 17 bright and 148 dwarf candidate galaxies remain. The ensure the group membership, spectroscopic observations will be carried out. 26 galaxies of all candidates were previously known, 139 are new objects. Only two (NGC 3665 and NGC 3658) have known redshifts. The luminosity function of the group members follows the shape of the Schechter function with a dip around $M_{\rm B} = -18$ which is in good agreement with luminosity functions of groups presented in recent publications, e.g. Miles *et al.* (2004).

Acknowledgements

The research project is supported by the Austrian Science Fund (FWF project P14783).

References

Bertin, E. & Arnouts, S. 1996, *A&AS* 117, 393
Garcia, A.M. 1993, *A&AS* 100, 47
Geller, M.J. & Huchra, J.P. 1983, *ApJS* 52, 61
Helsdon, S.F., Ponman, T.J. & Mulchaey, J.S. 2005, *ApJ* 618, 679
Mamon, G.A. 1994, in: F. Durret *et al.* (eds.) *Clusters of Galaxies*, Gif-sur-Yvette, Edition Frontières, p. 291
Miles, T.A., Raychaudhury, S., Forbes, D.A., Goudfrooij, P., Ponman, T.J. & Kozhurina-Platais, V. 2004, *MNRAS* 355, 785
Mulchaey, J.S., Davis, D.S., Mushotzky, R.F. & Burstein, D. 1996, *ApJ* 456, 80
Ponman, T.J., Bourner, P.D.J., Ebeling, H. & Bohringer, H. 1996, *MNRAS* 283, 690
Tran, K-V.H., Simard, L., Zabludoff, A.I. & Mulchaey, J.S. 2001, *ApJ* 549, 172
Tully, R.B. & Fisher, J.R. 1988 *Catalog of Nearby Galaxies*, Cambridge University Press
Zabludoff, A.I. & Mulchaey, J.S. 1998, *ApJ* 496, 73

Near-Field Cosmology with Dwarf Elliptical Galaxies
Proceedings IAU Colloquium No. 198, 2005
H. Jerjen and B. Binggeli, eds.

doi:10.1017/S1743921305004126

Charting the neighbourhood

Rami Rekola[1], Helmut Jerjen[2] and Chris Flynn[1]

[1]Tuorla Observatory, VISPA, University of Turku, Piikkiö, FI-21500, Finland
email: rareko@astro.utu.fi

[2]Research School of Astronomy and Astrophysics, The Australian National University, Mt Stromlo Observatory, Cotter Road, Weston ACT 2611, Australia

Abstract. We present Surface Brightness Fluctuation (SBF) distances of fifteen early-type dwarf galaxies and the S0 galaxy NGC 4150 in the Local Volume based on deep *B*- and *R*-band CCD images obtained with the 2.56 metre Nordic Optical Telescope (NOT). Distances have been determined from Fourier analysed fields selected from optimal locations on the galaxies. Distances were found to range from 3 to 16 Mpc.

The SBF method has been demonstrated to efficiently measure relatively accurate distances from the ground with mid-aperture telescopes. The SBF method to measure distances to dwarf elliptical (dE) galaxies has also been empirically calibrated using three dEs with known tip of the red giant branch (TRGB) distances.

Keywords. galaxies: clusters: individual (M81 group, M101 group, CVn I cloud, CVn II cloud, Leo I group, Coma I group), galaxies: dwarf, galaxies: stellar content, galaxies: structure, galaxies: distances and redshifts

---

1. Introduction

Recent imaging surveys of the Local Group (LG) neighbourhood have found and identified a large number of low surface brightness galaxy candidates, which potentially could be nearby dwarf galaxy systems (e.g. Côté, Freeman, Carignan & Quinn (1997); Karachentseva & Karachentsev (1998); Jerjen, Binggeli & Freeman (2000)). Many are dEs located in galaxy groups as satellites of giant galaxies. Whether dEs are local or in the background has cosmological implications. For example Moore, Ghigna, Governato, *et al.* (1999) have numerically studied galactic and cluster halo substructures in a hierarchical universe, and found that simulations predict an abundance of Milky Way dwarf satellites currently unobserved. This conflict may be partly addressed by catalogueing and obtaining accurate distances of dwarf galaxy populations in the local universe. As a new and numerous target group in extragalactic studies dEs serve also to test the peculiar linearity and smoothness of the Hubble flow in the local volume. As such they may help to define the dark energy solution to the problem (Baryshev, Chernin & Teerikorpi (2001)).

2. Distances

We have measured SBF distances to 16 galaxies in the vicinity of the LG (Jerjen, Rekola, Takalo, *et al.* (2001); Rekola, Jerjen & Flynn (2005)). Dwarf ellipticals DDO 44, KK98 77 and DDO 71 were used to calibrate the SBF method for dEs. Our distances of 27.52 ± 0.15 mag (or 3.2 ± 0.2 Mpc) for DDO 44, 27.71 ± 0.15 mag (or 3.5 ± 0.3 Mpc) for KK98 77 and 27.72 ± 0.15 mag (or 3.5 ± 0.3 Mpc) for DDO 71 agreed with existing

TRGB distances confirming previously established group memberships of these galaxies (in groups of NGC 2403, M81 and M81 respectively). Similarily KDG 61 and UGC 5442 had existing TRGB distances and BTS 128 an SBF distance; our distances of 27.61 ± 0.17 mag (or 3.3 ± 0.3 Mpc) for KDG 61, 27.74 ± 0.18 mag (or 3.5 ± 0.3 Mpc) for UGC 5442 and 31.02 ± 0.25 mag (or 16.0 ± 1.9 Mpc) for BTS 128 agreed with all of them (placing these galaxies in groups of M81, M81 and Coma I respectively). UGC 7639 had a rough distance based on brightest blue and red stars. We found a slightly smaller distance of 29.27 ± 0.16 mag (or 7.1 ± 0.6 Mpc), with no consequence to its assumed group membership in Canes Venatici II cloud.

The S0 galaxy NGC 4150 had an earlier SBF distance, which we were able to confirm. The distance of 30.79 ± 0.11 mag (or 14.4 ± 0.7 Mpc) places NGC 4150 to the outskirts of Virgo cluster. Its Local Group barycentric radial velocity of $v_{LG} = 207$ km s^{-1} is, however, very low for its distance. Solanes, Sanchis, Salvador-Solé, *et al.* (2002) have found that most Virgo galaxies in the region closest to us, corresponding to our distance to NGC 4150, have high radial velocities outward from the cluster centre. The low radial velocity of NGC 4150 can thus be explained with group dynamics.

We produced completely new distances to the remaining galaxies. The distances are 30.11 ± 0.18 mag (or 10.5 ± 0.9 Mpc) for UGC 4998 (in the background of M81 group), 27.44 ± 0.18 mag (or 3.1 ± 0.3 Mpc) for DDO 113 (in Canes Venatici I cloud), 29.12 ± 0.18 mag (or 6.7 ± 0.6 Mpc) for UGC 7356 (in Canes Venatici I cloud), 28.11 ± 0.15 mag (or 4.2 ± 0.3 Mpc) for UGC 1703 (a companion of NGC 784), 29.01 ± 0.27 mag (or 6.3 ± 0.8 Mpc) for UGCA 200 (possibly a field galaxy), 30.22 ± 0.17 mag (or 11.1 ± 0.9 Mpc) for UGC 5944 (in Leo I group), 30.19 ± 0.23 mag (or 10.9 ± 1.2 Mpc) for UGC 8799 (in Virgo cluster) with an alternative distance of 30.61 ± 0.26 mag (or 13.2 ± 1.7 Mpc), and 29.60 ± 0.20 mag (or 8.3 ± 0.8 Mpc) for UGC 8882 (in M101 group).

3. Conclusions

The SBF distances we have presented continue to support the understanding of the distribution of dwarf galaxies in galaxy groups and intermediate space in the Local Group neighbourhood. They also demonstrate well the feasibility of the Surface Brightness Fluctuation method in determining accurate distances with 2 metre class ground-based telescopes out to the Virgo cluster.

Acknowledgements

Based on observations made with the Nordic Optical Telescope, operated on the island of La Palma jointly by Denmark, Finland, Iceland, Norway, and Sweden, in the Spanish Observatorio del Roque de los Muchachos of the Instituto de Astrofísica de Canarias.

References

Baryshev, Yu.V., Chernin, A.D. & Teerikorpi, P. 2001, *A&A* 378, 729
Côté, S., Freeman, K.C., Carignan, C. & Quinn, P.J. 1997, *AJ* 114, 1313
Jerjen, H., Binggeli, B. & Freeman, K.C. 2000, *AJ* 119, 593
Jerjen, H., Rekola, R., Takalo, L., Coleman, M. & Valtonen, M. 2001, *A&A* 380, 90
Karachentseva, V.E. & Karachentsev, I.D. 1998, *A&AS* 127, 409
Moore, B., Ghigna, S., Governato, F., Lake, G., Quinn, T., Stadel, J. & Tozzi, P. 1999, *ApJL* 524, 19
Rekola, R., Jerjen, H. & Flynn, C. 2005, *A&A*, 437, 823
Solanes, J. M., Sanchis, T., Salvador-Solé, E., Giovanelli, R. & Haynes, M. P. 2002, *AJ* 124, 2440

Near-Field Cosmology with Dwarf Elliptical Galaxies
Proceedings IAU Colloquium No. 198, 2005
H. Jerjen and B. Binggeli, eds.

doi:10.1017/S1743921305004138

Infra-red studies of dwarf galaxies in the Fornax cluster

Rachel Smith, Steve Phillipps and Roberto De Propris

H.H. Wills Physics Laboratory, University of Bristol, Tyndall Avenue, Bristol, BS8 1TL
email: rachel.smith@bristol.ac.uk

1. Introduction

Dwarf galaxies are the most numerous type of galaxy in the local universe and are extremely important in the understanding of galaxy formation and evolution. A better understanding of the physics of dwarf galaxies is central to several current issues in observational cosmology. For example, simulations of galaxy evolution based on the Cold-Dark-Matter (CDM) model predict far more dwarfs than are observed locally. In the last 20 years there has been a huge amount of interest in the study of dwarf galaxies, partly due to these unanswered puzzles and also because of the improvements in the technology available.

One area into which we are researching involves analysing the follow-up infra-red (IR) data from the Fornax Cluster Spectroscopic Survey (FCSS). The FCSS is an all-object survey of a region around the Fornax Cluster of galaxies (Drinkwater *et al.* 2000 and Phillipps *et al.* 2001). The aim of the FCSS was to obtain spectra for a complete sample of all objects within the magnitude range $16.5 \leqslant b_j \leqslant 19.7$, irrespective of their image morphology. We have used this to obtain a well defined sample of cluster dwarfs.

2. Infra-red Observations

IR observations are a very useful tool in studying the astrophysical nature of dwarf galaxies. Near IR luminosities are generally regarded as a better surrogate for mass, than those at shorter wavelengths, which are more influenced by recent star formation. In particular this allows us to consider global properties such as the overall baryonic (or at least stellar) mass and the mass surface density.

We are interested in answering several different questions about Fornax Cluster dwarfs: Are they all old? Are nucleated dwarfs older than non-nucleated dwarfs? Are the ultra-compact dwarfs (UCDs) a different age to the other dwarfs, and what would this tell us about their evolution? We are also interested in the metallicity of these galaxies. For a few of the dwarfs we have spectroscopic data, however most will be investigated using broad-band colours.

The IR data we are using consists of observations of a sample of Fornax cluster dwarf galaxies in the K' band, taken in November 2000 using the Cryogenic Array Spectrometer/Imager (CASPIR) on the 2.3m telescope at Siding Spring Observatory and K_s band observations of a separate sample taken in December 2002 using the 3.6m ESO-New Technology Telescope (NTT) equipped with the Son of ISAAC (SOFI) infrared spectrograph and imaging camera.

3. Stellar Population Models

There are complications in using broad-band colours to model galaxy properties. As well as stellar evolution, chemical evolution in stars also affects the colour of the galaxy, which causes an age/metallicity degeneracy problem. Stars with high metallicities appear redder than those which are under abundant in heavy elements. It is therefore difficult to separate populations which have a combination of age and metallicity effects; the optical colours of composite stellar populations are virtually identical if the percentage change in age or metallicity (Z) follows Δage/ΔZ $\sim$3/2 (Worthey 1994). IR data is invaluable in disentangling age and metallicity effects in the star formation histories. By combining NIR data with optical colours we can constrain stellar populations by using stellar population synthesis (SPS) models.

Several different SPS models are available and their individual strengths and weaknesses have been discussed in the literature (e.g. Worthey 1994 and Bell & de Jong 2000). We are working with collaborators at the Astrophysics Research Institute, Liverpool John Moores University, who are developing their own predictions of the optical - NIR colours of stellar populations as a function of age and metallicity, using models developed by M. Salaris and S. Cassisi. They have identified the (B-K) - (J-K) colour-colour plane as particularly sensitive for breaking age-metallicity degeneracy. It is found that the integrated colour (J-K) appears to be a good tracer of the initial metallicity of a simple stellar population (SSP) and is weakly affected by age, whereas the integrated (B-K) is a good age indicator, mildly affected by the SSP metallicity (James *et al.* in prep.).

4. Current Progress and Future Work

We have high signal/noise reduced images for 9 Fornax Cluster dwarfs. Surface brightness profiles and K-band magnitudes will be measured from these images. In most cases these can be calibrated directly from the final image, as the majority of the galaxies have 2MASS stars within the object frame.

Using this IR data, alongside existing optical data, we will investigate morphological and structural properties, star formation rates and histories, and stellar populations and their metallicities in dwarf galaxies. A comparison of the broad-band colours and spectal methods of obtaining metallicities can be made, as both types of data are available for a few of our objects.

References

Bell, E.F. & de Jong, R.S. 2000, *MNRAS* 312, 497

Drinkwater, M.J., Phillipps, S., Jones, J.B., Gregg, M.D., Deady, J.H., Davies, J.I., Parker, Q.A., Sadler, E.M. & Smith, R.M. 2000, *A&A* 355, 900

James, P.A., Salaris, M., Davies, J.I., Phillipps, S. & Cassisi, S. In prep.

Phillipps, S., Drinkwater, M.J., Gregg, M.D. & Jones, J.B. 2001 *ApJ* 560, 201

Worthey, G. 1994, *ApJS* 95, 107

Near-Field Cosmology with Dwarf Elliptical Galaxies
Proceedings IAU Colloquium No. 198, 2005
H. Jerjen and B. Binggeli, eds.

doi:10.1017/S174392130500414X

NIR observations of dEs in the Virgo cluster: a structural continuity with giant ellipticals

Stefano Zibetti[1] and the GOLDMiNe Research Team[2]

[1]Max-Planck-Institut für Extraterrestrische Physik
Gießenbachstraße, D-85748, Garching bei München, Germany – email: szibetti@mpe.mpg.de
[2]`http://goldmine.mib.infn.it/`

Abstract. The structural properties of a sample of 50 dEs in the Virgo cluster are here derived from Near InfraRed (NIR, H-band 1.65 μm) surface photometry and analysed. One-dimensional surface brightness (SB) profiles are extracted using elliptical isophote fitting. They are characterised by means of structural parameters, namely the half light radius R_e, the average surface brightness within R_e (μ_e), and a concentration index (c_{31}). We show that typical dEs have close-to-exponential NIR SB distributions.

The relations between dEs and giant ellipticals (Es) are investigated by comparing the NIR structural parameters of 273 Es in nearby clusters. Further analysis is conducted using the optical-NIR colour $B - H$ and by studying the relationships between structural and dynamical parameters (fundamental plane) for the two classes of galaxies. The transition between the two regimes is smooth and no dichotomy is seen.

Keywords. galaxies: dwarf; elliptical and lenticular, cD; fundamental parameters; structure; kinematics and dynamics; galaxies: clusters: individual (Virgo, Coma, A1367)

1. Introduction and sample

Dwarf elliptical galaxies are **not** just a rescaled version of their giant counterparts. Optical observations have shown that, as opposed to Es, dEs have exponential rather than de Vaucouleurs SB profiles, and do not follow the same scaling relations linking luminosity, the effective radius R_e and the mean effective surface brightness μ_e.
In this work we re-assess the structural properties of dEs using NIR surface photometry, which is a good tracer of the bulk of the stellar mass, while being much less affected than optical bands by dust attenuation and metallicity effects. This project is part of a large NIR survey of galaxies in nearby clusters (Gavazzi *et al.* 2000, 2003). This has allowed to compare the derived properties with those of E galaxies in the full range of luminosities.

The 50 dEs of this work have been selected from the VCC (Binggeli, Sandage & Tammann, 1985), limited to photographic magnitude $m_{\mathrm{p}} < 16$, and observed at the 3.6m telescopes ESO/NTT and TNG (Gavazzi *et al.* 2001). Our sample is representative of $\sim$30% of the dEs more luminous than $10^{8.5}\mathrm{L}_{\odot}$ (H). 48 out of these 50 have B-band photometry available, and 6 have also velocity dispersion measurements. 273 Es (including 35 in Virgo and the complete sample of 217 Es in Coma and Abell 1367) have been used to investigate the relations between dwarves and giants.

2. Results

The SB profile of each galaxy is extracted by fitting elliptical isophotes with standard IRAF procedures. The decomposition of the 1-D profile with simple analytical functions (exponential, de Vaucouleurs and their combinations) is used to extrapolate an asymptotic magnitude (extrapolations are of the order of 0.1–0.2 mag). *From the measured*

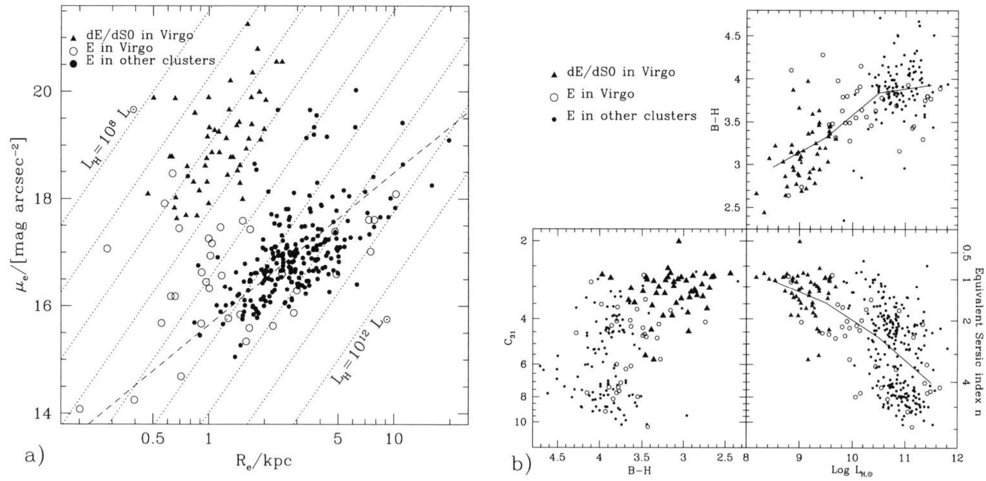

Figure 1. a) The H-band $\mu_e - R_e$ plane. The thick dashed line is the "Kormendy" relation for the Es. Dotted lines are isoluminosity lines assuming a fixed ellipticity=0.35. **b)** The "photometric cube", alias $c_{31} - (B - H) - L_H$ space. Trends are indicated by the thick lines, representing the median of c_{31} and $B - H$ in 4 bins of L_H. The right-side scale of the $c_{31} - L_H$ panel reports the value of the Sérsic index n that reproduces the same c_{31}.

SB profile we extract the effective radius R_e enclosing half of the asymptotic flux, and the average SB μ_e within R_e. The distribution of dEs and Es in this plane (Fig. 1a) does not display any significant gap between the two populations. The "Kormendy" relation for Es (thick dashed line) represents the upper limit for the SB at given R_e, with dEs having lower SB than Es. The multi-dimensional photometric parameter space $c_{31} - (B - H) - L_H$† (Fig 1b, see also Scodeggio *et al.* 2002) shows very clearly the smooth transition from blue, low-concentration (nearly exponential) dEs to red, concentrated (roughly de Vaucouleurs) Es.

We have analysed the Fundamental Plane relations for the galaxies with a velocity dispersion measure . Using NIR structural parameters, dEs lie on the same relations as Es. Systematic variations of M/L in the NIR have been investigated by means of the κ-space formalism (Bender, Burstein & Faber, 1992): although a trend with mass is clearly seen at high masses, dEs span the same range in M/L as low-mass Es. Adopting a more sophisticated formalism, in which profile deprojection is used to take non-homology and spectroscopic aperture effects into account (Zibetti *et al.* 2002), we show that the data are consistent with constant M/L throughout the sequence of Es and dEs.

References

Bender, R., Burstein, D. & Faber, S.M. 1992, *ApJ* 399, 462

Binggeli B., Sandage A. & Tammann G.A. 1985, *AJ* 90, 1681

Gavazzi, G., Franzetti, P., Scodeggio, M., Boselli, A. & Pierini, D. 2000, *A&A* 361, 863

Gavazzi, G., Zibetti, S., Boselli, A., Franzetti, P., Scodeggio, M. & Martocchi, S. 2001, *A&A* 372, 29

Gavazzi, G., Boselli, A., Donati, A., Franzetti, P. & Scodeggio, M. 2003, *A&A* 400, 451

Scodeggio, M., Gavazzi,G., Franzetti,P., Boselli,A., Zibetti,S. & Pierini,D. 2002, *A&A* 384, 812

Zibetti, S., Gavazzi, G., Scodeggio, M., Franzetti, P. & Boselli, A. 2002, *ApJ* 579, 261

† The concentration index c_{31} is the ratio of the radii enclosing 75% and 25% of the asymptotic flux.

Near-Field Cosmology with Dwarf Elliptical Galaxies
Proceedings IAU Colloquium No. 198, 2005
H. Jerjen and B. Binggeli, eds.

doi:10.1017/S1743921305004151

Globular clusters and nuclei in dwarf ellipticals

Bryan W. Miller

Gemini Observatory, Casilla 603, La Serena, Chile
email: bmiller@gemini.edu

Abstract. Globular clusters (GCs) and nuclei in dEs are important probes for for studying cosmology and galaxy evolution. GCs and nuclei are usually formed in the high pressures of starbursts, therefore resolved stellar color-magnitude diagrams and integrated colors, magnitudes, or spectra yield ages and metallicities which are snapshots of the conditions during the most violent and important star forming episodes of the galaxies. Most are older, more metal-poor, and more enriched in α-elements than the underlying dE field stars. About 25% have formed within the last $\sim$6 Gyr from enriched gas. Many nuclei have the properties of brights GCs but some can be distinguished by large sizes and composite stellar populations. The relatively large numbers of GCs in dEs shows that they are an important source of GCs to galaxies that accrete them.

Keywords. galaxies: dwarf, galaxies: star clusters, galaxies: nuclei

1. Introduction

The aim of near-field cosmology is to use local objects which can be studied in great detail to try to unravel the earliest periods of galaxy formation and evolution. What is discovered should be consistent with the results from high redshift galaxy studies and the values of the cosmological parameters. Globular clusters (GCs) are very useful tools for this endeavor: they are normally made of simple (single-age) stellar populations so they can be dated relatively easily; they are compact and luminous so they can be detected at large distances; and they are fairly numerous in dwarf elliptical (dE) galaxies. They have also been found to be among the oldest stellar systems known. Therefore, they provide an important view back to the earliest periods of star and galaxy formation and they have played a fundamental role in determining the formation history of the galaxy (e.g. Searle & Zinn 1978).

Nuclei are identified as distinct compact structures in the centers of galaxies. Nuclei are found in the centers of disk galaxies as well as in over half of bright dE galaxies in galaxy clusters. Nucleated dwarfs are the dominant galaxy population by number in rich galaxy clusters but their fraction decreases in lower density environments (Ferguson & Sandage 1989). A detailed comparison of GCs and nuclei may help determine how nuclei can form. Also, a comparison of the nuclei in spirals and dEs can help determine if some nucleated dEs are the harassed remnants of spiral galaxies.

This paper will review the basic properties of GCs and nuclei in nearby dE galaxies. We use the term dE to refer both to brighter systems like NGC 205 and the low surface-brightness systems like Fornax that are sometimes referred to as dwarf spheroidals (dSph). Detailed information about the stars within GCs in the Local Group will help us interpret the larger samples of GCs in more distant galaxies where only integrated properties can be measured. Finally, some inferences will be drawn about the formation of GCs in the context of our current understanding of the conditions in the early universe.

Galaxy	M_V	N_{GC}	Nuc	$\langle V-I\rangle_{GC}$	$[Fe/H]_{GC}$	$[Fe/H]_{FS}$	δt_{GC}
Fornax	−13.1	5	Yes?	0.83	−1.8	−0.9	3 Gyr
Sagittarius	−13.5?	6	Yes?	0.89	−1.6	−0.5	6 Gyr
NGC 147	−15.2	2	No	0.75	−2.0	−0.9	< 2 Gyr
NGC 185	−15.5	5	No	0.86	−1.7	−1.2	< 2 Gyr
NGC 205	−16.6	7	Yes	0.95	−1.4	−0.8	10 Gyr

Table 1. Globular cluster systems in Local Group dE galaxies. Data are taken from Lotz *et al.* (2004), Forbes *et al.* (2004), and references therein. δt_{GC} is the age spread of the GCs.

2. The Local Group

Of the approximately 20 dE/dSph galaxies known in the Local Group only 5 have massive star clusters. The properties of these star cluster systems are given Table 1. Some have proposed that between 4 and 6 GCs may also be associated with the newly discovered Canis Major dwarf (Martin *et al.* 2004; Forbes *et al.* 2004). However, Martínez-Delgado (2005) reports that new proper motions exclude these clusters from being associated with this galaxy. Therefore, we will not consider the Canis Majoris system.

Three of the five galaxies in Table 1 are possibly nucleated. NGC 205 has a clear nucleus but this nucleus is less than 1 Gyr old. Both Fornax and Sagittarius have globular clusters near the peaks in their background light distributions and so these are considered candidate nuclei (Da Costa & Armandroff 1995; Strader *et al.* 2003).

A defining characteristic of GCs is their small size. Half-light radii (R_h) are typically 3–5 pc but they can be as large as 30 pc (Figure 1). The sizes of GCs associated with the Sagittarius dE are the same as typical Galactic GCs. There may be an upper envelop to R_h at a given M_V for typical GCs that is indicated by the solid line in (Figure 1. The few objects that lie above this line are unusually large for their luminosities. M54 is the central GC in the Sagittarius dwarf and it may have a spread in internal abundances (Da Costa & Armandroff 1995). Omega Cen also has multiple stellar populations and it has been speculated that it is the former nucleus of a disrupted dwarf (Hilker & Richtler 2000). The one unambiguous dE nucleus in the Local Group belongs to NGC 205 (Butler & Martínez-Delgado 2005). It also falls on or just above the line in Figure 1. Therefore, dE nuclei have compact sizes similar to those of GCs but they may be distinguishable by having sizes that are relatively large for their luminosities.

The bane of stellar population studies is breaking the degeneracy between age and metallicity that is always present in the observations. Nearby GCs that can be resolved into stars are the among best environments for breaking this degeneracy and do consistency checks of different techniques. We will use the Fornax GCs as an example of the different methods and their uncertainties.

Figure 2 shows the color-magnitude diagrams (CMDs) of two GCs in the Fornax dwarf from images taken with *HST* (Buonanno *et al.* 1998). The narrow giant branch and single, well-defined main sequence turnoff are indicative of a simple (single age and metallicity) stellar population. Isochrones from stellar evolutionary models are fit to the CMDs to determine the ages, metallicities, reddenings, and distances. This analysis shows that they have low metallicities, ranging between [Fe/H] $= -1.8$ and [Fe/H] $= -2.2$, and age differences of less than 1 Gyr. These clusters are coeval with the Galactic GC M68 and perhaps 1–2 Gyr younger than the prototypical old GC M92. Similar analysis of the centrally-located cluster H4 shows that is has the same metallicity as the other Fornax clusters but that it is about 3 Gyr younger (Buonanno *et al.* 1999).

Ages and metallicities derived from spectroscopy should be consistent with the results from resolved stellar photometry, but this is not always the case. The analysis of low

The criteria for selecting GC candidates in these surveys are size (FHWM < 20 pc at Virgo for the WFPC2 sample), integrated color similar to that of Galactic GCs, and number density greater than that of a field sample. Both the mean $V - I$ color and the color dispersion of dE GCs increases with galaxy absolute magnitude (Strader *et al.* (2004); Lotz *et al.* 2004; see Figure 4). The colors and color trend for GCs in dEs matches those of GCs in the *blue* GC populations of giant Es and S0s. The blue GSs in E/S0s have properties similar to Galactic GSs, so this is consistent with GCs in Local Group dEs being similar to Galactic halo GCs. The red peak in the bimodal E/S0 GC color distributions is thought to be due to a second period of GC formation, perhaps during mergers (Ashman & Zepf 1992).

The mean colors of the field stars in the WFPC2 dE sample are 0.1 to 0.2 mag redder than the mean GC colors. In nucleated dEs there is a trend that the field-star colors become redder with galaxy M_B. There is a shallower trend for the non-nucleated dwarfs.

At a given absolute magnitude there is no color difference between nuclei and GCs. However, brighter galaxies have brighter and redder nuclei (Graham & Guzmán 2003; see Figure 5). From WFPC2 the nuclei are ~0.05 mag redder in $V - I$ than the GCs and between 0.0 and 0.1 mag bluer than the underlying stars. Other studies have tended to find no color differences between the nuclei and their galaxies (Caldwell & Bothun 1987).

Integrated $V - I$ colors suffer from age/metallicity degeneracy so we must make some assumptions about one property before deriving the other. The mean GC colors are the same as the integrated colors of old, metal-poor GCs in Sagittarius and Fornax (Table 1). Also, spectroscopy of the field-star light gives a mean age of ~6 Gyr and [Fe/H] ~ −0.3 (Geha *et al.* 2003). We know that in the Local Group most GCs in dEs are older and more metal-poor than the mean field-star population. Finally, the GCLF is consistent with the GCLF of an old, metal-poor population (Miller 2003). Therefore, a consistent and reasonable conclusion is that most GCs in the WFPC2 sample are older than 10 Gyr.

For an old stellar population one can convert $V - I$ to metallicity using $[Fe/H] = -4.50 + 3.27(V - I)$ (Kissler-Patig *et al.* 1998). Therefore, the observed trend in GC color with galaxy M_B gives $Z_{GCs} \propto L_B^{0.2}$. This is shallower than $Z_{FS} \propto L_B^{0.4}$ found for

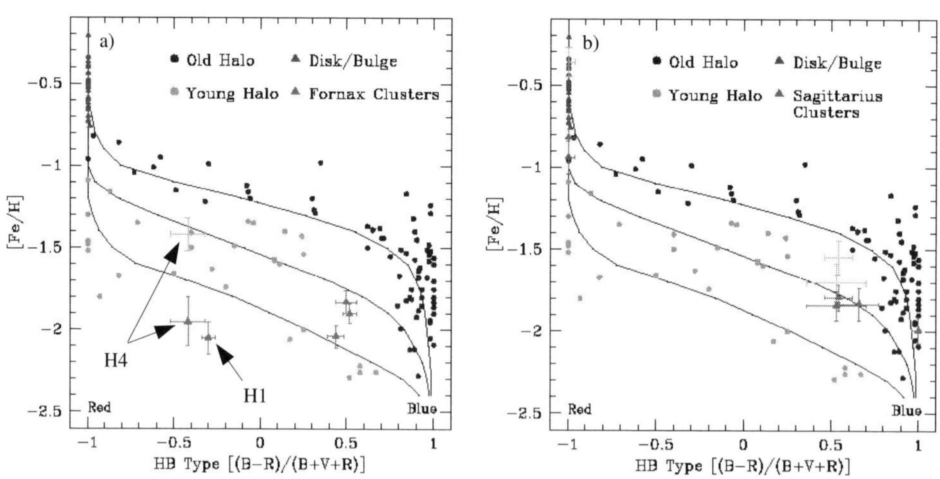

Figure 3. Horizontal-branch index versus metallicity for the Fornax and Sagittarius dwarfs in comparison with different populations of Galactic GCs (from Mackey & Gilmore 2004). Cyan (light) triangles with errors bars represent higher-metallicity measurements for a given cluster. The curves are isochrones from the models of Rey *et al.* (2001). The middle and lower isochrones are 1.1 Gyr and 2.2 Gyr younger than the top isochrone.

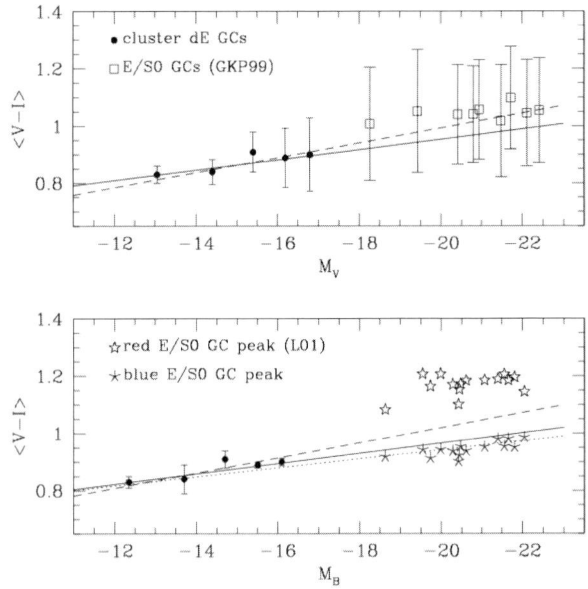

Figure 4. GC color versus galaxy magnitude (from Lotz *et al.* 2004). *Top*: Mean $V-I$ vs. host galaxy M_V. The error bars give the dispersion in the color. *Bottom*: Mean $V-I$ for the dEs and red and blue peak colors of 16 E/S0s versus host galaxy M_B. Error bars on the dE points are now the uncertainties in the mean. The solid line is fit to $\langle V-I\rangle - M_V$ for the WFPC2 sample only while the dashed line includes dEs from the literature. The dotted line is the fit to the blue E/S0 points from Larsen *et al.* (2001).

nearby field star populations (Dekel & Silk 1986) but consistent with the trends in field-star colors seen in the WFPC2 data. The shallower slopes can be explained if adiabatic winds remove gas on timescales comparable to the crossing times (Dekel & Silk 1986).

The increase in color dispersion with galaxy M_B also implies an increased metallicity spread. A simple closed-box chemical evolution model matches the color distribution for all the dEs fairly well if the effective yield is $p = 0.03Z_\odot$. This is lower than typical supernovae yields and implies the closed-box model is not appropriate, a significant amount of gas must be lost due to winds (Lotz *et al.* 2004).

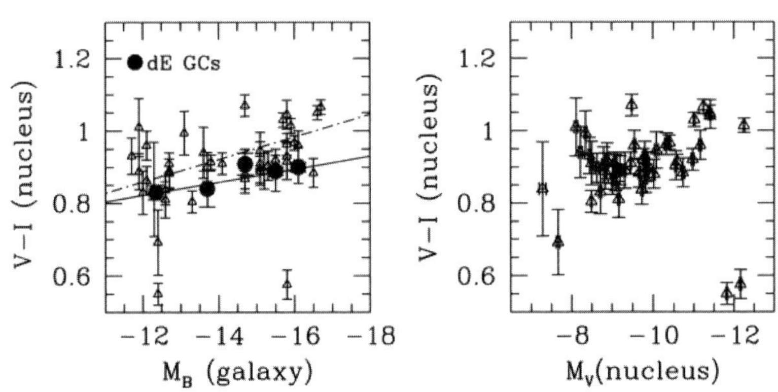

Figure 5. Colors of dE nuclei vs. galaxy M_B and the nuclei's M_V (from Lotz *et al.* 2004). Filled circles in the left panel are the mean GC colors and the dot-dashed line is the fit to the $(V-I)_{nuc} - M_B$ trend.

4. The Formation of GCs and Nuclei

The picture that emerges from several decades of work on GCs in dEs within 20 Mpc is that the GCs are older, more metal-poor, and more α-enhanced than the majority of the field stars within the galaxies. Both the GCs and the field stars show shallow trends of increasing mean metallicity with galaxy luminosity. The majority of the GCs in dEs have ages greater than 10 Gyr while about 25% have ages of about 6 Gyr or younger.

Mackey & Gilmore (2004) have done a detailed comparison of the GCs in Sagittarius and Fornax with the different GC populations of the Galaxy. Based on spatial distributions, kinematics, metalliticities, HB morphologies, core-radii distributions, and age distributions, the GCs in the dEs are most closely related to the "Young Halo" (YH) Galactic population. These are characterized by an extended spatial distribution, high velocity dispersion and no rotation, and low mean metallicities but HB morphologies that are too red for their metallicities (see Figure 3). Some "Young Halo" GCs are as old as the oldest "Old Halo" clusters but a high fraction have ages 2–4 Gyr younger than the oldest GCs, consistent with age being the dominant 2nd HB parameter. Based on these similarities Mackey & Gilmore (2004) argue that up to 75% of the "Young Halo" GCs and 25% of the "Old Halo" GCs have origins in accreted dwarf galaxies.

We can use this information on GCs to develop a timeline of the early universe within the CDM paradigm. The first stellar system to form would be GCs within gas-rich fragments in the most massive dark matter halos. Over the next 1–3 Gyr GC formation begins in lower-mass halos that eventually become today's dwarf galaxies. The vigorous star formation at this time produces clusters with high $[\alpha/\mathrm{Fe}]$. Between reionization and supernovae winds that occur at this time vigorous star formation in dwarfs is squelched and cluster formation stops. After reionization at $z \sim 1$ ($t \sim 8$ Gyr) it becomes easier for gas to cool and star formation to resume. Random density fluctuations or interactions due to dwarfs falling into galaxy groups or clusters can then form a few new GCs that are enriched in [Fe/H] and have lower $[\alpha/\mathrm{Fe}]$.

It has been shown that dE nuclei are closely related to GCs and so are likely form by similar processes. Nuclei can be just bright GCs but some can be distinguished by having large sizes for their luminosities and by not having simple stellar populations. Large GCs/nuclei may form near the centers of dwarfs because the densest gas collects at the centers of the potentials. As above, they would be likely to form early and so would be more metal-poor than most field stars. Massive GCs that don't form in the centers may quickly sink there as a result of dynamical friction (Lotz *et al.* 2001). Multiple populations can result from GC-GC mergers due to dynamical friction or from additional star formation from enriched gas.

More observational and theoretical work are needed to test and validate this picture and flesh out the details. Observationally, we should aim to get high signal-to-noise and high-resolution spectroscopy of all the remaining GCs in Local Group dEs in order to study abundance patterns. We are also working to get spectroscopy of the GCs in the Virgo and Fornax dEs in order to test our assumptions that they are old and compare them to Local Group dEs. GCs promise to continue to be some of our best fossils of the conditions in the early universe.

Acknowledgements

Supported by the Gemini Observatory, which is operated by the Association of Universities for Research in Astronomy, Inc., on behalf of the international Gemini partnership of Argentina, Australia, Brazil, Canada, Chile, the United Kingdom, and the United States of America.

References

Ashman, K.M. & Zepf, S.E. 1992, *ApJ* 384, 50
Beasley, M.A. *et al.* 2005, *AJ* 129, 1412
Brown, J.A., Wallerstein, G. & Zucker, D. 1997, *AJ* 114, 180
Brown, J.A., Wallerstein, G. & Gonzalez, G. 1999, *AJ* 118, 1245
Buonanno, R. *et al.* 1998, *ApJ* 501, L33
Buonanno, R. *et al.* 1999, *AJ* 118, 1671
Butler, D.J. & Martínez-Delgado, D. 2005, *AJ* 129, 2217
Caldwell, N. & Bothun, G.D. 1987, *AJ* 94, 1126
Carney, B.W. 1996, *PASP* 108, 900
Cohen, J.G. 2004, *AJ* 127, 1545
Da Costa, G.S. & Armandroff, T.E. 1995, *AJ* 109, 2533
Dekel, A. & Silk, J. 1986, *ApJ* 303, 39
Forbes, D.A., Strader, J. & Brodie, J.P. 2004, *AJ* 127, 3394
Ferguson, H.C. & Sandage, A. 1989, *ApJ* 346, L53
Geha, M., Guhathakurta, P. & van der Marel, R.P. 2003, *AJ* 126, 1794
Graham, A.W. & Guzmán, R. 2003, *AJ* 125, 2936
Hilker, M. & Richtler, T. 2000, *A&A* 362, 895
Kissler-Patig, M., *et al.* 1998, *AJ* 115, 105
Larsen, S.S., Brodie, J.P., Huchra, J.P., Forbes, D.A. & Grillmair, C.J. 2001, *AJ* 121, 2974
Layden, A.C. & Sarajedini, A. 2000, *AJ* 119, 1760
Lotz, J.M. *et al.* 2001, *ApJ* 552, 572
Lotz, J.M., Miller, B.W. & Ferguson, H.C. 2004, *ApJ* 613, 262
Mackey, A.D. & Gilmore, G.F. 2004, *MNRAS* 355, 504
Mackey, A.D. & van den Bergh, S. 2005, *MNRAS* in press (astro-ph/0504142)
Martin, N.F. *et al.* 2004, *MNRAS* 355, L33
Martínez-Delgado, D. 2005, *private communication*
Miller, B.W. 2003, *Extragalactic Globular Cluster Systems* 199
Pont, F., Zinn, R., Gallart, C., Hardy, E. & Winnick, R. 2004, *AJ* 127, 840
Rey, S., Yoon, S., Lee, Y., Chaboyer, B. & Sarajedini, A. 2001, *AJ* 122, 3219
Searle, L. & Zinn, R. 1978, *ApJ* 225, 357
Strader, J., Brodie, J.P., Forbes, D.A., Beasley, M.A. & Huchra, J.P. 2003, *AJ* 125, 1291
Strader, J., Brodie, J.P. & Forbes, D.A. 2004, *AJ* 127, 3431
Tautvaišienė, G., *et al.* 2004, *AJ* 127, 373
Worthey, G. 1994, *ApJS* 95, 107

Discussion

DRINKWATER: You mentioned that some of the nuclei were bluer. Could you elaborate on that and describe them?

MILLER: Two of the nuclei in Figure 5 have $V - I < 0.6$. This color is too blue to be explained by difference in metallicity so they must be extremely young, like the nucleus of NGC 205. One of these systems is FCC 46 which is discussed by Debattista in these proceedings.

MASHCHENKO: Is there a similar correlation between the metallicity of the GCs and the blue luminosity of the host galaxy for giant galaxies?

MILLER: Yes, Figure 4 shows that the color (or metallicity) - M_B trend for GCs dwarfs is consistent with the trend for the *blue* GCs in giant galaxies.

Near-Field Cosmology with Dwarf Elliptical Galaxies
Proceedings IAU Colloquium No. 198, 2005
H. Jerjen and B. Binggeli, eds.

doi:10.1017/S1743921305004163

Age and metallicity estimations in old stellar populations from Strömgren photometry

K. Rakos[1], J. Schombert[2] & A. Odell[3]

[1]Institute for Astronomy, University of Wien
[2]Department of Physics, University of Oregon
[3]Department of Physics, Northern Arizona University

Abstract. We present a new technique to determine age and metallicity of old stellar populations (globular clusters and elliptical galaxies) using an iterative principal component analysis on narrow band (Strömgren) colors. Our technique is capable of reproducing globular cluster [Fe/H] values to 0.02 dex and CMD ages to 1.0 Gyrs. We also present preliminary results on the application of our technique to a sample of high mass, field ellipticals and low mass, cluster dwarf ellipticals. We confirm the results of earlier studies which find that globular clusters increase in metallicity with age and that age and metallicity increase with galaxy mass. However, we find that dwarf ellipticals deviate from the elliptical sequence by having little to no correlation between age and metallicity.

Keywords. techniques: photometric, galaxies: abundances, galaxies: evolution

1. Introduction

The two primary processes that determine the characteristics of a stellar population are its star formation history and its chemical evolution. For an actively star forming system, such as the disk of our Galaxy, these two process are intertwined and will display a feedback loop as star formation continues and, thus, understanding an active system requires detailed HR diagrams and individual stellar spectroscopy. However, a simple stellar population (SSP), one formed in a single event from a single cloud of gas (e.g. a globular cluster), will have a fixed metallicity and age that may be derived from the color-magnitude diagram (CMD). A burst stellar population, i.e. one derived from a extended star formation event, will be composed of a combination of SSP's and the evolutionary processes are reflected into the population's age and metallicity by the luminosity weighted mean of the various SSP's. It is possible to characterize a burst population if the duration of the burst is short and distribution of metallicities is uniform (Renzini & Buzzoni 1986).

Early studies of composite stellar populations focused on broadband colors of spiral bulges and elliptical galaxies (Sandage & Visvanathan 1978, Tinsley 1980, Frogel 1985) and these datasets supported the hypothesis that red galaxies are composed, primarily, of old, metal-rich stellar populations under the burst hypothesis. Unfortunately, it was quickly realized that detailed interpretation of broadband colors with respect to age and metallicity are complicated by several factors. Foremost was the assumption that old stellar systems are composed of a uniform population in age and metallicity. It was soon demonstrated by population synthesis techniques (O'Connell 1980) that a young population quickly reddens to similar integrated colors as an old population and that the change in color is abrupt even while the differences in age may be quite large (greater than 5 Gyrs). In addition, it was identified through the use of stellar population models that slight changes in age and metallicity operate in the same direction of spectroevolutionary parameter space (Worthey 1994). This coupling of age and metallicity (known as age-metallicity degeneracy, Worthey 1999) is due to competing contributions from main sequence turn-off stars (sensitive to age) and red giant branch (RGB) stars (sensitive to metallicity) near 5000Å. Filters that bracket this region of a galaxy's spectrum will require increasingly accurate values for metallicity to determine a unique age and vice-versa.

To avoid the age-metallicity degeneracy problems, a majority of recent stellar population studies have focused on the determination of age and metallicity through the use of various spectral signatures, such as Hβ for age (Kuntscher 2000, Trager *et al.* 2001). This approach provides a finer comparison to stellar population models, but requires assumptions about the relationship between metallicity indicators (e.g. Mg_2) and the [Fe/H] value of the population as reflected into the behavior of the red giant branch. In other words, spectral lines provide the value of that element's abundance, but what is really required is the temperature of the RGB which is a function of the total metallicity, Z. Varying ratios of individual elements to Z complicates the interpretation of line studies (Ferreras, Charlot & Silk 1999). In addition, these techniques have limitations due to the required high S/N for the data that make them problematic for the study of high redshift systems.

An alternative approach to spectral line studies is to examine the shape of specific portions of a spectral energy distribution (SED) using narrow band filters centered on regions sensitive to the mean color of the RGB (metallicity) and the main sequence turnoff point (mean age) without the overlap that degrades broadband colors. The type of galaxy examined, for example stellar systems with ongoing star formation where a mix of different age populations may be present, will still limit this technique. However, for systems that have exhausted their gas supply many Gyrs ago (i.e. old and quiescent), it may be possible to resolve the underlying population with some simple assumptions on their star formation history and subsequent chemical evolution. Thus, we have the expectation, guided by the results of evolution models, that objects composed of SSPs or a composite of SSPs (e.g. ellipticals) present special circumstances where the age-metallicity degeneracy can be resolved and allow the study of the evolution of stellar populations.

In a series of earlier papers, we have examined the modified Strömgren uz, vz, bz, yz colors of globular clusters and used a combination of their colors and SED models to derive the mean age and metallicity of dwarf, bright and field ellipticals (Rakos *et al.* 2001, Odell, Schombert & Rakos 2002, Rakos & Schombert 2004). While spectroscopic data is superior for age and metallicity estimations in high S/N datasets, our goal has been to develop a photometric system that can be used for galaxies of low surface brightness and/or high redshift, where spectroscopy is impractical or impossible. In addition, we believe it is important to compare continuum based age and metallicity estimates to line indice estimates to expand the range of testable observables for our spectroevolutionary models.

Our past technique has been to relate the $vz - yz$ color index to mean metallicity, since the vz filter is centered on the absorption line region near 4100Å and the $bz - yz$ color index, whose filters are centered on continuum regions of the spectral energy curve, measures the mean stellar age. These results, as guided by comparison to multi-metallicity SED models, was crude since metallicity changes will move the effective temperature of the RGB and, thus, the continuum $bz - yz$ colors. In our more recent work, we have included photometry through the uz filter which provides an additional handle on age and metallicity effects to the other two color indices. In addition, we have applied a

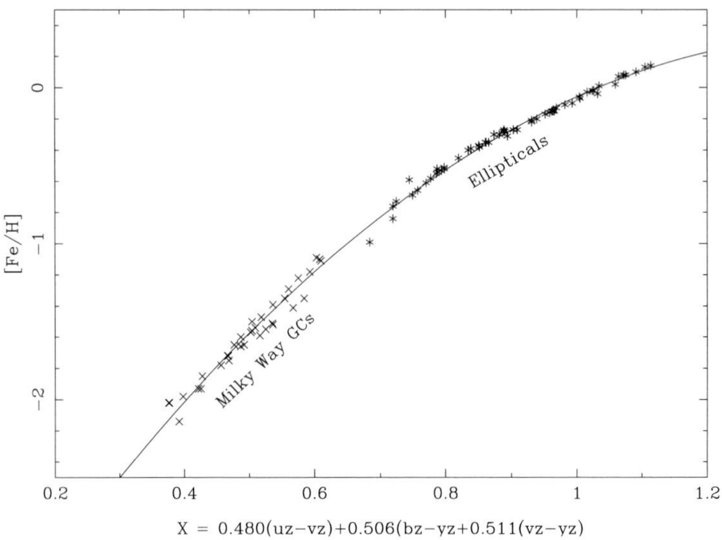

Figure 1. The mean metallicity of the globular cluster and elliptical samples as determined by the iterative technique versus the multi-color term that incorporates all of the PC terms. This figure demonstrates how the use of the full parameter space reduces the scatter in [Fe/H]. Also shown are the iterative results for a sample of dwarf and giant ellipticals in the Coma cluster with full *uvby* photometry. The relationship become less linear for galaxies due to the fact that they are composed of a integrated population of SSP's. The low scatter indicates that metallicity is the primary driver of galaxy color (see also Smolcic *et al.* 2004).

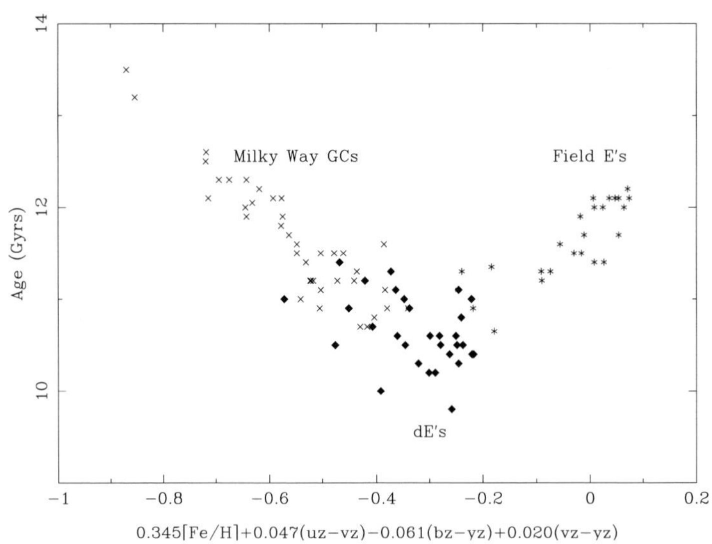

Figure 2. Mean stellar age, as determined by an iterative PC fit, versus an "ageless" color term. While any portion of this diagram can be occupied by a range of age and metallicities, in fact, old, metal-poor objects (globulars) and old, metal-rich objects (ellipticals) follow two separate linear correlations. Dwarf ellipticals bridge region between globulars and ellipticals as transition objects with a smaller range of metallicity in their underlying stellar populations.

principal component (PC) analysis on the multi-color data (Steindling, Brosch & Rakos 2001) that more fully isolates metallicity from age effects and the changes due to recent star formation. In the end, we believe we have isolated a reliable technique for estimating age and metallicity for non-starforming objects such as ellipticals and S0's. This paper

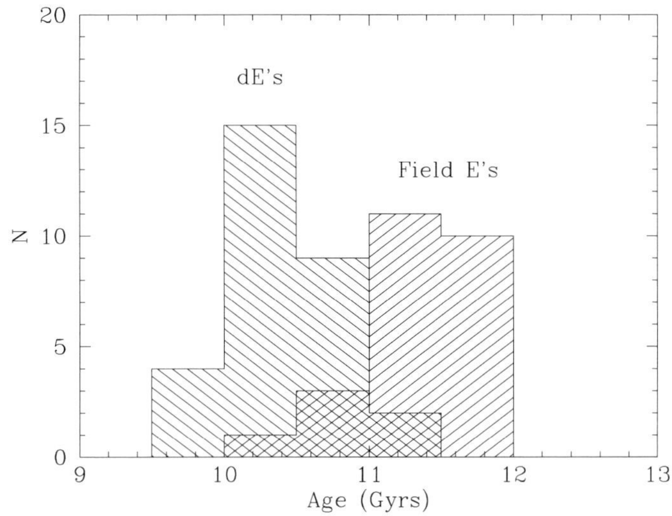

Figure 3. The distribution of mean stellar age for dwarf ellipticals (dEN's from Fornax) and field ellipticals (a bright, nearby sample). The mean dE age is 1.0 Gyrs younger than the mean for field ellipticals. This difference can reflect an earlier star formation epoch for field ellipticals or a prolonged initial stage of star formation for dE's.

will describe some of our preliminary results in comparing the star formation history of globular clusters, field ellipticals and dwarf ellipticals.

2. Age and Metallicity Calibration

The technique we have developed to apply to our narrow band filter system is full described in Rakos & Schombert (2005). In brief, our method hinges using principal component (PC) analysis on the multi-color phase space generated by a unique set of SED models (Schulz *et al.* 2002). A PC analysis in any n-dimensional space, calculates the axis along which the model points present the largest, most significant scatter. This is called the first principal component (PC1). PC analysis then proceeds to calculate PC2, the axis of the second most significant spread in the remaining $n-1$ dimensional space orthogonal to the first PC, and so on. While scientific data in the astronomy are usually neither linear nor orthogonal at the same time. Often, for small regions of data, linearity and orthogonality can be reached through the use of PC analysis. Close inspection of Schulz *et al.* theoretical models has shown acceptable linearity for ages larger than 3 Gyrs over the full range of model metallicities. In this restricted region, it is possible to apply PC analysis to; 1) separate the age and the metallicity of a stellar population, 2) select the most correlated variables and 3) determine linear combinations of variables for extrapolation.

The advantage of the smooth behavior for age and metallicity in the PC plane is that if one has all three narrow band colors ($uz-vz$, $vz-yz$ and $bz-yz$) then knowledge of the correct PC1 and PC2 values allows for the unique determination of age and metallicity from the PC equations. If the PC values are unknown, then an iterative search scheme could select a range of values for mean age and [Fe/H], determine PC1 and PC2 from the model fits, then compare how well those values compare with the results from the PC equations. To test this iterative procedure, we ran a sample of 40 globular clusters with known ages (CMD fitting) and metallicities (spectral) where we have obtained high precision *uvby* photometry. The resulting fits re-captured the correct ages and [Fe/H] values to an accuracy of 0.5 Gyrs in age and 0.05 dex in [Fe/H]. Additional analysis with the iterative solutions finds that for both globular clusters (simple SSP's) and ellipticals (composite SSP's), it is possible to find a highly accuracy metallicity simply from a linear combination of narrow colors, by themselves (see Figure 1).

This technique differs from previous line indice work in several regards. First, the narrow band filters used in our study are specifically selected to avoid major line regions of the spectrum. Thus, they provide a measure of the integrated continuum of a stellar population and the value for Fe/H determined by this method will be based on stellar atmospheric temperatures rather than individual line depth. While this may be less accurate than direct determination of the strength of Fe in a spectrum, it has the advantage of being the quantity you wish to know about a stellar population (its mean metallicity Z as it effects the colors of the stars) and does not suffer from calibration shifts when going from one line feature to Fe/H (e.g. changing ratios of Mg/Fe in ellipticals). The metallicity values produced by this method will also be luminosity weighted means for the entire stellar population. This is in contrast to line indice studies which are surface brightness selected and will be strongly biased by central core light.

3. Application to Globular Clusters and Ellipticals

We have three new sources of photometry in which to test our technique, Milky Way globular clusters (described in the last section), 25 nearby field ellipticals and a sample of 40 dwarf ellipticals in Fornax. From our previous work with ellipticals, we know that their narrow band colors cannot be described by a single metallicity/age SSP (Rakos *et al.* 2001). However, we found that models which combine a range models (roughly gaussian) of varying metallicities in a linear fashion accurately reproduce the color of ellipticals. Each set of models can be parameterized by a luminosity weighted mean metallicity and that decreasing metallicity with mass (the color-magnitude) effect reflects a decreasing spread or range in underlying population metallicities. Thus, at high luminosities, ellipticals are a composite of many stellar populations, but at low luminosities (mass), the colors of ellipticals approach the colors of SSP's.

We, therefore, have the expectation that dwarf ellipticals would bridge the gap between ellipticals and globular clusters in the PC space defined by our three narrow band colors, age and mean metallicity. This can be seen in Figure 2, a plot of galaxy age, as determined by our iterative PC technique and an "ageless" PC term given by the linear combination of metallicity and the three narrow band colors (note that the value of [Fe/H] is the one determined by the PC fits, although a value of [Fe/H] from relationship in Figure 1 would have produced identical results).

There are several important points to note about Figure 2. The first is that any region of this diagram can be occupied by various combinations of high or low metallicity versus old or young mean stellar age. Our samples divided into old, low metallicity objects (globular clusters) and old, high metallicity objects (high mass ellipticals). The addition of the dwarf elliptical data joins the globular and elliptical sequences at low metallicity, younger age. As found in our previous work on dwarf ellipticals (Rakos & Schombert 2004). Nucleated dwarfs (dEN's) track the globular sequence colors to higher metallicities and, as in Figure 2, younger age.

A direct comparison of the ages of field and dwarf elliptical is found in Figure 3, a histogram of their PC calculated ages. While the initial interpretation of this Figure is that dwarf ellipticals are younger than field ellipticals (hierarchical models of galaxy formation predict that low mass galaxies are older than high mass systems, but also

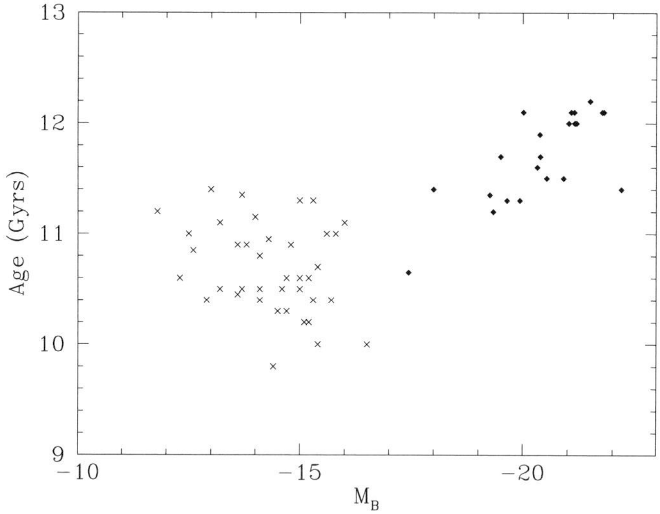

Figure 4. Mean stellar age versus galaxy blue luminosity (mass). There is a clear correlation for field ellipticals of decreasing age with decreasing mass, but dwarf ellipticals little to no correlation with age (a slight tendency to find increasing age with decreasing mass). Combined with the color-magnitude (mass-metallicity) relation for bright ellipticals, there is no anti-correlation between age and metallicity as predicted by hierarchical models of galaxy formation.

predict that field galaxies are older than cluster galaxies), "age" from these calculations refers to mean stellar age. A "younger" mean age can be obtained through a later epoch of initial star formation (not necessary when the dark+baryonic matter lump formed) or through a prolonged era of initial star formation. And, in the case of dwarf galaxies, later weak bursts of star formation would also reflect into a younger mean age.

While as a group the dE's in Fornax have lower mean stellar ages than field ellipticals, this trend is not as linear with luminosity (mass) as one would expect. Figure 4 displays the run of age versus absolute blue luminosity (M_B). The trend for field ellipticals is a clear one of decreasing age with decreasing mass. This parallels the well-known mass-metallicity relation such that older galaxies have higher metallicities (as predicted by closed box models of chemical evolution, Matteucci 2003). However, the dwarfs display little correlation with mass (similar to the breakdown of the narrow band color-magnitude relation for dwarf galaxies, Odell, Schombert & Rakos 2002) which may signal that higher mass ellipticals are constructed in a uniform fashion, but that low mass dwarf ellipticals have episodic past histories of star formation.

Given the trend α/Fe ratios in ellipticals (Thomas *et al.* 2005), which measures the duration of the initial star formation burst, such that high mass ellipticals have shorter durations, part of the trend in Figure 4 is from the prolonged star formation for low mass ellipticals. Between a luminosity of −17 and −22, Thomas *et al.* estimate a decrease in duration of approximately 0.75 Gyrs, which is close to the age of 1.5 Gyrs for the same mass range in Figure 4. Thus, bright ellipticals may be coeval, but with a slowly varying duration of initial star formation that reflects into younger mean age. Testing this hypothesis to a larger sample of cluster ellipticals will be complicated by recent mergers of star-forming galaxies which artificially lower the mean stellar age (Trager *et al.* 2001)

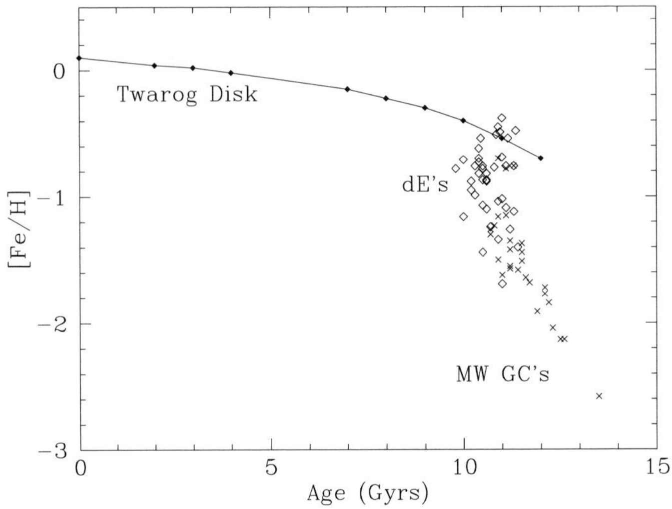

Figure 5. The age-metallicity diagram for Milky Way globular clusters and Fornax dwarf ellipticals. The age-metallicity relation for solar neighborhood F dwarfs (Twarog 1980) is also shown.

4. Chemical Evolution of Old Stellar Populations

Armed with a handful of age and metallicity values for Milky Way globular clusters and dwarf ellipticals, we can investigate the earliest epoch of the construction of the age-metallicity relation (AMR, Twarog 1980). Figure 5 displays the age and [Fe/H] values for the low metallicity region of the PC phase space. Also shown in Figure 5 is the AMR for the solar neighborhood as determined by Strömgren plus Hβ photometry of F dwarfs. The steeper slope to the galaxy data reflects the expected rapid enrichment process predicted by closed box models of chemical evolution (Bond 1981).

The era where the dE data ends corresponds roughly to a [Fe/H] ~ -0.6 which is the transition point between the Galactic disk, where the closed models for chemical evolution are inadequate, and the halo component (Gilmore & Reid 1983). While the dwarf elliptical data lacks the clear correlation between age and metallicity seen in the globular cluster data, it is obvious that the low metallicity dE's follow the globular clusters and that above [Fe/H] $= -0.8$ the dE's decouple from a simple model of chemical evolution (i.e. a closed box with instantaneous recycling). This is consistent with our earlier interpretation that the lower metallicity dwarfs have a very narrow range of internal metallicities and that their underlying stellar populations, to first order, mimic the colors of an SSP, such as a globular cluster.

Acknowledgements

We wish to thank all the observatories (NOAO, Steward, ESO) which have granted us time for this project. And we are grateful to the organizers of this Colloquium for a chance to present our newest results.

References

Bond, H. 1981, *ApJ* 248, 606
Ferreras, I., Charlot, S. & Silk, J. 1999, *ApJ* 521, 81
Frogel, J. 1985, *ApJ* 298, 528

Gilmore, G. & Reid, N. 1983 *MRNAS* 202 1025
Kuntschner, H. 2000 *MRNAS* 315, 184
Matteucci, F. 2003 *ASP Conf. 317* p. 337
O'Connell, R. 1980 *ApJ* 236, 430
Odell, A., Schombert, J. & Rakos, K. 2002 *AJ* 124, 3061
Rakos, K. & Schombert, J. 2004 *AJ* 127, 1502
Rakos, K. & Schombert, J. 2005 *PASP* 117, 245
Rakos, K., Schombert, J., Maitzen, H., Prugovecki, S. & Odell, A. 2001 *AJ* 121, 1974
Renzini, A. & Buzzoni, A. 1986 *Spectral Evolution of Galaxies, eds. C. Chiosi and A. Renzini (Dordrecht: Reidel)* p. 195
Sandage, A. & Visvanathan, N. 1978 *ApJ* 223, 707
Schulz, J., Fritze-v, Alvensleben,, Moumller, C. & Fricke, K. 2002 *A&A* 392, 1
Steindling, S., Brosch, N. & Rakos, K. 2001 *ApJS* 132, 19
Thomas, D., Maraston, C., Bender, R. & de Oliveira, C. 2005 *ApJ* 621, 673
Tinsley, B. 1980 *Fund. of Cosmic Physics, Vol. 5* p. 287
Trager, S., Faber, S., Worthey, G. & Gonzlez, J. *AJ* 120, 165
Twarog, B. 1980, *ApJ* 242, 242
Worthey, G. 1994 *ApJS* 95, 107
Worthey, G. 1999 *ASP Conference Proceedings, Vol. 192* p. 283

Discussion

De Rijcke: The metallicities you derive for Fornax dEs using Strömgren photometry seem to agree well with spectroscopic measurements: they are metal-poor, with [Fe/H] around -1. Virgo dEs appear to be more metal-rich. Do you plan to apply your method also to Virgo dEs?

Rakos: As soon as I get the telescope time for such a proposal. This is my greatest problem.

Near-Field Cosmology with Dwarf Elliptical Galaxies
Proceedings IAU Colloquium No. 198, 2005
H. Jerjen and B. Binggeli, eds.

doi:10.1017/S1743921305004175

Ultra-compact dwarf galaxies: a new component of galaxy clusters and groups

M.J. Drinkwater[1], E. Evstigneeva[1], M.D. Gregg[2], J.B. Jones[3] S. Phillipps[4] and R. Jurek[1]

[1]Department of Physics, University of Queensland, QLD 4072, Australia
email: mjd,katya,jurek@physics.uq.edu.au

[2]Physics Dept., U.C. Davis, and IGPP, Lawrence Livermore National Laboratory, L-413, Livermore, CA 94550, USA
email: gregg@igpp.ucllnl.org

[3]Astronomy Unit, School of Mathematical Sciences, Queen Mary University of London, Mile End Road, London E1 4NS, United Kingdom
email: bryn.jones@qmul.ac.uk

[4]H. H. Wills Physics Laboratory, University of Bristol, Tyndall Avenue, Bristol BS8 1TL, UK
email: S.Phillipps@bristol.ac.uk

Abstract. We recently reported the discovery of a new galaxy type, ultra-compact dwarf (UCD) galaxies, in both the Fornax and Virgo clusters. Here we present new detections of a larger population of UCDs in the Fornax Cluster, as well as the first detections of UCDs in two galaxy groups. We extended our original survey of the Fornax Cluster 1 magnitude fainter, finding a further 46 new compact objects. This means that UCDs outnumber normal galaxies in the central region of the Fornax Cluster. We used this large UCD sample to test the tidal stripping model: simulations indicate that only about half of the UCD population may have formed by disruption of the observed population of nucleated dwarf galaxies. As the UCDs are strongly clustered towards the cluster centre, we conclude that the remaining UCDs must be associated with the much earlier assembly of the central galaxy.

Keywords. globular clusters: general, galaxies: star clusters, galaxies: dwarf, galaxies: formation

1. Introduction

Our study of compact dwarf galaxies in the Fornax Cluster was originally motivated by the desire to determine if the cluster contained any "M32-like" compact elliptical galaxies as suggested by Ferguson (1989). We used the FLAIR-II multi-object spectrograph on the 48-inch U.K. Schmidt Telescope to measure redshifts of some 500 compact galaxies in the direction of the cluster (Drinkwater *et al.* 2001). None of the M32 candidates turned out to be cluster members so we concluded that whatever process resulted in M32 is not important in the cluster environment. There were other surprises however, notably that 8 of the previously-classified background galaxies were actually cluster members.

The detection of these new cluster members motivated us to start a much more extensive spectroscopic study to improve our knowledge of the dwarf galaxy population of the Fornax Cluster. The *Fornax Cluster Spectroscopic Survey* (FCSS; Drinkwater *et al.* 2000) has used the 2dF 400-fibre spectrograph on the 3.9m Anglo-Australian Telescope to perform the first *all-object* survey of a galaxy cluster. Rather than select resolved objects (i.e. "galaxies"), we have observed all objects (i.e. "stars" and "galaxies") in our magnitude range ($16.5 < b_J < 19.8$) in three 2dF fields around the cluster centre. The targets were selected from U.K. Schmidt Telescope photographic survey plates: 65% were morphologically classified as stars and 35% as galaxies.

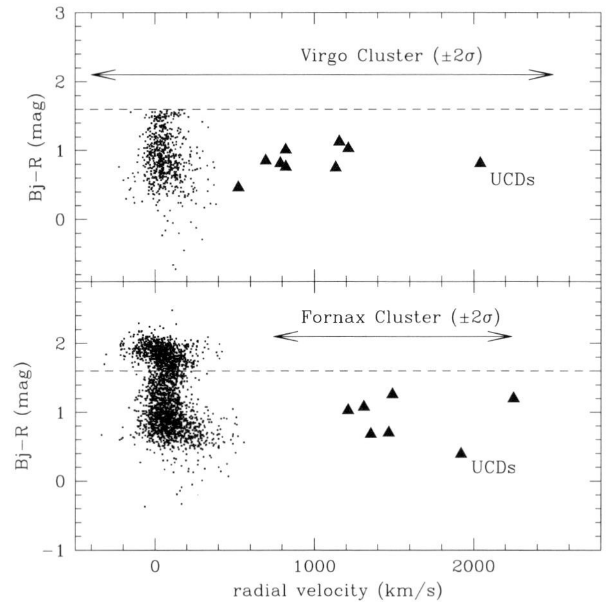

Figure 1. Comparison of the Virgo (upper) and Fornax (lower) Cluster UCDs (triangles) with foreground Galactic stars (points). For the Virgo, only stellar objects with $B_J - R < 1.6$ (dashed line) were observed. For each cluster the 2σ velocity range about the mean redshift is shown.

2. A new type of galaxy; also in the Virgo Cluster

To our surprise, six of the "stars" in the FCSS were found to have redshifts consistent with cluster membership: they were not Galactic stars. Two of these had previously been identified in a smaller survey by Hilker *et al.* (1999). As these objects were unlike any known objects in the cluster we described them as "ultra-compact dwarf" (UCD) galaxies (Phillipps *et al.* 2001).

In subsequent investigations we have shown that the UCDs are unlike any known type of galaxy or star cluster and therefore form a new class of galaxy (Drinkwater *et al.* 2003). They do not appear to be isolated globular clusters because they are much larger (r_e = 10–20 pc) than Galactic globulars (r_e = 3–5 pc), they do not follow the same relation between velocity dispersion and luminosity as globulars, and they have higher mass-to-light ratios (M/L = 2–4) than globulars (M/L = 1–2). They are not extreme cases of known galaxy types as we have shown they are very widely separated from known types in surface brightness as a function of luminosity. Our initial hypothesis is that UCDs are the remnant nuclei of nucleated dwarf elliptical galaxies tidally disrupted by the cluster potential. We refer to this process as galaxy threshing (Bekki *et al.* 2001).

We predicted that UCDs should be found in similar numbers in the Virgo Cluster around the central galaxy M87 as it is similar in size to NGC 1399 in Fornax and is surrounded by a similar surface density of dwarf elliptical galaxies. We used 2dF again to measure stellar objects towards the centre of the Virgo Cluster, but this time, informed by the properties of the Fornax UCDs we did not observe the reddest stellar objects. As a result we were able to test our hypothesis with only a few hours of service observations: in just 3 hours were were able to detect 9 UCDs in the Virgo Cluster. The results for both the Fornax and Virgo clusters are compared in Fig. 1 which also indicates our revised colour selection for Virgo.

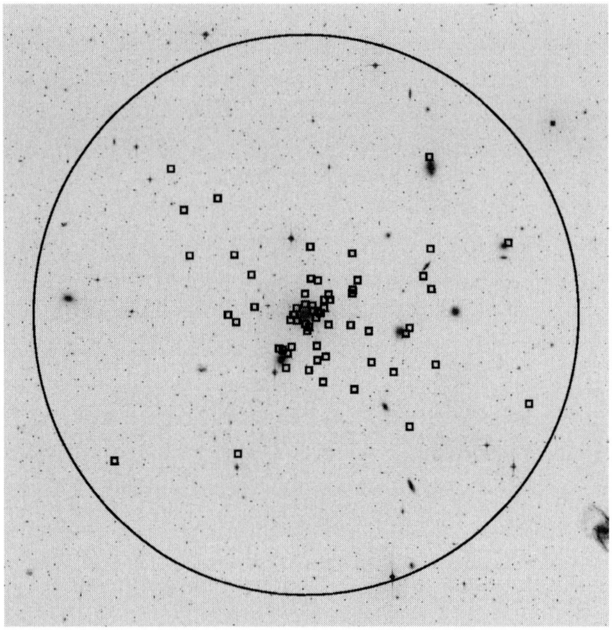

Figure 2. The distribution of the large population of UCDs (squares) in the Fornax Cluster. The circle shows the 2-degree diamater 2dF field searched for UCDs.

3. Large UCD Population

Based on our galaxy threshing hypothesis for UCD formation we predicted that the luminosity function of UCDs should approximately follow that of the nuclei of dE,N galaxies (e.g. Binggeli & Cameron 1991). On that basis we used 2dF to search the central region of Fornax for UCDs to a fainter limit of $b_J < 21.5$ and avoiding the red objects. We expected to triple our sample, but actually found 10 times as many objects making a total of 63. The locations of the objects are shown in Fig. 2. We note that these objects outnumber normal galaxies in the cluster centre.

In Fig. 3 we compare the magnitude and velocity distributions of the new UCDs with normal dE,N galaxies (Ferguson 1989) and globular clusters (Dirsch *et al.* 2004). It is clear that the new objects overlap the distribution of globular clusters, however there are some significant differences in their dynamical properties. First we note that the UCDs ($\sigma_v = 262 \pm 29\mathrm{kms}^{-1}$) are more relaxed than the dE,N galaxies in the same region ($\sigma_v = 417 \pm 63\mathrm{kms}^{-1}$) at the 99% confidence level. Conversely, the mean velocity of the UCDs ($< v >= 1556 \pm 40\mathrm{kms}^{-1}$) is larger than that of the globular clusters ($< v >= 1445 \pm 16\mathrm{kms}^{-1}$) at the 99% confidence level.

4. Origin of UCDs

We have used the larger population of UCDs to make further tests of the galaxy threshing hypothesis by making numerical simulations of the process, taking the parent population of UCDs to be the total of all current UCDs and dE,N galaxies. We then simulate their possible orbits to estimate the fraction of parent orbits at a given radius which will be threshed to form UCDs. In Fig. 4 we compare this to the observed fraction. The simulations predict at most half the current population of UCDs. This is basically because the UCDs are significantly more centrally-concentrated than the dE,N galaxies.

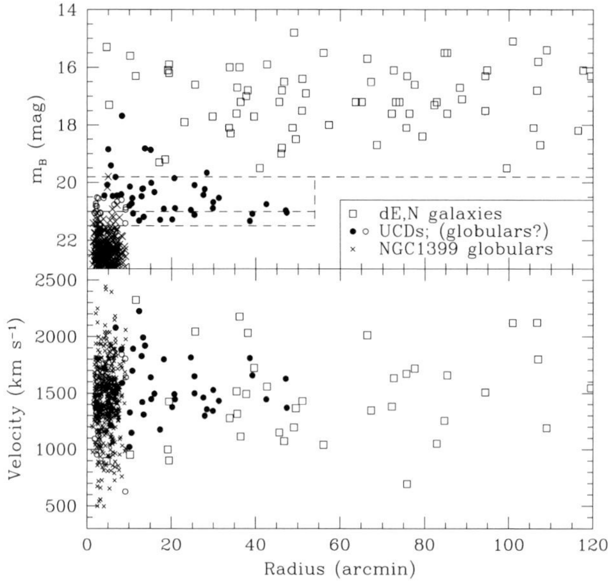

Figure 3. Radial distributions of dwarf elliptical galaxies, UCDs and globular clusters in the Fornax Cluster. Upper panel: apparent magnitude as a function of radius. The dashed lines indicate the parameter range we have searched for UCDs with completeness levels of 96% ($m < 19.8$), 82% ($19.8 < m < 21$) and 36% ($21 < m < 21.5$). Lower panel: velocity distributions.

There is no significant friction involved in the threshing process, so threshed objects will continue on the same orbits as their progenitors. We therefore infer that a significant fraction of the UCDs formed in close orbits around the central cluster galaxy. This implies an earlier formation age in hierarchical structure formation scenarios.

5. Summary

The main result we have presented here is that there is a large population of fainter UCDs in the Fornax Cluster. Whilst there is undoubtedly some overlap between these objects and true globular clusters associated with the central galaxy of the cluster, the UCD population has a significantly higher mean velocity than the globulars. The UCDs are also distributed over a much larger region of intra-cluster space than the globular clusters. The large sample of UCDs has enabled us to make a new test of the "galaxy threshing" hypothesis for their formation. We find that disruption of dE,N galaxies can only account for about half the observed UCD population. This is basically because the UCDs are much more concentrated to the cluster centre than the dE,N galaxies; they also have a much lower velocity dispersion. We have just begun searches for UCDs in a number of galaxy groups to investigate if their formation is dependent on the cluster environment. Our preliminary results suggest they are also present in groups, but in lower numbers: we have found one each so far in the Dorado and NGC 1400 groups.

Acknowledgements

M.J.D. is supported by an Australian Research Council Discovery Project grant.

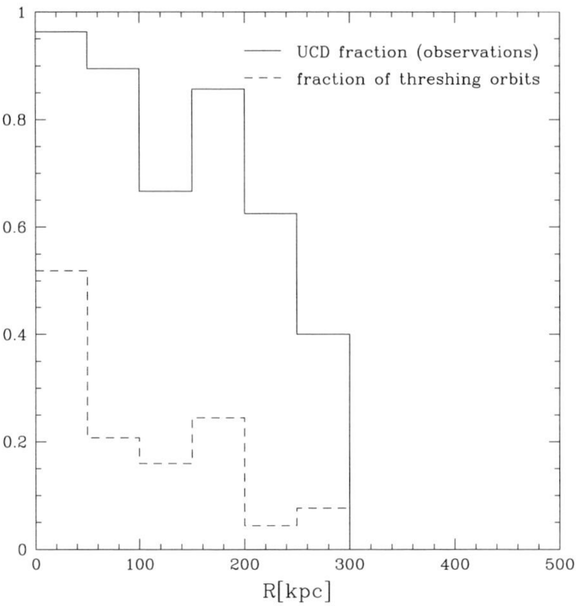

Figure 4. Predicted fraction of UCDs formed from threshed dE,N galaxies as a function of radius based on simulations of the threshing process. This is compared to the actual fraction of UCDs defined as the number of UCDs divided by the total number of UCDs plus dE,Ns. We assume a distance of 20 Mpc to Fornax.

References

Bekki, K., Couch, W.J. & Drinkwater, M.J. 2001, *ApJ* 552, L105

Binggeli, B. & Cameron, L.M. 1991, *A&AS* 252, 27

Dirsch, B., Richtler, T., Geisler, D., Gebhardt, K., Hilker, M., Alonso, M.V., Forte, J.C., Grebel, E.K., Infante, L., Larsen, S., Minniti, D. & Rejkuba, M. 2004, *AJ* 127, 2114

Drinkwater, M.J., Phillipps, S., Jones, J.B., Gregg, M.D., Deady, J.H., Davies, J.I., Parker, Q.A., Sadler, E.M. & Smith, R.M. 2000, *A&A* 355, 900

Drinkwater, M.J., Gregg, M.D., Holman, B.A. & Brown, M.J.I. 2001, *MNRAS* 326, 1076

Drinkwater, M.J., Gregg, M.D., Hilker, M., Bekki, K., Couch, W.J., Ferguson, H.C., Jones, J.B. & Phillipps, S. 2003, *Nature* 423, 519

Ferguson, H.C. 1989, *AJ* 98, 367

Hilker, M., Infante, L., Vieira, G., Kissler-Patig, M. & Richtler, T. 1999, *A&AS* 134, 75

Phillipps, S., Drinkwater, M.J., Gregg, M.D. & Jones, J.B. 2001, *ApJ* 560, 201

Discussion

CALDWELL:: In the mean, the UCDs are more massive and have higher M/L than GCs, but is there an overlap in the distributions such that some of the fainter UCDs in Fornax might be GCs?

DRINKWATER:: Yes, there is clearly an overlap in luminosity. We hope to use internal properties to separate objects in this range: internal velocity dispersion and M/L.

MIESKE:: 1) The GC system in Fornax (NGC1399, Dirsch *et al.* 2003) has been shown to extend out to 30' and possibly beyond. Therefore one has to be careful in postulating that the fainter UCDs which extend out to about 40' cannot be GCs. 2) In your comparison between the LF of UCDs and the GCLF of NGC1399, you should remark that the GCLF has been scaled down by a factor of ca. 50 to fit into the plot.

DRINWATER:: Yes, I agree on both your points and I will investigate the numbers from the Dirsch *et al.* sample. Thank you.

JORDAN:: A population of intra cluster globular clusters has been hypothesized for a long time and recently detected. West *et al.* (1995) presented a model that predicts the number of intra cluster globulars based on the mass of the cluster. Have you compared your number counts with the prediction of their models?

DRINKWATER:: I haven't, but that is something I'd like to do.

SAVIANE:: You said that UCDs were predicted by Disney. On what theoretical grounds?

DRINKWATER:: Disney's work wasn't theoretical really. He noted that current galaxy surveys were only sensitive to a small range of surface brightness so that low- and high-surface brightness objects would be missed. therefore he inferred large undiscovered populations of such objects, of which UCDs are one example.

KROUPA:: The formation of UCDs may be seen now in the Antenna galaxies as the young "knots", or star-cluster complexes, as described by Whitmore *et al.* (1999) and Kroupa (1998).

DRINKWATER:: Yes, this is one formation scenario we should consider, especially as the long axis of the UCD distribution in Fornax points towards the infalling subcluster.

DIEMAND:: In the "galaxy threshing" model: why did you start with an initial radial distribution similar to *todays* UCDs + dEs? CMD models motivate a much more concentrated (r^{-3}) initial population. In this case the larger efficiency of tidal destruction and stripping at smaller radii could explain the different concentrations of ICL + GCs, UCDs and dEs today.

DRINKWATER:: Our initial hypothesis was intentionally naive as we didn't have a good understanding of CDM-type predictions. We are now collaborating with people from the Virgo Consortium to make more realistic models.

READ:: What is the implication of these results for the missing satellite problem? If you go deeper, do you expect to find yet more of these things?

DRINKWATER:: We certainly expect to find more as we go both fainter and wider in our searches. My guess is they don't have enough dark mass to contribute much to the missing satellite problem, but others here are more expert in that area.

KARACHENTSEV:: Did you try to search for UCDs in the Leo I Group? This is the nearest E-dominated group, situated in the zone of the SDSS. It can be reached with ACS in the snap-shot survey, too.

DRINKWATER:: Thank you; that is an excellent suggestion.

Near-Field Cosmology with Dwarf Elliptical Galaxies
Proceedings IAU Colloquium No. 198, 2005
H. Jerjen and B. Binggeli, eds.

doi:10.1017/S1743921305004187

UCDs in Fornax and Abell 1689

S. Mieske[1], M. Hilker[1] and L. Infante[2]

[1]Sternwarte der Universität Bonn, Auf dem Hügel 71, 53121 Bonn, Germany
email: smieske@astro.uni-bonn.de, mhilker@astro.uni-bonn.de

[2]Departamento de Astronomía y Astrofísica, P. Universidad Católica, Casilla 104, Santiago 22, Chile
email: linfante@astro.puc.cl

Abstract. We review some properties of ultra compact dwarf galaxies (UCDs) in Fornax and of their brighter counterparts in Abell 1689, among which are two M32 twins, and discuss various possible UCD formation scenarios. We note that it is indispensable to carefully take into account the bright end of the globular cluster luminosity function when estimating the number density of UCDs. It is suggested that the search for more luminous UCDs in dense and rich galaxy clusters is the best way towards establishing UCDs as a new class of galaxies.

Keywords. Galaxies: nuclei, star clusters, formation, peculiar

1. Introduction

Ultracompact dwarf galaxies (UCDs) have recently been proposed as a new class of galaxies, populating the central regions of the Fornax and Virgo clusters (Hilker *et al.* 1999, Drinkwater *et al.* 2000, Hasegan *et al.* 2005). UCDs are placed between the sequence of globular clusters and dwarf elliptical galaxies in the fundamental plane of stellar systems (Drinkwater *et al.* 2003). One much discussed hypothesis on their origin is that they are remnant nuclei of dwarf galaxies stripped in the potential well of their host cluster (Bekki *et al.* 2003). It has also been proposed that UCDs actually are no "galaxies" but simply very bright globular clusters (GCs) (Mieske *et al.* 2002), or that they are stellar super clusters created in merger events (Fellhauer & Kroupa 2002). Hasegan *et al.* (2005) report on the discovery of several UCDs in the Virgo cluster with high M/L ratios and propose the M/L ratio as a criterion to distinguish between overluminous GCs and UCDs.

In this contribution, we review the luminosity and colour distribution of UCDs in the Fornax cluster (Sect. 2) with respect to the various proposed formation scenarios. Furthermore, we report on the search for UCDs and the subsequent discovery of two M32 twins in the very massive cluster Abell 1689 (Sect. 3).

2. UCDs in Fornax

In Fig. 1 we show the distribution of Fornax compact objects in luminosity and colour, as derived in the course of the Fornax Compact Object Survey FCOS (Mieske *et al.* 2004b). We would like to draw the attention of the reader to the following findings:

1. The luminosity distribution of compact objects in Fornax with $M_V \geqslant -11.4$ mag is consistent both in shape and total number of objects with the extrapolated globular cluster luminosity function (GCLF) as determined at fainter magnitudes. That is, *it is not necessary to invoke the existence of UCDs with* $M_V \geqslant -11.4$ *to explain the luminosity distribution of compact objects. Any survey mapping the luminosity distribution of compact objects fainter than* $M_V \simeq -11.4$ *mag in Fornax is bound to be totally dominated by the globular cluster population.*

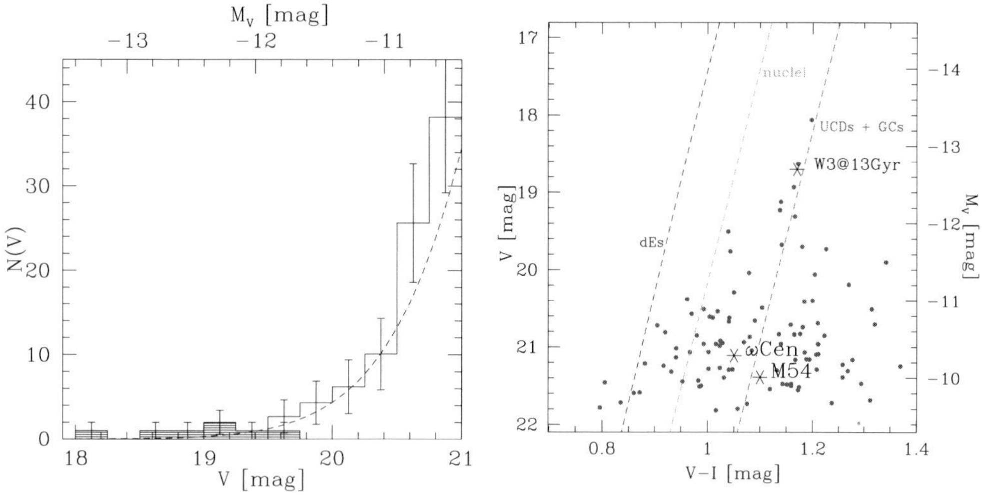

Figure 1. Left: *Solid histogram:* incompleteness corrected luminosity distribution $N(V)$ of the compact objects detected in the FCOS (Mieske *et al.* 2004a). *Shaded histogram:* luminosity distribution of the UCDs detected by Drinkwater *et al.* (2003), photometry taken from the FCOS. *Dashed line:* Gaussian LF for the GCLF of NGC 1399, taken from Dirsch *et al.* (2003). Note that the Gaussian LF and the luminosity distribution of compact objects are equally scaled and refer to the same total area. **Right:** Colour-magnitude (CM) diagram of the objects in the left panel plus fainter globular clusters with $V > 21$ mag from Dirsch *et al.* (2003). The asterisks indicate the positions of ω Cen, M54 – the two brightest MW globular clusters – and that of the massive young star cluster W3 (Maraston *et al.* 2004) if aged to 13 Gyrs (passive evolution), using BC03 models (Bruzual & Charlot 2003) and assuming solar metallicity. Dashed lines are from left to right fits to the CM distribution of: all objects plotted except the asterisks; Fornax and Virgo dE,N nuclei (Lotz *et al.* 2004; Fornax dEs (Hilker *et al.* 2003).

2. The UCDs are significantly redder than dE,N nuclei. This may be consistent with the stripping scenario if the star formation in stripped nuclei gets suppressed due to gas removal while being able to continue over more extended periods in the surviving dE,Ns. UCDs plus bright GCs define a colour-magnitude (CM) sequence offset to, but of similar slope as that of dEs. This is consistent with the stripping scenario, because stripped nuclei are expected to trace the colour-magnitude relation of their host galaxies.
3. The massive young cluster W3 (Maraston *et al.* 2004) matches the colour and luminosity of Fornax UCDs if aged to 13 Gyrs. This is consistent with a scenario in which UCDs are evolved stellar super clusters created in galaxy mergers (Fellhauer & Kroupa 2002).

3. UCDs in Abell 1689

In Fig. 2 we present the main results of our ACS/VLT search for UCDs in Abell 1689 (Mieske *et al.* 2004b and 2005). The following findings are obtained:
1. The luminosity distribution of UCD candidates with $M_V < -13$ mag cannot be explained by the GCLF of Abell 1689, while for $M_V > -13$ mag one cannot reject the hypothesis that all UCD candidates are genuine GCs (see also point 1 in Sect. 2).
2. The colours of UCD candidates in Abell 1689 are redder than normal dwarf galaxies and define a similar slope as the CM relation of normal dwarfs. This is consistent with the stripping scenario (see also point 2 in Sect. 2).
3. The brighter UCD candidates ($M_V < -13$ mag) form a separate sequence in the size-luminosity plane as compared to normal dwarf galaxies. There is a significant gap in

measured size between both populations. This is inconsistent with the hypothesis that UCDs are simply the most compact dEs in a continuous size distribution.
4. Two of the three brightest UCD candidates are spectroscopically confirmed as cluster members. According to their total luminosity and surface brightness profiles, they are twins of M32. Finding two of these extremely rare objects in a massive cluster like Abell 1689 then suggests that tidal forces are responsible for their creation. If the fainter UCD candidates are also confirmed as cluster members by spectroscopy, Abell 1689 would host a continuous sequence of compact objects extending from bright compact ellipticals (cEs) to equivalents of the Fornax UCDs. That is, there might not be a conceptual difference between cEs and UCDs. Both classes of objects might be created by tidal stripping of normal galaxies, leaving only the compact core "alive".

4. Conclusions

We suggest that in order to estimate the number of bona fide UCDs detected in spectroscopic surveys, one needs to subtract the bright end of the GCLF as extrapolated from fainter magnitudes from the luminosity function of UCD candidates. If no significant overpopulation is detected, it is not possible to confirm the presence of UCDs. Derivation of sizes and masses of UCD candidates (see e.g. Hasegan *et al.* 2005, Drinkwater *et al.* 2003) seem appropriate ways to distinguish between GCs and UCDs in luminosity regimes where the GCLF has not yet dropped to zero. The existence of a CM relation (see Fig. 1) may be a criterion to favour a galaxian origin, since no GC system has been found to exhibit such a relation, yet. The search for very luminous and massive UCDs in the densest and most massive galaxy clusters (Mieske *et al.* 2004b and 2005) appears the most promising way to definitely establish UCDs as a new class of galaxies.

Acknowledgements

We thank the ESO User Support Group for carrying out the DDT spectroscopy (program 273.B-5008) in service mode. ACS was developed under NASA contract NAS 5-32864.

References

Bekki, K., Couch, W.J., Drinkwater, M.J. & Shioya, Y. 2003, *MNRAS* 344, 399
Bruzual, G. & Charlot, S. 2003, *MNRAS* 344, 1000
Czoske, O. 2004, *IAUC 195*, ed. A. Diaferio *et al.*, astro-ph/0403650
Dirsch, B., Richtler, T., Geisler, D. *et al.* 2003, *AJ* 125, 1908
Drinkwater M.J., Jones J.B., Gregg M.D. & Phillipps S. 2000, *PASA* 17, 227
Drinkwater, M.J., Gregg, M.D., Hilker, M. *et al.* 2003, *Nature* 423, 519
Fellhauer, M. & Kroupa, P., 2002, *MNRAS* 330, 642
Graham, A.W. 2002, *ApJ* 568, L13
Hasegan M., Jordán A., Cote P. *et al.* 2005, *ApJ* in press, astro-ph/0503566
Hilker, M., Infante, L. & Richtler, T., 1999, *A&AS* 138, 55
Hilker, M., Mieske, S., & Infante, L. 2003, *A&A*, 397, L9
Kundu, A. & Whitmore, B.C. 2001, *AJ* 121, 2950
Lotz, J.M., Miller, B.W. & Ferguson, H.C. 2004, *ApJ* 613, 262
Maraston, C., Bastian, N., Saglia, R.P. *et al.* 2004, *A&A* 416, 467
Mieske, S., Hilker, M. & Infante, L. 2002, *A&A* 383, 832
Mieske, S., Hilker, M. & Infante, L. 2004a, *A&A* 418, 445
Mieske, S., Infante, L., Benítez, N., *et al.* (ACS GTO team) 2004b, *AJ* 128, 1529
Mieske, S., Infante, L., Hilker, M. *et al.* 2005, *A&A*, 430, 25

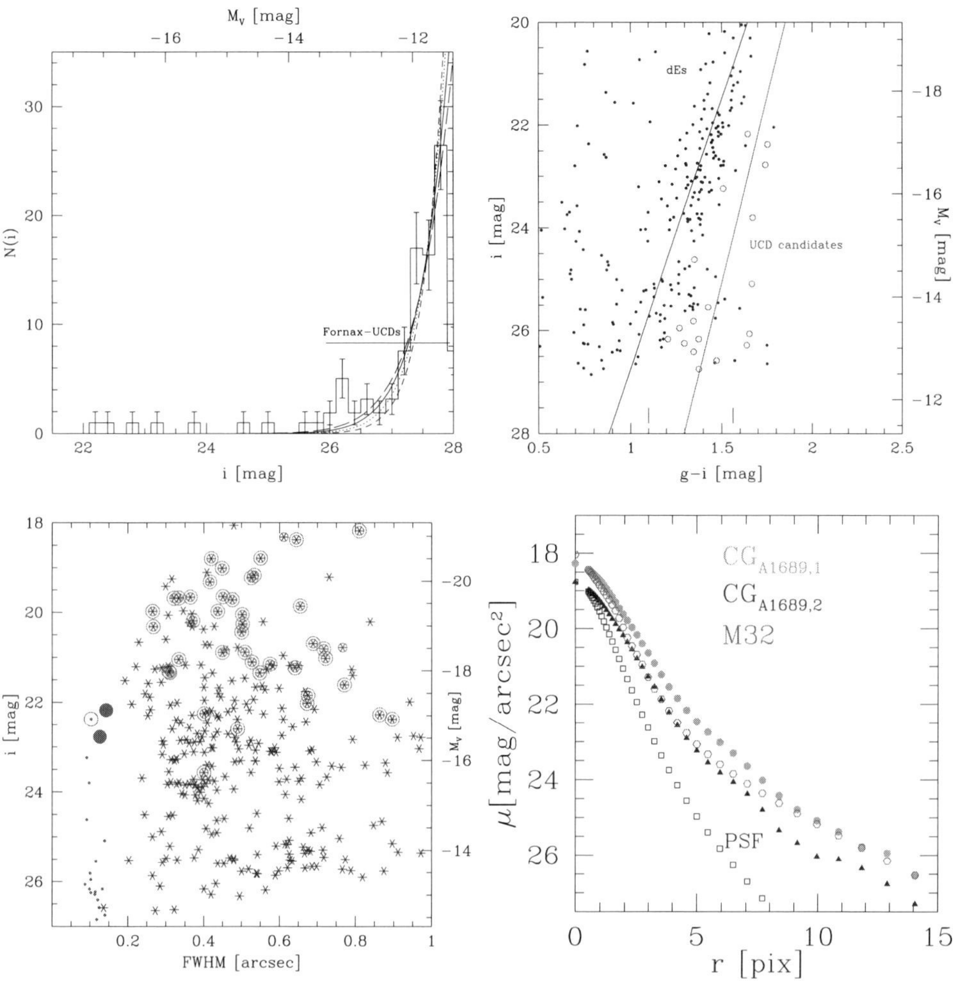

Figure 2. Top left: Luminosity distribution of UCD candidates in Abell 1689 (Mieske *et al.* 2004b). The magnitude range of the UCDs in Fornax is indicated by the horizontal tick. The long dashed, solid, dotted, short dashed lines correspond to a Gaussian GCLF at Abell 1689's distance with $\sigma = 1.5$, 1.4, 1.3, 1.2 mag respectively. **Top right:** CMD in i, $(g-i)$ of UCD candidates in A1689 with $i < 27$ mag and $z_{\rm phot} < 0.5$ indicated as circles, and resolved objects with $z_{\rm phot} < 0.5$ indicated as dots (Mieske *et al.* 2004b). The lines indicate the fitted CM relation to the resolved sources (dashed) and the UCD candidates (solid). The vertical ticks denote the expected position of the blue and red peak of the GCLF (Kundu & Whitmore 2001). **Bottom left:** Size-luminosity distribution of normal, i.e. resolved, dwarf galaxies (asterisks) and (unresolved) UCD candidates (dots) in Abell 1689 (Mieske *et al.* 2005). The larger filled circles indicate the two spectroscopically confirmed M32 twins $\mathrm{CG_{A1689,1}}$ (at $i = 22.2$ mag) and $\mathrm{CG_{A1689,2}}$ (at $i = 22.7$ mag), the UCD candidate with an open circle is a foreground star. All other UCD candidate have not been observed spectroscopically, yet, but observations are underway. The open circles around dwarf galaxy candidates indicate a measured spectroscopic redshift (Czoske 2004), double circles are members of Abell 1689. **Bottom right:** Comparison of surface brightness (SB) profiles, with scale $0.05''$/pixel, or 155 pc/pixel at Abell 1689's distance. Filled circles are $\mathrm{CG_{A1689,1}}$, filled triangles are $\mathrm{CG_{A1689,2}}$, open hexagons are M32's profile from Graham (2002), projected to A1689's distance and PSF convolved. M32 is of intermediate luminosity between $\mathrm{CG_{A1689,1}}$ and $\mathrm{CG_{A1689,2}}$. Squares indicate the PSF profile.

Discussion

BINGGELI: I think there cannot be a *physical* continuity between M32-types and globular clusters, as M32s contain DM, while GCs don't.

MIESKE: That might be correct, although I don't know the M/L ratio of M32. As it is that compact, it will be hard to derive a DM profile far out. I have not heard that M32 is DM dominated. If it is a stripped spiral bulge, it need not be. My point regarded the photometric parameters, not necessarily the underlying physical ones.

FERGUSON: The statistical distinction of UCDs from GCs based on the LF is a very weak argument, since the GCLF functional form need not be Gaussian in the tails (there is no physical theory that predicts that).

MIESKE: I agree to that. My point was that you don't have to postulate UCDs as long as the number counts can be explained by the GCLF. Comparison of number counts can then give an upper limit on the number of UCDs, assuming that the GCLF as fitted at fainter luminosities holds for the bright end as well. (Neither there is a theory that predicts that the GCLF is non-Gaussian. As in the VCS, GCLFs are very well fit by Gaussians, it seems the appropriate way to assume a Gaussian at the bright end and make first order conclusions based on that. We keep in mind that it need not be the whole truth.)

KARACHENTSEV: Do you expext any lensing phenomena which could be seen around UCDs?

MIESKE: I think that resolution and photometry limits don't allow to measure such an effect. Furthermore, UCDs are very few and you need a perfect alignment for lensing. This makes it improbable.

Near-Field Cosmology with Dwarf Elliptical Galaxies
Proceedings IAU Colloquium No. 198, 2005
H. Jerjen and B. Binggeli, eds.

doi:10.1017/S1743921305004199

Globular clusters in low mass galaxies

Margarita E. Sharina[1], T. H. Puzia[2], V. L. Afanasiev[1], D. I. Makarov[1], A. E. Dolphin[3] and G. M. Karataeva[4]

[1]Special Astrophysical Observatory, Russian Academy of Sciences, N. Arkhyz, KChR, 369167, Russia
emails: sme@sao.ru, vafan@sao.ru, dim@sao.ru

[2]Space Telescope Science Institute, 3700 San Martin Drive, Baltimore, MD21218, USA
email: tpuzia@stsci.edu

[3]Steward Observatory, University of Arizona, 933 North Cherry Avenue, Tucson, AZ 85721, USA
email: adolphin@as.arizona.edu

[4]Astronomical Institute, St. Petersburg State University, Universitetsky pr., 28 Petrodvoretz, St. Petersburg 198504, Russia
email: narka@astro.spbu.ru

Abstract. We study properties of globular cluster candidates (GCCs) and the diffuse stellar populations in nearby low surface brightness dwarf galaxies, using HST WFPC2 photometry in the V and I bands. Our sample consists of 18 dwarf spheroidal (dSph), 36 dwarf irregular (dIrr), and 3 transition-type galaxies, with projected linear diameters less than 3.5 kpc and mean blue surface brightness > 23 mag/arcsec^2 situated at the distance 2–6 Mpc in the field and in nearby groups. Our sample dwarf spheroidal galaxies were not detected in HI and are located at the distances of up to $\sim$1 Mpc away from a nearby bright galaxy. Transition-type galaxies, which are distributed like dSph galaxies, form a very rare class of galaxies. dIrr galaxies show a weaker concentration to the nearest massive neighbors than dSph and transition-type galaxies. At a given surface brightness and luminosity, they exhibit lower mean metallicities than dSphs. In contrast to dIrr galaxies, the majority of dSph galaxies at a similar mean surface brightness contains GCCs. The percentage of GCCs located near the centers of dSph galaxies is much higher than that for dIrr galaxies. The composite population of GCCs in dSphs is spatially more concentrated than in dIrr galaxies. The color distributions of GCCs in dSph and dIrr galaxies show major differences. While the latter shows obvious bimodality with the peaks near $(V-I)_0 \sim 0.5$ and $\sim$1.0 mag, the GCC color distribution in dSph galaxies shows only one peak with a mean color $(V-I)_0 \sim 1.0$ mag. There is a tendency of increasing half-light radii with increasing projected galactocentric distances for a large number of GCCs in dSph galaxies, which is also observed in the Galactic globular cluster system. We embarked on a spectroscopic survey of GCCs in low-mass galaxies to obtain their chemical compositions and relative ages.

Keywords. photometry, dwarf galaxies, star clusters, fundamental parameters

1. Properties of our sample galaxies

We select low surface brightness dwarf galaxies with accurately measured distances situated in the field and in the nearby groups: M81, Centaurus A, Sculptor, Canes Venatici I cloud. More than 50% dSph are located within the projected radius of 250 kpc from the nearest massive galaxy (see Fig. 1), which is well reproduced by CDM models (e.g. Kravtsov *et al.* 2004).

The galaxies were surveyed with the HST WFPC2 (GO-8192, GO-8601) with 600-s. expositions taken in the F606W and F814W filters. Photometry and artificial star tests were performed using DOLPHOT program (Dolphin 2000). We estimate the metallicity

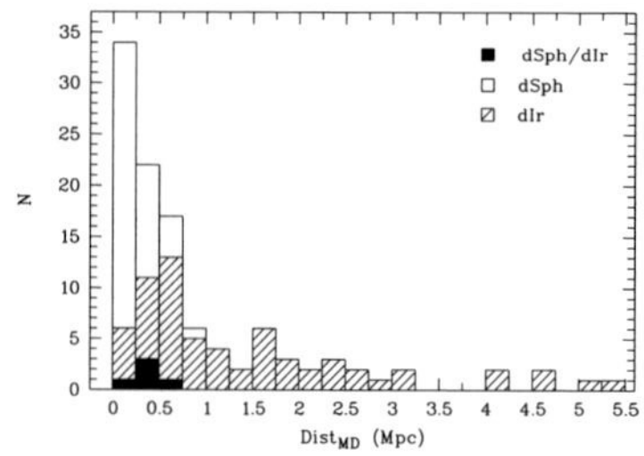

Figure 1. Distribution of nearby galaxies (our sample + LG) according to their projected distances from the nearest massive galaxy, $Dist_{MD}$(Mpc), from Karachentsev *et al.* (2004)

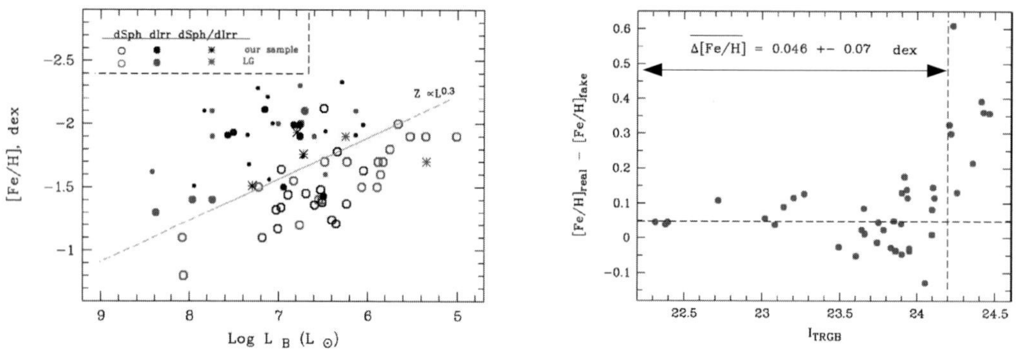

Figure 2. **(a)** B-band luminosity versus metallicity of red giant branch stars in our sample galaxies. Isolated dIrr galaxies (with tidal index, $TI < 0$, according to Karachentsev *et al.* (2004)) are marked by small solid dots. The data for the LG galaxies were taken from Grebel *et al.* (2003). **(b)** Comparison of metallicity estimated from real and artificial RGB stars in dependence of apparent I magnitude of TRGB in our sample galaxies.

of RGB stars from the mean color of the red giant branch measured at an absolute magnitude $M_I = -3.5^m$ (Lee *et al.* 1993): $[\mathrm{Fe/H}] = 12.64 + 12.6(V-I)_{-3.5} - 3.3(V-I)^2_{-3.5}$. We corrected our measurements using artificial star tests (see Fig. 2b). Fig. 2a shows that dSph galaxies follow a galactic metallicity-luminosity relation ($Z \sim L^{0.3}$) fairly well, but dIrr galaxies have higher metallicity for their luminosity as compared to dSphs.

2. Properties of globular cluster candidates

Globular cluster candidates were selected and photometry was performed using the FIND task of the DAOPHOT-II Stetson(1987) package. To convert instrumental magnitudes F606W and F814W into the standard Johnson-Cousins system we used the surface photometry recipes and equations presented by Holtzman *et al.* (1995). The magnitudes of GCCs were corrected for Galactic extinction using reddening maps from Schlegel *et al.* (1998). We select GCCs with half-light radii of $2 <$ FWHM < 9 pix (corresponds to $3 < r_h < 20$ pc, which is typical for Galactic GCs), absolute magnitudes $M_V < -5.0^m$ and color $0.3 < (V-I)_0 < 1.5$. We fit surface brightness profiles of our GCCs with the King law (King 1962). About 10% of the sample was removed from our GCCs list due to uncertain structural parameters. The final list contains 105 GCCs. In general, 50% of all surveyed galaxies contain globular cluster candidates, irrespective of morphological type.

Figure 3a shows a color-magnitude diagram for all GCCs. The red peak GCC subpopulation in dIrrs and most GCCs in dSph galaxies cover the same $(V-I)_0$ color range

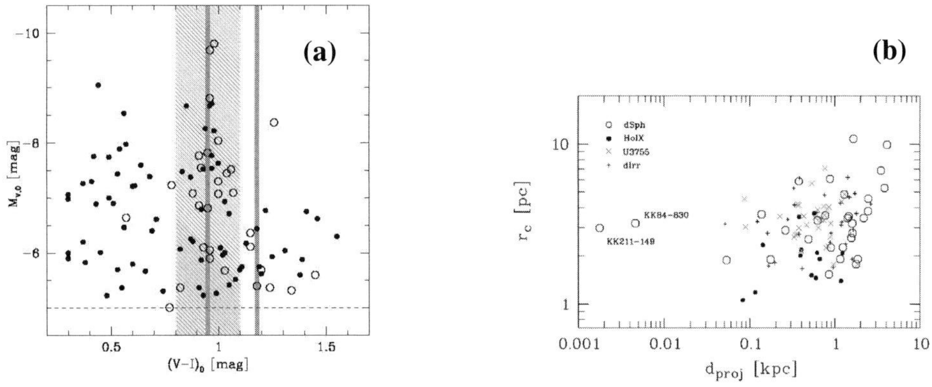

Figure 3. **(a)** Color-magnitude diagram of globular cluster candidates in dSph (*open circles*) and dIrr galaxies (*solid dots*). The dashed line indicates the limit of our photometry The hatched region shows the color range for Galactic and M31 globular clusters (Puzia *et al.* 2004). The two vertical lines indicate colors of blue and red sub-populations in massive early-type galaxies. **(b)** Core radius of GCCs versus their projected galactocentric distance. The data for the two richest GCC systems, UGC 3755 and Holmberg IX, are marked by different symbols.

as the ancient Galactic and M31 globular clusters and are similar to *blue* globular clusters in early-type galaxies, which implies similar ages and metallicities. For GCCs in dSphs the LF shows a turnover at around $M_V \approx -7.4^m$. We find indications for an excess population of faint GCCs with $M_V > -6.5$ in both dSph and dIrr, reminiscent of excess populations of faint GCs in the LG spiral galaxies. Figure 3b shows a trend of increasing half-light and core radius as a function of increasing galactocentric distance. We find that most GCCs have structural parameters similar to extended outer halo globular clusters in the Milky Way and M31, as well as the recently discovered population of "faint fuzzy" clusters in nearby lenticular galaxies (Brodie & Larsen 2002).

We find that the surface density profiles of globular cluster systems in dIrr galaxies are "flatter" ($x \approx 1.1$) and ($x \approx 1.85$) than those in dSph galaxies ($x \approx 2$) (see Fig. 4).

It should be mentioned, that in contrast to dIrr galaxies, ~60% of dSphs with GCCs have cluster candidates located near their galaxy center. These GCCs are bright and compact ($-7 < M_{V,0} < -10$, $r_c \approx$ 2–3 pc), similar to those found at the center of nucleated Virgo dE galaxies (Lotz *et al.* 2004).

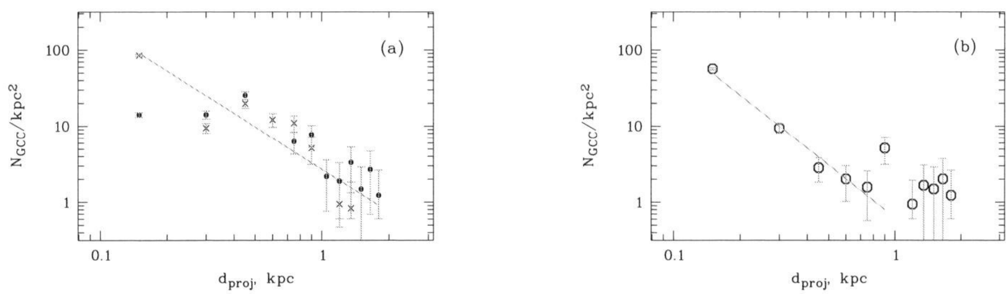

Figure 4. Radial distribution of GCCs in dIrr galaxies *(panel a)* and dSph galaxies *(panel b)*. The sample of GCCs in dIrrs is divided into blue GCCs with $(V-I)_0 < 0.75$ *(dots)* and red GCCs $(V-I)_0 > 0.75$ *(crosses)*. Plotted here is the logarithm of the surface density of GCCs per square kpc (evaluated in 0.15-kpc bins) vs. the logarithmic projected distance from the galaxy center, in kpc.

3. Conclusions

Our findings do not contradict a hypothesis that dSph galaxies turned their gas reservoirs early and more efficiently into stars than dIrrs of similar luminosity. dSph galaxies have higher metallicity of red giant stars for their luminosity as compared to dIrr galaxies. The detection rate of GCCs in dSph galaxies is ~2 times higher than in dIrr galaxies in the same surface brightness range $\mu_B \geqslant 24.4$ mag/arcsec^2. GCC systems of dSph galaxies are more spatially concentrated than in dIrr galaxies. The fraction of GCCs located near the center of dSph galaxies is 2 times higher than that for dIrrs. However, it is worth to note that dSph, dIrr and transition type galaxies host GCCs with $(V-I)_0 \approx 1.0^m$, which might indicate that at least early globular cluster formation took place irrespective of the host galaxy's morphological type.

If accretion of globular clusters plays an important role in the assembly of outer globular cluster systems in massive galaxies, one should expect similarities in magnitudes and structural parameters between clusters in our sample dwarf galaxies and the outskirts of the Local Group spirals (Harris 1996). For GCCs in dSphs the luminosity function shows a turnover at around $M_V = -7.4$. Roughly half of our globular cluster candidates have structural parameters similar to extended outer halo globular clusters in the Local group spiral galaxies.

An ongoing spectroscopic survey of globular clusters in low-mass nearby galaxies will help us to reject background galaxies from our list and study the chemical composition and relative ages of globular clusters in low surface brightness dwarf galaxies.

Acknowledgements

M. E. S. would like to acknowledge IAU grant, RFBR travel grant 05-02-26517 and the conference organizers for a stimulating meeting.

References

Dolphin, A.E. 2000, *PASP* 112, 1383
Brodie, J.P. & Larsen, S.S. 2002, *AJ* 124, 1410
Grebel, E.K., Gallagher, J.S. & Harbeck, D. 2003, *AJ* 125, 1926
Harris, W.E. 1996, *AJ* 112, 1487
Holtzman, J.A., Burrows, C.J., Casertano, *et al.* 1995, *PASP* 107, 1065
Karachentsev, I.D., Karachentseva, V.E., Huchtmeier, W.K. & Makarov, D.I. 2004, *AJ* 127, 2031
King, I. 1962, *AJ* 67, 471
Kravtsov,A.V., Gnedin, O.Y. & Klypin, A.A. 2004, *ApJ* 609, 482
Lee, M.G., Freedman, W.L. & Madore, B.F. 1993, *ApJ* 417, 553
Lotz, J.M., Miller, B.W. & Ferguson, H.C. 2004, *ApJ* 613, 262
Puzia, T.H. *et al.* 2004, *A&A* 415, 123
Schlegel, D.J., Finkbeiner, D.P & Davis M. 1998, *ApJ* 500, 525
Stetson, P.B. 1987 *PASP* 99, 191

Discussion

MILLER: What are the specific frequencies for globular cluster candidates in dIrrs?

SHARINA: A mean value of the specific frequency for GCCs in our sample dIrrs is ~15.

Near-Field Cosmology with Dwarf Elliptical Galaxies
Proceedings IAU Colloquium No. 198, 2005
H. Jerjen and B. Binggeli, eds.

doi:10.1017/S1743921305004205

The origin of ultra-compact dwarf galaxies in the Virgo cluster

E.A. Evstigneeva[1], M.D. Gregg[2]† and M.J. Drinkwater[1]

[1]Department of Physics, University of Queensland, QLD 4072, Australia
e-mail: katya@physics.uq.edu.au

[2]Department of Physics, University of California, Davis, CA 95616, USA

Abstract. We present the internal velocity dispersions and spectral line indices for six Virgo ultra-compact dwarf (UCD) galaxies obtained from spectroscopy with the Keck II 10-m telescope. We use these results: 1) to compare the Virgo UCDs with the Fornax UCD population; 2) to compare UCDs with globular clusters (GCs) to determine if UCDs and the most luminous GCs are the same or distinct systems; 3) to further test the nucleated dwarf elliptical (dE,N) stripping hypothesis for UCD formation.

Keywords. galaxies: dwarf, galaxies: kinematics and dynamics, galaxies: abundances, galaxies: clusters: individual (Virgo Cluster)

1. Introduction

UCDs are a new type of galaxy recently discovered in the center of the Fornax, Virgo, and other clusters of galaxies (Hilker *et al.* 1999; Drinkwater *et al.* 2003,2004; Jones *et al.* 2005). Possible origins of UCDs include the following hypotheses: 1) they are luminous intra-cluster globular clusters; 2) they are extremely luminous star clusters formed from the amalgamation of stellar super-clusters that were created in galaxy interactions; 3) they are the remnant nuclei of stripped dwarf galaxies which have lost their outer parts in the course of tidal interaction with the galaxy cluster potential; 4) they are highly compact galaxies formed in the early Universe.

Here we present new results of the spectroscopic analysis from the Keck II telescope for six Virgo UCDs (Table 1) to explore and test hypotheses of their formation.

2. Internal Dynamics

Figure 1 compares UCDs to other types of stellar systems, on the luminosity - velocity dispersion plane. The internal velocity dispersions for the UCDs (and GCs) were measured from CaT region (8400–8750Å) using the direct-fitting method (van der Marel 1994). We include the most massive and luminous known GCs: G1, ω Cen, and NGC5128 GCs as well as GCs in the Milky Way and M31, dE,Ns and their nuclei, and giant ellipticals. We find: 1) The Virgo UCDs have the same dynamical properties as the Fornax ones (Drinkwater *et al.* 2003). 2) There is no gap between luminous GCs and UCDs in this plot. We obtained velocity dispersions for the two brightest M87 GCs from Hanes *et al.* (2001) list: they lie in the same part of $M_V - \sigma$ plane as UCDs. 3) The UCDs follow approximately the same relation between luminosity and velocity dispersion as GCs, except for the two brightest ones. More data on the velocity dispersions for bright GCs or fainter UCDs are required to confirm this. 4) There is an overlap in luminosities and

† Also at: Institute for Geophysics and Planetary Physics, Lawrence Livermore National Laboratory, L-413, Livermore, CA 94550, USA

Table 1. Targets. M_V – the V band absolute magnitude. Strom547 and Strom417 are the two brightest globular clusters in M87 (the central galaxy in Virgo) according to Hanes *et al.* (2001) list. Strom547 has luminosity comparable to that of UCDs and was therefore included into UCD list.

Name	Object type	R.A.(J2000)	Dec.(J2000)	M_V
UCD1/Strom547	UCD/GC	12h30m57.40s	+12°25′44.8″	−12.3
UCD2	UCD	12h30m07.61s	+12°36′31.1″	−12.2
UCD3	UCD	12h31m04.51s	+11°56′36.8″	−12.1
UCD4	UCD	12h31m28.41s	+12°25′03.3″	−12.0
UCD5	UCD	12h31m52.93s	+12°15′59.5″	−13.9
UCD6	UCD	12h31m11.90s	+12°41′01.2″	−12.1
Strom417	GC	12h31m01.29s	+12°19′25.6″	−11.6

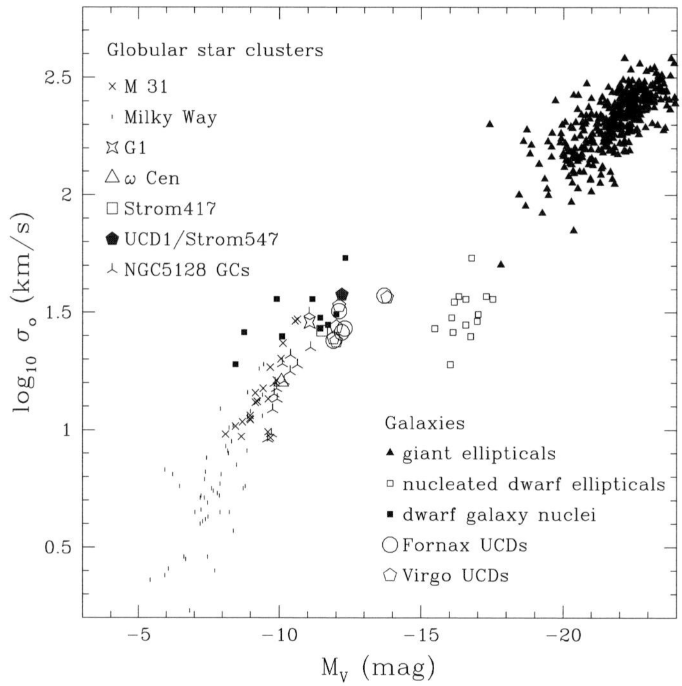

Figure 1. Comparison of the internal dynamics of UCDs (our data) with globular clusters and nucleated dwarf ellipticals (literature data).

velocity dispersions of the nuclei of dE,Ns and the properties of bright GCs and UCDs . It supports the dE,N tidal stripping hypothesis for UCD and GC formation.

3. Ages and Chemical Compositions

We estimate ages, metallicities, and abundances of UCDs using Lick/IDS indices (Worthey *et al.* 1994) and single-aged stellar population models of Thomas *et al.* (2003).

In *Figure 2a* we plot abundance ratios of the Virgo UCDs. To estimate [α/Fe], we plot Mgb (an indicator of α-elements) versus <Fe> (an average of the indices Fe5270 and Fe5335). [α/Fe] traces the timescale of star formation activity in galaxies. Most α-elements are produced rapidly by Type II supernovae, while Fe is produced by Type Ia

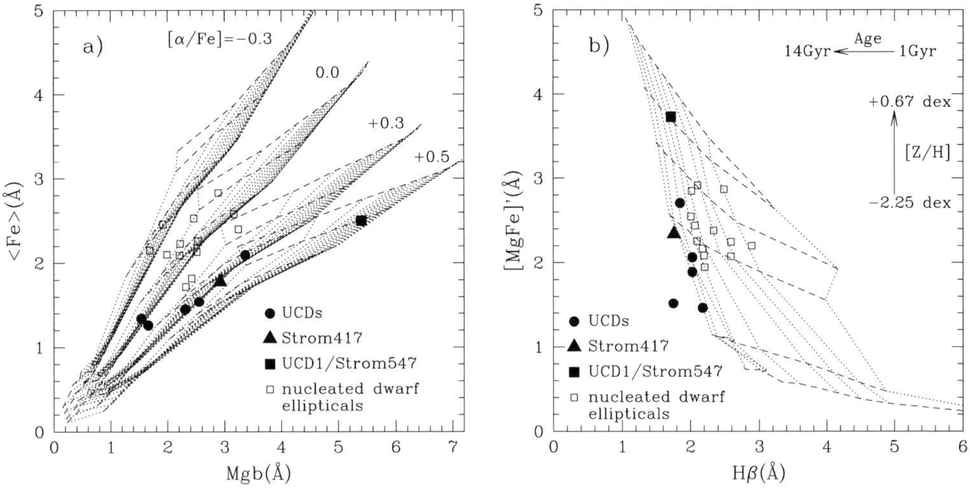

Figure 2. Comparison of our data with model grids from Thomas *et al.* (2003).
a) Mgb – <Fe>. Thomas *et al.* models with variable $[\alpha/\mathrm{Fe}]$ are shown for ages 1,2,3,...,15 Gyr (dotted lines, from left to right) and metallicities −2.25,−1.35,−0.33,0.0,+0.35,+0.67 dex (dashed lines, from bottom to top). **b)** $\mathrm{H}\beta$ – [MgbFe]'. Thomas *et al.* models are shown for ages 1,2,3,4,6,8,10,12,14 Gyr (dotted lines, from right to left) and metallicities −2.25,−1.35,−0.33,0.0,+0.35,+0.67 dex (dashed lines, from bottom to top).

SNe on longer timescales. We find: 1) Four UCDs and Strom417 (GC) have super-solar abundance ratio, $[\alpha/\mathrm{Fe}] \sim +0.3 - +0.5$, and two UCDs appear to have solar abundances, $[\alpha/\mathrm{Fe}] \sim 0.0$. The super-solar abundances are typical for old stellar populations (in globular clusters and elliptical galaxies). 2) Virgo dE,Ns (Geha *et al.* 2003) are also shown in the same plot. The majority of dE,Ns are consistent with solar abundance ratios, while the majority of UCDs have super-solar abundances. This can be the evidence that UCDs and dE,Ns have different star formation histories.

Figure 2b shows ages and metallicities of the Virgo UCDs. We plot the age-sensitive $\mathrm{H}\beta$ index versus the metallicity sensitive [MgFe]' index (Thomas *et al.* 2003). We find: 1) The UCDs are old and metal-poor, except UCD1 which appears to be intermediate-aged and metal-rich. 2) The ages and metallicities of UCDs (except UCD1) are similar to Strom417 (GC) and to those found for M87 globular clusters by Cohen *et al.* (1998). 3) The UCDs have older integrated stellar populations than present-day dE,Ns. It is consistent with the stripping formation scenario, when the ISM is removed from the UCD progenitor and star formation activity ceases.

References

Cohen, J.G., Blakeslee, J.P. & Ryzhov, A. 1998, *ApJ* 496, 808
Drinkwater, M.J., Gregg, M.D., Hilker, M., *et al.* 2003, *Nature* 423, 519
Drinkwater, M.J., Gregg, M.D., Couch, W.J., *et al.* 2004, *PASA* 21, 375
Geha, M., Guhathakurta, P. & van der Marel, R.P. 2003, *AJ* 126, 1794
Hanes, D.A., Cote, P., Bridges, T.J., *et al.* 2001, *ApJ* 559, 812
Hilker, M., Infante, L., Viera, G., *et al.* 1999, *A&AS* 134, 75
Jones, J.B., Drinkwater, M.J., Jurek, R., *et al.* 2005, *MNRAS* in press
van der Marel, R.P. 1994, *MNRAS* 270, 271
Thomas, D., Maraston, C. & Bender, R. 2003, *MNRAS* 339, 897
Worthey, G., Faber, S.M., Gonzalez, J.J., *et al.* 1994, *ApJS* 94, 687

Near-Field Cosmology with Dwarf Elliptical Galaxies
Proceedings IAU Colloquium No. 198, 2005
H. Jerjen and B. Binggeli, eds.

doi:10.1017/S1743921305004217

Faint stellar systems in the Virgo cluster: ultracompact dwarfs and globular clusters

Monica Haşegan[1,2] **and Patrick Côté**[1,3]

[1]Department of Physics and Astronomy, Rutgers, The State University of New Jersey, Piscataway, New Jersey 08854, USA
email: mhasegan@physics.rutgers.edu

[2]Institute for Space Sciences, P.O.Box MG-23, RO 77125, Bucharest-Magurele, Romania

[3]Herzberg Institute of Astrophysics, National Research Council of Canada, 5071 West Saanich Road, Victoria, BC, V9E 287, Canada
email: Patrick.Cote@nrc-cnrc.gc.ca

Abstract. We examine the properties of ten compact objects in the vicinity of M87 using *HST* imaging and Keck spectroscopy. We find three objects to be larger and brighter than typical globular clusters; they resemble the nuclei of nucleated dEs in Virgo, having similar mass-to-light ratios, luminosities and colors. Their properties are consistent with models of tidally stripping of nucleated dEs. One object could be an old "stellar supercluster", formed through the amalgamation of multiple young massive clusters. Two other objects appear to be massive versions of otherwise "normal" globular clusters.

Keywords. galaxies: clusters: individual (Virgo) – galaxies: star clusters – galaxies: dwarf – galaxies: formation – galaxies: kinematics and dynamics

1. Introduction and Sample Selection

We are carrying out a program to study the structural, photometric and dynamical properties of faint stellar systems in the Virgo Cluster. Here we present results from a recent paper (Haşegan *et al.* 2005) where we have examined the properties of ultracompact dwarf (UCD) galaxies in the Virgo Cluster. UCDs are a potentially new type of faint and compact stellar system which were first identified in the Fornax Cluster (see Drinkwater *et al.* 2003 and references therein). They may represent a missing link between dwarf elliptical galaxies and globular clusters.

Based on a combination of ground- and space-based imaging surveys, we identified several compact objects with $-11.8 \leqslant M_V \leqslant -10.8$ in the vicinity of M87, which we provisionally classify as "Dwarf Globular Transition Objects" (DGTOs). Our analysis is based on imaging from *HST* — taken as part of the ACS Virgo Cluster Survey (ACSVCS; Côté *et al.* 2004) or from the *HST*/WFPC2 archive — and high-resolution spectroscopy from Keck/ESI.

2. Results and conclusions

The left panel of Figure 1 shows a finding chart for three DGTOs (S928, S999 and H8005) near M87, along with two additional candidates without ESI spectra. The $0.75'' \times 20''$ ESI slit used for the Keck observations is shown by the rectangles.

The right panel of Figure 1 compares the properties of these DGTOs with those of other low-mass stellar systems. From top to bottom, central velocity dispersion, half-light radius and mass surface density averaged within the half-light radius are plotted against total mass. Among the sample of six DGTOs shown in this figure (squares), we find two objects which are likely massive M87 globular clusters (i.e., those lying on the solid lines which represent the scaling relations of Galactic globular clusters), three objects which are probable UCDs (i.e., those clustered together with the Fornax UCDs and with the simulated tidally stripped cores of dE,Ns; Bekki *et al.* 2003), and one *possible* UCD (i.e., the object near the simulated merged massive young clusters; Fellhauer & Kroupa 2002).

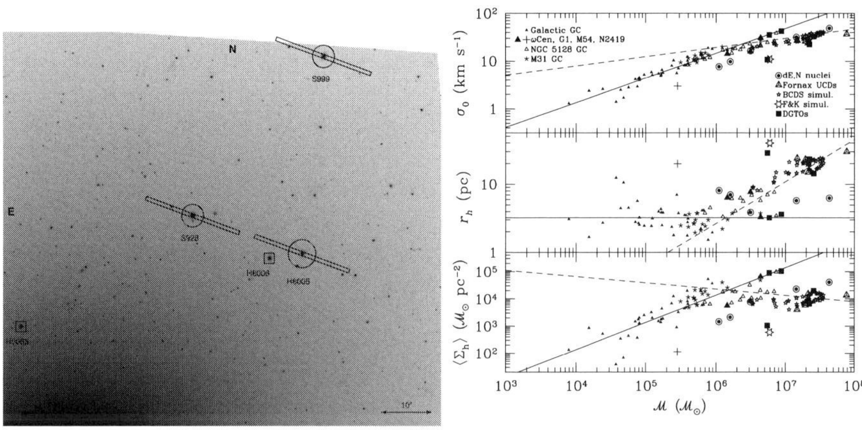

Figure 1. Left: ACSVCS image of M87 showing the location of several Dwarf Globular Transition Objects (DGTOs). Right: Scaling relations for low-mass, hot stellar systems. (Figures taken from Haşegan *et al.* 2005)

The UCDs are larger than typical globular clusters (with half-light radii of $r_h \approx 20$ pc rather than $\approx$3 pc,) yet much more compact than normal dEs. Moreover, with mass-to-light ratios in the range $6 \leqslant M/L_V \leqslant 9$, they appear to contain *dark matter*, which would suggest a fundamental difference from globular clusters. They also bear a strong resemblance in terms of size, luminosity and mass-to-light ratio to the *nuclei of dwarf elliptical galaxies* in Virgo (Geha *et al.* 2002), supporting the identification of at least some UCDs with the tidally-stripped remains of nucleated dEs (Bekki *et al.* 2003) although other scenarios can not be completely ruled out. Future analysis of an extended sample of UCDs from the ACSVCS, together with nucleated dwarfs and their nuclei, should allow a greater insights into the origin of these faint stellar systems.

Acknowledgements

These results are part of a collaborative effort by the ACSVCS team, whom we thank.

References

Bekki, K., Couch, W.J., Drinkwater, M.J. & Shioya, Y. 2003, *MNRAS* 346, 11

Côté, P., Blakeslee, J.P., Ferrarese, L., Jordán, A., Mei, S., Merritt, D., Milosavljević, M., Peng, E., Tonry, J.L. & West, M.J. 2004, *ApJS* 153, 223

Drinkwater, M.J., Gregg, M.D., Hilker, M., Bekki, K., Couch, W.J., Ferguson, H.C., Jones, J.B. & Phillipps, S. 2003, *Nature* 423, 519

Fellhauer, M. & Kroupa, P. 2002, *ApJ* 124, 2006

Geha, M., Guhathakurta, P. & van der Marel, R.P. 2002, *AJ* 124, 3073

Haşegan, M., Jordan, A., Côté, P., Djorgovski, S.G., McLaughlin, D.E., Blakeslee, J.P., Mei, S., West, M.J., Peng, E.W., Ferrarese, L., Milosavljević, M., Tonry, J.L. & Merritt, D. 2005, *ApJ* 627, 203

Near-Field Cosmology with Dwarf Elliptical Galaxies
Proceedings IAU Colloquium No. 198, 2005
H. Jerjen and B. Binggeli, eds.

doi:10.1017/S1743921305004229

Abundances from a large spectroscopic survey in ω Centauri

A. Kayser[1], M. Hilker[2], T. Richtler[3] and P. Willemsen[2]

[1]Astronomisches Institut der Universität Basel, Venusstrasse 7, 4102 Binningen, Switzerland
[2]Sternwarte der Universität Bonn, Auf dem Hügel 71, 53121 Bonn, Germany
[3]Universidad de Concepción, Departamento de Física, Casilla 106-C, Concepción, Chile

ω Centauri, the largest globular cluster of our Milky Way, is an outstanding object in many aspects. Studies on the red giant branch (RGB) revealed a large spread in iron abundance, which ranges from [Fe/H]~ -2.0 to -0.4 dex and the existence of multiple sub-populations that do not only differ in their element abundances but also in their spatial and kinematic distributions (e.g. Norris et al. 1997). Furthermore more recent photometric studies suggested the existence of an age spread between these subpopulations (e.g. Hilker & Richtler 2000).

In order to draw further conclusions on the origin of ω Centauri we analysed a large sample of spectra with medium resolution of stars near the age sensitive subgiant branch and main sequence turnoff region observed with FORS2/MXU at the VLT. For the metallicity calibration our dataset comprises spectra with known [Fe/H] values: of the chemically homogeneous globular cluster M55, standard stars and a grid of synthetic spectra taken from Bailer-Jones (2000). To quantify the strength of the absorption lines we defined line indices (mostly for Fe, but also Ca, Mg and H lines) with central widths adjusted to the resolution of our spectra. As a measure for the Fe abundance we combined several line indices to a mean Fe indictor. We corrected for the temperature effect by using a temperature dependent mean Balmer index. From the derived metallicities and in combination with photometric data kindly provided by Rey *et al.* (2004), we were able to disentangle the age-metallicity degeneracy. We determined the ages by interpolating in a grid of Yonsei-Yale isochrones taken from Kim *et al.* (2002). We assumed $[\alpha/\text{Fe}] = 0.3$

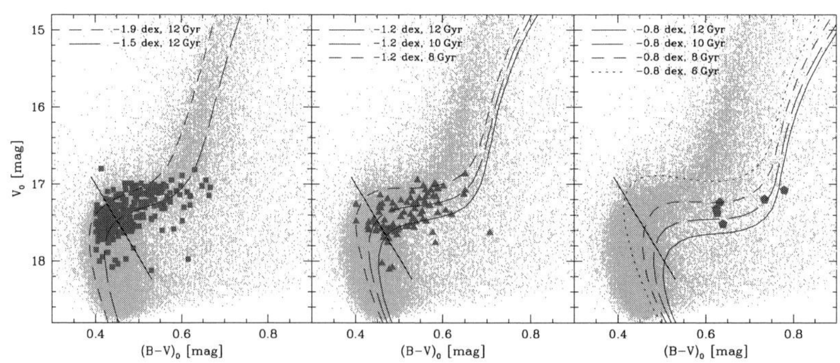

Figure 1. The position of stars of different abundances compared with isochrones of different ages and metallicities. The abundance ranges for the highlighted stars are: left: $-2.0 <$[Fe/H]< -1.4, middle:$-1.4 <$[Fe/H]< -1.0, right:$-1.0 <$[Fe/H]. The dominant metal poor population is best fit by 12 Gyr isochrones. More metal-rich stars scatter around isochrones that are up to 2 Gyr younger. The most metal-rich stars are distributed between the 9 to 11 Gyr isochrones.

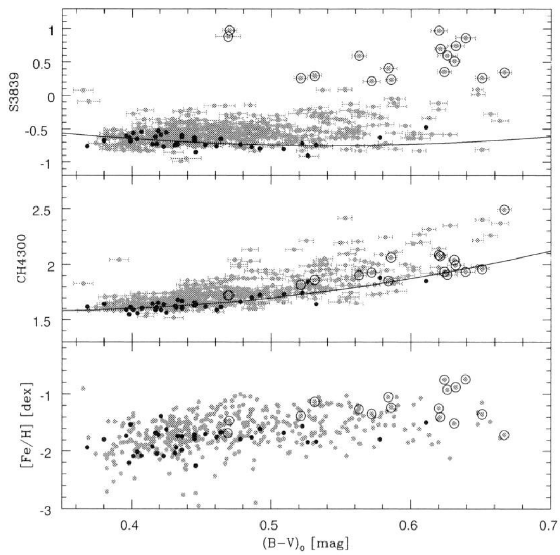

Figure 2. From top to bottom, S(3839), CH(4300) and [Fe/H] vs. $(BV)_0$ for stars in ω Cen and M55 (black). The fit curves of the M55 data indicated the temperature effect in the indices. Highlighted by circles are the most CN-rich stars. These stars are reidentified in the CH(4300) vs. $(B-V)_0$ and [Fe/H] vs. $(B-V)_0$ diagrams.

dex for stars with [Fe/H]< -1.0 dex and $[\alpha/\mathrm{Fe}] = 0.1$ dex for stars with [Fe/H]> -1.0 dex (Pancino *et al.* 2002). The resulting age-metallicity dependencies are shown in Fig. 1. Our results confirm an age spread of up to 4 Gyr in the sense that the metal-rich stars are on average younger than the dominant metal-poor population (Hilker *et al.* 2004).

The CN and CH absorption bands were measured by the commonly used indices S(3839) and CH(4300) (e.g. Norris *et al.* 1981). Again, M55 served as a comparison object for estimating the strength of the CN and CH variations and to estimate the effect of temperature on these indices. Specifically, we found that the line strength of CH(4300) is strongly anticorrelated with temperature, while the strength of S(3839) shows a very week temperature effect. The analysis of star-to-star CN and CH variations revealed, as for the RGB, the existence of a large scatter toward very high CN and CH abundances. From an inspection of the location of the most CN-rich stars in the corresponding CH(4300) and [Fe/H] vs. $(B-V)_0$ diagrams, we found that the CN-rich stars tend to be more CH-weak, and mainly belong to the metal-rich populations (Fig. 2).

Our study shows that ω Centauri definitely has experienced a complex chemical enrichment history with yields of different types of stars contributing. This has not been observed in any other globular cluster. The results provide further evidence for a prolonged star formation history as would be expected for a dwarf galaxy.

References

Hilker, M., Kayser, A., Richtler, T. & Willemsen., P. 2004, *A&A* 422, L9
Hilker, M. & Richtler, T. 2000, *A&A* 362, 895
Kim, Y., Demarque, P., Yi, S. K. & Alexander, D.R. 2002, *AJ* 143, 499
Norris, J., Cottrell, P.L., Freeman, K.C. & da Costa, G.S. 1981, *ApJ* 244, 205
Norris, J., Freeman, K.C., Mayor, M. & Seitzer, P. 1997, *ApJL* 487, 187
Panino, E., Pasquini, L., Hill, V., Ferraro, F.R. & Bellazzini, M. 2002, *ApJL* 568, L101
Rey, S., Lee, Y., Ree, C.H., Joo, J. & Sohn, Y. 2004, *AJ* 127, 958

Near-Field Cosmology with Dwarf Elliptical Galaxies
Proceedings IAU Colloquium No. 198, 2005
H. Jerjen and B. Binggeli, eds.

doi:10.1017/S1743921305004230

Summary from an observational perspective

John S. Gallagher, III

Dept. Astronomy, U. Wisconsin, 475 N. Charter St. Madison, WI 53760, USA
email: jsg@astro.wisc.edu

Abstract. The meeting "Near Field Cosmology with Dwarf Elliptical Galaxies" covered a wide range of topics relating the general issues associated with galaxy formation and evolution as well as problems which are specific to spheroidal dwarfs. This paper summarizes these presentations in the form of an annotated guide to the presenters and their subjects. The results of the meeting demonstrate the value of focused, quantitative research on small galaxies as a way to enhance our understanding of how internal processes, cosmological foundations, and environmental conditions effect galaxy evolution.

Keywords. galaxies: dwarf, galaxies: evolution, galaxies: formation, galaxies: stellar content, galaxies: structure, galaxies:clusters: general, Local Group, dark matter

---

1. Introduction

The extreme nature of spheroidal dwarfs became immediately clear with the discovery of the Sculptor and Fornax dSph systems in the late 1930s (see Shapley 1939, Baade & Hubble 1939). Early studies of the Virgo cluster with the Lick astrograph led G. Reaves (1953) to recognize the surprising existence of a substantial population of low luminosity Sculptor-type galaxies in the cluster and in the process provide the first appearance of the term "dwarf galaxy" in a title listed by the NASA ADS. However, it was only with the advent of fast Schmidt telescopes that the extensive populations of low luminosity dwarfs became clear. Swiss astronomer Fritz Zwicky used the 18-inch Palomar Schmidt to find large number densities of dwarf galaxies and thus demonstrated their cosmic importance (Zwicky 1957). Holding this meeting in Switzerland then was especially appropriate for many reasons, including as a reminder (especially on the centennial of Einstein's 1905 paper) that size and importance are not always correlated.

Soon after their discovery, the value of spheroidal dwarfs for stellar population studies was recognized by Baade (1944). Because of their low stellar densities and deceptively simple central structures, the dE and dSph galaxies provided insights into the characteristics of old stellar populations and formed a basis for defining the properties of Population II stars (Baade 1951). By the time the DDO catalog was published (van den Bergh 1959) dwarf galaxies were a recognized component of the nearby universe. In his influential review at the Yale symposium on stellar populations, van den Bergh (1977) further remarked on the division of dwarfs into two major groups–the old spheroidal systems and star-forming dIs. This provided an additional stimulus to understand how the populations of dwarf galaxies evolved (see also the pioneering review by Hodge 1971).

The direction of dwarf galaxy research took a new twist with the advent of models in which galaxies contain dark matter. Precision stellar radial velocity dispersion measurements by Aaronson (1983) showed that some dSphs have higher apparent mass-to-light ratios than can be accounted for by normal stellar populations, thereby opening the way for observational support of cold dark matter (CDM) models. The companion paper by Lin & Faber (1983) initiated explorations of the wider astrophysical implications of dwarf galaxies whose mass is domnated by CDM.

In this meeting we have heard discussions that center on spheroidal dwarfs but extend over topics ranging from star formation to cosmology. There is no easy way to coherently summarize the diversity of subjects. In §II I present the main topics of discussions followed by lists of contributors, so that this summary might serve as an annotated guide to the program and the proceedings. A few concluding remarks are included in §III.

2. Major Discussion Points

2.1. *Dwarfs in the Youthful Universe*

Dwarfs & Cosmology – *Colafrancesco, Zhao:* The survival of dSph galaxies with their low stellar densities as neighbors of giant galaxies is a long standing problem. One possibility is that the existence of such objects is a signature of non-Newtonian gravity, such as the MOND models of gravitational interactions. It therefore was useful to hear a discussion of the issues that arise and paths towards their solution in placing MOND in a cosmological context by including relativity. On the other side, the current standard model for low density dwarfs is that they are bound within CDM halos. In this case these objects have high central phase space densities of CDM. This and the proximity of the Galactic dSphs make them interesting targets for efforts to directly detect particle emission associated with possible annihilation processes in CDM.

Formation of Dwarf Galaxies – *Diemand, Mashchenko, Ricotti:* Numerical models of mass particles which interact only via gravity now have sufficient resolution to describe the evolution of moderate mass bound structures within a cosmological context. These models, when run for standard density perturbation spectra in ΛCDM cosmologies, predict large numbers of low mass 'mini-halos' clustered around the large halos which presumably contain giant galaxies. How one interprets this excess is a key question: If dSphs occupy what originally were relatively low mass DM halos, then the number of luminous dwarfs per giant galaxy is far less than the expected number of halos. This in turn leads to a fundamental issue that is illustrated in Figure 1: How do we connect the observable stellar galaxies to the models of DM halos; are we even talking about the same objects?

Impact of Reionization – *Grebel, Suza:* DSph galaxies contain ancient stars and yet evolved to the point where they produce mean metal abundances above 1% of solar. Since such galaxies, or related classes of low mass objects that merged rapidly into giant galaxies, should have been very common, they are potentially important contributors to the reionization of the universe. On the other hand, in some models reionization profoundly influences the evolution of low mass galaxies to the point where it can cut off or cut down the available amount of ISM and thus truncate star formation. Unfortunately, modern stellar population studies have yet to detect any signature of a preferred epoch for the cessation of star formation in nearby dSphs. Instead the star formation histories appear to vary with galaxy location, suggesting that local factors determine the evolutionary paths of these systems.

Dwarfs in the Distant Universe? – *Barazza, Ferguson, Pettini:* In the past decade we gained a remarkable level of access to the youthful universe. Galaxies now have been reliably observed up to $z \approx 6$, or at a point where only ∼1 Gyr had elapsed since the big bang. The compact sizes of galaxies observed with *HST* at $z \geqslant 2$ suggest that some connections with present day dwarfs is possible if some dwarfs are born via extreme starbursts, where a significant fraction of their stars are produced in a galactic crossing time of ∼100 Myr. However, the star formation histories of most nearby dwarfs suggest they did not evolve in this way, but rather formed stars more gradually over a time scale

Figure 1. A key point is how to compare theoretical models and observations. While the same words are used, observers and theorists are not necessarily describing objects in the same ways, as illustrated by the above descriptions of Swiss cows by observers and theorists.

of several Gyr. Unfortunately, the present studies of galaxies at high redshifts have yet to turn up any objects that are unambiguously dwarfs. Perhaps studies of low luminosity Lyα emission line sources might be one way to find the high redshift ancestors of present day dwarf galaxies?

2.2. *The Stellar Populations*

Modeling the Birth and Evolution of Stellar Populations – *Dekel, Hensler, Lin, Marcolini, Michielsen, Weidner:* We face a fundamental difficulty in that galaxies, with the exception of cases where an object is found via gravitational lensing alone, are the only known tracers of moderate mass scale cosmic structures that are bound by gravity. Galaxies contain baryons and these in turn make stars. Thus the complex processes of gas capture into galaxies and the subsequent feedback from stellar populations, as well as external affects, must be taken into account when producing models of galaxies vs. models of dark halos. The ability of gas to change phases and interact with the host galaxy in many ways, including gravity, collisional processes, photoionization, and via magnetic fields complicates the critical problem of modeling the baryons. In dealing with dSphs we face the further difficulty that some mechanisms must exist to cause star formation to cease and the remaining system to be almost free of any cool gas. Further progress in "gastrophysics," the study of the flow of mass and energy through the ISM and IGM associated with galaxies, remains essential. However, even in the absence of a full physical understanding of the many physical links between gas and stars, models are developing in predictive power and providing important guides to observations, such as the increased emphasis in understanding the role of mass outflows *and* inflows of both gentle and violent types.

Reading the Stellar Populations – *Arimoto, Da Costa, Gallart, Gullieuszik, Harbeck, Held, Hempel, Koch, Miller, Rejkuba, Saviane, Tolstoy:* The advent of the *Hubble Space Telescope* in combination with large ground-based telescopes has allowed us to extend "Galactic astronomy," the study of galaxies through the properties and dynamics of their stellar populations, throughout the Local Group and beyond. One key set of observations show that the lower stellar mass function is relatively normal in Galactic dSphs; there is little room for extra stellar mass in dSphs in the form of very low mass stars.

The interpretation of Hess diagrams, stellar densities on color-magnitude planes, has become a developed art that yields information on stellar abundances and star formation histories. These studies reveal the complex nature of star formation in even the dSphs dominated by old stellar populations. As is appropriate for galaxies, a range in metallicity exists, radial stellar population gradients are seen, and extended periods of active star formation are derived. This work also highlights differences between the Galactic and M31 dSph satellites; the Milky Way has the younger, more diverse companions while the M31 dSphs are predominantly older stellar fossils. It also demonstrates the existence of a smooth evolutionary continuum between the various classes of dSphs, and "transition" dwarf galaxies, such as Phoenix or LGS3, which appear to share stellar population features with the younger dSphs, like Fornax. On the other hand, the dSphs and transition dwarfs appear to have had higher initial star formation rates than the dIs, indicating that some of the differences among dwarfs go back to early times.

Work on the chemical evolution of dwarfs also is experiencing a boom powered largely by high and medium resolution spectroscopic studies carried out at the VLT and Keck Observatories. Consistent with models based on Hess diagrams, the Galactic dSphs show depressed [α/Fe] relative to Galactic halo stars, implying that enrichment took place over a longer time. Evidently the Galactic halo is not primarily composed of stars that were stripped from mature dSph galaxies. On the other hand, the detection of very extended old stellar populations around some irregular galaxies supports the idea that basic structural similarities exist among all classes of dwarfs.

The Gas Crises – *Dekel, Marcolini, Skillman:* This topic was widely discussed across the meeting, but was not the main focus of an individual contribution. The main point is that dSphs are the most gas-deficient class of galaxy; they have *no* detectable ISM (see Young 2000, Gallagher *et al.* 2003). How the ISM in these galaxies was lost is not yet established, but a combination of internal (shell break-outs, photoionization) and external (tides, ram pressure stripping) factors are likely to be involved. The issue of gas retention connects to one distinction between star-forming dwarfs, which tend to be widely distributed within and beyond galaxy groups, versus the dSphs/dEs that are clustered within groups. No field dSph galaxies currently are known, suggesting that the conditions leading to gas removal are intimately tied to galaxy groups and clusters.

2.3. *Environmental Connections*

Satellite Galaxies – *Chilingarian, Coleman, Evans, Kroupa, Lee, Mateo, Read, Prugniel, Wilkinson, Zaggia:* A key problem in dSph astronomy is how these objects manage to survive, and what their original nature might have been. Are dSphs tracers of the full population of low mass dark halos, tidally trimmed remnants of galaxies in moderate mass halos, or systems that are not in simple equilibrium and therefore have no dark matter halos? The answers to these questions go beyond dwarf galaxy research and constrain the nature and amount of dark matter. The difficulty in understanding the structures of dwarfs on the observational side arise from the problems in studying their outer regions, and from our inability to measure internal proper motions, thus leaving us with only projected radial velocities. Here we saw that the outermost radial velocity dispersion profiles are still being debated, as are the forms of the density profiles. The structural issues are illustrated by the rather smooth Draco system, and clumpy, somewhat distorted Ursa Minor dSph, and yet both galaxies have similar kinematics. Thus the amount and radial distribution of dark matter–cusp or core structure–remains uncertain.

Dwarf-Giant Connections – *Bailin, Font, Gallagher, Ibata, Kroupa, Martinez-Delgado, Sohn:* There is no longer any doubt that dwarfs feed the halos of giants. The Sagittarius stream is one such long-lived remnant of tracing stellar contributions to the Milky Way's halo. The issue is how common could such streams be, and how many should we expect to observe if the bulk of the halo was produced from dissolved dwarfs? Realistic models show that early tidal debris is randomized as a giant galaxy forms around it, while the later streams can survive for several wraps. We also need to keep in mind the baryonic mass contributions in the form of gas that is being made to the Milky Way by the Magellanic Clouds, and was likely also captured from dSphs in their youth. However, for a dwarf companion that is in a prograde orbit near the disk plane of a spiral, there also is the possibility of substantial gas capture from a tidally induced spiral arm. Thus the gas flows between dwarfs and their hosts may go two ways, as can gravitational interactions, e.g., by dwarfs exciting warps in their spiral hosts.

Another feature in the dwarf giant connection is the Holmberg effect, where companion galaxies are in either the disk or polar directions. Furthermore, within the the polar cone, where most of the Milky Way's dSphs lie, the systems seem to be nearly coplanar. Its not clear how this feature of the Galactic dSphs fits with CDM dwarf models.

Locating Dwarfs – *Dunn, Jerjen, Karachentsev, Rizzi, Rekola:* While the low stellar densities of NGC 147 and NGC 185 allowed Baade to resolve their red giants in the mid-1940s, these properties now permit accurate distance measurements to dwarfs out to and beyond 10 Mpc from the luminosity of the tip of the red giant branch. This is done directly via resolved stars with *HST* and indirectly via surface brightness from the ground. The results presented here show that we are entering an era where dwarf galaxies can be prized distance markers, a process that soon should be extended as adaptive optics (AO) techniques improve on large ground-based telescopes.

Membership in Galaxy Groups – *Da Costa, Grebel, Karachentsev, Mateo, McConnachie, Skillman, Taylor:* The populations of dwarf galaxies in the Local Group are statistically sharply divided with the early-type systems mainly (but not entirely!) clumped around the Milky Way and M31, while the dI galaxies spread throughout the Local Group. Within the Local Group we can look for finer details; how do the properties of distant and nearby dSphs compare in terms of star formation histories, gas content, structures, etc.? Is there a "typical dSph galaxy"? The general spatial distribution of dwarfs seems to be repeated elsewhere and so such detailed studies are likely to apply beyond the Local Group. Yet the M31 dSphs, as well as those in other galaxy groups have different mean properties than the Milky Way satellites. Whether these properties are imprinted through initial conditions or the later environment is difficult to sort out as dwarf galaxies are mostly cosmically immobile; they stay near where they were formed.

Evolutionary Transitions – *Debattista, Drinkwater, Evistigneeva, Graham, Hwang, Kayser, Knezek, Mayer, Prugniel, Scannapieco, Read, Vaduvescu:* At one level we are asking whether the evolutionary paths of dwarfs are primarily set by properties at formation or by later events–a classic nature vs. nurture issue. On another we explore how, when, and why low mass galaxies should cease being active star formers and drop into the inactive dwarf category. A third issue is the connection, if any, between rotationally supported and non-rotating classes of dwarfs. Models show that a variety of opportunities exist in nature to modify the structures of galaxies: tides, gas loss or accretion, dynamical heating through collisions, mergers, and the effects of external ionizing radiation fields, to name a few. We therefore expect that some fraction of present day dwarfs will have been shaped by their environments.

What about the observations? There is no doubt that the population of dwarfs continues to experience secular evolution. This can be seen in features such as tidal and gas streams from satellite dwarfs to their wide range in stellar population ages. We also find galaxies whose properties place them between the disky Irrs and triaxial dEs, such as the apparently non-rotating NGC 5253 starburst dwarf. Ultracompact dwarfs (and possibly closely related peculiar globular stars clusters, such as ω Cen) may be the ultimate examples of galaxies whose natures have been changed via interactions; it is very unlikely that the structures of these types of galaxies bear much resemblance to their initial states.

Galaxy Clusters – *Caldwell, Conselice, Coté, DeRijcke, Hilker, Lisker, Kovac, Mastropietro, Matkovic, Popesso, Rakos, Tully, van Zee, Zibetti:* Galaxy clusters are the densest galactic environments and also the places with the largest populations of dwarfs. The two broad explanations for this phenomenon are that small halos containing stars were better able to survive in clusters and thus give us a true picture of dwarf galaxy numbers (which are closer to the expectations of CDM models), or that dwarfs in clusters of galaxies are remnants of a variety of types of galaxies that were trimmed down through interactions within the cluster. This is another area where the past decade has seen a substantial shift in position as the inhomgeneous nature of cluster dE/dSphs populations was recognized. Some spheroidal dwarfs are rotationally flattened, and others are evidently intrinsically triaxial; a range of stellar population ages exist; and at a given luminosity there also are spreads in metallicity and surface brightness.

It seems unlikely that either the dwarf by birth or dwarf by transformation will explain everything and some combination of processes is likely. For example, differing rates of the intrinsic evolutionary clocks can lead to stellar metallicity and age differences due solely to internal factors. If the majority of dwarfs are perturbed descendants of other classes of galaxies, then large numbers of these objects must have fallen into galaxy clusters. These were primordial star forming dwarfs in pictures where the transformations are from late-to-early type dwarfs, or moderate luminosity disk systems if the transformations were more severe. In either case the process of whittling down galaxies to make dwarfs would contribute intracluster baryonic matter, gas and stars. Properties of the intracluster stars (ages, metallicities, spatial distributions) will be very useful in constraining dwarf-by-transformation models for galaxy clusters.

The comparison of luminosity functions as a function of galaxy density suggests the existence of a trend for steeper luminosity functions in denser locations. However, even the steepest luminosity functions for spheroidal dwarfs are not as steep as the predictions for illuminated low mass halos from standard CDM models.

3. Concluding Thoughts

Bruno Binggeli introduced the meeting by briefly touching on several key issues: the degree of unity of dSph/dE galaxies as a structural class, the degree to which the fair samples from the local universe complement far field cosmological studies, the apparently prominent position of dark matter in dwarf galaxies, and the role that such systems play in the overall process of galaxy formation. That we can even ask and quantify these questions is a measure of the degree of progress in this field.

While most of the questions rate a "work in progress" status, we can answer the point about the homogeneity of the dSph/dE classes. Neither is homogeneous as both classes show a range of kinematic and stellar population properties. Perhaps we can make progress by viewing these types of galaxies as being structural endpoints that are reached by a variety of evolutionary paths accessible to low density, gas poor galaxies?

We also are still in a difficult triangle bounded by the expectations for large numbers of dwarf satellite halos (and therefore dwarf galaxies?) from CDM models, the uncertain dynamical states and structures of satellite dwarf galaxies, and the difficulties in dealing with the evolution of baryons from capture into galaxies and eventual formation of stars. We can look for continued improvements in several key areas, e.g., as more accurate radial velocities and proper motions become available (and a potential leap forward via GAIA and SIM). Furthermore, in addressing these issues through studies of dwarf galaxies we have the advantages of numerous objects spanning a range properties that are nearby. We therefore expect work on these visually unimpressive objects will continue to contribute in major ways to our understanding of galaxies and the state of matter in the universe.

Acknowledgements

I would like to thank the organizers for producing an excellent meeting and for providing a full week of Alpine photometric weather. My research on dwarf galaxies is supported by NASA through grants administered by the Space Telescope Science Institute in conjunction with U.S. general observer programs, and by the National Science Foundation through grant AST-9803018 to the University of Wisconsin.

References

Aaronson, M. 1983, *ApJ* 266, L11

Baade, W. 1944, *ApJ* 100, 147

Baade, W. 1951, *Publ. U. Michigan Obs.*, vol. 10, p. 33

Baade, W. & Hubble, E. 1991, *PASP* 51, 40

Bergh, S. van den 1959, *Publ. David Dunlop Obs.*, vol. 2, p. 147 (No. 5)

Bergh, S. van den 1977, in: B. M. Tinsley & R. B. Larson (eds.), *The Evolution of Galaxies and Stellar Populations* (New Haven: Yale Univ. Obs.), p. 19

Gallagher, J.S., Madsen, G.J., Reynolds, R.J., Grebel, E.K. & Smecker-Hane, T. A. 2003, *ApJ* 588, 326

Hodge, P. 1971, *ARA&A* 9, 35

Lin, D.N.C. & Faber, S.M. 1983, *ApJ* 266, L21

Reaves, G. 1953, PhD thesis, University of California, Berkeley

Shapley, H. 1939, *Proc. Nat. Acad. Sci.* 25, 565

Young, L. 2000, *AJ* 119, 188

Zwicky, F. 1957, *Morphological Astronomy* (Berlin: Springer)

Author Index